Geophysical Monograph Series

Geophysical Monograph Series

193 **Abrupt Climate Change: Mechanisms, Patterns, and Impacts** *Harunur Rashid, Leonid Polyak, and Ellen Mosley-Thompson (Eds.)*

194 **Stream Restoration in Dynamic Fluvial Systems: Scientific Approaches, Analyses, and Tools** *Andrew Simon, Sean J. Bennett, and Janine M. Castro (Eds.)*

195 **Monitoring and Modeling the Deepwater Horizon Oil Spill: A Record-Breaking Enterprise** *Yonggang Liu, Amy MacFadyen, Zhen-Gang Ji, and Robert H. Weisberg (Eds.)*

196 **Extreme Events and Natural Hazards: The Complexity Perspective** *A. Surjalal Sharma, Armin Bunde, Vijay P. Dimri, and Daniel N. Baker (Eds.)*

197 **Auroral Phenomenology and Magnetospheric Processes: Earth and Other Planets** *Andreas Keiling, Eric Donovan, Fran Bagenal, and Tomas Karlsson (Eds.)*

198 **Climates, Landscapes, and Civilizations** *Liviu Giosan, Dorian Q. Fuller, Kathleen Nicoll, Rowan K. Flad, and Peter D. Clift (Eds.)*

199 **Dynamics of the Earth's Radiation Belts and Inner Magnetosphere** *Danny Summers, Ian R. Mann, Daniel N. Baker, and Michael Schulz (Eds.)*

200 **Lagrangian Modeling of the Atmosphere** *John Lin (Ed.)*

201 **Modeling the Ionosphere-Thermosphere** *Jospeh D. Huba, Robert W. Schunk, and George V. Khazanov (Eds.)*

202 **The Mediterranean Sea: Temporal Variability and Spatial Patterns** *Gian Luca Eusebi Borzelli, Miroslav Gacic, Piero Lionello, and Paola Malanotte-Rizzoli (Eds.)*

203 **Future Earth – Advancing Civic Understanding of the Anthropocene** *Diana Dalbotten, Gillian Roehrig, and Patrick Hamilton (Eds.)*

204 **The Galápagos: A Natural Laboratory for the Earth Sciences** *Karen S. Harpp, Eric Mittelstaedt, Noemi d'Ozouville, and David W. Graham (Eds.)*

205 **Modeling Atmospheric and Oceanic Flows: Insightsfrom Laboratory Experiments and Numerical Simulations** *Thomas von Larcher and Paul D. Williams (Eds.)*

206 **Remote Sensing of the Terrestrial Water Cycle** *Venkat Lakshmi (Ed.)*

207 **Magnetotails in the Solar System** *Andreas Keiling, Caitriona Jackman, and Peter Delamere (Eds.)*

208 **Hawaiian Volcanoes: From Source to Surface Rebecca** *Carey, Valerie Cayol, Michael Poland, and Dominique Weis (Eds.)*

209 **Sea Ice: Physics, Mechanics, and Remote Sensing** *Mohammed Shokr and Nirmal Sinha (Eds.)*

210 **Fluid Dynamics in Complex Fractured-Porous Systems** *Boris Faybishenko, Sally M. Benson, and John E. Gale (Eds.)*

211 **Subduction Dynamics: From Mantle Flow to Mega Disasters** *Gabriele Morra, David A. Yuen, Scott King, Sang Mook Lee, and Seth Stein (Eds.)*

212 **The Early Earth: Accretion and Differentiation** *James Badro and Michael Walter (Eds.)*

213 **Global Vegetation Dynamics: Concepts and Applications in the MC1 Model** *Dominique Bachelet and David Turner (Eds.)*

214 **Extreme Events: Observations, Modeling and Economics** *Mario Chavez, Michael Ghil, and Jaime Urrutia-Fucugauchi (Eds.)*

215 **Auroral Dynamics and Space Weather** *Yongliang Zhang and Larry Paxton (Eds.)*

216 **Low-Frequency Waves in Space Plasmas** *Andreas Keiling, Dong-Hun Lee, and Valery Nakariakov (Eds.)*

217 **Deep Earth: Physics and Chemistry of the Lower Mantle and Core** *Hidenori Terasaki and Rebecca A. Fischer (Eds.)*

218 **Integrated Imaging of the Earth: Theory and Applications** *Max Moorkamp, Peter G. Lelievre, Niklas Linde, and Amir Khan (Eds.)*

219 **Plate Boundaries and Natural Hazards** *Joao Duarte and Wouter Schellart (Eds.)*

220 **Ionospheric Space Weather: Longitude and Hemispheric Dependences and Lower Atmosphere Forcing Timothy Fuller-Rowell,** *Endawoke Yizengaw, Patricia H. Doherty, and Sunanda Basu (Eds.)*

221 **Terrestrial Water Cycle and Climate Change Natural and Human-Induced Impacts** *Qiuhong Tang and Taikan Oki (Eds.)*

222 **Magnetosphere-Ionosphere Coupling in the Solar System** *Charles R. Chappell, Robert W. Schunk, Peter M. Banks, James L. Burch, and Richard M. Thorne (Eds.)*

223 **Natural Hazard Uncertainty Assessment: Modeling and Decision Support** *Karin Riley, Peter Webley, and Matthew Thompson (Eds.)*

224 **Hydrodynamics of Time-Periodic Groundwater Flow: Diffusion Waves in Porous Media** *Joe S. Depner and Todd C. Rasmussen (Auth.)*

225 **Active Global Seismology** *Ibrahim Cemen and Yucel Yilmaz (Eds.)*

226 **Climate Extremes** *Simon Wang (Ed.)*

227 **Fault Zone Dynamic Processes** *Marion Thomas (Ed.)*

228 **Flood Damage Survey and Assessment: New Insights from Research and Practice** *Daniela Molinari, Scira Menoni, and Francesco Ballio (Eds.)*

229 **Water-Energy-Food Nexus – Principles and Practices** *P. Abdul Salam, Sangam Shrestha, Vishnu Prasad Pandey, and Anil K Anal (Eds.)*

230 **Dawn–Dusk Asymmetries in Planetary Plasma Environments** *Stein Haaland, Andrei Rounov, and Colin Forsyth (Eds.)*

231 **Bioenergy and Land Use Change** *Zhangcai Qin, Umakant Mishra, and Astley Hastings (Eds.)*

232 **Microstructural Geochronology: Planetary Records Down to Atom Scale** *Desmond Moser, Fernando Corfu, James Darling, Steven Reddy, and Kimberly Tait (Eds.)*

233 **Global Flood Hazard: Applications in Modeling, Mapping and Forecasting** *Guy Schumann, Paul D. Bates, Giuseppe T. Aronica, and Heiko Apel (Eds.)*

234 **Pre-Earthquake Processes: A Multidisciplinary Approach to Earthquake Prediction Studies** *Dimitar Ouzounov, Sergey Pulinets, Katsumi Hattori, and Patrick Taylor (Eds.)*

235 **Electric Currents in Geospace and Beyond** *Andreas Keiling, Octav Marghitu, and Michael Wheatland (Eds.)*

236 **Quantifying Uncertainty in Subsurface Systems** *Celine Scheidt, Lewis Li, and Jef Caers (Eds.)*

237 **Petroleum Engineering** *Moshood Sanni (Ed.)*

238 **Geological Carbon Storage: Subsurface Seals and Caprock Integrity** *Stephanie Vialle, Jonathan Ajo-Franklin, and J. William Carey (Eds.)*

239 **Lithospheric Discontinuities** *Huaiyu Yuan and Barbara Romanowicz (Eds.)*

240 **Chemostratigraphy Across Major Chronological Eras** *Alcides N.Sial, Claudio Gaucher, Muthuvairavasamy Ramkumar, and Valderez Pinto Ferreira (Eds.)*

241 **Mathematical Geoenergy:Discovery, Depletion, and Renewal** *Paul Pukite, Dennis Coyne, and Daniel Challou (Eds.)*

242 **Ore Deposits: Origin, Exploration, and Exploitation** *Sophie Decree and Laurence Robb (Eds.)*

243 **Kuroshio Current: Physical, Biogeochemical and Ecosystem Dynamics** *Takeyoshi Nagai, Hiroaki Saito, Koji Suzuki, and Motomitsu Takahashi (Eds.)*

Geophysical Monograph 244

Geomagnetically Induced Currents from the Sun to the Power Grid

Jennifer L. Gannon
Andrei Swidinsky
Zhonghua Xu
Editors

This Work is a co-publication of the American Geophysical Union and John Wiley and Sons, Inc.

AGU 100
ADVANCING EARTH AND SPACE SCIENCE

WILEY

This Work is a co-publication between the American Geophysical Union and John Wiley & Sons, Inc.

This edition first published 2019 by John Wiley & Sons, Inc., 111 River Street, Hoboken, NJ 07030, USA and the American Geophysical Union, 2000 Florida Avenue, N.W., Washington, D.C. 20009

Published under the aegis of the AGU Publications Committee

Brooks Hanson, Executive Vice President, Science
Lisa Tauxe, Chair, Publications Committee
For details about the American Geophysical Union visit us at www.agu.org.

Wiley Global Headquarters
111 River Street, Hoboken, NJ 07030, USA

For details of our global editorial offices, customer services, and more information about Wiley products visit us at www.wiley.com.

Library of Congress Cataloging-in-Publication Data is available.

ISBN: 9781119434344

Cover image: (c) Sunday Stock/Shutterstock, (c) Color4260/Shutterstock
Cover design by Wiley

Set in 10/12pt Times New Roman by SPi Global, Pondicherry, India

V10013526_091219

CONTENTS

CONTRIBUTORS

Christopher Balch
Space Weather Prediction Center, National Oceanic and Atmospheric Administration, Boulder, CO, USA

Paul A. Bedrosian
Geology, Geophysics, and Geochemistry Science Center, United States Geological Survey, Denver, CO, USA;
Crustal Geophysics, United States Geological Survey, Denver, CO, USA

Stephen W. Cuttler
Department of Geophysics, Colorado School of Mines, Golden, CO, USA

Robyn A. D. Fiori
Geomagnetic Laboratory, Canadian Hazards Information Service, Natural Resources Canada, Ottawa, Ontario, Canada

Kevin F. Forbes
The Busch School of Business and Department of Economics, The Catholic University of America, Washington, DC, USA

Maryam Kazerooni
Department of Electrical and Computer Engineering, University of Illinois at Urbana-Champaign, Urbana, IL, USA

Anna Kelbert
Geomagnetism Program, United States Geological Survey, Denver, CO, USA;
College of Earth, Ocean, and Atmospheric Sciences, Oregon State University, Corvallis, OR, USA

Jeffrey J. Love
Geomagnetism Program, United States Geological Survey, Denver, CO, USA

Greg M. Lucas
Geomagnetism Program, United States Geological Survey, Denver, CO, USA

Esteban Bowles-Martinez
College of Earth, Ocean, and Atmospheric Sciences, Oregon State University, Corvallis, OR, USA

Benjamin S. Murphy
College of Earth, Ocean, and Atmospheric Sciences, Oregon State University, Corvallis, OR, USA

Chigomezyo M. Ngwira
Department of Physics, The Catholic University of America, Washington, DC, USA;
Goddard Space Flight Center, Space Weather Laboratory, National Aeronautics and Space Administration, Greenbelt, MD, USA

Thomas J. Overbye
Department of Electrical and Computer Engineering, Texas A&M University, College Station, TX, USA

Antti A. Pulkkinen
Goddard Space Flight Center, Space Weather Laboratory, National Aeronautics and Space Administration, Greenbelt, MA, USA

E. Joshua Rigler
Geomagnetism Program, United States Geological Survey, Golden, CO, USA

Adam Schultz
College of Earth, Ocean, and Atmospheric Sciences, Oregon State University, Corvallis, OR, USA

Komal S. Shetye
Department of Electrical and Computer Engineering, Texas A&M University, College Station, TX, USA

O. C. St. Cyr
Goddard Space Flight Center, National Aeronautics and Space Administration, Greenbelt, MD, USA

D. R. Weimer
Bradley Department of Electrical and Computer Engineering and Center for Space Science and Engineering Research, Virginia Tech, Blacksburg, VA, USA;
National Institute of Aerospace, Hampton, Virginia, USA

Daniel Welling
University of Texas at Arlington Physics Department, Arlington, TX, USA

Michael Wiltberger
High Altitude Observatory, National Center for Atmospheric Research, Boulder, CO, USA

L. M. Winter
Los Alamos National Laboratory, Los Alamos, NM, USA

J. R. Woodroffe
Space Science and Applications, Los Alamos National Laboratory, Los Alamos, NM, USA

PREFACE

When beginning work in the study of Geomagnetically Induced Currents (GICs), one of the most common difficulties experienced by a researcher is in overcoming, and eventually embracing, the interdisciplinary nature of the field. To understand the GIC process from start to end, we must understand a little bit of solar physics, space weather, geophysics, and power engineering – fields with completely different applications, expertise, and language. At a typical university, these disciplines are often in different departments, and likely in completely different schools. Overcoming disciplinary barriers takes a willingness to leave behind the high degree of specialization that the academic path requires and become a novice in at least one new field. This is daunting for most of us.

GIC research is also highly applied. Because of the potential for extreme GICs to damage critical power grid infrastructure, there is a great deal of societal importance. New research may have a significant effect on industry and government. This real-world relevance is often what draws us to the field, but means that there are also practical considerations. Given the wide range of backgrounds and priorities among participants in the field, it is no wonder that there are sometimes disagreements, but also great potential for cross-pollination of ideas and information.

Some of these differences will be apparent in this book. The book is presented in three parts – Space Weather, Geomagnetic Induction, and Power System Impacts. Each section is curated by an editor in the particular subfield, with authors invited from among the leading experts in the field. We have endeavored to include a range of expertise and backgrounds, and each of the authors has provided a perspective on their topic aimed at experts in adjacent fields. This book would not exist without the hard work of these authors, for which we are extremely grateful and humbled that they would join us in this effort. We hope that the result is a practical, interdisciplinary introduction to a full range of topics relevant to the GIC problem.

We hope that you, as a reader of this book, find it useful as a bridge to a new discipline within the broad range of GIC topics, or perhaps even find something new within your own specialty.

Jennifer L. Gannon
Andrei Swidinsky
Zhonghua Xu

Part I
Space Weather

1
An Introduction to Geomagnetically Induced Currents

Chigomezyo M. Ngwira[1,2,3] **and Antti A. Pulkkinen**[2]

ABSTRACT

Earth-directed space weather is a serious concern that is recognized as one of the top priority problems in today's society. Space weather-driven geomagnetically induced currents (GICs) can disrupt operation of extended electrically conducting technological systems. This threat to strategic technological assets, like power grids, oil and gas pipelines, and communication networks, has rekindled interest in extreme space weather. To improve national preparedness, it is critical that we understand the physical processes related to extreme events in order to address key national and international objectives. This paper serves to provide a basic introduction to space weather and GICs, and highlights some of the major science challenges the GIC community continues to face.

Key Points

- Geomagnetically induced currents (GICs) is a space weather-driven phenomena.
- It is a threat to strategic technological assets, such as power grids, oil and gas pipelines, and communication networks.
- This paper serves to provide basic introduction on space weather and GICs, and the major science challenges the GIC community continues to face.

1.1. INTRODUCTION

Space weather is a serious natural threat to national security, and is recognized as one of the top priority problems today. The term "space weather" generally refers to dynamic conditions on the Sun, in the solar wind, and in the near-Earth space environment that can influence the performance of man-made technology, and can also affect human health and activities. Space weather is a multi-faceted phenomenon, thus the scientific community is faced with a challenge to better understand this natural hazard in order to enhance preparedness.

Geomagnetically induced currents (GICs), a space weather-driven phenomena, have received increased international policy, science, industry, and public interest. GICs flowing on ground-based electrically conducting systems can disrupt operation of critical infrastructure such as power grids, pipelines, telecommunication cables, and railway systems (e.g., Barlow, 1849; Davidson, 1940; Boteler and Jansen van Beek, 1999; Molinski et al., 2000; Pirjola, 2000; Pulkkinen et al., 2001; Eroshenko et al., 2010, and references therein). The majority of community efforts focus on extreme forms of space weather which not only have severe impact on our technology and human space travel, but also challenge our understanding of the space weather phenomena.

Scientific investigations are critical for understanding the basic physics and predicting the potential impact of extreme space weather. Public opinions on the topic of extreme space weather include wide ranging views. This

[1] *Department of Physics, The Catholic University of America, Washington, DC, USA*
[2] *Goddard Space Flight Center, Space Weather Laboratory, National Aeronautics and Space Administration, Greenbelt, MD, USA*
[3] *Now at Atmospheric and Space Technology Research Associates, Louisville, CO, USA*

Geomagnetically Induced Currents from the Sun to the Power Grid, Geophysical Monograph 244, First Edition.
Edited by Jennifer L. Gannon, Andrei Swidinsky, and Zhonghua Xu.

chapter provides a high-level summary of space weather and GICs. While some of the topics touched on cover a broad range of space weather domains, the discussions are oriented/biased towards the geophysical facet of GICs. For more insight on specific GIC aspects, the reader is urged to consult other sections of this volume.

1.2. THE SPACE WEATHER CHAIN

The Sun is the primary source of all space weather in the heliosphere. Sudden, violent eruptions of solar material from the Sun's atmosphere (the corona) called coronal mass ejections (CMEs), mark the beginning of major space weather events that eventually produce geomagnetic storms (disturbances) in the Earth's upper atmosphere. The Sun's activity is closely governed by the solar activity cycle, which has an average length of about 11 years. The cycle is defined by the number of visible active sunspots on the solar surface.

During solar maximum period when solar activity is high, the Sun can launch multiple CMEs towards Earth per day. A CME can be perceived as a cloud of plasma with the solar magnetic field known as the interplanetary magnetic field (IMF) embedded within it. Upon arriving at Earth, CMEs interact with the magnetosphere, a low-density partially ionized region around the upper atmosphere dominated by Earth's magnetic field. This interaction then triggers geomagnetic disturbances (GMDs) that lead to violent global magnetic field variations.

Orientation of the IMF varies with time and is important for interaction between the solar wind and the magnetosphere. Historically, the most intense disturbances have been recorded when the IMF Bz component, which is parallel to the solar rotation axis is oppositely directed to the Earth's magnetic field, a condition often referred to as a southward or negative IMF. Under southward condition, the coupling between the solar wind and the magnetosphere is enhanced and the transfer of CME plasma, momentum, and energy into the near-Earth space environment is increased. This enhanced energy flow stimulates a chain of complex processes within the magnetosphere–ionosphere (M–I) coupled system that regulate phenomena such as storm enhanced density, ionospheric irregularities, substorms, GICs, and auroral displays at high-latitude locations. In addition to these effects, space weather can also compromise the integrity and performance of our technology (Lanzerotti, 2001). Figure 1.1 highlights some of the key technological assets affected by space weather. Per the purpose of this book, we now focus our discussions exclusively on GICs that occur at the end of the space weather chain.

1.3. GEOMAGNETICALLY INDUCED CURRENTS

Overall the GIC problem can be categorized by a two-step approach (Pirjola, 2000, 2002a). In step 1, the geophysical facet involving the estimation of the geoelectric field based on M-I currents and the ground conductivity is considered. Step 1 is fundamentally a science piece and the connection to space weather phenomenon. In step 2 ("engineering piece") the current flowing on the system is calculated based on the estimated geoelectric field and detailed information about the particular ground system (e.g., Lehtinen and Pirjola, 1985; Molinski et al., 2000; Pirjola, 2000). In other words, the magnitude of GICs flowing through a network is generally determined by a combination of the horizontal surface geoelectric field, the geology, and elements of a given network (e.g., Molinski et al., 2000; Pirjola, 2000). We now briefly examine each of these three components.

1.3.1. The Geoelectric Field

The ground geoelectric field is the actual link to space weather through M-I processes. The primary feature of geomagnetic storms that pertains to GICs is the variation of electric currents in the M-I mode. Intense time-varying magnetosphere and ionosphere currents lead to rapid fluctuation of the geomagnetic field on the ground. Faraday's law of induction is the basic principle related to the flow of GICs on ground networks: a changing magnetic field induces an electric field through geomagnetic induction in the earth. In turn this electric field is responsible for currents that flow on ground conductors, such as power grids, according to Ohm's law $\mathbf{J} = \sigma \mathbf{E}$, where $\mathbf{J}$ is the current density, σ is the conductivity, and $\mathbf{E}$ is the electric field. The key processes for the creation and flow of GICs are illustrated in Figure 1.2.

Mathematically, Faraday's law of induction can be expressed as:

$$\nabla \times \mathbf{E} = -\frac{\partial \mathbf{B}}{\partial t}. \tag{1.1}$$

Hence, the observed induced surface geoelectric field depends only on geomagnetic field variations and electromagnetic induction in the earth determined by the local geology (e.g., Pirjola, 1982). It follows, therefore, that this induced geoelectric field is completely independent of any technological system but is determined by M-I currents that are a function of space weather conditions and the ground conductivity, as discussed before.

To calculate the geoelectric field, a simple but illustrative 1-dimensional (1-D) model that assumes a plane wave propagating vertically downwards and a uniform

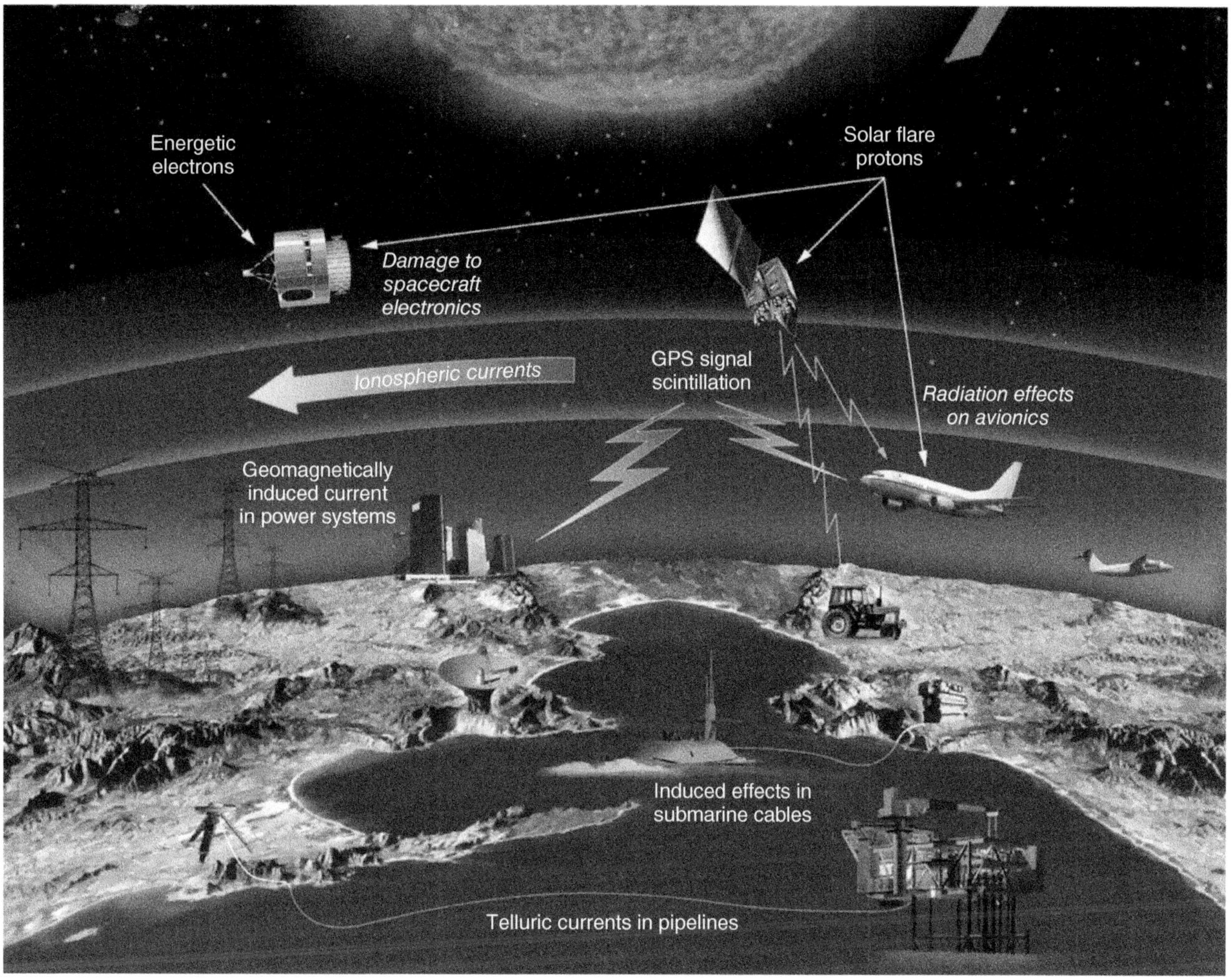

Figure 1.1 Technological infrastructure affected by space weather events at the Earth. Source: Courtesy of NASA: https://www.nasa.gov/mission_pages/rbsp/science/rbsp-spaceweather.html. (*See electronic version for color representation of this figure.*)

half-space earth with conductivity σ is traditionally used (Cagniard, 1953; Pirjola, 1982). The fields are all presumed to be horizontally uniform to simplify the modeling. Adopting a single frequency ω, then the geoelectric field E_x and E_y components can be deduced in terms of the perpendicular geomagnetic field component B_y and B_x as:

$$E_{x,y}(\omega) = \pm\sqrt{\frac{\omega}{\mu_0 \sigma}} e^{\frac{i\pi}{4}} B_{y,x} \tag{1.2}$$

where μ_0 is permeability of free space, whereas the layer of air between the ground and the ionosphere is taken to have zero conductivity to limit significant attenuation of external electromagnetic fields. Since Equation (1.2) outlines the basis for deriving the Earth's conductivity using geoelectric and geomagnetic field measurements recorded at the surface, it is considered as the "basic equation of magnetotellurics."

1.3.2. Ground Conductivity

The Earth's geology is another key ingredient in the geomagnetic induction process. Penetration of the geomagnetic field into the Earth's crust is determined by the ground conductivity and frequency of the geomagnetic field variations. That is to say, the rate of attenuation of the induced electric field is dependent on the vertical distribution of the resistivity of the ground, and the period considered. Upper layers generate stronger influences at short periods, and deeper layers are more bearing at long periods, as depicted in Figure 1.3. It should be noted that the geomagnetic induction process is not fully discussed in the present paper. Here, we mostly emphasize the different conductivity models used for GIC applications. However, Part 2 of this volume is dedicated to discussions on geomagnetic induction.

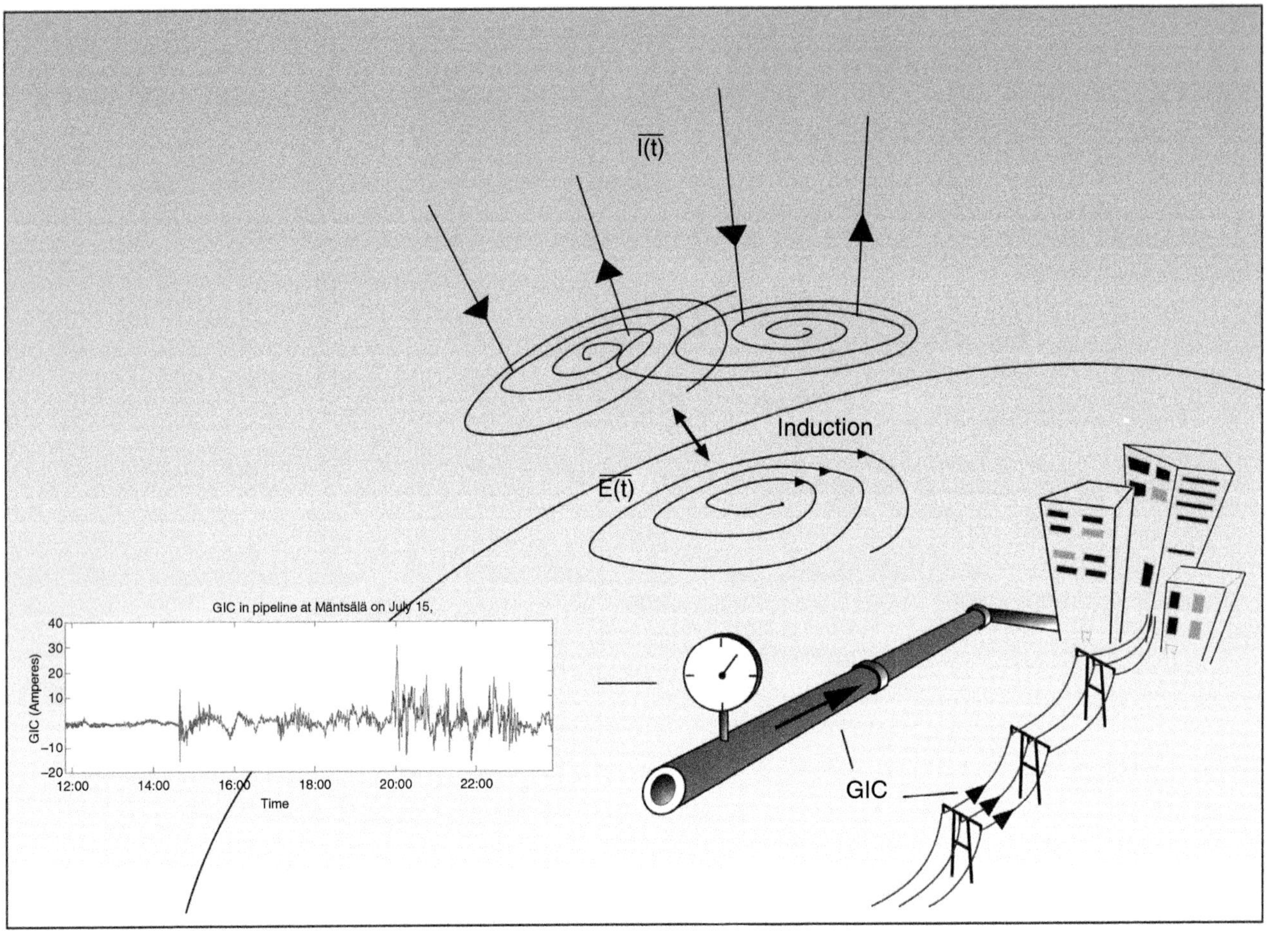

Figure 1.2 The basic principle for the generation of GICs: variations of the ionospheric currents ($I(t)$) generate an electric field ($E(t)$) through geomagnetic induction in the earth. This electric field then drives GICs on ground conductors. Also shown are actual GIC recordings from the Finnish natural gas pipeline. Image credit Wikipedia. (*See electronic version for color representation of this figure.*)

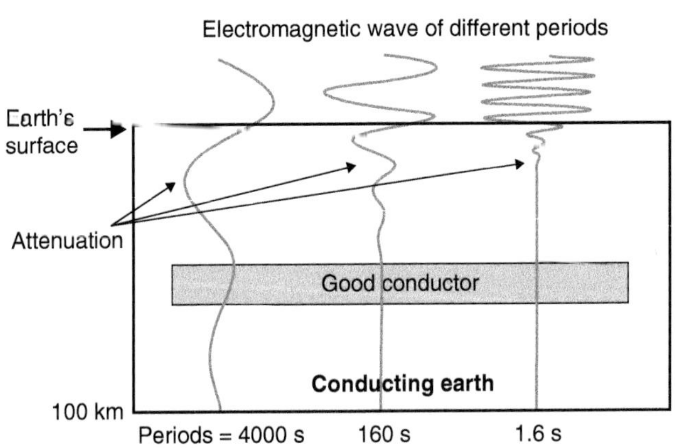

Figure 1.3 Depiction of electromagnetic signal penetration at different periods. Long-period signals penetrate deeper into the underground than short periods. Source: Adopted with modification: http://userpage.fu-berlin.de/mtag/MT-principles.html.

The "plane wave" approach described above is a firmly-established and simplest procedure for calculating GICs (see e.g., Pirjola, 2002a; Viljanen et al., 2006; Ngwira et al., 2008; Liu et al., 2009; da Silva Barbosa et al., 2015; Pulkkinen et al., 2015). This approach has also been applied to extreme events with much success (Wik et al., 2009; Pulkkinen et al., 2012; Ngwira et al., 2013, 2015). Historically, the most widely used ground structure has been the 1-D layered conductivity (where σ is depth dependent) applied to specific or given location (e.g., Boteler and Pirjola, 1998; Viljanen et al., 2006; Pulkkinen et al., 2007; Ngwira et al., 2008; Fernberg, 2012; Zhang et al., 2012). In the United Kingdom, however, past studies have calculated GICs flowing on the high-voltage transmission system using the "thin-sheet" approximation (Beamish et al., 2002; McKay, 2003; Thomson et al., 2005; Beggan, 2015). The thin-sheet approach uses a spatially varying conductance on a 2-D surface covering the region of interest, combined with a 1-D layered conductivity of upper lithosphere conductance (McKay, 2003). Thus, a thin-sheet model incorporates the effect of lateral conductivity variations on redistribution of regional currents induced elsewhere (e.g., oceans or shelf seas).

Several recent studies emphasize the use of 3-D conductivity that more accurately represent the true 3-D earth response (e.g., Love, 2012; Bonner IV and Schultz, 2017; Kelbert et al., 2017, and references therein). Unfortunately, these 3-D conductivity models are not readily available in many areas, thus their application is quite limited. In the United States, data from magnetotelluric (MT) campaigns such as EarthScope USArray MT program (http://ds.iris.edu/spud/emtf) are improving the conductivity models (e.g., Schultz, 2009). So far, nearly 60% of the continental United States has already been covered [Adam Schultz, personal communication]. In Figure 1.4 is a map showing the locations and current status of the NSF-funded EarthScope USArray MT project.

With the development of 3-D ground models, one challenge is to pinpoint exactly when and where the 1-D case fails and the 3-D case becomes necessary for GIC purposes. This is partly due to the data limitation mentioned before but with more 3-D models becoming available, the picture is beginning to change. Take for instance the recent Mid-Atlantic region case study by Love et al. (2018). They estimate that geoelectric fields calculated from 3-D conductivity models could be a few orders of magnitude larger than the fields estimated from 1-D models. If indeed the 1-D models under-estimate the induced fields (the order of magnitude might vary), then this could raise significant concern for power system operators in affected regions. However, the actual impact of such fields on any system is a matter requiring more detailed analysis that consider all sides of the problem, including the coupling of space weather processes to the grid.

1.3.3. Engineering Considerations

Generally, information about the geoelectric field produced on the ground during GMD events is acquired as described above. Once this information is obtained, determining the level of GICs flowing through a given node for any ground system is relatively straightforward. The GIC can be calculated by considering the geoelectric field to be uniform in the near vicinity of the network using the expression

$$GIC(t) = aE_x(t) + bE_y(t) \tag{1.3}$$

where a and b are the network coefficients specific to each network node depending only on the resistance and geometrical composition of a system (Viljanen and Pirjola, 1994). This is a purely engineering task that requires a full description of the system under consideration, which is beyond the scope of this paper. Nevertheless, readers can turn to Lehtinen and Pirjola (1985) or Viljanen and Pirjola (1994), and more recently Boteler (2014), for more information concerning this procedure. In addition, Part 3 of this volume contains several discussions on GICs and the power system.

1.4. EXTREME EVENTS

While mild and moderate space weather is fairly "common," relatively speaking, it is often the extreme events that gather the most attention because they are "infrequent" but pose the highest risk. A truly extreme and rare space weather event could have produced large GICs that can seriously disrupt technology. Policy makers, the general public, industry, and the science community are all interested to know "how bad can space weather really get?" Recent policy action at the White House level in terms of development of the National Space Weather Strategy and National Space Weather Action Plan (SWAP) has sparked renewed interest on this topic (National Science and Technology Council, 2015a,b). It is worth noting at this point that goal 1 of the SWAP calls for extracting information about extreme 1-in-100 year geoelectric fields and theoretical maximums. While GICs are not the only space weather hazard highlighted in these policies, the phenomenon does play an important role in them. In this section, a general view of GICs, extreme events, and impact are covered.

1.4.1. General View of GIC Studies

Space weather is a global phenomenon, however, most notable effects tend to occur locally, that is, isolated area, as is the case for GICs. For this reason, many GIC studies focus on specific regions or networks. Nevertheless, there are several examples of studies that have a wider scope and provide a global snapshot of events (e.g., Pulkkinen et al., 2012; Ngwira et al., 2013, 2015; Fiori et al., 2014; Kataoka and Ngwira, 2016; Carter et al., 2016; Moldwin and Tsu, 2017; de Villiers et al., 2017; Barbosa et al., 2017; Oliveira et al., 2018, and references therein).

The largest number of GIC studies have come from high-latitude regions because of the proximity to the auroral zone (Pirjola, 1982; Lehtinen and Pirjola, 1985; Viljanen and Pirjola, 1994; Boteler et al., 1998; Boteler, 2001; Pulkkinen et al., 2001; Pirjola, 2002b,c; Pulkkinen et al., 2003; Trichtchenko and Boteler, 2004; Thomson et al., 2005; Viljanen et al., 2006; Wintoft, 2005; Wik et al., 2009; Myllys et al., 2014; Beggan, 2015; Ngwira et al., 2018a, and references therein). The most interesting

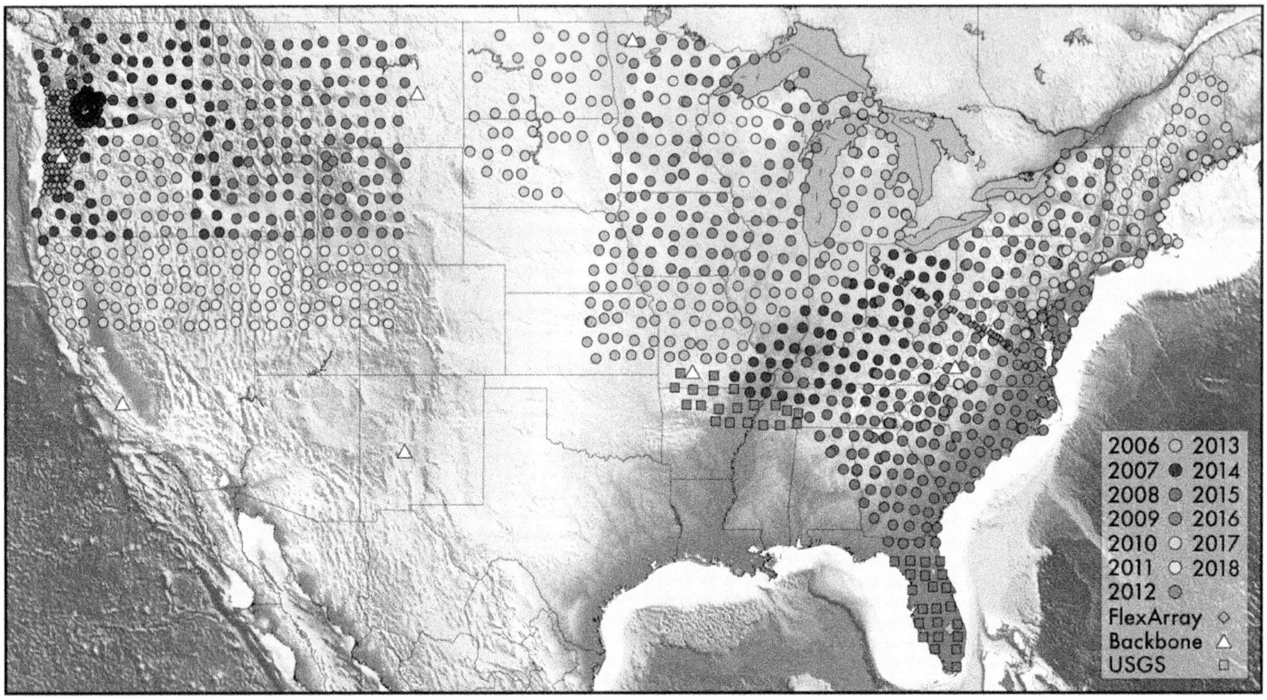

Figure 1.4 EarthScope USArray MT status map across the lower-48 U.S. The stations are spaced in an approximate 70 km grid. http://www.usarray.org/researchers/obs/magnetotelluric. (*See electronic version for color representation of this figure.*)

feature about the auroral zone is associated with auroral electrojet current flowing in the ionosphere. During storms, this current system can be strongly intensified mostly by magnetospheric substorms, thereby causing large GICs on the ground.

For many years, it was believed that GICs were a high-latitude phenomena, thus mid-low latitudes were generally not regarded to be susceptible to adverse impact by GICs. But this picture changed after evidence in South Africa revealed that GICs may have contributed significantly to the failure of several transformers (see reports by Koen, 2002; Gaunt and Coetzee, 2007). Since then, the GIC community has experienced a major growth in the number of studies focusing on mid-latitude locations such as, Australia, China, France, Greece, Hungary, Ireland, Japan, New Zealand, South Africa, Spain, and the United States (Kappenman, 2006; Bernhardi et al., 2008; Ngwira et al., 2008, 2009; Watari et al., 2009; Turnbull et al., 2009; Liu et al., 2009; Ngwira et al., 2011; Love, 2012; Torta et al., 2012; Zois, 2013; Marshall et al., 2013; Lotz and Cilliers, 2014; Fujii et al., 2015; Blake et al., 2016; Matandirotya et al., 2016; Lotz and Danskin, 2017; Kelly et al., 2017; Love et al., 2018, and references therein).

In general, low-latitude geomagnetic variations tend to be relatively smaller than those experienced at mid- and high-latitudes, thus the region has largely been overlooked and is the least studied area in terms of GICs. One of the earliest studies on low-latitude networks was conducted in Brazil by Trivedi et al. (2007). However, on examining the March 1989 and October 2003 extreme geomagnetic storms, Pulkkinen et al. (2012) first showed that induced surface geoelectric fields can be strongly amplified at the magnetic equator, thus could pose a higher threat to power systems at low-latitudes than at mid-latitudes. Then, Ngwira et al. (2013) extended study of extreme storms not only confirmed the findings by Pulkkinen et al. (2012), but also associated the effect to amplification of equatorial electrojet (EEJ) current by high-latitude penetration electric fields. Penetration electric fields are attributed to sudden changes in the strength of field-aligned currents, which are required for shielding the inner magnetosphere and the low-mid-latitude ionosphere from the dawn–dusk magnetospheric convection electric field (Fejer et al., 2007; Maruyama and Nakamura, 2007). After the extended study of extreme storms, Carter et al. (2015) investigated the potential effects of interplanetary shocks on the equatorial region and further demonstrated that their magnetic signature was amplified by the EEJ. Partly due to the investigation by Pulkkinen et al. (2012), we have witnessed an increased interest in GICs at low-latitudes during the last 5 years (e.g., Liu et al., 2014; da Silva Barbosa et al., 2015; Barbosa et al., 2015; Adebesin et al., 2016; Moldwin and Tsu, 2017; Oliveira et al., 2018, and references therein).

1.4.2. Extreme GICs

In the past three decades, the space weather field has observed significant progress that has strengthened insight on the central processes driving GICs. However, rare but extremely intense geomagnetic storms continue to challenge our understanding of space weather (Kataoka and Ngwira, 2016; Pulkkinen et al., 2017; Ngwira and Pulkkinen, 2018; Tsurutani et al., 2018; Ngwira et al., 2018b). As noted in our introduction, scientific investigations are critical for raising awareness and predicting the impact of extreme space weather.

The space weather community is well aware that during extreme geomagnetic storms, intense high-latitude currents can expand into the mid-latitudes (e.g., Kappenman, 2005; Ngwira et al., 2013, 2014). However, understanding how deep into the lower latitudes the high-latitude ionospheric currents can extend is still a challenge. Figure 1.5 shows global maximum geomagnetic field *dB*/*dt* distribution computed from ground magnetometer recordings. The distribution in this figure comprises of data from 12 from historical extreme geomagnetic storms that occurred between the years 1989 and 2005 (see report by Ngwira et al., 2013). Firstly, the figure clearly illustrates the impact of extreme storms on geomagnetic field perturbations at the geomagnetic equator, as seen by the amplified response near zero geomagnetic latitude. Secondly, the dark gray dashed lines mark the geomagnetic latitude boundary (GLB) location, a dynamic transition zone between high- and mid-latitudes, while the thick solid curve is the sixth order polynomial fit.

The GLB is crucial because it identifies the latitude band where *dB*/*dt* (or the geoelectric field) experience roughly an

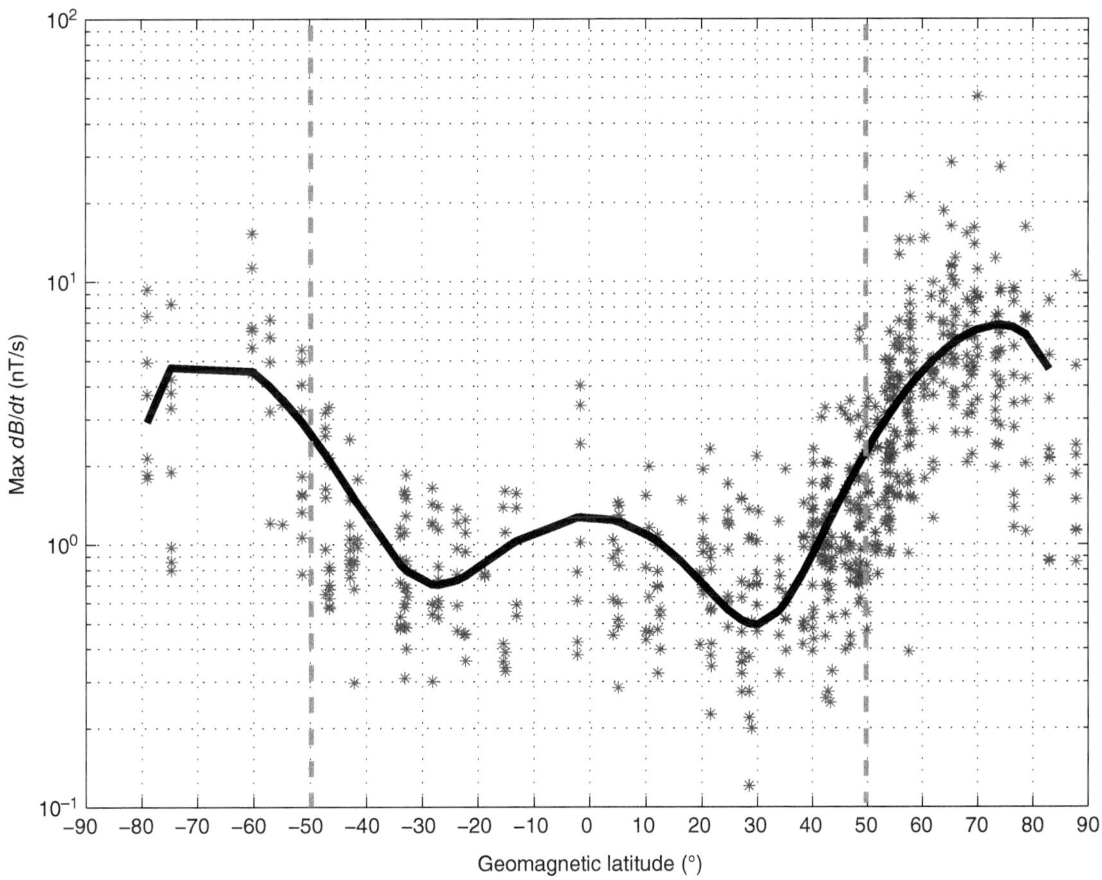

Figure 1.5 Geomagnetic latitude distributions comprising 12 extreme events that occurred between 1989 and 2005. Plot shows the maximum time derivative of the horizontal magnetic field, *dB/dt*, at specific ground sites represented by the "*" symbol. Source: Image credit *Ngwira et al.* (2013). Reproduced with permission of John Wiley and Sons. (*See electronic version for color representation of this figure.*)

order of magnitude drop/jump, and hence helps to define locations most exposed to the GIC hazard. Some investigators have determined the GLB location to be around 50–55°. geomagnetic latitude (e.g., Thomson et al., 2011; Pulkkinen et al., 2012; Ngwira et al., 2013). However, Ngwira et al. (2014) suggest that under extremely strong space weather conditions ("Carrington-type" event), the GLB location can be substantially displaced deeper (~40° geomagnetic latitude) into the lower latitudes. Such a location is much lower than previously determined for observed extreme geomagnetic storms, including the March 1989 and the October 2003 Halloween storm.

1.4.3. Impact

Reducing the Nation's vulnerability to space weather is identified as a national priority, and GICs are also identified as the top threat (National Science and Technology Council, 2015b). The telegraph system was the first technology to report disruption (1847) by space weather-driven GICs (Barlow, 1849). But perhaps in todays society, the power grid is the most critical infrastructure affected by GICs due to the wide-spread demand for electrical power. The first reported disruption on power grids was in 1940 (Davidson, 1940; McNish, 1940).

However, the most widely known impact of space weather on any system is the collapse of the Hydro-Quebec power network grid in Canada during the 13 March 1989 superstorm. Intense GICs produced during the superstorm triggered a complete blackout of the entire Hydro-Quebec network in a relatively short time interval (Boteler, 2001; Bolduc, 2002, and references therein). It is believed that after a large substorm, system stability was lost thereby cutting off main load points from a major generation source (see Bolduc, 2002, for a more detailed discussion). During the same March 1989 event, a generator step-up power transformer was damaged in New Jersey, USA.

Other, more recent but perhaps not well-known GIC impacts on power grids include, failure of a high-voltage power transmission system in Sweden (e.g., Pulkkinen et al., 2005; Wik et al., 2009), and possible transformer damages in South Africa (Gaunt and Coetzee, 2007). These two event are both associated with the "Halloween storm" of October 2003. GICs affect a wide range of technologies, as noted above. An extensive record of GIC impact on different systems within the last 80 years was recently compiled by Ngwira and Pulkkinen (2018).

1.5. CONCLUDING REMARKS

Space weather is an interesting but complex phenomenon. One of the top priorities of the science community today is extending our current understanding of the space weather phenomena. Extreme events in particular not only challenge our understanding of the physics of this phenomena, but can also cause deleterious effects. With several on-going national and international efforts, our awareness of the problem is growing, thus could help to resolve outstanding challenges. More detailed discussion on specific areas of the GIC problem are provided in different chapters of this volume.

ACKNOWLEDGMENTS

We thank Peter Schuck at NASA Goddard Space flight Center and the two anonymous reviewers for assisting this effort. The discussion presented in this chapter relies on the data collected at magnetic observatories. The authors thank the national institutes that support their operation and INTERMAGNET for promoting high standards of magnetic observatory practice (www.intermagnet.org). Research effort at The Catholic University of America (CUA) was supported via NASA Grants NNG11PL10A 670.135 and NNG11PL10A 670.157 to CUA/IACS.

REFERENCES

Adebesin, B. O., A. Pulkkinen, and C. M. Ngwira (2016), The interplanetary and magnetospheric causes of extreme db/dt at equatorial locations, *Geophysical Research Letters*, *43*, 11501–11509, doi:10.1002/2016GL071526.

Barbosa, C., L. Alves, R. Caraballo, G. A. Hartmann, A. R. R. Papa, and R. J. Pirjola (2015), Analysis of geomagnetically induced currents at a low-latitude region over the solar cycles 23 and 24: Comparison between measurements and calculations, *Journal of Space Weather and Space Climate*, *55*, A35, doi:10.1051/swsc/2015036.

Barbosa, C. S., R. Caraballo, L. R. Alves, G. A. Hartmann, C. D. Beggan, A. Viljanen, C. M. Ngwira, A. R. R. Papa, and R. J. Pirjola (2017), The Tsallis statistical distribution applied to geomagnetically induced currents, *Space Weather*, *15*, 9, 1094–1101, doi:10.1002/2017SW001631.

Barlow, W. H. (1849), On the spontaneous electrical currents observed in the wires of the electric telegraph, *Philosophical Transactions of the Royal Society of London*, *139*, 61–72.

Beamish, D., T. G. D. Clark, E. Clarke, and A. W. P. Thomson (2002), Geomagnetically induced currents in the UK: Geomagnetic variations and surface electric fields, *Journal of Atmospheric and Solar Terrestrial Physics*, *64*, 1779–1792.

Beggan, C. D. (2015), Sensitivity of geomagnetically induced currents to varying auroral electrojet and conductivity models, *Earth, Planets, and Space*, *67*, 24, doi:10.1186/s40623-014-0168-9.

Bernhardi, E., P. J. Cilliers, and C. T. Gaunt (2008), Improvement in the modeling of geomagnetically induced currents in South Africa, *South African Journal of Science*, *104*, 265–272.

Blake, S. P., P. T. Gallagher, J. McCauley, A. G. Jones, C. Hogg, J. Campanya, C. Beggan, A. W. P. Thomson, G. S. Kelly, and

D. Bell (2016), Geomagnetically induced currents in the Irish power network during geomagnetic storms, *Space Weather*, *14*, 1136–1154, doi:10.1002/2016SW001534.

Bolduc, L. (2002), GIC observations and studies in the Hydro-Québec power system, *Journal of Atmospheric and Solar Terrestrial Physics*, *64*(16), 1793–1802.

Bonner IV, L. R., and A. Schultz (2017), Rapid prediction of electric fields associated with geomagnetically induced currents in the presence of three-dimensional ground structure: Projection of remote magnetic observatory data through magnetotelluric impedance tensors, *Space Weather*, *15*, doi:10.1002/2016SW001535.

Boteler, D. H. (2001), Space weather effects on power systems, In *Space Weather*, Edited by D. Song, H.J. Singer, and G.L. Siscoe, *AGU Geophysical Monograph 125*, AGU, Washington, DC, pp. 347–352.

Boteler, D. H. (2014), Methodology for simulation of geomagnetically induced currents in power systems, *Space Weather and Space Climate*, *4*, A21, doi:10.1051/swsc/2014018.

Boteler, D. H., and G. Jansen van Beek (1999), August 4, 1972 revisited: A new look at the geomagnetic disturbance that caused the L4 cable system outage, *Geophysical Research Letters*, *26*(5), 577–580.

Boteler, D. H., and R. J. Pirjola (1998), Modeling geomagnetically induced currents produced by realistic and uniform electric fields, *IEEE Transactions on Power Delivery*, *13*(4), 13031308.

Boteler, D. H., R. J. Pirjola, and H. Nevanlinna (1998), The effects of geomagnetic disturbances on electrical systems at the Earth's surface, *Advances in Space Research*, *22*(1), 17–27.

Cagniard, L. (1953), Basic theory of the magneto-telluric methods of geophysical prospecting, *Geophysics*, *18*(3), 605.

Carter, B. A., E. Yizengaw, R. Pradipta, A. J. Halford, and K. Zhang (2015), Interplanetary shocks and the resulting geomagnetically induced currents at the equator, *Geophysical Research Letters*, *42*, doi:10.1002/2015GL065060.

Carter, B. A., E. Yizengaw, R. Pradipta, J. M. Weygand, M. Piersanti, A. Pulkkinen, M. B. Moldwin, R. Norman, and K. Zhang (2016), Geomagnetically induced currents around the world during the 17 March 2015 storm, *Journal of Geophysical Research*, *121*, 10496–10507, doi:10.1002/2016JA023344.

da Silva Barbosa, C., G. A. Hartmann, and K. J. Pinheiro (2015), Numerical modeling of geomagnetically induced currents in a brazilian transmission line, *Advances in Space Research*, *55*, 1168–1179, doi:10.1016/j.asr.2014.11.008.

Davidson, W. F. (1940), The magnetic storm of March 24, 1940–Effects in the power system, *Edison Electric Institute Bulletin*, *8*, 365–366.

de Villiers, J. S., M. Kosch, Y. Yamazaki, and S. Lotz (2017), Influences of various magnetospheric and ionospheric current systems on geomagnetically induced currents around the world, *Space Weather*, *17*, 403–417, doi:10.1002/2016SW001566.

Eroshenko, E. A., A. V. Belov, D. Boteler, S. P. Gaidash, S. L. Lobkov, R. Pirjola, and L. Trichtchenko (2010), Effects of strong geomagnetic storms on Northern railways in Russia, *Advances in Space Research*, *46*, 1102–1110, doi:10.1016/j.asr.2010.05.017.

Fejer, B. G., J. W. Jensen, T. Kikuchi, M. A. Abdu, and J. L. Chau (2007), Equatorial ionospheric electric fields during the November 2004 magnetic storm, *Journal of Geophysical Research*, *112*, A10304, doi:10.1029/2007JA012376.

Fernberg, P. (2012), One-dimensional Earth resistivity models for selected areas of continental United States and Alaska. EPRI Technical Update 1026430, Palo Alto, CA.

Fiori, R. A. D., D. H. Boteler, and D. M. Gillies (2014), Assessment of GIC risk due to geomagnetic sudden commencements and identification of the current systems responsible, *Space Weather*, *12*, 76–91, doi:10.1002/2013SW000967.

Fujii, I., T. Ookawa, S. Nagamachi, and T. Owada (2015), The characteristics of geoelectric fields at Kakioka, Kanoya, and Memambetsu inferred from voltage measurements during 2000 to 2011, *Earth, Planets, and Space*, *67*, 62, doi:10.1186/s40623-015-0241-z.

Gaunt, C. T., and G. Coetzee (2007), Transformer failure in regions incorrectly considered to have low gic-risks, *IEEE Power Tech.*, Conference *Paper 445*, Lausanne, July, pp. 807–812.

Kappenman, J. G. (2005), An overview of the impulsive geomagnetic field disturbance and the power grid impacts associated with the violent Sun-Earth connection events of 29-30 October 2003 and a comparative evaluation with other contemporary storms, *Space Weather*, *3*, S08C01, doi:10.1029/2004SW000128.

Kappenman, J. G. (2006), Great geomagnetic storms and extreme impulsive geomagnetic field disturbance events – An analysis of observational evidence including the great storm of May 1921, *Advances in Space Research*, *38*, 188–199, doi:10.1016/j.asr.2005.08.055.

Kataoka, R., and C. Ngwira (2016), Extreme geomagnetically induced currents, *Progress in Earth and Planetary Science*, *3*, 23, doi:10.1186/s40645-016-0101-x.

Kelbert, A., C. C. Balch, A. Pulkkinen, G. D. Egbert, J. J. Love, E. J. Rigler, and I. Fujii (2017), Methodology for time-domain estimation of storm time geoelectric fields using the 3-D magnetotelluric response tensors, *Space Weather*, *15*, 874–894, doi:10.1002/2017SW001594.

Kelly, G. S., A. Viljanen, C. D. Beggan, and A. W. P. Thomson (2017), Understanding GIC in the UK and French high-voltage transmission systems during severe magnetic storms, *Space Weather*, *15*, 99–114, doi:10.1002/2016SW001469.

Koen, J. (2002), Geomagnetically induced currents in the Southern African electricity transmission network, Ph.D. thesis, University of Cape Town, March.

Lanzerotti, L. J. (2001), Space weather effects on technologies, In *Space Weather*, Edited by D. Song, H.J. Singer, and G.L. Siscoe, *AGU Geophysical Monograph 125*, AGU, Washington, DC, pp. 11–22.

Lehtinen, M., and R. Pirjola (1985), Currents produced in earthed conductor networks by geomagnetically-induced electric field, *Annales Geophysicae*, *3*(4), 479–484.

Liu, C.-M., L.-G. Liu, R. Pirjola, and Z.-Z. Wang (2009), Calculation of geomagnetically induced currents in mid- to low-latitude power grids based on the plane wave method: A preliminary case study, *Space Weather*, *7*, S04005, doi:10.1029/2008SW000439.

Liu, C., Y. Li, and R. Pirjola (2014), Observations and modeling of GIC in the Chinese large-scale high-voltage power networks, *Journal of Space Weather and Space Climate*, *4*, A03, doi:10.1051/swsc/2013057.

Lotz, S. I., and P. J. Cilliers (2014), A solar wind-based model of geomagnetic field fluctuations at a mid-latitude station, *Advances in Space Research*, *55*, 220–230, doi:10.1016/j.asr.2014.09.014.

Lotz, S. I., and D. W. Danskin (2017), Extreme value analysis of induced geoelectric field in South Africa, *Space Weather*, *15*, doi:10.1002/2017SW001662.

Love, J. J. (2012), Credible occurrence probabilities for extreme geophysical events: Earthquakes, volcanic eruptions, magnetic storms, *Geophysical Research Letters*, *39*, L10301, doi:10.1130/G32655.1.

Love, J. J., G. M. Lucus, A. Kelbert, and P. A. Bedrosian (2018), Geoelectric hazard maps for the mid-atlantic United States: 100 Year extreme values and the 1989 magnetic storm, *Geophysical Research Letters*, *45*, 5–14, doi:10.1002/2017GL076042.

Marshall, R. A., H. Gorniak, T. V. D. Walt, C. L. Waters, M. D. Sciffer, M. Miller, M. Dalzell, T. Daly, G. Pouferis, G. Hesse, and P. Wilkinson (2013), Observations of geomagnetically induced currents in the Australian power network, *Space Weather*, *11*, 1–11.

Maruyama, T., and M. Nakamura (2007), Conditions for intense ionospheric storms expanding to lower midlatitudes, *Journal of Geophysical Research*, *112*, A05310, doi:10.1029/2006JA012226.

Matandirotya, E., P. J. Cilliers, R. R. Van Zyl, D. T. Oyedokun, and J. de Villiers (2016), Differential magnetometer method applied to measurement of geomagnetically induced currents in Southern African power networks, *Space Weather*, *14*, 221–232, doi:10.1002/2015SW001289.

McKay, A. J. (2003), Geoelectric fields and geomagnetically induced currents in the United Kingdom, Ph.D. thesis, University of Edinburgh, UK.

McNish, A. G. (1940), The magnetic storm of March 24, 1940, *Terrestrial Magnetism and Atmospheric Electricity*, *45*(3), 359–364.

Moldwin, M. B., and J. S. Tsu (2017), Stormtime equatorial electrojet ground-induced currents: Increasing power grid space weather impacts at equatorial latitudes, In *Ionospheric Space Weather: Longitude and Hemispheric Dependences and Lower Atmosphere Forcing*, Edited by T. Fuller-Rowell et al., *AGU Geophysical Monograph*, *220*, pp. 33–40, doi:10.1002/9781111892916.ch3.

Molinski, T. S., W. E. Feero, and B. L. Damsky (2000), Shielding grids from solar storms, *IEEE Spectrum*, *37*, 55, doi:10.1109/6.880955.

Myllys, M., A. Viljanen, ø. A. Rui, and T. M. Ohnstad (2014), Geomagnetically induced currents in Norway: the northernmost high-voltage power grid in the world, *Journal of Space Weather and Space Climate*, *4*, A10, doi:10.1051/swsc/2014007.

National Science and Technology Council (2015a), National Space Weather Strategy, Executive Office of the President (EOP), USA. Available at https://www.whitehouse.gov/sites/default/files/microsites/ostp/final_nationalspace- weather-strategy_20151028.pdf.

National Science and Technology Council (2015b), National Space Weather Action Plan, Executive Office of the President (EOP), USA. Available at https://www.whitehouse.gov/sites/default/files/microsites/ostp/final_nationalspace- weatheractionplan_20151028.pdf.

Ngwira, C. M., and A. Pulkkinen (2018), An overview of science challenges pertaining to our understanding of extreme geomagnetically induced currents, In *Extreme Events in Geospace: Origins, Predictability and Consequences*, Edited by N. Buzulukova et al., Elsevier, pp. 187–208.

Ngwira, C. M., A. Pulkkinen, L. A. McKinnell, and P. J. Cilliers (2008), Improved modeling of geomagnetically induced currents in the South African power network, *Space Weather*, *6*, S11004, doi:10.1029/2008SW000408.

Ngwira, C. M., L. A. McKinnell, P. J. Cilliers, A. Viljanen, and R. Pirjola (2009), Limitations of the modelling of geomagnetically induced currents in the South African power network, *Space Weather*, *7*, S10002, doi:10.1029/2009SW000478.

Ngwira, C. M., L.-A. McKinnell, and P. J. Cilliers (2011), Geomagnetic activity indicators for geomagnetically induced current studies in South Africa, *Advances in Space Research*, *48*, 529–534.

Ngwira, C. M., A. Pulkkinen, F. D. Wilder, and G. Crowley (2013), Extended study of extreme geoelectric field event scenarios for geomagnetically induced current applications, *Space Weather*, *11*, 121–131, doi:10.1002/swe.20021.

Ngwira, C. M., A. Pulkkinen, M. M. Kuznetsova, and A. Glocer (2014), Modeling extreme "Carrington–type" space weather events using three-dimensional MHD code simulations, *Journal of Geophysical Research*, *119*, 4456–4474, doi:10.1002/2013JA019661.

Ngwira, C. M., A. Pulkkinen, E. Bernabeu, J. Eichner, A. Viljanen, and G. Crowley (2015), Characteristics of extreme geoelectric fields and their possible causes: Localized peak enhancements, *Geophysical Research Letters*, *42*, doi:10.1002/2015GL065061.

Ngwira, C. M., D. Sibeck, M. V. D. Silveria, M. Georgiou, J. M. Weygand, Y. Nishimura, and D. Hampton (2018a), A study of intense local dB/dt variations during two geomagnetic storms, *Space Weather 16*, 676–693, doi:10.1029/2018SW001911.

Ngwira, C. M., A. Pulkkinen, M. M. Kuznetsova, and A. Glocer (2018b), Reply to comments by Tsurutani et al. on "Modeling extreme 'Carrington–type' space weather events using three-dimensional MHD code simulations", *Journal of Geophysical Research*, *123*, doi:10.1002/2017JA024928.

Oliveira, D. M., D. Arel, J. Raeder, E. Zesta, C. M. Ngwira, B. A. Carter, E. Yizengaw, A. J. Halford, B. T. Tsurutani, and J. W. Gjerloev (2018), Geomagnetically induced currents caused by interplanetary shocks with different impact angles and speeds, *Space Weather*, *16*, 636–647 doi:10.1029/2018SW001880.

Pirjola, R. (1982), Electromagnetic induction in the Earth by a plane wave or by fields of line currents harmonic in time and space, *Geophysica*, *18*(1–2), 1–161.

Pirjola, R. (2000), Geomagnetically induced currents during magnetic storms, *IEEE Transactions on Plasma Science*, *28*(6), 1867–1873.

Pirjola, R. (2002a), Review on the calculation of the surface electric and magnetic fields and geomagnetically induced currents in ground based technological systems, *Surveys in Geophysics*, *23*, 71–90.

Pirjola, R. (2002b), Space weather effects on technological systems on the ground, *Asia-Pacific Conference on Environmental Electromagnetics*, CEEM'2000 (IEEE Cat. No.00EX402), Shanghai, China, 7 May 2002, IEEE. doi:10.1109/CEEM.2000.853936

Pirjola, R. (2002c), Fundamentals about the flow of geomagnetically induced currents in a power system applicable to estimating space weather risks and designing remedies, *Journal of Atmospheric and Solar Terrestrial Physics*, *64*, 1967–1972.

Pulkkinen, A., A. Viljanen, K. Pajunpaa, and R. Pirjola (2001), Recordings and occurrence of geomagnetically induced currents in the Finnish natural gas pipeline network, *Journal of Applied Geophysics*, *48*, 219–231.

Pulkkinen, A., A. Thomson, E. Clarke, and A. McKay (2003), April 2000 geomagnetic storm: Ionospheric drivers of large geomagnetically induced currents, *Annales Geophysicae*, *21*, 709–717.

Pulkkinen, A., S. Lindahl, A. Viljanen, and R. Pirjola (2005), Geomagnetic storm of 29-31 October: Geomagnetically induced currents and their relation to problems in the Swedish high-voltage power transmission system, *Space Weather*, *3*, S08C03, doi:10.1029/2004SW000123.

Pulkkinen, A., R. Pirjola, and A. Viljanen (2007), Determination of the ground conductivity and system parameters for optimal modeling of geomagnetically induced current flow in technological systems, *Earth Planets and Space*, *59*, 999–1006.

Pulkkinen, A., E. Bernabeu, J. Eichner, C. Beggan, and A. W. P. Thomson (2012), Generation of 100-year geomagnetically induced current scenarios, *Space Weather*, *10*, S04003, doi:10.1029/2011SW000750.

Pulkkinen, A., E. Bernabeu, J. Eichner, A. Viljanen, and C. M. Ngwira (2015), Regional-scale high-latitude extreme geoelectric fields pertaining to geomagnetically induced currents, *Earth, Planets, and Space*, *67*, 93, doi:10. 1186/s40623-015-0255-6.

Pulkkinen, A., E. Bernabeu, A. Thomson, A. Viljanen, R. Pirjola, D. Boteler, J. Eichner, P. J. Cilliers, D. Welling, N. P. Savani, R. S. Weigel, J. J. Love, C. Balch, C. M. Ngwira, G. Crowley, A. Schultz, R. Kataoka, B. Anderson, D. Fugate, J. J. Simpson, and M. MacAlester (2017), Geomagnetically induced currents: Science, engineering and applications readiness, *Space Weather*, *15*, 28–856, doi:10.1002/2016SW001501.

Schultz, A. (2009), EMScope: A continental scale magnetotelluric observatory and data discovery resource, *Data Science Journal*, *8*, PIGY6–IGY20.

Thomson, A. W. P., E. Clarke, A. McKay, and S. J. Reay (2005), Surface electric fields and geomagnetically induced currents in Scottish Power grid during 30 October 2003 geomagnetic storm, *Space Weather*, *3*, S11002, doi:10.1029/2005SW00156.

Thomson, A. W. P., E. B. Dawson, and S. J. Reay (2011), Quantifying extreme behavior in geomagnetic activity, *Space Weather*, *9*, S10001, doi:10.1029/2011SW000696.

Torta, J. M., L. Serrano, J. R. Regué, A. M. Sánchez, and E. Roldán (2012), Geomagnetically induced currents in a power grid of northeastern Spain, *Space Weather*, *10*, S06002, doi:10.1029/2012SW000793.

Trichtchenko, L., and D. H. Boteler (2004), Modeling geomagnetically induced currents using geomagnetic indices and data, *IEEE Transactions on Plasma Science*, *32*(4), 1459–1469.

Trivedi, N. B., I. Vitorello, W. Kabata, S. L. G. Dutra, A. L. Padilha, M. S. Bologna, M. B. de Pádua, A. P. Soares, G. S. Luz, F. de A. Pinto, R. Pirjola, and A. Viljanen (2007), Geomagnetic conjugate observations of large-scale traveling ionospheric disturbances using GPS networks in Japan and Australia, *Space Weather*, *5*, S04004, doi:10.1029/2006SW000282.

Tsurutani, B. T., G. S. Lakhina, E. Echer, R. Hajra, C. Nayak, A. Mannucci, and X. Meng (2018), Comment on "Modeling extreme 'Carrington-type' space weather events using three-dimensional global MHD simulations by C. M. Ngwira, A. Pulkkinen, M. M. Kuznetsova, and A. Glocer", *Journal of Geophysical Research*, *123*, doi:10.1002/2017JA024779.

Turnbull, K. L., J. A. Wild, F. Honary, A. W. P. Thomson, and A. J. McKay (2009), Characteristics of variations in the ground magnetic field during substorms at mid latitudes, *Annales Geophysicae*, *27*, 3421–3428.

Viljanen, A., and R. Pirjola (1994), Geomagnetically induced currents in the Finnish high-voltage power system, *Surveys in Geophysics*, *15*, 383–408.

Viljanen, A., A. Pulkkinen, R. Pirjola, K. Pajunpää, P. Posio, and A. Koistinen (2006), Recordings of geomagnetically induced currents and a nowcasting service of the Finnish natural gas pipeline, *Space Weather*, *4*, S10004, doi:10.1029/2006SW000234.

Watari, S., M. Kunitake, K. Kitamura, T. Hori, T. Kikuchi, K. Shiokawa, N. Nishitani, R. Kataoka, Y. Kamide, T. Aso, Y. Watanabe, and Y. Tsuneta (2009), Measurements of geomagnetically induced current (GIC) in a power grid in Hokkaido, Japan, *Space Weather*, *7, S03002*, S03002, doi:10.1029/2008SW000417.

Wik, M., R. Pirjola, H. Lundstedt, A. Viljanen, P. Wintoft, and A. Pulkkinen (2009), Space Weather events in July 1982 and October 2003 and the effects of geomagnetically induced currents on Swedish technical systems, *Annales of Geophysicae*, *27*, 1775–1787.

Wintoft, P. (2005), Study of solar wind coupling to the time difference horizontal geomagnetic field, *Annales Geophysicae*, *23*, 1949–1957.

Zhang, J. J., C. Wang, and B. B. Tang (2012), Modeling geomagnetically induced electric field and currents by combining a global MHD model with a local one-dimensional method, *Space Weather*, *10*, S05005, doi:10.1029/2012SW000772.

Zois, I. P. (2013), Solar activity and transformer failures in the Greek national electric grid, *Journal of Space Weather and Space Climate*, *3*, A32, doi:10. 1051/swsc/2013055.

2

Interpolating Geomagnetic Observations: Techniques and Comparisons

E. Joshua Rigler[1], Robyn A. D. Fiori[2], Antti A. Pulkkinen[3], Michael Wiltberger[4], and Christopher Balch[5]

ABSTRACT

Five geomagnetic vector interpolation techniques are reviewed and compared by analyzing their performance when applied to realistic inputs simulated by a state-of-the-art geospace general circulation model. The availability of synthetic "ground truth" allows meaningful estimates of relative interpolation error as a two-dimensional function of separation between geographically sparse input coordinates. Three of these techniques – nearest neighbor, triangular barycentric, and Gaussian Process regression – are entirely based on the input data, and do not benefit from any knowledge of physics that might improve predictions in unsampled regions. Two of the techniques – spherical cap harmonic analysis and spherical elementary current system inversion – incorporate simple physical understanding into their basis functions and generally provide better predictions even when far removed from input measurements. Spherical elementary currents generate fewer interpolation artefacts in the spatial domain.

Key Points

- Five geomagnetic vector interpolation techniques are reviewed and compared by analyzing their performance when applied to realistic inputs simulated by a state-of-the-art geospace general circulation model.
- The nearest neighbor, triangular barycentric, and Gaussian Process regression techniques are based on the input data and do not benefit from any knowledge of physics that might improve predictions in unsampled regions.
- Spherical cap harmonic analysis and spherical elementary current system inversion techniques incorporate simple physical understanding into their basis functions and generally provide better predictions even when far removed from input measurements.

2.1. INTRODUCTION

The magnetic field measured at Earth's surface is a superposition of magnetic fields generated by electrical currents flowing both internal and external to the Earth. The Earth's predominant "main field" is the quasi-static consequence of an electromagnetic dynamo within its

[1] *Geomagnetism Program, United States Geological Survey, Golden, CO, USA*

[2] *Geomagnetic Laboratory, Canadian Hazards Information Service, Natural Resources Canada, Ottawa, Ontario, Canada*

[3] *Goddard Space Flight Center, Space Weather Laboratory, National Aeronautics and Space Administration, Greenbelt, MA, USA*

[4] *High Altitude Observatory, National Center for Atmospheric Research, Boulder, CO, USA*

[5] *Space Weather Prediction Center, National Oceanic and Atmospheric Administration, Boulder, CO, USA*

Geomagnetically Induced Currents from the Sun to the Power Grid, Geophysical Monograph 244, First Edition.
Edited by Jennifer L. Gannon, Andrei Swidinsky, and Zhonghua Xu.

molten, mostly iron, outer core. The main field typically comprises over 95% of the total magnetic field intensity measured at any given time, but it evolves very slowly, with variation time scales that range from decades to millennia. Much of the data analysis and effort extended to create main field models (e.g., Alken et al., 2015; Finlay et al., 2015; Hamilton et al., 2015; Sabaka et al., 2015; Thébault et al., 2015) goes into filtering out the remaining ~5% of the total measured magnetic field intensity, which (1) tends to vary on time scales of less than a second to years, (2) is mostly due to changes in the electromagnetic environment above Earth's surface, and (3) is considered noise by those primarily concerned with the Earth's main field.

However, this geomagnetic noise, or disturbance, is a primary concern in space physics, and its careful study led to some of the earliest and most profound discoveries in this field, as summarized nicely by Cliver (1994). Nishida (1978) provides a concise, but fairly complete, synopsis of the advances in our understanding of the Earth's magnetosphere made possible by using geomagnetic data to remotely sense electrical currents flowing above the atmosphere using Maxwell's equations. This remote sensing capability has allowed the electromagnetic geospace environment to be monitored nearly continuously since the mid-twentieth century, largely in the form of various geomagnetic indices (Mayaud, 1980). More recently, geomagnetic disturbance data have become critical as "ground truth" for validating increasingly complex global geospace models used for space weather forecasting (e.g., Pulkkinen et al., 2010, 2011, 2013; Rastätter et al., 2013).

Historically, both basic and applied research have relied largely on point measurements of geomagnetic disturbance on the Earth's surface, or on carefully constructed indices that combine a few point measurements into a global proxy for some geospace phenomenon of interest. However, there is a growing, even urgent, need to reliably and accurately determine magnetic disturbance at arbitrary geographic coordinates. As technological systems become more sensitive to the effects of space weather (or more accurately, as we learn of their sensitivities), much more localized estimates of the disturbance field are necessary to accurately assess geomagnetic hazards, and to help mitigate related impacts. One obvious example is directional drilling, used commonly in the petroleum extraction industry, where magnetometers located on drill heads must be referenced to a known surface field in order to determine and track their orientation (e.g., Buchanan et al., 2013; Gleisner et al., 2006).

More germane to this monograph is the requirement for local estimates of geomagnetic disturbance that induce geoelectric fields known to adversely impact electrical power grids (e.g., Boteler, 2003; Boteler et al., 1998; Pirjola, 2002). Such estimates are required at arbitrary locations along a given power grid segment, where they are combined with local estimates of ground conductivity in order to model induced electric field, which is then integrated to obtain the voltage drop between grid nodes (e.g., Bonner and Schultz, 2017; Kappenman, 2004; Lucas et al., 2018). It is not feasible to install a sufficient number of magnetometers to satisfy such spatial resolution requirements without invoking interpolation of one form or another. Fortunately, spatial scales associated with geomagnetic disturbance are generally long enough that a much smaller number of magnetometers is required if a decent interpolation scheme is used (although, as we shall see, there currently are not nearly enough magnetometers in operation over most of North America).

The purpose of this document is to review several common methods for two-dimensional (2D) vector field interpolation, to compare their relative strengths and limitations, and to consider how each might be adapted to better support specific power grid requirements. A practical definition of "interpolation" is used here: to specify a value at an arbitrary and unsampled location, given measurements at sampled locations, and an assumed mathematical relationship between these locations. The reader will notice that no attempt is made to differentiate curve or function fitting, which is still interpolation, but which acknowledges the possibility for error in the mathematical relationship and/or error in the measurements. Extrapolation is also included, although it should be acknowledged that optimal methods might not be identical to those used for interpolation. Formal data assimilation is excluded, but only because the underlying mathematical relationships between observation points can be very complicated and warrant separate study beyond the scope of this document. Also excluded are regression models based on independent variables that differ from those being predicted, although we acknowledge that most interpolation schemes can be improved by using such exogenous information to refine mathematical relationships between sample locations. The different interpolation techniques studied here are not applied to real geomagnetic observations, but rather realistic synthetic observations simulated by Lyon–Feddar–Mobarry (LFM; Lyon et al., 2004), a well-known geospace general circulation model (GGCM), in order to evaluate performance far away from input observations.

2.2. SYNTHETIC GEOMAGNETIC DISTURBANCE OVER NORTH AMERICA

If geomagnetic field interpolation is to support studies of geomagnetically induced currents (GICs), there is little value in modeling Earth's main magnetic field. Rather, it is the disturbance to the main field caused by magnetospheric and ionospheric currents (and their induced telluric currents) that is of primary concern. Removal of the quasi-static main field from real observations is not especially straightforward, nor is it standardized. Although numerous time series detrending techniques have been implemented over the years that attempt to do so (e.g., Gjerloev, 2012; Love and Gannon, 2009; Rigler, 2017; Sugiura, 1964), none of these are guaranteed to fully filter out the main field from observations and leave only a disturbance field to interpolate. Practically speaking, this uncertainty is relatively low, and most of these techniques work well enough for operational use.

However, when validating and comparing geomagnetic disturbance interpolation tools, this source of uncertainty can be completely eliminated by employing synthetic observations that are known, a priori, to have no main field component. More importantly, synthetic observations provide a "ground truth" that may not be perfectly accurate in an absolute sense, but allows a relative uncertainty to be estimated far from samples used to constrain the interpolation. This kind of analysis is not uncommon. For example, Amm and Viljanen (1999) constructed a simple Cowling channel and employed the Biot–Savart relationship to estimate a ground truth against which to compare interpolated magnetic fields generated using both spherical elementary currents and spherical cap harmonics. Pulkkinen et al. (2003) did something similar using even more complicated ionospheric current models designed to mimic ionospheric configurations thought to manifest above Scandinavia during geomagnetically active intervals.

Still, these models represented highly idealized ionospheric configurations with limited spatial extent, and are generally understood to be isolated to high latitudes. For the purpose of this study, namely the demonstration and comparison of geomagnetic disturbance interpolation techniques over North America, a more general model was required that could generate realistic disturbance for a broad range of locations in a physically self-consistent manner. The LFM model, using the MIX (Merkin and Lyon, 2010) ionospheric potential solver, and coupled with the Rice Convection Model (RCM; Toffoletto et al., 2003), is a modern magnetohydrodynamics-based GGCM that provides some of the most realistic simulations available to date (Wiltberger et al., 2017).

LFM+MIX+RCM does not model geomagnetic disturbance directly. Rather, it simulates magnetospheric, ionospheric, and field-aligned electrical current distributions that, through the well-known Biot–Savart relationship, are used to synthesize magnetic disturbance at arbitrary locations on Earth's surface. We note briefly the LFM+MIX+RCM cannot presently simulate telluric currents below Earth's surface. Approximations can be made to emulate such currents without requiring a full first-principles electromagnetic model. A long-known technique is to assume a virtual thin superconducting spherical shell at some depth below Earth's surface, and invoke so-called "image" dipoles and/or currents (e.g., Bonnevier and Boström, 1970; Boteler and Pirjola, 1998; Kisabeth and Rostoker, 1977; Pirjola and Viljanen, 1998; Pulkkinen et al., 2003) to correct estimated surface fields in a physically self-consistent manner. Image dipoles/currents are roughly consistent with real observations, enhancing the horizontal disturbance, and reducing the vertical disturbance, but since synthetic data are already being used to compare interpolation methods, there is freedom to ignore telluric currents entirely in this analysis, thus reducing the variables that must be considered when making comparisons between interpolation techniques. We revisit this issue later as we discuss practical considerations and possible future studies.

While there are obvious advantages to using a state-of-the-art GGCM to generate synthetic observations, there is also substantial processing time required, even on large supercomputers. Therefore, for this study, the scope of the simulation was limited to a relatively short interval that captures a single strong geomagnetic storm. This event occurred on 17 March 2015, lasted ~2 days, and has come to be referred to as the 2015 "St. Patrick's Day Storm." It exhibited a strong and somewhat extended sudden commencement, a complicated main phase, and a gradual recovery phase that was still not finished by midnight UTC on March 19. All of this can be seen in Figure 2.1, which shows a time series of simulated horizontal magnetic field components at the RES, BLC, FRD, and SJG magnetic observatories for 17–18 March 2015. The B_X component is geographically northward, while B_Y is eastward, following conventions in the geomagnetism research community. The locations of these, and 22 additional synthetic observatories, are provided in Table 2.1. It is worth noting that, while these plots present simulated output, they resemble real observations quite closely over multi-hourly time scales, and even capture most shorter period variations in a statistical sense. One exception is that expected substorm signatures are muted, and not as periodic in the simulated output as in reality. It should be emphasized, however, that this report is not intended as a validation study for the LFM+MIX+RCM model, which may be found in Lyon et al. (2004), Merkin and Lyon (2010), Toffoletto et al. (2003), or Wiltberger et al. (2017), and demonstrated by the multitude of studies that use this model for magnetospheric research. As such, a direct comparison between the synthetic data and real measurements will not be provided here.

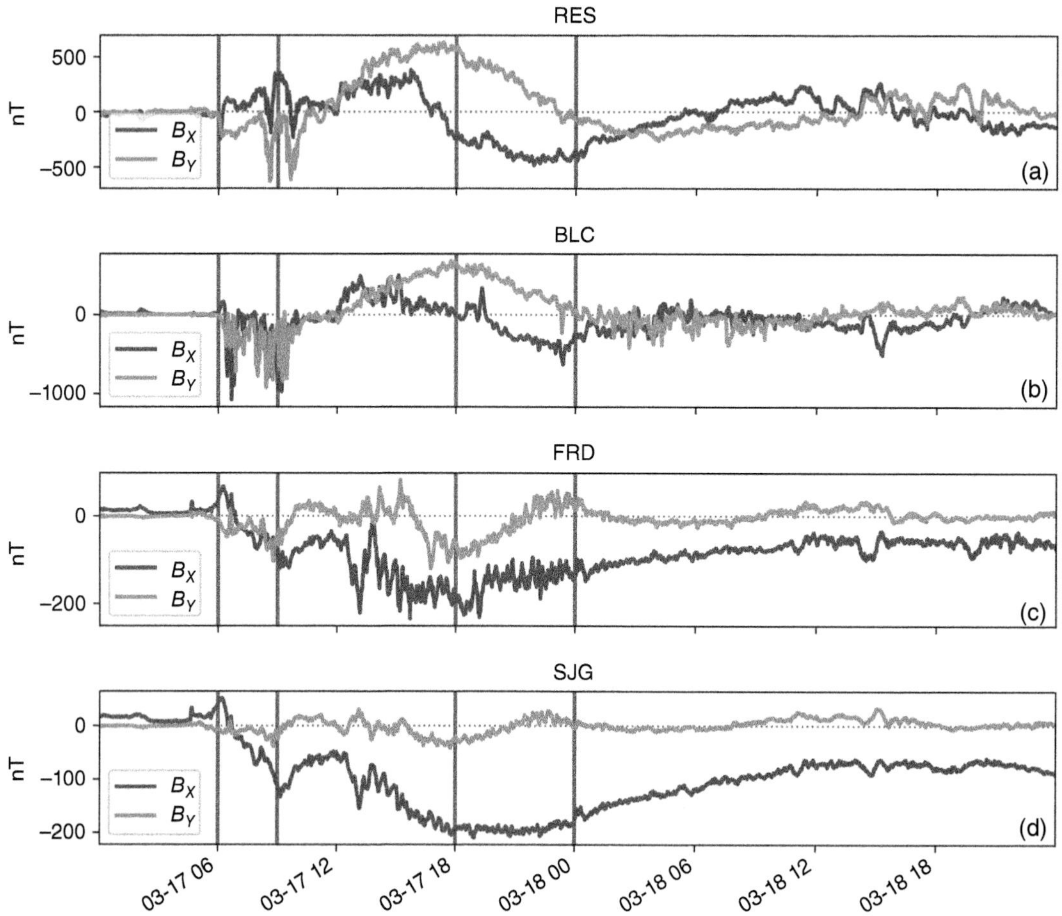

Figure 2.1 Simulated "St. Patrick's Day" geomagnetic storm horizontal vector component time series in geographically northward (X) and eastward (Y) directions for RES, BLC, FRD, and SJG synthetic magnetic observatories. The vertical lines correspond to 2D vector field snapshots presented in subsequent figures. (*See electronic version for color representation of this figure.*)

Table 2.1 Names, Abbreviations, and Locations of All Synthetic Magnetic Observatories Used in This Study.

IAGA ID (common name)	Latitude	Longitude	IAGA ID (common name)	Latitude	Longitude
BLC (Baker Lake[a])	64.32	−96.01	IQA (Brandon[a])	63.75	−68.52
BOU (Boulder[b])	40.14	−105.24	KAK (Kakioka[c])	36.23	140.18
BRD (Brandon[a])	49.87	−99.97	MEA (Meanook[a])	54.62	−113.35
BRW (Barrow[b])	71.32	−156.62	NEW (Newport[b])	48.27	−117.12
BSL (Stennis[b])	30.35	−89.64	OTT (Ottawa[a])	45.40	−75.55
CMO (College[b])	64.87	−147.86	RES (Resolute Bay[a])	74.69	−94.89
DED (Deadhorse[b])	70.36	−148.79	SHU (Shumagin[b])	55.35	−160.46
FRD (Fredericksburgh[b])	38.20	−77.37	SIT (Sitka[b])	57.06	−135.33
FRN (Fresno[b])	37.09	−119.72	SJG (San Juan[b])	18.11	−66.15
GUA (Guam[b])	13.59	144.87	STJ (St. John's[a])	47.59	−52.68
HAD (Hartland[d])	51.00	−4.48	TUC (Tucson[b])	32.18	−110.73
HER (Hermanus[e])	−34.43	19.23	VIC (Victoria[a])	48.52	−123.42
HON (Honolulu[b])	21.32	−158.00	YKC (Yellowknife[a])	62.48	−114.48

[a] Natural Resources Canada (NRCan)
[b] United States Geological Survey (USGS)
[c] Japan Meteorological Agency (JMA)
[d] British Geological Survey (BGS)
[e] South African National Space Agency (SANSA)

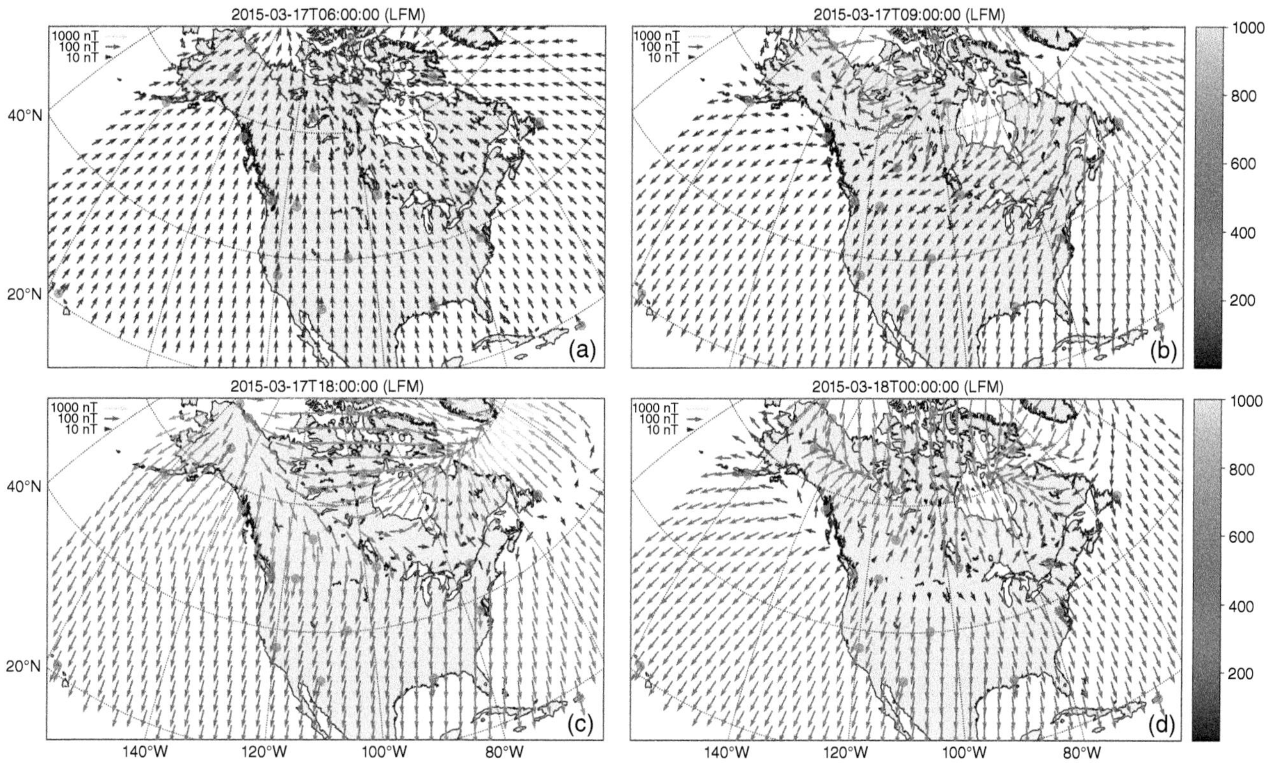

Figure 2.2 Snapshots of the North American distribution of simulated horizontal magnetic field vectors during the 2015 "St. Patrick's Day Magnetic Storm". Panel (a) is during the sudden commencement. Panel (b) is the main phase. Panel (c) is the transition from main to recovery phase. Panel (d) is near the heart of the recovery phase. *(See electronic version for color representation of this figure.)*

Representative snapshots of the simulated 2D gridded geomagnetic vector field are shown in Figure 2.2. Light gray spots are the locations of synthetic magnetic observatories used to interpolate to the same grid as the simulated ground truth, and the vectors overlaying these spots represent the horizontal magnetic vectors used as input for these interpolation schemes. All vector lengths are log-scaled, while their color corresponds to a more familiar linear scale. This is intended to help visualize a vector field that exhibits a range of variation of more than an order of magnitude.

Panel (a) occurs during the storm's sudden commencement, which corresponds to an enhancement in large-scale magnetopause currents that cause a global increase in fields parallel to the Earth's magnetic axis. With the exception of minor deviations in the polar cap that are barely visible in this plot, the geomagnetic disturbance field across North America is almost uniformly positive. Panel (b) takes place during the storm's main phase, while North America is in the midnight-dawn local time sector. The large geomagnetic disturbance seen at high latitudes is due to an enhancement in the westward auroral electrojet, with the signature of a possible substorm occurring over the northernmost portions of Canada's Yukon and Northwest Territories. Panel (c) occurs during the transition from storm main phase to recovery, but more interestingly, North America is centered on local noon, allowing one to see both westward and eastward auroral electrojet geomagnetic signatures, plus that of a polar cap current associated with strong magnetospheric convection. Panel (d) falls in the heart of the recovery phase of this magnetic storm, while North America is centered on the dusk meridian. There is a characteristic geomagnetic signature of the eastward auroral electrojet falling roughly along the United States–Canada border, whereby horizontal vectors diverge as the influence of nearby ionospheric currents gives way to the more global influence of the storm-enhanced ring current.

These four snapshots by no means capture all the dynamic variations in the geomagnetic disturbance over North America. They do, however, offer a representative set of horizontal vector field configurations that can be used to visually compare and contrast the performance of the different interpolation techniques studied in the following sections. Subsequent statistical analyses of the interpolation results were done using a time series of all 2880 1-min frames.

2.3. INTERPOLATION METHODOLOGIES AND RESULTS

We define interpolation in a very general sense: a process of specifying a useful value at an unsampled location given measurements at sampled locations, and an assumed relationship between these locations. There are countless such relationships that might be assumed, and the choice of which to use involves a trade-off between computational speed, precision, accuracy, robustness, and in some cases, physical interpretability. Five different cases were considered that represent a spectrum of such complexities for 2D geospatial interpolation: nearest neighbor (NN), triangle barycentric interpolation (TRI), Gaussian Process regression (GPR, a.k.a. Kriging), spherical cap harmonic analysis (SCHA), and spherical elementary currents (SECs).

Evaluation of the five interpolation schemes is performed through a straightforward comparison of the interpolated field against the expected field, as represented by the GGCM. Synthesized magnetic observations are generated at the location of ground observatories indicated by light gray dots in Figure 2.2. These synthetic data are then used as inputs to the interpolation schemes to model the geomagnetic disturbance field on a 2°×2° grid from 20° to 80° geographic latitude and −170° to −30° geographic longitude for comparison with simulated data at these same grid locations. Comparison maps are generated for the entire 17–18 March 2015 period at 1-min increments resulting in 2880 maps, and a statistical analysis performed for each interpolation scheme.

2.3.1. Nearest Neighbor

Perhaps the simplest form of interpolation is to use the nearest available observation as a best estimate of the physical quantity of interest at a given location. Of course, finding the nearest neighbor still requires a measure of the spatial separation between neighboring points. Ideally, this would be the distance along a great circle arc between points on a sphere, but the optimized algorithms found in many scientific software libraries cannot be easily modified to use this distance. Meanwhile, a naive application of such algorithms to the latitude and longitude of points of interest will lead to substantial distance errors for points near the poles. A compromise was made for this study whereby all location coordinates were projected onto a Lambert azimuthal equal area (LAEA) surface, with a "true scale" central coordinate of 50°N, and −110°E. Most points of interest were within 80° longitude, and 30° latitude, of this location, so the mapping distortion was minimal, even as one approached the North Pole. To be clear, only the location coordinates were transformed; directional vectors remained in their original local spherical coordinates to be interpolated.

Figure 2.3 contains representative snapshots of the NN-interpolated horizontal magnetic vector field over North America during the St. Patrick's Day storm. These are the same instants in time presented in Figure 2.2. Panel (a) is during the storm sudden commencement, and the uniform ground truth field is well represented by the nearest neighbor interpolation. However, the limitations of NN interpolation for such sparse input data become very obvious in panels (b–d). The spatial structure of known ionospheric features like the auroral electrojets is completely lost because only a single observatory actually sampled this in panel (b). Worse still, the substorm signature previously noted over the northern Yukon and Northwest Territories is entirely missing because there was no observatory sampling this feature; likewise for the eastward auroral electrojet noted in panel (c). Finally, while the characteristic divergence of horizontal fields associated with the eastward auroral electrojet might be visible in panel (d), its orientation is completely dependent on the geographic locations of the nearest observatories, and therefore looks nothing like the ground truth.

Next, we present accumulated statistics of temporal deviations between synthetic truth and interpolated geomagnetic vector fields. Rather than treat horizontal vector components as two independent variables, a horizontal vector correlation is considered, as described by Hanson et al. (1992), and summarized in the appendix of this report. This is similar to the familiar Pearson correlation for scalar series but uses complex numbers to represent vectors. Figure 2.4, panel (a), presents ρ^2, the squared magnitude of the complex vector correlation. Like the traditional r^2, this scalar metric represents the fraction of the synthetic truth's variation that can be explained by the interpolated series after an optimal (in a least-squares sense) linear transformation. For scalar sequences, that linear transformation is a simple scale factor, which is presented in panel (b). This should be interpreted as the average scale factor required to inflate/deflate the interpolated vector's magnitude to match the synthetic truth. A value greater than 1 means the interpolation scheme tends to systematically underestimate truth and must be inflated; a value less than 1 means the scheme tends to systematically overestimate synthetic truth and must be deflated. It should be noted that scaling tends toward unity as one moves closer to an observatory used for the interpolation. For 2D vector time series, there is an additional possible linear transformation – rotation. Panel (c) presents the average rotation required to align interpolated vectors with synthetic truth. Like the scaling factor, the rotation angle tends toward 0 as one moves

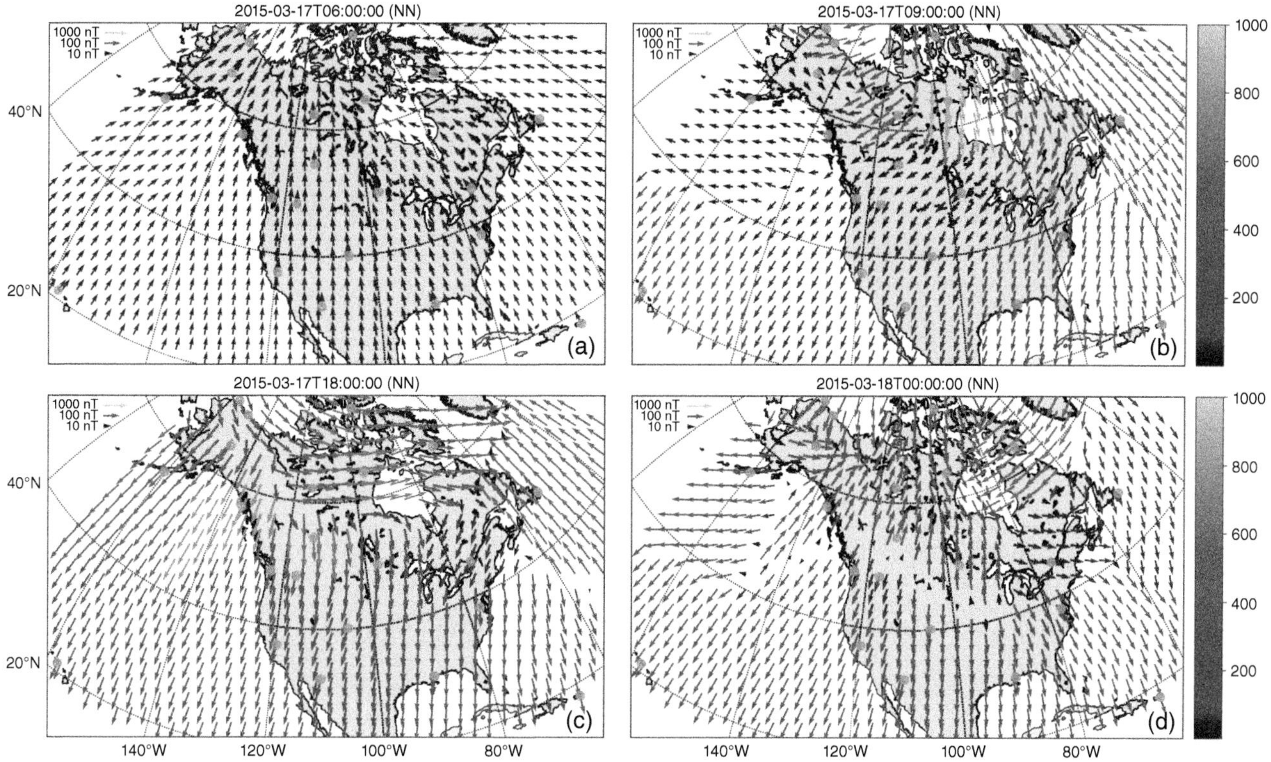

Figure 2.3 Snapshots of the North American distribution of nearest neighbor-interpolated horizontal magnetic field vectors during the 2015 "St. Patrick's Day Magnetic Storm." Panel (a) is during the sudden commencement. Panel (b) is the main phase. Panel (c) is the transition from main to recovery phase. Panel (d) is near the heart of the recovery phase. (*See electronic version for color representation of this figure.*)

closer to an observatory used in the interpolation. The range of rotation angles here is −180 to +180°, which leads to some contour plotting artefacts far away from observations, but we judge this to be a benefit, as it draws attention to regions where interpolation performs exceedingly poorly.

Nearest-neighbor interpolation performs reasonably well over much of our region of interest, especially the southern half of the continental United States (CONUS), where more than 90% of the synthetic truth variation is captured during the 2015 St. Patrick's Day storm. This is indicative of longer spatial scales expected for geomagnetic disturbance at lower latitudes, which is driven primarily by magnetospheric currents flowing farther away. This high squared correlation deteriorates quickly at locations southeastward of the Hudson Bay, in Canada, and in the mid-to-north Atlantic. The low squared correlations near Hudson Bay are mostly associated with inflated magnetic disturbance according to panel (b). These higher latitudes exhibit shorter disturbance spatial scales (due to ionospheric currents flowing only 100–200 km above the ground), which means that the nearest neighbors must be closer for accurate NN interpolation. The mid-to-north Atlantic's low squared correlation is mostly associated with rotation error, as indicated by panel (c). Spatial scales at these middle latitudes are expected to be relatively long because they are mostly out from under the high-latitude ionospheric currents, but because the nearest observatory available for NN-interpolating in this mid-latitude region is St. Johns, on Newfoundland, the interpolated disturbance fields behave as if they are in fact being driven by high-latitude ionospheric currents. For much of this interval, the actual high-latitude disturbance fields pointed in the opposite direction to mid-to-north Atlantic disturbance fields (see panel (c) of Figures 2.2 and 2.3 for a representative snapshot). In the northern portions of CONUS, we see that ρ^2 also drops, if not quite as significantly as the regions just described. Like southeastward of Hudson Bay, however, this drop is associated with inflationary error, not rotational error, suggesting that magnetic disturbance here exhibits shorter spatial scales than are achievable with NN interpolation.

It should be noted that nearest-neighbor interpolation will never produce a value that was not already measured.

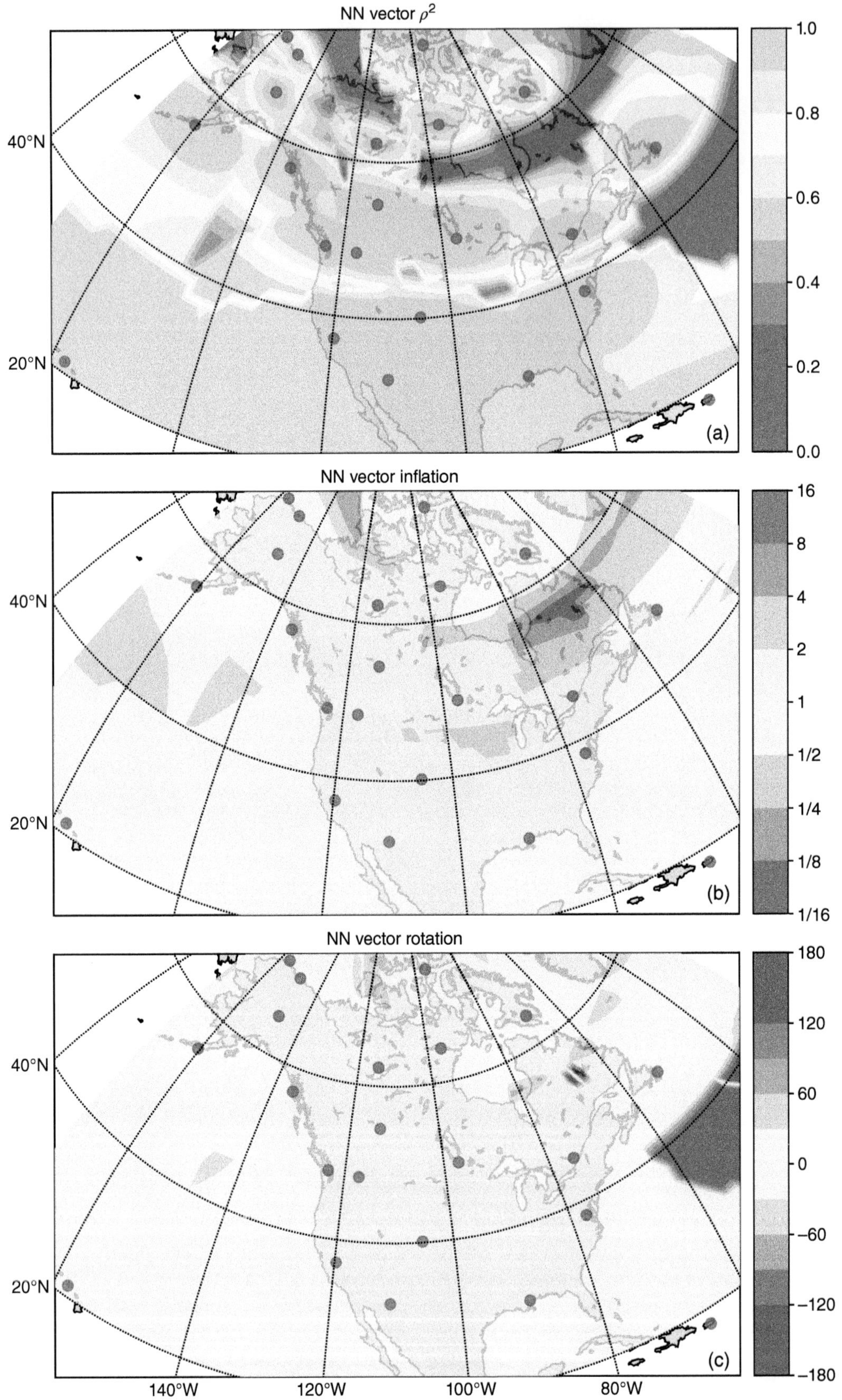

Figure 2.4 Comparison of synthetic truth and nearest neighbor-interpolated vector fields over North America. Panel (a) presents squared temporal vector correlations. Panel (b) presents the average vector inflation required to correct interpolation to match synthetic truth. Panel (c) presents the average vector rotation required to correct interpolation to match truth. (*See electronic version for color representation of this figure.*)

This makes it robust, and possibly a safe extrapolation tool in the absence of any other information. Otherwise, it is not recommended, and it is included here mostly to round out the spectrum of interpolation possibilities.

2.3.2. Triangle Barycentric Interpolation

The next logical level of interpolation complexity is a linear function between two points. In the one-dimensional case (i.e., a straight line between two sampled points runs through an unsampled point of interest), this reduces to a weighted average of the two sampled points, where weights are just the inverse of the normalized distance from the unsampled point to the sampled points:

$$f(x) = f(x_0)\left(1 - \frac{x - x_0}{x_1 - x_0}\right) + f(x_1)\left(1 - \frac{x_1 - x}{x_1 - x_0}\right). \quad (2.1)$$

In the 2D case where sampled points fall on a rectilinear grid, one simply applies this process thrice: once for each sampled pair above and below the point of interest, and parallel to the primary axis of the grid; then again using these two interpolated values, and orthogonal to the primary axis of the grid. This bilinear interpolation can be readily optimized, and in spite of the three linear fits involved, produces a smooth surface that is quadratic along any lines but the original rectilinear grid axes.

For unstructured data coordinates, bilinear interpolation is not possible. However, the concept of a weighted average may still be invoked if there are at least three sampled locations whose coordinates create a triangle that encompasses the unsampled coordinate of interest. This internal point can be written as a unique convex combination of the triangle's three vertices and so-called barycentric coordinates (α, β, γ) (i.e., $\mathbf{r} = \alpha\mathbf{r}_1 + \beta\mathbf{r}_2 + \gamma\mathbf{r}_3$, where $\mathbf{r} \equiv \{x, y\}$, and $\alpha + \beta + \gamma = 1$; see Farin, 1993, for a comprehensive review). Triangular barycentric coordinates are simply the normalized areas of the three sub-triangles constructed from the unsampled interior point and the three vertices constructed from sampled points:

$$\begin{aligned}
\alpha &= \frac{(y_2 - y_3)(x - x_3) + (x_3 - x_2)(y - y_3)}{(y_2 - y_3)(x_1 - x_3) + (x_3 - x_2)(y_1 - y_3)}, \\
\beta &= \frac{(y_3 - y_1)(x - x_3) + (x_1 - x_3)(y - y_3)}{(y_2 - y_3)(x_1 - x_3) + (x_3 - x_2)(y_1 - y_3)}, \quad (2.2) \\
\gamma &= 1 - \alpha - \beta.
\end{aligned}$$

These weights are applied to values at each corresponding vertex to interpolate to the unsampled interior location:

$$f(\mathbf{r}) = \alpha f(\mathbf{r}_1) + \beta f(\mathbf{r}_2) + \gamma f(\mathbf{r}_3). \quad (2.3)$$

Many modern scientific software libraries provide optimized interpolation functions based on barycentric weights, and these in turn often rely on so-called Delauney triangulation to generate the tessellation required to perform TRI interpolation. Much like NN interpolation, these algorithms are optimized for Cartesian coordinates, so an LAEA map projection was applied prior to determining the Delaunay tessellation required for TRI interpolation; this is presented in Figure 2.5.

Figure 2.6 contains representative snapshots of the TRI-interpolated horizontal magnetic vector field over

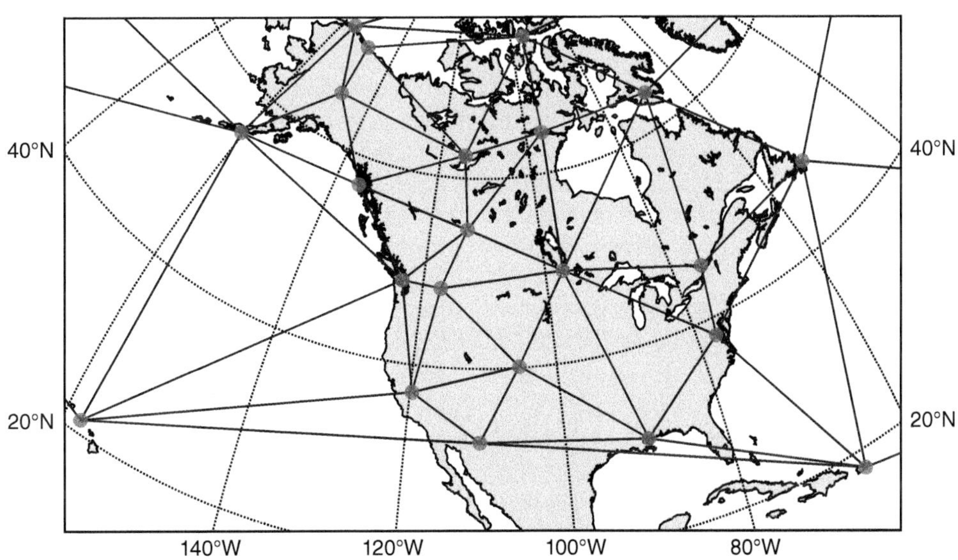

Figure 2.5 Delaunay tessellation for the set of geomagnetic observatory coordinates listed in Table 2.1. *(See electronic version for color representation of this figure.)*

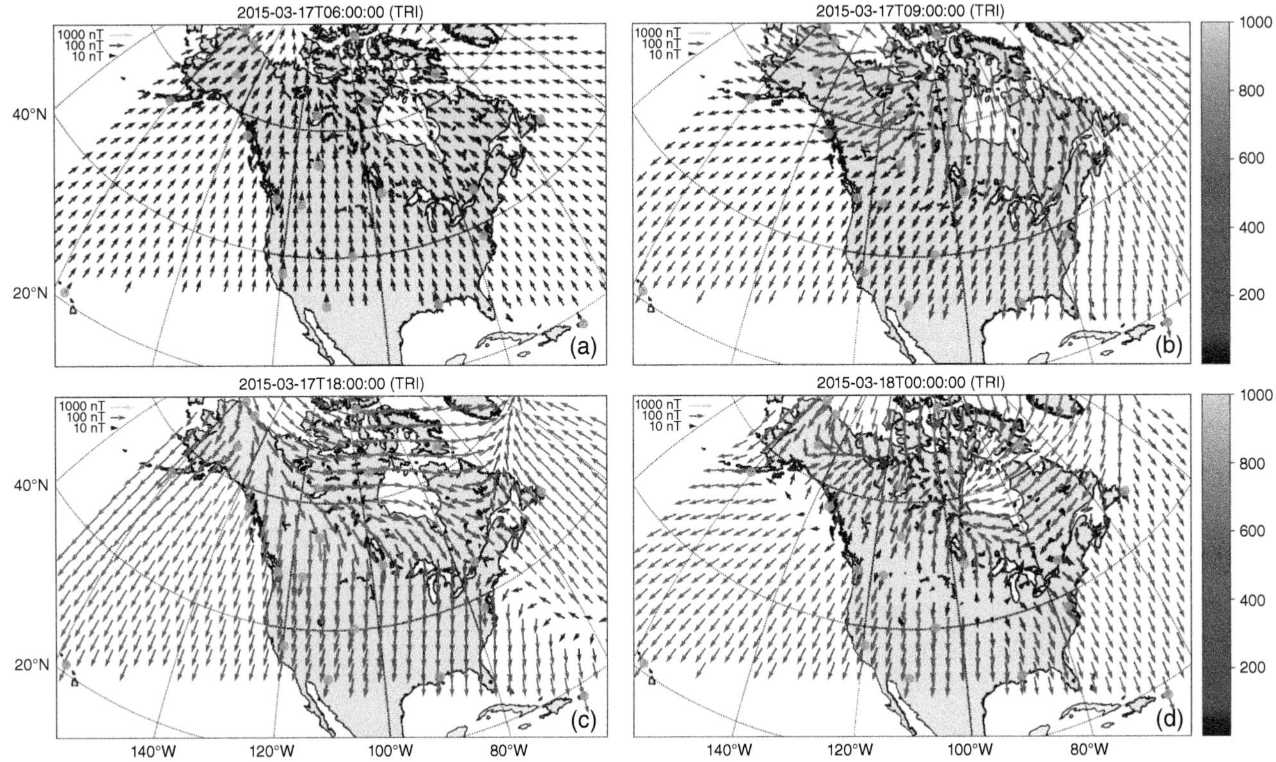

Figure 2.6 Snapshots of the North American distribution of TRI-interpolated horizontal magnetic field vectors during the 2015 "St. Patrick's Day Magnetic Storm." Panel (a) is during the sudden commencement. Panel (b) is the main phase. Panel (c) is the transition from main to recovery phase. Panel (d) is near the heart of the recovery phase. (*See electronic version for color representation of this figure.*)

North America during the St. Patrick's Day storm. These were generated with locations projected into LAEA coordinates and are the same instants in time presented in Figure 2.2. Panel (a) is during the storm sudden commencement, and the uniform ground truth field is well represented. However, the limitations of linear interpolation for such sparse input data become obvious in panels (b–d). In fact, the issues are very similar to those with NN interpolation, except that there is a smoother transition between observatories, rather than the abrupt boundaries that delineate the points of equal distance to a nearest observatory. The westward auroral electrojet and possible substorm noted in panel (b) of Figure 2.2 are very poorly reproduced; likewise for the eastward electrojet in panel (c). Linear interpolation does seem to capture at least some of the westward electrojet in panel (c), but only because, for this particular instant in time, there is a reasonable distribution of observatories available with which to interpolate. And again, there is an apparent divergence of geomagnetic vectors characteristic of an eastward electrojet in panel (d), but its orientation seems dictated by the observatory distribution more than anything else.

Perhaps the most glaring feature in Figure 2.6 is the lack of interpolated vectors across large swaths of the southern United States and northern Mexico, as well as the Caribbean and eastern Pacific Ocean between 20 and ~30° latitude. The reason is that enclosing triangles required for barycentric interpolation could not be constructed for these regions. To force such a triangle, say one with vertices at SJG, TUC, and HON, would lead to an improper tessellation, and gaps on the interior of the region of interest, which is even less acceptable than the noted lack of exterior coverage. A hybrid approach, perhaps using NN for exterior regions like this, might be employed to extrapolate, but we left this feature as-is to demonstrate a very typical complication with TRI interpolation.

Figure 2.7 presents accumulated statistics of temporal deviations between synthetic truth and TRI-interpolated geomagnetic vector fields. They look somewhat similar to results obtained from NN interpolation, especially at higher latitudes like the low squared correlation region southeastward of Hudson Bay. One exception is over the middle Atlantic Ocean, where TRI interpolation leads to a larger area of even poorer performance than NN. This is largely the result of using the HER (South Africa) magnetic observatory as a vertex of two different triangles used for interpolation. HER is particularly

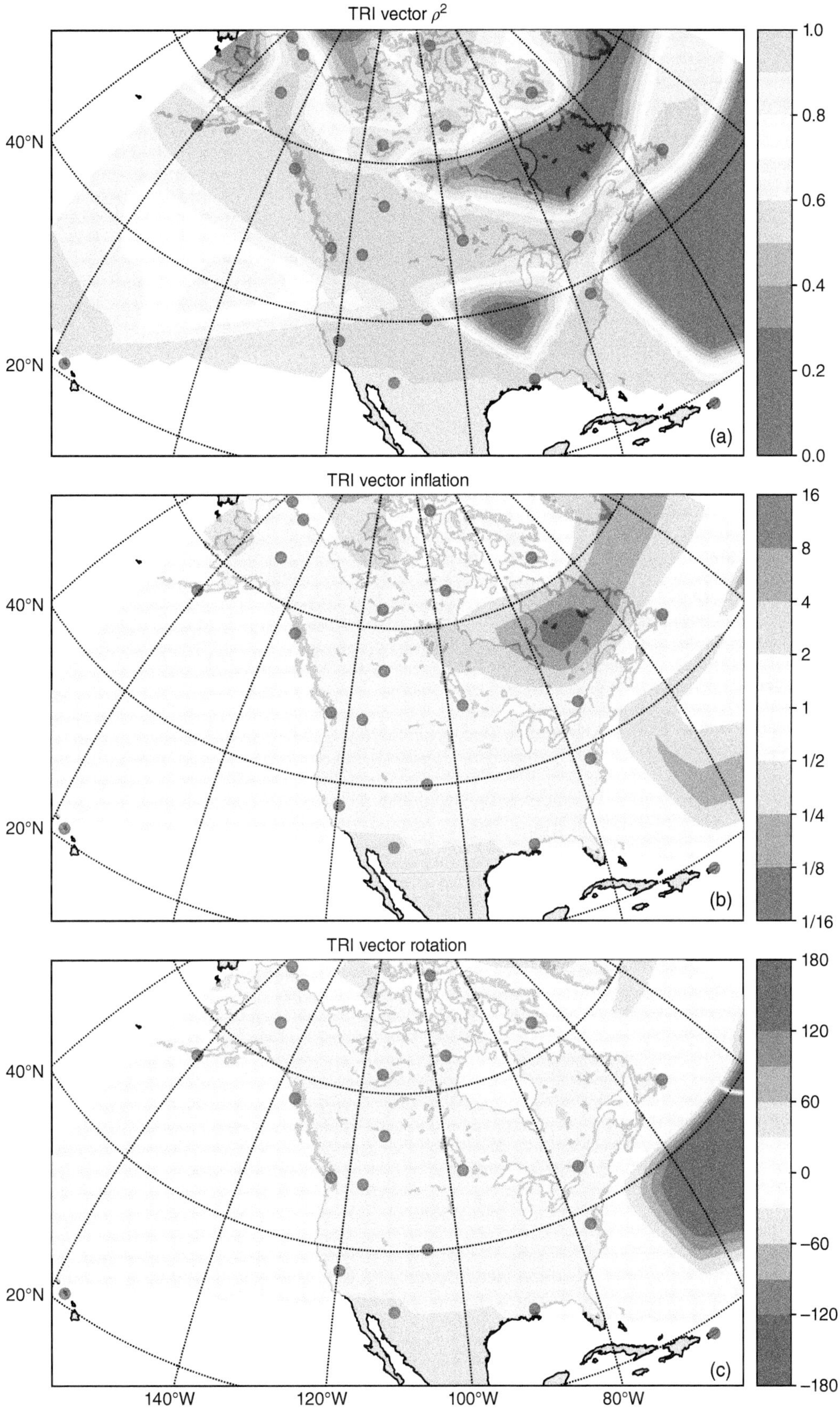

Figure 2.7 Comparison of synthetic truth and triangle barycentric-interpolated vector fields over North America. Panel (a) presents squared temporal vector correlations. Panel (b) presents the average vector inflation required to correct interpolation to match synthetic truth. Panel (c) presents the average vector rotation required to correct interpolation to match truth. (*See electronic version for color representation of this figure.*)

non-representative of this region, and yet it is given equal weight to the other two observatories when interpolating over the middle of the Atlantic Ocean. A similar argument might be made against using the HAD (United Kingdom) observatory when interpolating to locations in the mid-to-north Atlantic. However, it turns out that, at least along the line from STJ (Newfoundland) to HAD, the interpolation results are quite good. This is due to the fact that magnetic disturbance organizes itself roughly latitudinally. Upon reexamining the Delauney triangles used for this interpolation, we quickly see that triangles which are limited in latitudinal extent lead to relatively good interpolations, while triangles that traverse many latitudes tend to provide poor interpolations. Often there is no alternative triangular tessellation, especially near the edges of the spatial domain. Occasionally, however, the tessellation generated by standard techniques violates these physically motivated assumptions to create triangles that span large ranges of latitude when just a slight modification to the tessellation would result in significantly improved interpolations. This is what caused the very low squared correlations seen over the state of Missouri in the United States.

2.3.3. Gaussian Process Regression

The two preceding interpolation techniques can be problematic because they either: (1) consider only the nearest observations; or (2) assume inflexible scaling relationships between measurement locations. These can lead to artefacts in interpolated geomagnetic vector fields that are not only non-physical, but are often not even statistically plausible. One powerful tool to help address these issues is Gaussian Process Regression (GPR; Rasmussen and Williams, 2006), or as it is commonly referred to in geostatistics, Kriging, after the technique's originator Danie G. Krige (although its theoretical formalization was left largely to others like Georges Matheron, considered by many to be the founder of modern geostatistics).

GPR can be summarized as a weighted average of multiple input observations, where weights are a function of the relative positions of the observations and the point of interest. The manner in which GPR addresses those problems identified above becomes evident when one recognizes that this weighted average is just a multivariate linear regression, which itself provides the most likely value to be drawn from an N-dimensional Gaussian distribution conditioned on available observations. So, the interpolated value is statistically plausible by construction, and while GPR cannot by itself shed any light on underlying physics, it can be readily tuned to incorporate physical assumptions if they can be stated as a position-dependent weighting function.

Some basic definitions are needed before moving on. The covariance of two variables is the expected product of the deviations of each variable from their respective expected values:

$$\sigma_{XX'} = \mathrm{E}\left[\left(X - \mathrm{E}[X]\right)\left(X' - \mathrm{E}\left[X'\right]\right)\right]. \tag{2.4}$$

The covariance of a variable with itself is simply its variance, which is typically represented as σ_X^2. A variance–covariance matrix is square and symmetric, with diagonal elements composed of the univariate variances, and covariances for the off-diagonals. For a bivariate system, this is:

$$\Sigma_{XX'} \equiv \begin{bmatrix} \sigma_X^2 & \sigma_{XX'} \\ \sigma_{X'X} & \sigma_{X'}^2 \end{bmatrix}. \tag{2.5}$$

If linear regression of a set of dependent variables $\mathbf{Y} = [Y_1 \;\cdots\; Y_n]^\top$ on a set of independent variables $\mathbf{X} = [X_1 \;\cdots\; X_m]^\top$ is required, it may be convenient to consider the following:

$$\Sigma \equiv \begin{bmatrix} \Sigma_{\mathbf{XX}} & \Sigma_{\mathbf{XY}} \\ \Sigma_{\mathbf{YX}} & \Sigma_{\mathbf{YY}} \end{bmatrix}$$

where

$$\Sigma_{\mathbf{XX}} \equiv \begin{bmatrix} \sigma_{X1}^2 & \cdots & \sigma_{X1Xm} \\ \vdots & \ddots & \vdots \\ \sigma_{XmX1} & \cdots & \sigma_{Xm}^2 \end{bmatrix}, \Sigma_{\mathbf{YY}} \equiv \begin{bmatrix} \sigma_{Y1}^2 & \cdots & \sigma_{Y1Yn} \\ \vdots & \ddots & \vdots \\ \sigma_{YnY1} & \cdots & \sigma_{Yn}^2 \end{bmatrix},$$
$$\text{and } \Sigma_{\mathbf{XY}} \equiv \begin{bmatrix} \sigma_{X1Y1} & \cdots & \sigma_{X1Yn} \\ \vdots & \ddots & \vdots \\ \sigma_{XmY1} & \cdots & \sigma_{YmYn} \end{bmatrix} = \Sigma_{\mathbf{YX}}^\top. \tag{2.6}$$

With these definitions, a linear algebraic regression equation can be built that makes clear the weighted-average nature of GPR:

$$\hat{\mathbf{Y}} = \Sigma_{\mathbf{YX}}\Sigma_{\mathbf{XX}}^{-1}\mathbf{X} = \mathbf{WX}. \tag{2.7}$$

Perhaps equally useful to the regression itself is the robust estimate of the prediction error covariance that falls out of these relationships:

$$S_{\mathbf{YY}} = \Sigma_{\mathbf{YY}} - \Sigma_{\mathbf{YX}}\Sigma_{\mathbf{XX}}^{-1}\Sigma_{\mathbf{XY}} = \Sigma_{\mathbf{YY}} - \mathbf{W}\Sigma_{\mathbf{XY}}. \tag{2.8}$$

All of this is standard multivariate linear regression, if not necessarily presented in the form most familiar to the space physics community. While it is more typical to consider *sample* variances and covariances that are calculated from observations, there is no reason they cannot be constructed to exhibit some characteristic spatial relationship between points, so long as any resulting variance–covariance matrix remains positive definite. When combined with Equation (2.7), this is referred to as "simple" Kriging. One common approach is to define covariances that are a squared exponential function of distance:

$$\sigma_{XX'} = \exp\left(-\frac{d^2}{2\ell^2}\right) \tag{2.9}$$

where the distance between the points is d, and ℓ defines a length scale over which X's influence on X' diminishes by $1/e$ (i.e., $\exp(-1)$). In general, the larger ℓ, the smoother the interpolation; the smaller ℓ, the more localized the interpolation.

GPR is not so common as NN or TRI interpolation, and because it is more complicated algorithmically, implementation details differ between various scientific software libraries. To facilitate reproducibility, it is sensible to recommend a relatively mature and validated GPR library that is freely available as part of Python's SciPy machine learning toolkit scikit-learn (Pedregosa et al., 2011). This library refers to covariance functions as "kernels," and the squared exponential in particular as the radial basis function (RBF) kernel. Once again, in order to take advantage of these optimized libraries, and to ensure consistent comparisons between different interpolation techniques, latitudes and longitudes were projected into LAEA map coordinates prior to analysis. With that, an isotropic length scale of 1×10^6 pseudo-meters (1000 km) was specified. Input values are not transformed but remain in their original spherical coordinates.

Figure 2.8 contains representative snapshots of the GPR-interpolated horizontal magnetic vector field over North America during the St. Patrick's Day storm. These are the same instants in time presented in Figure 2.2. Panel (a) is during the storm sudden commencement, and the uniform ground truth field is well represented. However, some limitations of GPR interpolation for such sparse input data become obvious in panels (b–d). In fact, the issues are similar to those for both nearest neighbor and linear interpolation, but with fewer obvious interpolation artefacts. For example, there is none of the segmentation so obvious in the NN results, and the tessellation issues associated with exterior points in TRI

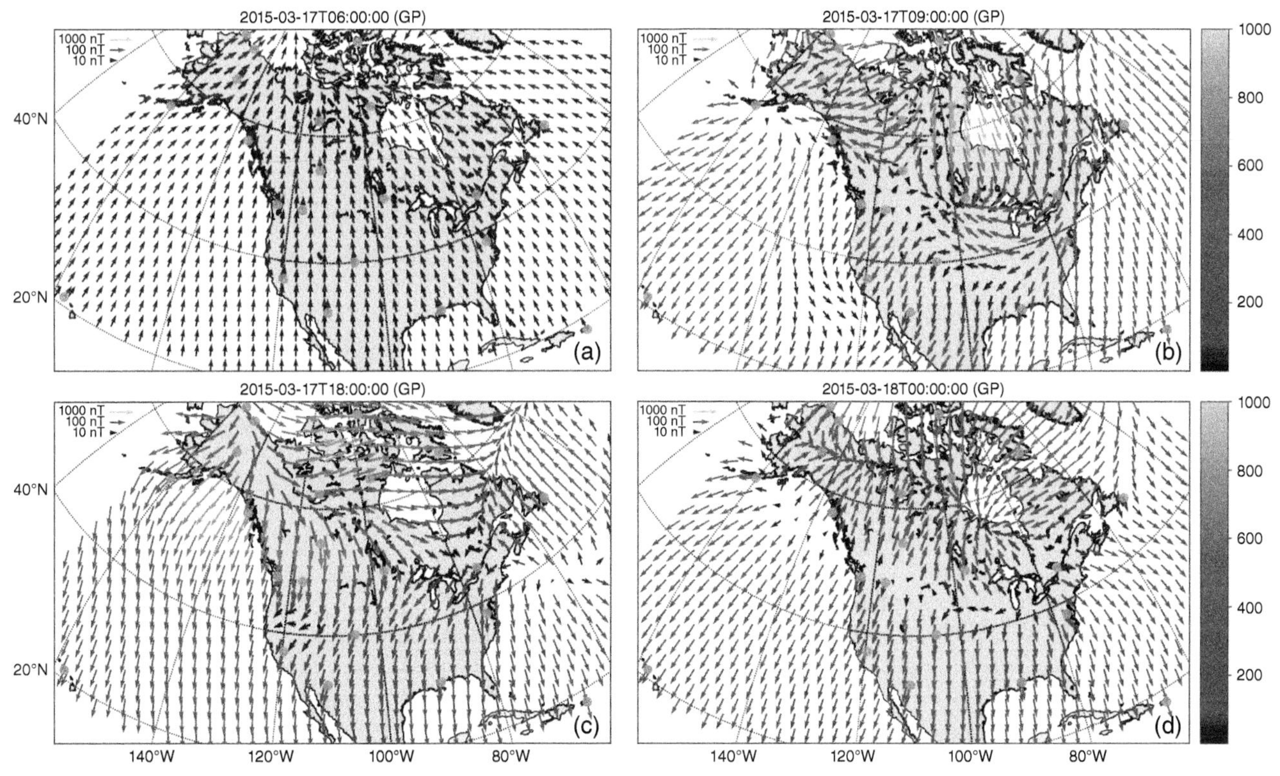

Figure 2.8 Snapshots of the North American distribution of Gaussian Process Regression-interpolated horizontal magnetic field vectors during the 2015 "St. Patrick's Day Magnetic Storm." Panel (a) is during the sudden commencement. Panel (b) is the main phase. Panel (c) is the transition from main to recovery phase. Panel (d) is near the heart of the recovery phase. (*See electronic version for color representation of this figure.*)

interpolation are nonexistent. However, the westward auroral electrojet and possible substorm noted in panel (b) are still poorly reproduced; likewise for the eastward electrojet in panel (c). GPR interpolation does seem to capture much of the westward electrojet in panel (c), but again, only because, for this particular instant in time, there is a reasonable distribution of observatories available with which to interpolate. Finally, the apparent divergence of geomagnetic vectors characteristic of an eastward electrojet in panel (d) is somewhat better captured by GPR than the other interpolation methods, but it is still problematic.

In panel (c) of Figure 2.8, there is an anomalous enhancement of eastward geomagnetic field, centered at ~ −110°E longitude, and 70°N latitude. This is not obvious in other interpolation results, and it is almost certainly not real. It turns out this is a less-than-intuitive consequence of the normalization of weights provided by Σ_{XX}^{-1} in Equation (2.7), which acts to deemphasize clusters of nearby observations, an intentional and usually valuable feature of GPR. This de-clustering works by generating weights for neighboring observatories that partially screen each other (e.g., one observatory is assigned a weight of 1.5, while another one nearby is assigned a weight of −1.0, rather than just reducing the two observatory weights to ~0.25). This works well so long as the independent inputs **X** actually exhibit covariance behavior similar to that of the covariance/kernel function used by GPR. However, in this circumstance, the geomagnetic field reconfigured in a manner that violated length scale assumptions, and observatories that should have screened out the impact of clusters for the anomalous region did not. More specifically, the College (CMO) observatory in Fairbanks, Alaska, and the Meanook (MEA) observatory just north of Edmonton, Alberta, were assigned strong negative weights that should have partially screened the positive weights associated with other nearby observatories, but there was almost no measured eastward magnetic disturbance at these two locations at this time, so the other positive weights overpredicted eastward geomagnetic disturbance.

This kind of post-regression analysis is not practical for most real-world applications, especially for non-experts who simply require a reliable interpolation tool. One way to mitigate the problem is to reduce the kernel length scale for GPR so that clusters are effectively smaller. In this instance, a length scale of 500 km, or half that used to generate the results in Figure 2.8, was sufficient to eliminate the anomaly. The downside is that those areas that are more distant from observations revert to the global mean more quickly. This is especially apparent over the Pacific and Atlantic Oceans. Another possibility is to regularize the solution using simple ridge parameters to augment the covariance matrix Σ_{XX}. This is sometimes referred to as Kriging with a nugget and is equivalent to assigning an uncertainty to observations. The overall fit will be smoothed and may not match observations exactly at their respective locations. Multiplying the diagonal of Σ_{XX} by 0.1, and adding this back to its diagonal, proved to be sufficient to remove the anomaly, but it oversmoothed the global solutions. In the end, a 1000 km length scale was retained, and no ridge parameters were specified, for demonstration purposes.

Figure 2.9 presents accumulated statistics of temporal deviations between synthetic truth and GPR-interpolated geomagnetic vector fields. Many of the features noted in the earlier interpolation statistics remain, but have become muted, especially over the North American continent. Squared correlation coefficients over the Pacific Ocean are actually quite a bit worse than both nearest neighbor and linear interpolation as one moves farther away from observations. This is to be expected though because, by design, GPR regresses to the mean of the entire domain as the covariance functions tend toward 0 at large distances d. The simpler interpolation methods did not exhibit such low squared vector correlations because they do not regress to a global mean, but rather to the nearest neighbor, or 3. The geomagnetic disturbance field was uniform enough over the Pacific Ocean to ensure relatively high squared correlations with the simpler interpolation techniques. This could be mitigated in GPR with larger length scales ℓ, but then issues like the anomalous enhancement noted for panel (c) in Figure 2.8 would occur more frequently. In the end, these kind of metrics and validation results would be useful to inform decisions on where to install new geomagnetic observatories if and when resources allow.

2.3.4. Spherical Cap Harmonic Analysis

Spherical cap harmonic analysis (SCHA) was originally developed for regional modelling of magnetic variations (Haines, 1985, 1988, 2007). In a current-free region of the Earth, the vector magnetic field (**B**) is expressed as the negative gradient of a scalar potential (Φ_B), which satisfies the Laplace equation and can therefore be represented as a series of harmonic basis functions. For global modelling, the familiar spherical harmonic basis functions are applied. For the case presented here, where data are limited to a specific region (i.e., North America), a modified set of spherical cap harmonic basis functions are instead used.

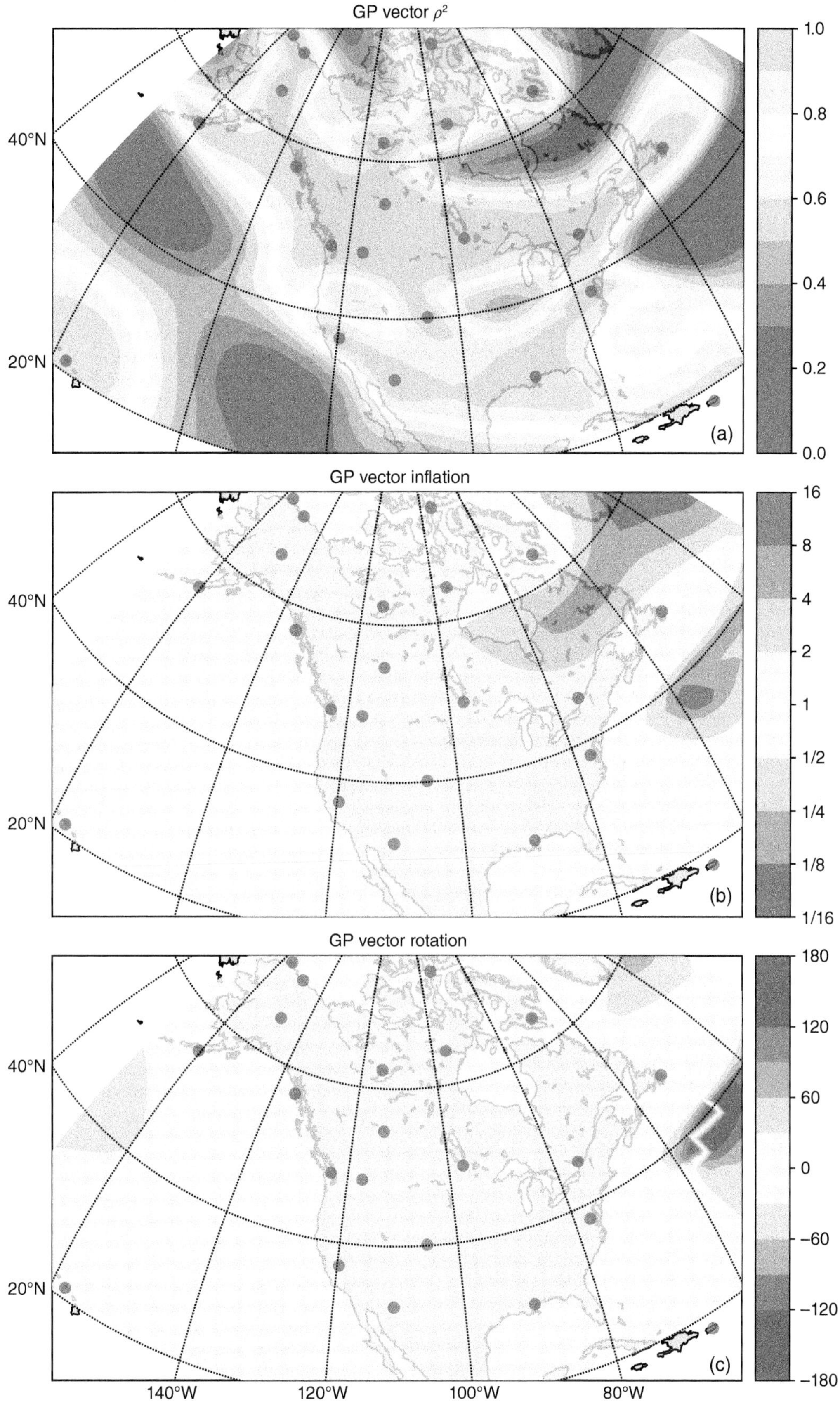

Figure 2.9 Comparison of synthetic truth and Gaussian Process Regression-interpolated vector fields over North America. Panel (a) presents squared temporal vector correlations. Panel (b) presents the average vector inflation required to correct interpolation to match synthetic truth. Panel (c) presents the average vector rotation required to correct interpolation to match truth. (*See electronic version for color representation of this figure.*)

To construct the spherical cap coordinate system, consider a conical section of the Earth having an angular radius of θ_c. The spherical cap is represented by the outer surface of that section which extends from a radius of r_1 to r_2. The coordinate system is defined with its pole located at the center of the spherical cap with co-latitude (θ) spanning from 0° to θ_c and longitude (φ) spanning 360°. The spherical cap harmonic expansion of Φ_B, separated for internal (int) and external (ext) sources, is given by

$$\Phi_B(r,\theta,\varphi) = R_E \sum_{k=0}^{K^i_{max}} \sum_{m=0}^{M} \left(\frac{R_E}{r}\right)^{n_k(m)+1} \left[g^{int}_{km}\cos(m\varphi) + h^{int}_{km}\sin(m\varphi)\right] P^m_{n_k(m)}(\cos\theta) + R_E \sum_{k=0}^{K^e_{max}} \sum_{m=0}^{M} \left(\frac{r}{R_E}\right)^{n_k(m)} \left[g^{ext}_{km}\cos(m\varphi) + h^{ext}_{km}\sin(m\varphi)\right] P^m_{n_k(m)}(\cos\theta) \quad (2.10)$$

R_E is the radius of the Earth. $P^m_{n_k(m)}(\cos\theta)$ are the associated Legendre functions of the first kind having a non-integer degree $n_k(m)$ and order m, k is an integer degree-indexing term. K_{max} is the maximum degree-index which truncates the series expansion, M is the minimum of k and M_{max}, and M_{max} is the maximum order which truncates the series expansion. The coefficients g_{km} and h_{km} are constants for each combination of k and m and represent the amplitude of the harmonics. In the special case where $n_k(m) = k$, this would be identical to the familiar global spherical harmonic expansion.

Haines (1985, 1988, 2007) provides a thorough description of how to calculate the associated Legendre functions, how to find the gradients for estimating the magnetic field, and how to construct the system of equations required to invert for optimal coefficients g_{km} and h_{km} given measurements in the spherical cap's reference frame, which is not reproduced here. More importantly, Haines describes how to choose the appropriate non-integer $n_k(m)$ so that a spherical cap expansion can be obtained. In short, in order to ensure continuity in longitude, regularity of potential at the spherical cap pole, and that both Φ_B and $\frac{d\Phi_B}{d\theta}$ are arbitrary at the boundary of the spherical cap, $n_k(m)$ should be chosen such that:

$$P^m_{n_k(m)}(\cos\theta)\Big|_{\theta=\theta_c} = 0 \quad k-m=\text{odd} \quad (2.11)$$

$$\left.\frac{dP^m_{n_k(m)}(\cos\theta)}{d\theta}\right|_{\theta=\theta_c} = 0 \quad k-m=\text{even}. \quad (2.12)$$

Haines (1988) expands slightly on the earlier work and provides an algorithm to numerically solve for appropriate non-integer $n_k(m)$ values, and all other required steps of a spherical cap expansion if needed.

Before the SCHA algorithm can be applied in practice, mapping parameters must first be determined. As with any interpolation scheme, SCHA is limited by the distribution of input (measurement) coordinates. For the case presented in this work, data are most densely distributed throughout North America, primarily at high latitudes with five additional stations spread in longitude. Traditionally, a spherical cap would be chosen to fit over North America, and the area over the oceans would be neglected due to a lack of constraint. However, to be consistent with work done in the rest of this chapter, a large spherical cap was chosen centered at 65° latitude and 250° longitude having an angular radius of 60°. Next, a maximum degree-index and order must be chosen to truncate the series expansion. Ideally, a high K_{max} and M_{max} are chosen to maximize the resolution of the mapped geomagnetic disturbance field. For the sparse observational data distribution here, limited to only 26 stations which are significantly separated at the outer regions of the spherical cap, different considerations must be made. Choosing values of K_{max} and M_{max} that are too large will result in highly unphysical structures being mapped in regions not constrained by data. Following the recommendations of Walker (1989), and the examples of Weimer (1995, 2013), Weimer et al. (2010), and Pothier et al. (2015), the maximum degree-index and order are chosen such that $M_{max} < K_{max}$. Such a consideration is reasonable when mapping magnetic data as the mapped features have more structure in latitude than longitude. For this particular data set, $K_{max} = 9$ and $M_{max} = 4$ were chosen through trial and error to minimize the difference between input and output data sets. Although increasing these values does improve the agreement between input and output data sets, it compromises the smoothness of the fit in regions not constrained by data. Ideally, data should be uniformly distributed about the mapping region at a density sufficient to support the resolution of the structures being mapped. In this more realistic scenario, the data are only sparsely distributed. Therefore, it is not reasonable to expect a high-resolution output. To reduce the problem of a poor solution at the boundary of the spherical cap, only the $k-m$ = even basis functions were used Equation (2.12). Similar considerations are made by Richmond and Kamide (1988), Weimer et al. (2010), Weimer (2013), and Pothier et al. (2015).

Figure 2.10 contains representative snapshots of the SCHA-interpolated horizontal magnetic vector field over North America during the St. Patrick's Day storm. These are the same instants in time presented in Figure 2.2.

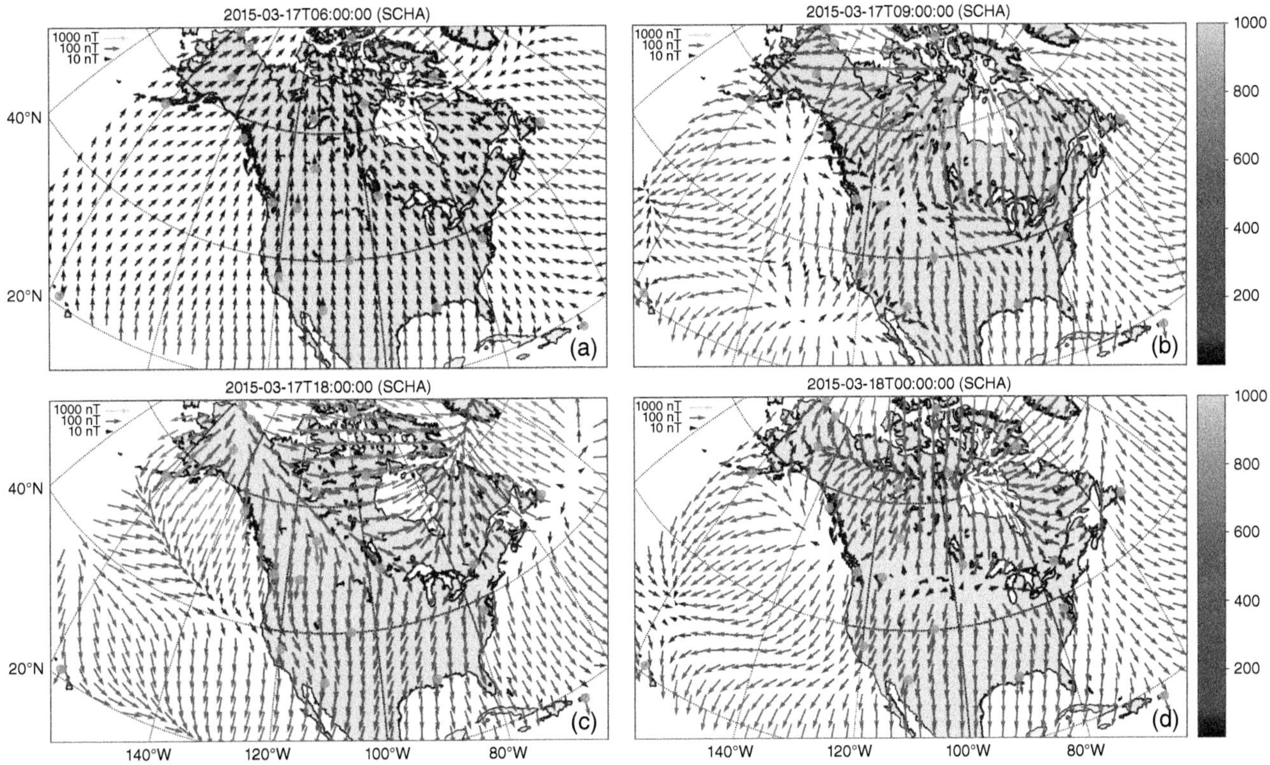

Figure 2.10 Snapshots of the North American distribution of spherical cap harmonic analysis-interpolated horizontal magnetic field vectors during the 2015 "St. Patrick's Day Magnetic Storm." Panel (a) is during the sudden commencement. Panel (b) is the main phase. Panel (c) is the transition from main to recovery phase. Panel (d) is near the heart of the recovery phase. (*See electronic version for color representation of this figure.*)

Panel (a) represents the geomagnetic storm sudden commencement, and the uniform ground truth field is fairly well represented. However, there are some slightly anomalous fields over Greenland, the Pacific Ocean, and just north of Puerto Rico in the Atlantic Ocean that all seem to hint at a systematic geometric bias. Panel (b) reproduces the eastward auroral electrojet signature fairly well, although somewhat diffused. This is probably to be expected given the low order/degree of the spherical harmonic fit employed here. What is more intriguing is that SCHA appears to capture the shape of this feature more accurately than the preceding interpolation techniques despite the lack of observatories directly under the electrojet. It even appears as though SCHA captured a weak signature of the substorm previously noted over the northernmost portions of Canada's Yukon and Northwest Territory, something that the preceding interpolation techniques failed at simply because there were effectively no observations sampling this region. In other words, the nominal physics wrapped into the SCHA equations appears to actually improve interpolations. This is further demonstrated in panel (c), where a clear eastward auroral electrojet signature can be seen where only a hint of one was visible in the preceding interpolation results. Panel (d) also does a good job of reproducing the diminished eastward auroral electrojet as North America rotates directly under it, locating the polar cap boundary and the separatrix between the electrojet and ring current signatures that falls rather neatly along the United States–Canada border.

However, after noting the advantages of a more physics-based interpolation scheme, we must acknowledge the anomalies/artefacts in the modelled field present over the Pacific Ocean, and to a lesser extent the Atlantic Ocean, north of Puerto Rico, especially during more geomagnetically active periods. These apparent convergence and divergence of geomagnetic vector features are ultimately the result of a model that is not flexible enough to capture the full range of spatial variability. A higher degree/order SCHA expansion could reproduce many if not all of the smaller spatial features more accurately, but given our sparse set of observations, it is impossible to constrain such a model far from the data. One typical, if generic, approach to dealing with this kind of problem is to regularize the solution via truncated singular value decomposition (SVD; see Mandel (1982), for a comprehensive review). This involves discarding singular values below a specified threshold when solving the linear system of equations for SCHA coefficients. We attempted as much here and found minor improvements over the results presented in Figure 2.10. But when the truncation threshold was set large enough to remove the offending artefacts, the entire solution became so smoothed as to be not useful. A more nuanced form of regularization that targets specific SCHA coefficients to be zeroed out could be employed. Smoothing through coefficient rejection as described in Haines and Fiori (2013) actually eliminates non-significant fitting coefficients. Coefficient rejection was successful in improving the agreement between the input model and the interpolated magnetic field in some cases but did not significantly influence the overall statistics.

In hindsight, this lack of significant improvement in the SCHA solution when using truncated SVD, or even coefficient rejection, is not entirely surprising. SCHA fits spatially periodic basis functions to observations of largely aperiodic spatial features. Regularization applied to these solutions amounts to a smoothing in the SCHA's wave-number domain, not the data's natural spatial domain. The consequence is a superposition of basis functions that, rather than smooth things out over poorly sampled regions, generates non-physical periodic features. Statistically speaking, these solutions may be no worse or better than a smooth solution over the unsampled regions, but they are not very visually satisfying, and potentially distracting if such maps are to be used by non-experts in an operational setting.

A different approach, one that increases overall complexity, but still falls short of our self-imposed threshold to avoid physics model-based data assimilation, is to use output from a statistical model of the geomagnetic disturbance field to supplement observatory data when fitting SCHA coefficients. For example, the empirical model described in Weimer (2013) is based on 8 years of observations and is parameterized according to a number of geophysical and solar parameters. The higher density of fitted data points allows a higher degree and order SCHA model to be specified that better captures the measured spatial variability, and the solution reverts to a statistical model over regions not actually sampled at the moment in question, a much more physically reasonable and visually satisfying result. This was not attempted for this study, which retains artefacts for demonstration purposes, but others, such as Waters et al. (2015), have demonstrated the validity of this approach.

Figure 2.11 presents accumulated statistics of temporal deviations between synthetic truth and SCHA-interpolated geomagnetic vector fields. Many of the features noted in the earlier interpolation statistics remain, although somewhat muted. SCHA does poorly near the edge of the spherical cap, and over the oceans, as expected. For example, the squared correlation coefficients are quite small just north of Puerto Rico, and east of Hawai'i, even though these regions are relatively near observations. This can be explained by the comparative isolation of these stations, since at least two observations must be made on a given wavelength to map a given structure, and it is not surprising that a single isolated point, particularly one near the edge of the spherical cap, cannot adequately constrain the fit. In contrast, SCHA performs well over the continental United States, Canada, and Alaska.

2.3.5. Spherical Elementary Currents

A different kind of basis function exists that, like SCHA, is grounded in a real physical understanding, but which is less constrained by rigid geometric considerations. Divergence-free spherical elementary currents (SECs) were first described by Amm (1997). In short, each SEC is characterized by its "pole" location in global spherical coordinates, and an electrical current amplitude. There is no poloidal component J_θ in the SEC's local spherical coordinate system, and its toroidal current J_ϕ diminishes from a maximum at the primary pole to zero at the opposite pole as a cotangent function.

$$J_\phi = \frac{I_0}{4\pi R_i}\cot\frac{\theta}{2} = \frac{I_0}{4\pi R_i}\frac{1+\cos\theta}{\sin\theta}. \tag{2.13}$$

Amm also described a curl-free elementary current basis function, largely building upon earlier work by Fukushima (1976), but because this cannot be sensed magnetically below its spherical shell of radius R_i, we do not consider it when interpolating geomagnetic disturbance. Figure 2.12 offers a visual representation of a single divergence-free SEC.

Amm and Viljanen (1999) demonstrated how this simple basis function could be combined with Ampere's Law in spherical coordinates to provide an analytic expression for radial and poloidal magnetic disturbance anywhere inside the spherical shell with radius R_i.

$$B_r = \frac{\mu_0 I_0}{4\pi r}\left(\frac{1}{\sqrt{1-\frac{2r\cos\theta}{R_i}+\left(\frac{r}{R_i}\right)^2}}-1\right) \tag{2.14}$$

$$B_\theta = \frac{\mu_0 I_0}{4\pi r}\left(\frac{\frac{r}{R_i}-\cos\theta}{\sqrt{1-\frac{2r\cos\theta}{R_i}+\left(\frac{r}{R_i}\right)^2}}+\cos\theta\right) \tag{2.15}$$

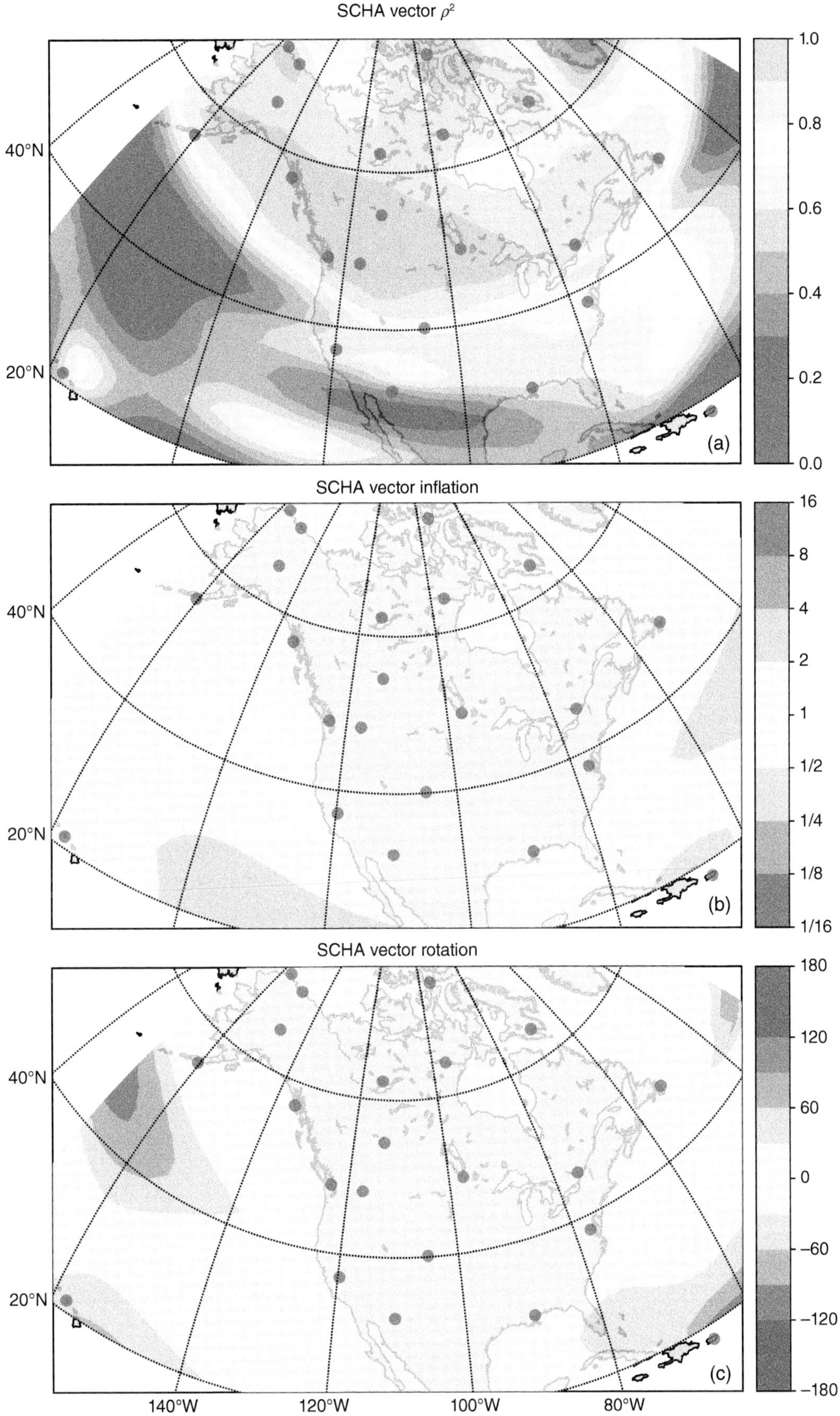

Figure 2.11 Comparison of synthetic truth and spherical cap harmonic analysis-interpolated vector fields over North America. Panel (a) presents squared temporal vector correlations. Panel (b) presents the average vector inflation required to correct interpolation to match synthetic truth. Panel (c) presents the average vector rotation required to correct interpolation to match truth. (*See electronic version for color representation of this figure.*)

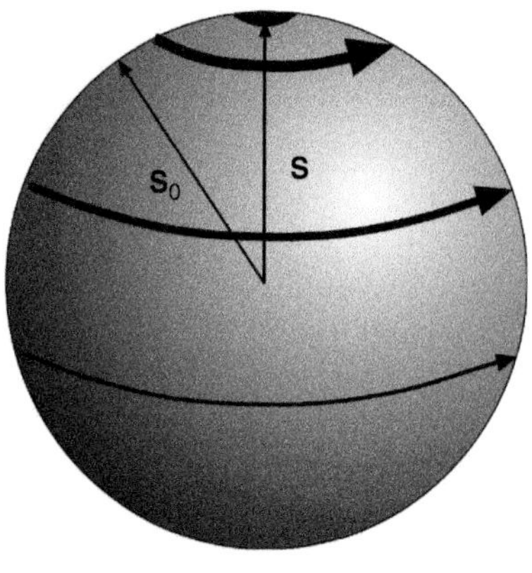

Figure 2.12 Divergence-free spherical elementary current (SEC). Vector **S** points to the SEC pole. Toroidal current density on a spherical shell scales as the cotangent of co-latitude, pointed to by vector $\mathbf{S}_0$, and is independent of longitude.

There is no toroidal magnetic vector component within an SEC's local spherical coordinate frame, but one is easily obtained through rotation into any other local spherical frame. If a distribution of SECs is constructed, the resulting magnetic vectors from each SEC can be superposed. This magnetic disturbance distribution is defined completely by the SEC pole locations and amplitudes, so with a fixed and finite set of pole locations, we can solve for the set of SEC amplitudes that result in magnetic disturbances that best matches available observations.

$$\mathbf{Z} = \mathbf{T}\mathbf{I}_0$$

$$\mathbf{I}_0 \equiv \begin{bmatrix} I_{01} \\ I_{02} \\ \vdots \\ I_{0m} \end{bmatrix}, \mathbf{T} \equiv \begin{bmatrix} T_{11,\theta} & T_{12,\theta} & \cdots & T_{1m,\theta} \\ T_{11,\phi} & T_{12,\phi} & \cdots & T_{1m,\phi} \\ T_{11,r} & T_{12,r} & \cdots & T_{1m,r} \\ \vdots & \vdots & \ddots & \vdots \\ T_{n1,\theta} & T_{n2,\theta} & \cdots & T_{nm,\theta} \\ T_{n1,\phi} & T_{n2,\phi} & \cdots & T_{nm,\phi} \\ T_{n1,r} & T_{n2,r} & \cdots & T_{nm,r} \end{bmatrix}, \mathbf{Z} \equiv \begin{bmatrix} Z_{1,\theta} \\ Z_{1,\phi} \\ Z_{1,r} \\ \vdots \\ Z_{n,\theta} \\ Z_{n,\phi} \\ Z_{n,r} \end{bmatrix} \tag{2.16}$$

Here, m is the number of SECs in our system, n is the number of observations. The vector **Z** holds measured magnetic disturbance vector components in their respective local coordinate frames, while the matrix **T** holds the magnetic disturbance vector components calculated using Equations (2.14) and (2.15), with I_0 equal to 1 Ampere, then rotated from each SEC frame into the local spherical coordinate frame of the respective observation. By inverting **T**, and multiplying by **Z**, the SEC amplitudes are solved for. With these, magnetic disturbance vectors can be calculated for any set of virtual observatories desired.

Following the example of Amm and Viljanen (1999), as well as most subsequent studies that have employed the SEC technique (e.g., Pulkkinen et al., 2003; Wei et al., 2013; Wygand et al., 2011), truncated SVD was used to solve for $\mathbf{I}_0$. This approach was chosen primarily because there are typically fewer available observations to fit to than the number of SECs; that is, the system of equations is underdetermined and setting the inverse of singular values below a finite positive threshold to zero ensures that the positive-definite requirement for matrix inversion is satisfied. This form of regularization also removes higher-frequency content in the final solution, basically assuming it was due to observational noise anyway. A threshold of 5% of the maximum singular value was sufficient to obtain reasonable solutions over the St. Patrick's Day storm.

The spatial distribution of SECs can take on any structure. In practice, however, a regular grid of SEC poles is typical. This grid should extend beyond the region in which interpolated magnetic disturbance values are desired, or else edge artefacts become a problem. The spacing of the grid should be no larger than the spatial scales over which a single SEC's influence remains significant, and smaller if the density of observations supports it. Amm and Viljanen (1999) showed that, for an SEC with radius $R_i \approx 100$ km larger than the Earth's surface, the radial component of a magnetic disturbance vector has an e-folding distance of ~2.5° on the Earth's surface, or ~280 km along a meridian. The polar component actually rises to a peak at an angular distance of just over 1°, before dropping back toward 0 at larger distances, with an e-folding distance from here of ~7.5°, or ~1000 km. A regular grid in LAEA coordinates was defined in order to be more consistent with the interpolation techniques covered previously. Spacing for this grid was ~260 km. Figure 2.13 shows the locations of the SECs used for this study. Predictions were made on the same 2 × 2° regular geographic grid as all other interpolation techniques.

Figure 2.14 contains representative snapshots of the SECs-interpolated horizontal magnetic vector field over North America during the St. Patrick's Day storm. These are the same instants in time presented in Figure 2.2. Panel (a) is during the geomagnetic storm sudden commencement, and the uniform ground truth field is very well represented, even far from any observatories. Also, in panel (b), the signature of the substorm previously noted over the northernmost portions of Canada's Yukon and Northwest Territory is reproduced quite well, in spite of the lack of any observatory in its immediate vicinity. In contrast to the SCHA method, it does this while managing not to introduce the anomalies/artefacts over the Pacific Ocean, and to a lesser extent the Atlantic Ocean, north of Puerto Rico, during more geomagnetically active periods.

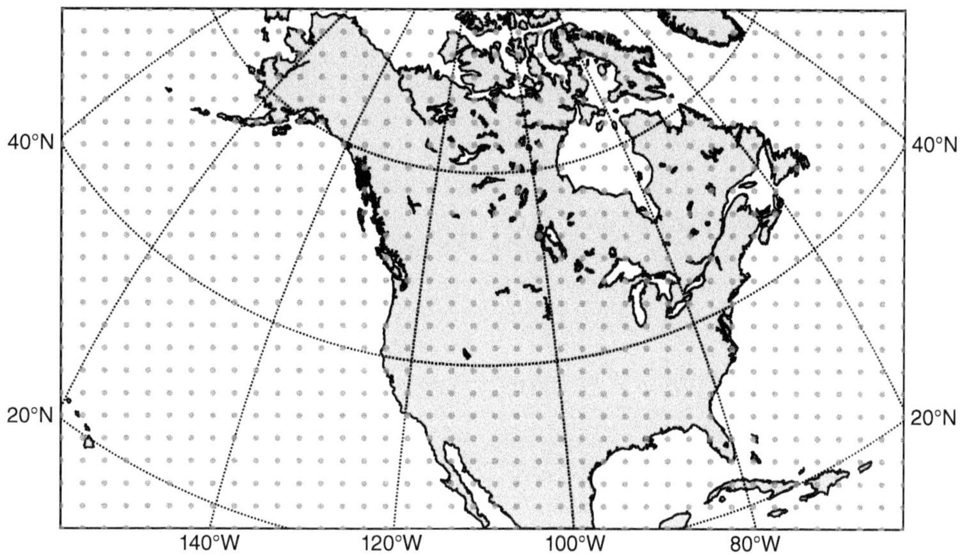

Figure 2.13 Distribution of spherical elementary current (SEC) poles used for this study.

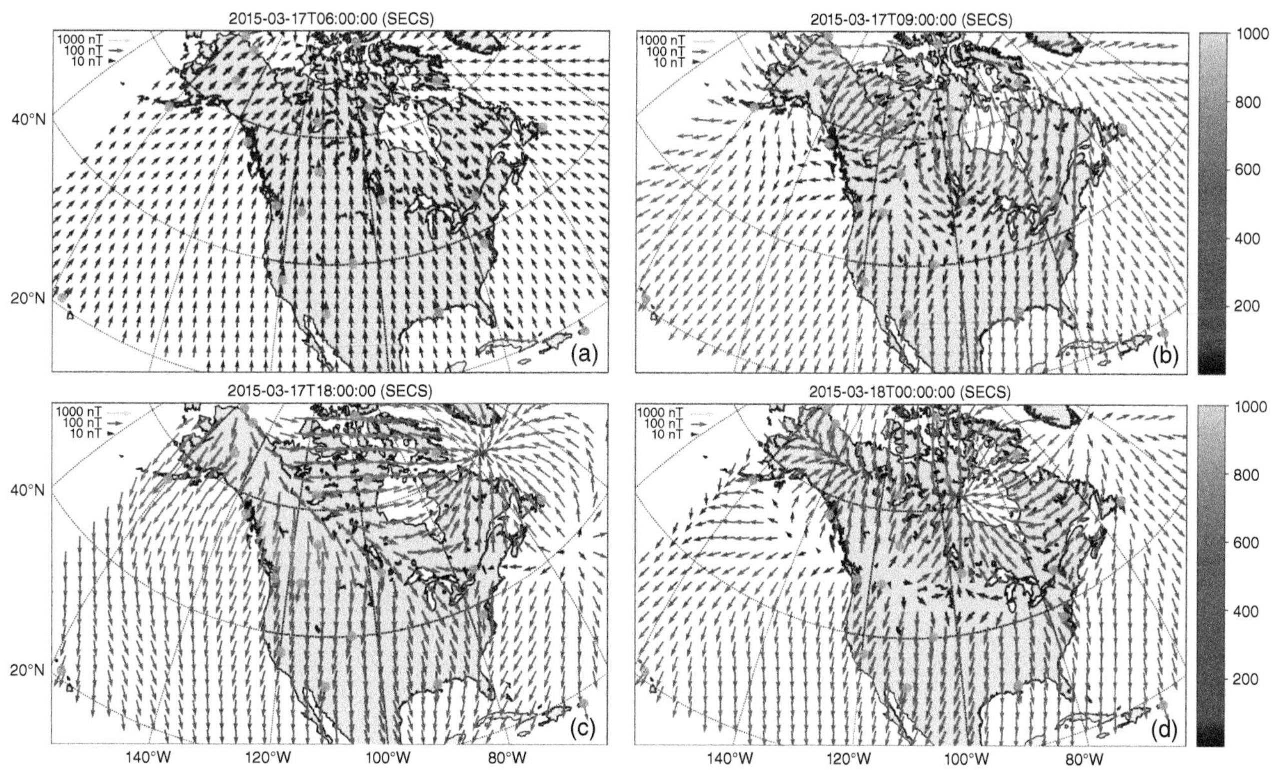

Figure 2.14 Snapshots of the North American distribution of spherical elementary current system-interpolated horizontal magnetic field vectors during the 2015 "St. Patrick's Day Magnetic Storm." Panel (a) is during the sudden commencement. Panel (b) is the main phase. Panel (c) is the transition from main to recovery phase. Panel (d) is near the heart of the recovery phase. (*See electronic version for color representation of this figure.*)

However, one should resist the natural inclination to assume that the reason for this improvement is that SECs offer a more accurate representation of real physics than SCHA. A more likely explanation is that, while SECs do provide a more sophisticated vector basis function than the squared exponential used for GPR, they still have a limited range of influence that falls off roughly exponentially with distance from their pole (recall e-folding distances for B_r and $B\theta$ mentioned above). Locations far from observations are not actually being physically modeled more accurately, but rather they are regressing to a regional, if not quite global, mean. An observation in support of this explanation is that SEC interpolation still does a fairly poor job capturing the signature of an enhanced eastward auroral electrojet south of Greenland in panel (c), where the reader will recall SCHA performed relatively well. SEC interpolation performs well during the recovery phase presented in panel (d), not unlike the other interpolation techniques.

Figure 2.15 presents accumulated statistics of temporal deviations between synthetic truth and SEC-interpolated geomagnetic vector fields. As expected, panel a's squared correlation coefficients over the mid-to-north Atlantic, Greenland, and the Hudson Bay are not exceptionally high, but they are better than any of the interpolation techniques analyzed thus far except SCHA. SEC interpolation does considerably better than SCHA over the Pacific Ocean, but is still poor, especially north and east of Hawai'i. It also does poorly in the sparsely sampled region south of the Great Lakes, although interestingly there is no systematic inflation or rotational bias here, just random error. There appears to be an overall degradation in interpolation accuracy from higher to lower latitudes. Some of this is due to a relative reduction in observation density, but it is also indicative of the increasing violation of Fukushima's theorem as magnetic field lines become less and less radial, a fundamental physical motivation for using SEC basis functions.

2.4. DISCUSSION AND RECOMMENDATIONS

Five techniques for interpolating geomagnetic disturbance vector fields were examined and compared: nearest neighbor (NN), triangle barycentric (TRI), Gaussian Process regression (GPR), spherical cap harmonic analysis (SCHA), and spherical elementary current (SEC) vector basis functions. While none of these techniques are nearly as computationally intensive as first-principal GGCM forward models, some do require considerably more computer cycles than the others. For this study, this was not much of an issue, but it is easy to imagine scenarios that include more observations, or a higher density grid of virtual observatories at which to predict geomagnetic disturbance, and where computational efficiency becomes a practical concern.

If the five interpolation techniques were to be ranked by computational complexity, they sort roughly as NN, TRI, GPR, SEC, and SCHA. It should be reemphasized that, for this study, all measurement coordinates were transformed into LAEA map coordinates. This was partly to take advantage of existing optimized numerical libraries, but also to make consistent comparisons between the different techniques feasible (i.e., to compare "apples-to-apples"). This latter consideration is not one a general practitioner should be concerned with, and to maximize accuracy, they should probably adapt the NN, TRI, and GPR techniques to work in more natural spherical coordinates. If this were done, the differences in computational complexity would lessen, and there would be little justification in using NN, TRI, or even GPR interpolation over SCHA or SECs. On the other hand, if the observation and prediction locations are not expected to change between observations, the added complexity of using true spherical coordinates is only required once, after which, the efficiencies afforded by NN, TRI, and GPR interpolation become considerable.

Unlike NN or TRI interpolation, GPR weights converge toward zero at large distances (at least when the squared-exponential kernel is used), and the solution reverts to a global mean rather than a weighted average of three points that may not be nearby at all. This global mean is generally preferable to the artefacts previously noted for NN and TRI interpolation. Also, while it is often less obvious than with NN interpolation, TRI interpolation still generates a discontinuous solution at tessellation boundaries. GPR with a squared-exponential kernel does not. These facts, combined with its relative computational efficiency when simple optimizations are applied, are probably sufficient to favor GPR over NN and TRI in almost all cases where the observation and prediction locations are constant, but the independent input values can change.

The preceding recommendation applies to interpolation of any dimensional vector field because each of these first three techniques treats vector components as orthogonal, and therefore independent. In other words, they interpolate each vector field's northward component, then its eastward component, and so on, using exactly the same functional relationships to describe the covariation between observations. SCHA does not share this characteristic because there is an additional functional relationship between the actual vector components that is motivated and justified by the assumption of a potential geomagnetic field, which can be approximated using a modified spherical harmonic expansion. This is why SCHA seems able to reproduce genuinely physical features with relatively little in the way of empirical evidence (recall the substorm signature noted in Section 2.3.4).

However, SCHA suffers in performance relative to GPR due to at least two factors: it forces a rigid geometric constraint on its basis functions when the degree/order is too small; and it does not, by default, de-cluster relatively dense concentrations of observations. The former means the model cannot accommodate strong localized enhancements without biasing results in distant regions that may not actually be physically connected. This can be at least

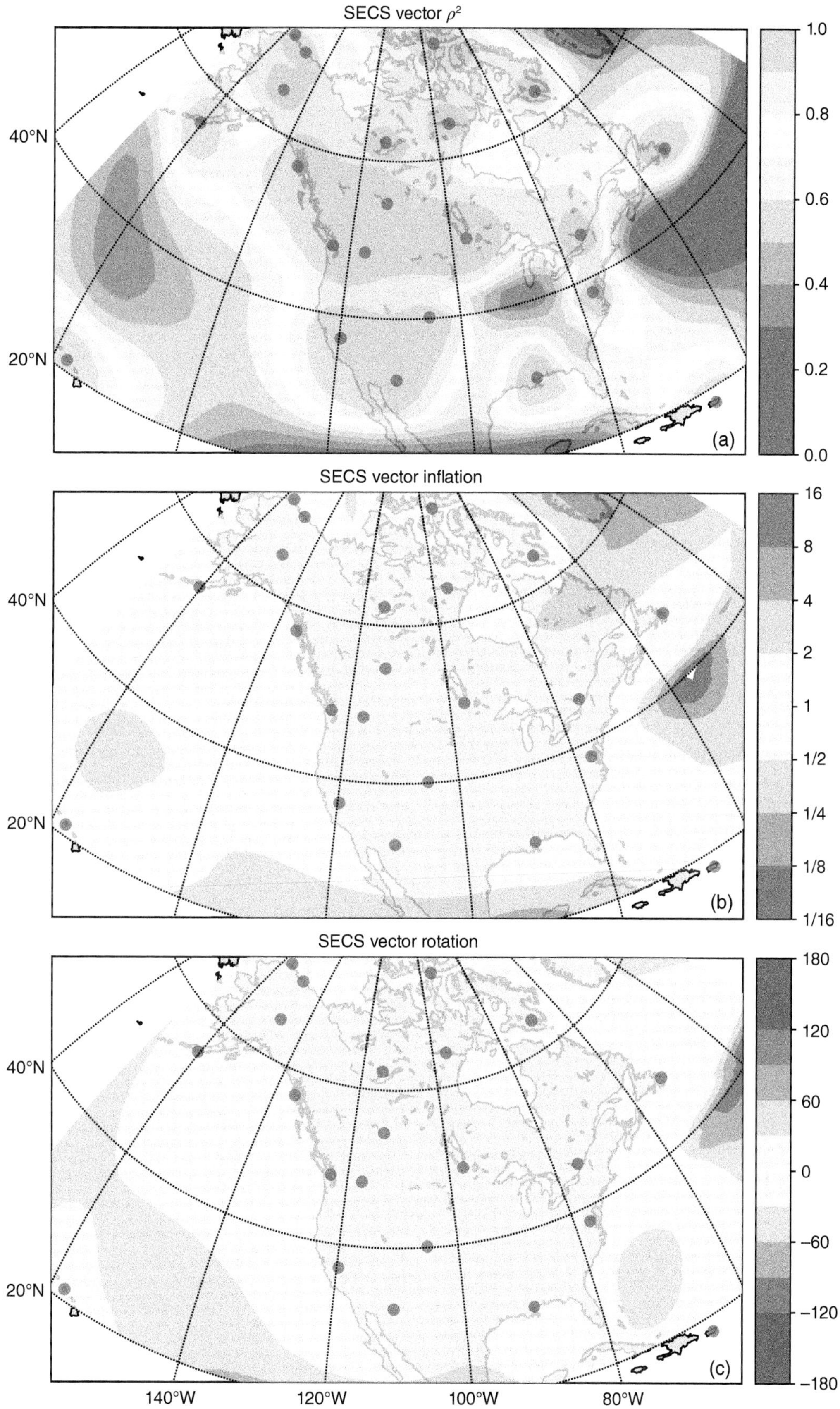

Figure 2.15 Comparison of synthetic truth and spherical elementary current system-interpolated vector fields over North America. Panel (a) presents squared temporal vector correlations. Panel (b) presents the average vector inflation required to correct interpolation to match synthetic truth. Panel (c) presents the average vector rotation required to correct interpolation to match truth. (*See electronic version for color representation of this figure.*)

partially addressed by increasing the degree/order of the harmonic expansion and applying standard regularization. However, the latter means that the model will always be biased to match regions with more observations. One solution to this would be to modify the truncation level of the series expansion so that it is not uniform across the spherical cap; this is analogous to applying the SECs algorithm with a non-uniform distribution of poles. Such a procedure has not yet been applied to SCHA but is the subject of future study. It is not clear if there exists an analog to the normalization in GPR provided by Σ_{XX}^{-1}, but a practical, if less mathematically elegant, alternative is to introduce additional statistically plausible inputs on a more regular grid. Similar procedures exist for mapping ionospheric plasma flow using a modified spherical harmonic expansion (Ruohoniemi and Baker, 1998; Shepherd and Ruohoniemi, 2000), and could similarly be applied to magnetic field interpolation with SCHA given an appropriate reference field.

SEC interpolation, also motivated and justified by being able to generate potential geomagnetic fields, is not immune from an uneven distribution of observations. But it is not nearly as geometrically constrained as SCHA. This means that an SEC whose pole is located relatively near an enhanced geomagnetic disturbance can adjust to fit that enhancement without strongly impacting an SEC amplitude fitted in a region whose geomagnetic disturbance vector is not enhanced, has a different sign, or even a different direction. Truncated SVD regularization, which is almost always required because the number of SEC basis functions typically exceeds the number of observations, automatically removes those principal components that do not contribute to a solution that best fits available observations, which often tend to correspond to be those SECs that are far away. The end result is similar to GPR's regression to a global mean far away from observations, which, as stated previously, is preferable to non-physical artefacts.

At this point, it is prudent to remind the reader that all data used herein, input observations and the ground truth used for validation, were synthesized using a geospace general circulation model. There may be more realistic GGCMs for specific scenarios, but LFM+MIX+RCM is currently considered among the state-of-the-art and reproduces most expected ionospheric and magnetospheric electrical current distributions, and their associated ground signatures. However, as noted earlier, LFM+MIX+RCM does not generate telluric currents, realistic or otherwise. If it did, it would be necessary to adjust the SCHA and SECs interpolation techniques to address this, since both were configured to assume telluric currents were nonexistent for this study. It is actually very common to solve simultaneously for two sets of spherical (cap) harmonic coefficients, one corresponding to internal, and one to external source current distributions (e.g., Haines, 1985; Olsen et al., 2010). Pulkkinen et al. (2003) demonstrated that an internal system of SECs could be fit simultaneously with an external system, and thus provide more accurate estimates of geomagnetic disturbance vectors, especially in the radial direction, which has so far been ignored in this analysis. However, both of these approaches will require significantly more observations to obtain a reliable fit. First, the number of unknowns doubles with respect to the number of available observations. But perhaps more relevant here, there is a whole additional degree of freedom that may vary in unpredictable ways as the solution moves far away from observations that might constrain it.

Even disregarding the problem of telluric currents, performance of all interpolation schemes is expected to diminish when using real observations because geomagnetic disturbance spatial scales can be quite a bit smaller than all but our most densely deployed magnetometer networks, never mind the operational geomagnetic observatories used for this study. However, this diminishment should be comparable for each interpolation technique in most situations, so we do not anticipate our relative comparisons to change substantially. One legitimate concern may be sporadic missing data, especially if interpolation is required for real time applications. How much can a single missing observation impact the overall solution? How far away from a missing observation's location is this effect significant? Is the interpolation more sensitive to certain observation locations, or clusters of observation locations? These are all important questions, but fall largely in the category of operations research, since the answers will always be specific to a given configuration and are therefore beyond the scope of this study.

Finally, the performance of the different interpolation techniques with respect to the time derivative, or spectral content, of geomagnetic disturbance, which are arguably more relevant variables for electric power grid applications, was not discussed. The answer to this question is complicated and warrants substantially more analysis and discussion than presented here. A straightforward extension of the analysis conducted above to *dB*/*dt* estimated from interpolated magnetic fields is probably not sufficient though, because these interpolation techniques will perform differently on the disturbance time derivative and/or its spectral content. The more empirical interpolation techniques (NN, TRI, and GPR) can easily be applied to *dB*/*dt* directly, or even to spectrally decomposed observations, and improve performance in the time derivative or frequency domains, respectively. While the physics-inspired techniques (SECs and SCHA) are both predicated on quasi-static approximations to Maxwell's equations, their empirical coefficients can also be made time-dependent. Indeed, this is common practice with global geomagnetic models in order to track secular variation in the Earth's main field, but what is not clear is whether this is appropriate for tracking geomagnetic disturbance relevant to GICs.

In closing, the reader is encouraged to consider carefully all the interpolation techniques studied here, or close relations, before selecting just one for their particular application. That application may very well involve real data that do not resemble our admittedly limited synthetic test data at all. But even if the data are similar, there are other practical considerations like computational resources, availability and timeliness of observations, spatial scales (local, regional, and global) of the desired interpolation, and the smoothness or physicality of the solution. In the end, the interpolated geomagnetic disturbance field is just a tool to facilitate analysis and must always be adapted to maximize its utility for a particular purpose.

ACKNOWLEDGMENTS

We thank Carol Finn, Greg Lucas, and Jill McCarthy, and Janet Slate at the USGS for reading drafts of this lengthy manuscript and providing valuable feedback. We also thank two anonymous reviewers for their critiques and helpful recommendations.

APPENDIX: 2D VECTOR CORRELATIONS

Rather than treat horizontal vector components as independent variables, the horizontal (2D) vector correlation is considered, as described by Hanson et al. (1992). This is similar to the familiar Pearson correlation for scalar series, but uses complex numbers to represent vectors. To summarize, if $\mathbf{z}_j = x_j + iy_j$ and $\mathbf{w}_j = u_j + iv_j$ are two series of j vectors, their vector variances and covariance may be defined as:

$$\sigma_{\mathbf{z}}^2 = \frac{1}{n}\sum_{j=1}^{n}(\mathbf{z}-\bar{\mathbf{z}})^*(\mathbf{z}-\bar{\mathbf{z}}) = \sigma_x^2 + \sigma_y^2$$

$$\text{and} \tag{A.1}$$

$$\sigma_{\mathbf{w}}^2 = \frac{1}{n}\sum_{j=1}^{n}(\mathbf{w}-\bar{\mathbf{w}})^*(\mathbf{w}-\bar{\mathbf{w}}) = \sigma_u^2 + \sigma_v^2$$

$$\sigma_{zw} = \frac{1}{n}\sum_{j-1}^{n}(\mathbf{z}-\bar{\mathbf{z}})^*(\mathbf{w}-\bar{\mathbf{w}}) = (\sigma_{xu}+\sigma_{yv}) + i(\sigma_{xv}-\sigma_{yu}) \tag{A.2}$$

where $\mathbf{z}_j^* = x_j - iy_j$ is the complex conjugate of $\mathbf{z}_j$, and an overbar represents the arithmetic mean. Note that the vector variances ($\sigma_{\mathbf{z}}^2$ and $\sigma_{\mathbf{w}}^2$) are actually scalar values, and the sum of the variances of each component. The covariance ($\boldsymbol{\sigma}_{zw}$) is itself a vector value, represented as a complex number where the real term describes the sum of covariances of the corresponding vector components, and the imaginary term describes a "twisting" of the vector components of the first variable into their orthogonal components in the second variable. With these definitions, a vector correlation analogous to the Pearson correlation is:

$$\rho_{zw} = \frac{\sigma_{zw}}{\sigma_z \sigma_w}. \tag{A.3}$$

Details are provided in Hanson et al. (1992), but much like the square of Pearson's correlation (i.e., R-squared, or the coefficient of determination), the squared magnitude of this complex value (i.e., $\rho_{zw}\rho_{zw}^*$) is a scalar that represents the proportion of variance in $\mathbf{w}_j$ that can be explained by $\mathbf{z}_j$ after an optimal (in a least-squares sense) linear transformation. For scalars, this transformation is a simple scale factor, but for 2D vectors, it also includes rotation.

The square root of the squared magnitude of the vector correlation is:

$$\rho_{zw} = \sqrt{\frac{(\sigma_{xu})^2 + (\sigma_{yv})^2 + (\sigma_{xv})^2 + (\sigma_{yu})^2 + 2(\sigma_{xu}\sigma_{yv} - \sigma_{xv}\sigma_{yu})}{(\sigma_x^2+\sigma_y^2)(\sigma_u^2+\sigma_v^2)}}. \tag{A.4}$$

This value will always fall between 0 and 1 (0 implies no correlation between the vectors; 1 is a perfect correlation). In addition, the optimal scale factor required to equalize the magnitude of the two vectors is:

$$B_{zw} = \rho_{zw}\sqrt{\frac{\sigma_u^2+\sigma_v^2}{\sigma_x^2+\sigma_y^2}}. \tag{A.5}$$

The optimal rotation required to align the direction of the two vectors is:

$$\Theta_{zw} = \tan^{-1}\left(\frac{\sigma_{xv}-\sigma_{yu}}{\sigma_{xu}+\sigma_{yv}}\right). \tag{A.6}$$

Hanson et al. also describe a translational element of the linear transformation, or the intercepts of a bivariate linear regression. We choose to ignore this since it could be forced to equal zero by simply removing the mean of each respective vector sequence and is of limited additional value for our purpose of comparing interpolation techniques.

REFERENCES

Alken, P., Maus, S., Chulliat, A., & Manoj, C. (2015). NOAA/NGDC candidate models for the 12th generation International Geomagnetic Reference Field. *Earth, Planets, and Space*, *67*(68), https://doi.org/10.1186/s40623-015-0215-1

Amm, O. (1997). Ionospheric elementary current systems in spherical coordinates and their application. *Journal of Geomagnetism and Geoeelctricity*, *49*(7), 947–955, https://doi.org/10.5636/jgg.49.947

Amm, O., & Viljanen, A. (1999). Ionospheric disturbance magnetic field continuation from the ground to the ionosphere using spherical elementary current systems. *Earth Planets Space*, *51*, 431–440, https://doi.org/10.1186/BF03352247

Bonner, L. R., & Schultz, A. (2017). Rapid prediction of electric fields associated with geomagnetically induced currents in the presence of three-dimensional ground structure: Projection of remote magnetic observatory data through magnetotelluric impedance tensors. *Space Weather*, *15*, 204–227, https://doi.org/10.1002/2016SW001535

Bonnevier, B., & Boström, R. (1970). A three-dimensional model current system for polar magnetic substorms. *Journal of Geophysical Research*, *75*(1), 107–122, https://doi.org/10.1029/JA075i001p00107

Boteler, D. H. (2003). Geomagnetic hazards to conducting networks. *Natural Hazards*, *28*, 537–561, https://doi.org/10.1023/A:1022902713136

Boteler, D. H., & Pirjola, R. J. (1998). The complex-image method for calculating the magnetic and electric fields produced at the surface of the Earth by the auroral electrojet. *Geophysical Journal International*, *132*, 31–40, https://doi.org/10.1046/j.1365-246x.1998.00388.x

Boteler, D. H., Pirjola, R. J., & Nevanlinna, H. (1998). The effects of geomagnetic disturbances on electrical systems at the Earth's surface. *Advances in Space Research*, *22*(1), 17–27, https://doi.org/10.1016/S0273-1177(97)01096-X

Buchanan, A., Finn, C. A., Love, J. J., Worthington, E. W., Lawson, F., Maus, S., … Poedjono, B. (2013). Geomagnetic referencing – The real-time compass for directional drillers. *Oilfield Review*, *25*(3), 32–47.

Cliver, E. W. (1994, December 6). Solar activity and geomagnetic storms: The first 40 years. *EOS, Transactions of the American Geophysical Union*, *75*(49), 569, https://doi.org/10.1029/94EO02060

Farin, G. (1993). *Curves and Surfaces for Computer Aided Geometric Design – A Practical Guide* (Third ed.). San Diego, CA: Academic Press, Inc.

Finlay, C. C., Olsen, N., & Tøffner-Claussen, L. (2015). DTU candidate field models for IGRF-12 adn the CHAOS-5 geomagnetic field model. *Earth, Planets, and Space*, *67*(114), https://doi.org/10.1186/s40623-015-0274-3

Fukushima, N. (1976). Generalized theorem for no ground magnetic effect of vertical currents connected with Pedersen currents in the uniform-conductivity ionosphere. *Report of Ionosphere and Space Research in Japan*, *30*(1/2), 35–40.

Gjerloev, J. W. (2012). The SuperMAG data processing technique. *Journal of Geophysical Research*, *117*, A09213, https://doi.org/10.1029/2012JA017683

Gleisner, H., Rasmussen, O., & Watermann, J. (2006). Large-magnitude geomagnetic disturbances in the North Sea region: Statistics, causes, forecasting. *Advances in Space Research*, *37*, 1169–1174, https://doi.org/10.1016/j.asr.2005.04.082

Haines, G. V. (1985). Spherical cap harmonic analysis. *Journal of Geophysical Research*, *90*(B3), 2583–2591, https://doi.org/10.1029/JB090iB03p02583

Haines, G. V. (1988). Computer programs for spherical cap harmonic analysis of potential and general fields. *Computers and Geosciences*, *14*(4), 413–447, https://doi.org/10.1016/0098-3004(88)90027-1

Haines, G. V. (2007). Spherical cap harmonics. In D. Gubbins, & E. Herrero-Bervera (Eds.), *Encyclopedia of Geomagnetism and Paleomagnetism* (pp. 395–397). Dordrecht, The Netherlands: Springer

Haines, G. V., & Fiori, R. A. (2013). Modeling by singular value decomposition and the elimination of statistically insignificant coefficients. *Computer and Geosciences*, *58*, 19–28, https://doi.org/10.1016/j.cageo.2013.04.021

Hamilton, B., Ridley, V. A., Beggan, C. D., & Macmillan, S. (2015). The BGS magnetic field candidate models for the 12th generation IGRF. *Earth, Planets, and Space*, *67*(69), https://doi.org/10.1186/s40623-015-0227-x

Hanson, B., Klink, K., Matsuura, K., Robeson, S. M., & Willmot, C. J. (1992). Vector correlation: Review, exposition, and geographic application. *Annals of the Association of American Geographers*, *82*(1), 103–116, https://doi.org/10.1111/j.1467-8306.1992.tb01900.x

Kappenman, J. G. (2004). Space weather and the vulnerability of electric power grids. In I. A. Daglis (Ed.), *Effects of Space Weather on Technology Infrastructure. NATO Science Series II: Mathematics, Physics and Chemistry*, vol. *176*, Dordrecht: Springer, pp. 257–286, https://doi.org/10.1007/1-4020-2754-0_14

Kisabeth, J. L., & Rostoker, G. (1977). Modelling of three-dimensional current systems associated with magnetospheric substorms. *Geophysical Journal of the Royal Astronomical Soceiety*, *49*, 655–683, https://doi.org/10.1111/j.1365-246X.1977.tb01310.x

Love, J. J., & Gannon, J. (2009). Revised Dst and the epicycles of magnetic disturbance: 1958-2007. *Annales Geophyicae*, *27*(1), 3101–3131, https://doi.org/10.5194/angeo-27-3101-2009

Lucas, G. M., Love, J. J., & Kelbert, A. (2018). Calculation of voltages in electric power transmission lines during historic geomagnetic storms: An investigation using realistic Earth impedances. *Space Weather*, *16*, https://doi.org/10.1002/2017SW001779

Lyon, J. G., Fedder, J. A., & Mobarry, C. M. (2004). The Lyon-Fedder-Mobarry (LFM) global MHD magnetospheric simulation code. *Journal of Atmospheric and Solar Terrestrial Physics*, *66*(1), 1333–1350, https://doi.org/10.1016/j.jastp.2004.03.020

Mandel, J. (1982). Use of the singular value decomposition in regression analysis. *The American Statistician*, *36*(1), 15–24.

Mayaud, P. N. (1980). *Derivation, Meaning, and Use of Geomagnetic Indices*. Washington, DC: American Geophysical Union.

Merkin, V. G., & Lyon, J. G. (2010). Effects of the low-latitude ionospheric boundary condition on the global magnetosphere. *Journal of Geophysical Research*, *115*(A), 10202, https://doi.org/10.1029/2010JA015461

Nishida, A. (1978). *Geomagnetic Diagnosis of the Magnetosphere*. New York: Springer-Verlag

Olsen, N., Glasmeier, K.-H., & Jia, X. (2010). Separation of the magnetic field into external and internal parts. *Space Science Reviews*, *152*, 135–157, https://doi.org/10.1007/s11214-009-9563-0

Pedregosa, F., Varoquaux, G., Gramfort, A., Michel, V., Thirion, B., Grisel, O., … Duchesnay, É. (2011). Scikit-learn: Machine learning in python. *Journal of Machine Learning Research*, *12*, 2825–2830.

Pirjola, R. J. (2002). Fundamentals about the flow of geomagnetically induced currents in a power system applicable to estimating space weather risks and design remedies. *Journal of Atmospheric and Solar Terrestrial Physics*, *64*, 1967–1972, https://doi.org/10.1016/S1364-6826(02)00228-6

Pirjola, R. J., & Viljanen, A. (1998). Complex image method for calculating electric and magnetic fields produced by an auroral electrojet of finite length. *Annales Geophysicae*, *16*, 1434–1444, https://doi.org/10.1007/s00585-998-1434-6

Pothier, N. M., Weimer, D. R., & Moore, W. B. (2015). Quantitative maps of geomagnetic perturbation vectors during substorm onset and recovery. *Journal of Geophysical Research*, *120*, 1197–1214, https://doi.org/10.1002/2014JA020602

Pulkkinen, A., Amm, O., Viljanen, A., & BEAR Working Group (2003). Separation of the geomagnetic variation field on the ground into external and internal parts using the spherical elementary current system method. *Earth Planets Space*, *55*, 117–129, https://doi.org/10.1186/BF03351739

Pulkkinen, A., Kuznetsova, M., Ridley, A., Raeder, J., Vapirev, A., Weimer, D., … Chulaki, A. (2011). Geospace environment modeling 2008-2009 challenge: Ground magnetic field perturbations. *Space Weather*, *9*, S02004, https://doi.org/10.1029/2010SW000600

Pulkkinen, A., Rastätter, L., Kuznetsova, M., Hesse, M., Ridley, A., Singer, H. J., & Chulaki, A. (2010). Systematic evaluation of ground and geostationary magnetic field predictions generated by global magnetohydrodynamic models. *Journal of Geophysical Research*, *115*, A03206, https://doi.org/10.1029/2009JA014537

Pulkkinen, A., Rastätter, L., Kuznetsova, M., Singer, H. J., Weimer, D., Toth, G., … Weigel, R. S. (2013). Community-wide validation of geospace model ground magnetic perturbation predictions to support model transition to operations. *Space Weather*, *11*, 369–385, https://doi.org/10.1002/swe.20056

Rasmussen, C. E., & Williams, C. K. (2006). *Gaussian Processes for Machine Learning*. Cambridge, MA: The MIT Press

Rastätter, L., Kuznetsova, M., Glocer, A., Welling, D., Meng, X., Raeder, J., … Gannon, J. (2013). Geospace environment modeling 2008-2009 challenge: Dst index. *Space Weather*, *11*(4), 187–205, https://doi.org/10.1002/swe.20036

Richmond, A. D., & Kamide, Y. (1988). Mapping electrodynamic features in the high-latitude ionosphere from localized observations: Technique. *Journal of Geophysical Research*, *93*(A6), 5741–5759, https://doi.org/10.1029/JA093iA06p05741

Rigler, E. J. (2017). Time-causal decomposition of geomagnetic time series into secular variation, solar quiet, and disturbance signals. *U.S. Geological Survey Open-File Report 2017-1037*, 26 p., https://doi.org/10.3133/ofr20171037

Ruohoniemi, J. M., & Baker, K. B. (1998). Large-scale imaging of high-latitude convection with Super Dual Auroral Network HF radar observations. *Journal of Geophysical Research*, *103*(A9), 20797–20811, https://doi.org/10.1029/98JA01288

Sabaka, T. J., Olsen, N., Tyler, R. H., & Kuvshinov, A. (2015). CM5, a pre-Swarm comprehensive geomagnetic field model derived from over 12 yr of CHAMP, Ørsted, SAC-C and observatory data. *Geophysical Journal International*, *200*, 1596–1626, https://doi.org/10.1093/gji/ggu493

Shepherd, S. G., & Ruohoniemi, J. M. (2000). Electrostatic potential patterns in the high-latitude ionosphere constrained by SuperDARN measurements. *Journal of Geophysical Research*, *105*(A10), 23005–23014, https://doi.org/10.1029/2000JA000171

Sugiura, M. (1964). Hourly values of equatorial Dst for the IGY. *Annals of the International Geophysical Year. Vol. 35*, pp. 9–45. Pergamon Press.

Thébault, E., Finlay, C. C., Alken, P., Beggan, C. D., Canet, E., Chulliat, A., … Rother, M. S. (2015). Evaluation of candidate geomagnetic field models for IGRF-12. *Earth, Planets, and Space*, *67*(112), https://doi.org/10.1186/s40623-015-0273-4

Toffoletto, F. R., Sazykin, S., Spiro, R. W., & Wolf, R. A. (2003). Inner magnetosphere modeling with the Rice Convection Model. *Space Science Reviews*, *107*, 175–196, https://doi.org/10.1023/A:1025532008047

Walker, J. K. (1989). Spherical cap harmonic modelling of high latitude magnetic activity and equivalent sources with sparse observations. *Journal of Atmospheric and Terrestrial Physics*, *51*(2), 67–80, https://doi.org/10.1016/0021-9169(89)90106-2

Waters, C. L., Gjerloev, J. W., Dupont, M., & Barnes, R. J. (2015). Global maps of ground magnetometer data. *Journal of Geophysical Research*, *120*, 9651–9660, https://doi.org/10.1002/2015JA021596

Wei, L. H., Homeier, N., & Gannon, J. L. (2013). Surface electric fields for North America during historical geomagnetic storms. *Space Weather*, *11*, 451–462, https://doi.org/10.1002/swe.20073

Weimer, D. R. (1995). Models of high-latitude electric potentials derived with a least error fit of spherical harmonic coefficients. *Journal of Geophysical Research*, *100*(A10), 19595–19607, https://doi.org/10.1029/95JA01755.

Weimer, D. R. (2013). An empirical model of ground-level geomagnetic perturbations. *Space Weather*, *11*, 107–120, https://doi.org/10.1002/swe.20030

Weimer, D. R., Clauer, C. R., Engebretson, M. J., Hansen, T. L., Gleisner, H., Mann, I., & YKumoto, K. (2010). Statistical maps of geomagnetic perturbations as a function of the interplanetary magnetic field. *Journal of Geophysical Research*, *115*, A10320, https://doi.org/10.1029/2010JA015540

Wiltberger, M., Merkin, V., Zhang, B., Toffoletto, F., Oppenheim, M., Wang, W., … Stephens, G. K. (2017). Effects of electrojet turbulence on a magnetosphere-ionosphere simulation of a geomagnetic storm. *Journal of Geophysical Research*, *122*, 5008–5027, https://doi.org/10.1002/2016JA023700

Wygand, J. M., Amm, O., Viljanen, A., Angelopoulos, V., Murr, D., Engebretsen, M. J., … Mann, I. (2011). Application and validation of the spherical elementary currents systems technique for deriving ionospheric equivalent currents with the North American and Greenland magnetometer arrays. *Journal of Geophysical Research*, *116*, A03305, https://doi.org/10.1029/2010JA016177

3

Magnetohydrodynamic Models of B and Their Use in GIC Estimates

Daniel Welling

ABSTRACT

Global magnetohydrodynamic (MHD) models, coupled with height-integrated ionospheric electrodynamic models, are important tools for GIC research and operations. This chapter reviews the physics behind global MHD models, recent advances in physics and inter-model coupling capabilities, and how surface magnetic perturbations are obtained from these models. Data-model validation studies are surveyed to illustrate the strengths and limitations of the MHD approach. Transition to operations and future directions for this modeling methodology are also discussed. While many MHD models are available in the field, emphasis is placed on the three most widely used and operationally ready today.

Key Points

- The capabilities of magnetohydrodynamics (MHD) models are reviewed in terms of their ability to contribute to solving the problem of geomagnetically induced currents (GICs).
- The validity of MHD as demonstrated through data-model comparisons is reviewed.
- Improving the predictive skill of MHD predictions and validation of MHD models is an ongoing process.

In order to successfully model and forecast geomagnetically induced currents, one must be able to capture the interaction between the solar wind and interplanetary magnetic field (IMF) with the Earth system, including the magnetosphere, ionosphere, and below. The electromagnetic and plasma response of the system must be translated into the perturbation of the field at ground level, both in terms of the perturbation itself (ΔB) and its time derivative (dB/dt). Further, to achieve forecast capabilities, any model of ΔB and dB/dt must be driven from upstream solar conditions only. Achieving these goals via a first-principles-based approach is a uniquely challenging task, numerically and computationally, as the scales and physics involved vary wildly from bow shock to ionosphere. Success can provide broad insight into the physics behind GICs and unlock operational predictive capability.

Though there are many first-principles-based approaches used in research, only one approach – global magnetohydrodynamics (MHD) with an embedded height-integrated electrostatic ionosphere – has shown the ability to be operationally viable for space weather applications. MHD models have the key characteristics necessary for both research and operations in GIC applications. First, the input set only includes upstream conditions of the solar wind (including IMF) and the F10.7 solar radio flux, enabling both forecasts which use solar wind observations, and simulations of hypothetical or idealized space weather storms. They have evolved to become computationally efficient, requiring only modest computational resources to achieve real-time execution speeds. Data-model comparisons show that they are capable of reproducing many sets of observations, often with higher skill than empirical counterparts. The output is directly relevant to GIC work as MHD calculates dB/dt anywhere about the globe. Though new models and

University of Texas at Arlington Physics Department, Arlington, TX, USA

Geomagnetically Induced Currents from the Sun to the Power Grid, Geophysical Monograph 244, First Edition.
Edited by Jennifer L. Gannon, Andrei Swidinsky, and Zhonghua Xu.

approaches are always being developed, MHD remains one of the most widely used in the field today for both research and applications.

This chapter reviews the capabilities of MHD models in terms of their ability to contribute to solving the problem of GICs. MHD theory and models are briefly introduced with a focus on applicability to the Earth system. The validity of MHD as demonstrated through data-model comparisons is reviewed. MHD's use in GIC research and operations is summarized, and the future directions of MHD modeling is discussed. While many MHD models are available in the field, emphasis is placed on the three most widely used and operationally ready today.

3.1. THEORY AND MODELS

Ideal magnetohydrodynamics combines the fluid dynamic equations with idealized Ohm's law and Maxwell's equations to describe the evolution of plasma as a single species charged fluid (Boyd and Sanderson, 2003; Chen, 2016). This yields four equations for two unknown scalar quantities (mass density, ρ, isotropic thermal pressure, P) and two unknown vector quantities (bulk velocity, $\bar{u}$, and magnetic field, $\bar{B}$):

$$\frac{\partial \rho}{\partial t}+\nabla\cdot(\rho\bar{u})=0 \quad (3.1)$$

$$\frac{\partial \rho\bar{u}}{\partial t}+\nabla\cdot\left[\rho\bar{u}\bar{u}+\bar{\bar{I}}\left(P+\frac{1}{2\mu_0}B^2\right)-\frac{1}{\mu_0}\bar{B}\bar{B})\right]=0 \quad (3.2)$$

$$\frac{\partial e}{\partial t}+\nabla\cdot\left[\bar{u}\left(e+P+\frac{1}{2\mu_0}B^2\right)-\frac{1}{\mu_0}\bar{u}\cdot\bar{B}\bar{B}\right]=0 \quad (3.3)$$

$$\frac{\partial \bar{B}}{\partial t}+\nabla\times(\bar{u}\times\bar{B})=0 \quad (3.4)$$

These equations represent conservation of mass (Equation (3.1)), momentum (3.2), and energy (3.3) as a function of time, t. The total energy, e, is defined as

$$e=\frac{P}{\gamma-1}+\frac{\rho u^2}{2}+\frac{B^2}{2\mu_0}. \quad (3.5)$$

The three terms on the right-hand side of Equation (3.5) represent the thermal, kinetic, and the magnetic energy, respectively. γ is the ratio of specific heats and is 5/3 for an ideal gas with three degrees of freedom. Equation (3.4) is known as the induction equation, which relates the motion of the plasma to that of the magnetic field. Without any resistive or Hall terms, Equation (3.4) is the "frozen flux" condition, meaning that the magnetic field and plasma must move together. Numerical MHD models solve a discretized version of these equations given some initial and boundary conditions for a variety of situations, including that of the Earth's magnetosphere (Raeder, 2003).

Using ideal MHD at Earth has a number of downsides that must be acknowledged. In the inner magnetosphere, energy dependent kinetic drifts become important (e.g., Daglis et al., 1999). The single temperature treatment in ideal MHD is poorly equipped to capture the diamagnetic drifts of the ring current (e.g., Heinemann and Wolf, 2001), yielding inaccurate solutions within geosynchronous distances. Reconnection is also a concern in ideal MHD. Without any explicit terms allowing the frozen flux condition to be broken (Equation (3.4)), all reconnection is the result of numerical resistivity. In many Riemann-type solvers, which dominate in the space physics community, numerical resistivity is proportional to the local fastest wave speed (Toro, 2006). Therefore, the reconnection rate scales with driving. In ideal MHD models, it has been shown that the reconnection rates scale as $0.1B_{in}V_A$, where B_{in} is the inflowing field strength and V_A is the local Alfven speed (Ouellette et al., 2013). The reconnection rate is still much slower than that of Hall MHD, hybrid, or fully kinetic approaches (Birn and Hesse, 2001) and approximately half what is expected by theory (Ouellette et al., 2014). These limitations are frequently a concern for magnetosphere modelers.

Despite these shortcomings, ideal MHD is still very capable at reproducing magnetospheric dynamics, as will be shown in Section 3.2. More physically complete treatments exist, but none have been validated to the extent that standard ideal MHD models have been. In magnetospheric physics, the ideal MHD formalism is still the go-to approach for global modeling.

3.1.1. Ideal MHD at Earth

Though MHD theory has been well established for a long time, it was not until the 1980s that it began to be applied to Earth (Brecht, 1985). Simple 2D simulations gave way to full 3D simulations using idealized solar wind conditions (e.g., Brecht et al., 1981; Leboeuf et al., 1981; Wu et al., 1981). Early simulations used simplistic representations of the Earth and ionosphere, such as imposing a simple azimuthal current to force a dipole field. Rudimentary attempts to include the ionosphere soon followed (Fedder and Lyon, 1984; Ogino and Walker, 1984). By the mid and early 1990s, new MHD advancements and scientific results were arriving with regularity (e.g., Fedder and Lyon, 1995; Raeder et al., 1995; Gombosi et al., 1998).

Throughout this period, modeling the solar-wind Earth system presented new challenges and solutions. Very high

Alfven wave speeds near the inner boundary (typically between 2 and 3 R_E, where the dipole field strength is strong) requires diminishingly small time steps in explicit schemes. Nearly all MHD codes have therefore adopted the Boris correction factor (Boris, 1970; Gombosi et al., 2002), which artificially lowers the speed of light and reduces the MHD wave speeds. Numerical grids for the Earth system required additional optimization as the magnetosphere combines regions where coarse resolution will suffice and other regions where fine spacing is required. New solvers and approaches to enforcing $\nabla \cdot B = 0$ were introduced and employed (Evans and Hawley, 1988; Koren, 1993; Tóth, 2000). Renewed consideration of inner boundary conditions has led to a better representation of the body's electrodynamic (Merkin and Lyon, 2010) and mass source (Welling and Liemohn, 2014) properties.

Over three decades of work has yielded many different MHD models, but three stand out as both widely used and immediately applicable to the problem of GICs:

• The Lyon-Fedder-Mobarry model has a long history of use within the field (Fedder and Lyon, 1995; Fedder et al., 1995; Mobarry et al., 1996). It can include an Magnetosphere-Ionosphere Coupler Solver (LFM-MIX (Merkin and Lyon, 2010)) which allows interfacing with a thermosphere-ionosphere model (Wiltberger et al., 2004), collectively referred to as the Coupled Magnetosphere Ionosphere Thermosphere model (CMIT). It employs a high-order solver on a stretched spherical grid that better aligns with terrestrial field lines than more traditional rectangular grids (Lyon et al., 2004). The default resolution is 53 radial, 24 azimuthal, and 32 polar points ("single" resolution); higher resolution configurations double the number of points in the azimuthal and polar directions ("double" resolution) or all directions ("quad" resolution). Double resolution is now the standard employed at NASA's Community Coordinated Modeling Center.

• The Open Global Generalized Circulation Model (OpenGGCM (Raeder et al., 1995, 1998, 2001a)) is another well established MHD model. It uses a stretched rectangular grid, which allows resolution to be concentrated across set ranges. This improves resolution where needed but also in areas where a coarser grid would suffice. OpenGGCM has long had the ability to include anomalous resistivity (Raeder, 1999). It is frequently used to study tail dynamics during substorm events (Raeder et al., 2008) as well as day side magnetopause conditions (Dorelli, 2004).

• The Block Adaptive Tree Solar wind Roe-type Upwind Scheme (BATS-R-US) is the youngest of the three models (Powell et al., 1999; De Zeeuw et al., 2000; Groth et al., 2000). It employs a flexible, block-based Cartesian grid that can adaptively refine to the physical conditions in the magnetosphere. One to two million cells are commonly used for Earth simulations, but much higher resolutions are possible (Glocer et al., 2016a). BATS-R-US is frequently used via the Space Weather Modeling Framework (SWMF (Tóth et al., 2005, 2007, 2012)), which allows it to be coupled to a host of other numerical models of the thermosphere, ionosphere, ring current, and more.

Though these models will be the focus of the rest of this chapter, several alternatives are also established in the field. Some notable codes include the Winglee and GUMICS models. The Winglee multi-fluid model (Winglee, 1998, 2000, 2003) was the first global model to include outflow of both light and heavy ions from the magnetosphere to the ionosphere. The Grand Unified Ionosphere-Magnetosphere Coupling Simulation (GUMICS) uses an adaptive grid with a unique 3D ionosphere model (Palmroth et al., 2004; Janhunen et al., 2012). These codes continue to make important scientific contributions to the field today.

3.1.2. Adding the Ionosphere

Early MHD simulations at Earth leveraged simplified representations of the conducting ionosphere. Given the importance of ionospheric dynamics to the magnetosphere, a more realistic, physics-based representation was required. The now-standard approach is to solve a Poisson-like equation for a height-integrated, two-dimensional electrostatic ionosphere (Goodman, 1995; Amm and Goodman, 1996):

$$J_R = \nabla_\perp \left(\Sigma \cdot \nabla_\perp \Phi \right) \tag{3.6}$$

Here, J_R is the radial currents as supplied by the MHD, Σ is the conductance tensor (comprising of the Hall and Pedersen conductances), Φ is the electric potential in the ionosphere, and derivatives are taken tangent to the spherical surface. The MHD currents are taken at some set altitude near but not on the inner boundary and are mapped to the ionosphere altitude along assumed dipole field lines. The gradient of the potential yields the $E \times B$ electric field, which is returned to the MHD model to set the tangential velocity about the inner boundary. Nearly all global MHD models follow this approach (Ridley et al., 2001, 2004; Wiltberger et al., 2009).

The outstanding challenge in including ionospheric electrodynamics is prescribing the conductance. In the real magnetosphere, there are two dominant sources for conductance: ionization due to solar extreme ultra-violet (EUV) radiation (Moen and Brekke, 1993) and ionization due to precipitating ions and electrons (Frahm et al., 1997; Newell et al., 2007, 2009). The former is included via an empirical formula that is a function of solar zenith angle. The latter requires knowledge of electron and ion precipitation into the ionosphere. This is a significant requirement, as it depends on the full particle distribution within the magnetosphere, how those distributions are being scattered into the loss cone by waves and other processes, and how the soft precipitation is accelerated into the ionosphere to produce discrete aurora. These requirements far exceed the capabilities of fluid-based representation found in MHD.

To include precipitation-driven conductance that is self-consistent with an MHD magnetosphere, significant compromises must be made. Early experiments used prescribed, constant and spatially uniform conductance, finding significant effects on magnetospheric dynamics (Siscoe, 2002; Siscoe et al., 2002; Merkine et al., 2003, 2005b; Ridley et al., 2004). The LFM model turned to a semi-empirical approach (Wiltberger et al., 2009). Electron precipitation is initially estimated by taking the near-boundary fluid number density and setting the thermal temperature via the local sound speed. The Knight (1973) relationship between precipitation, field-aligned current, and field-aligned electric potential drops is used to harden the precipitation spectrum. The precipitating energy flux is then converted into Hall and Pedersen conductances via the oft-used Robinson et al. (1987) empirical formulas. These relationships are based on a limited number of ionospheric radar observations (Vickrey et al., 1981), but have shown be be fairly robust (Kaeppler et al., 2015). This technique involves four parameters that were initially tuned to optimize comparisons between the model and observations (Fedder and Lyon, 1995; Slinker et al., 1999). A simpler empirical model is employed by the Ridley Ionosphere Model (RIM), which is coupled to the BATS-R-US MHD model. This model uses a conductance relationship obtained from results of Assimilative Mapping of Ionospheric Electrodynamics (AMIE), a technique for reconstructing ionospheric conditions via inverting a large number of ground- and space-based observations (Richmond and Kamide, 1988). From a full month (January 1997) of AMIE results, a simple relationship was constructed between the strength of FACs and Hall or Pedersen conductance:

$$\Sigma = \Sigma_0 e^{-A|J_{\parallel}|} \tag{3.7}$$

Σ_0 and A are functions of latitude and longitude. Two sets of coefficients are used, one for upward FACs and one for downward FACs. These two approaches represent the most widely used conductance specifications for MHD-based magnetosphere–ionosphere modeling today.

It should be noted that considerable effort is spent on improving the physics behind these models. Zhang et al. (2015) introduced a new approach for including broadband and cusp precipitation into the LFM model. Initial work has been performed to model the precipitation from the ring current into the ionosphere via first-principles-based model coupling (Yu et al., 2017; Perlongo et al., 2017). Including anomalous conductance as an additional conductance source in global models has yielded promising early results (Merkin et al., 2005a; Liu et al., 2016; Wiltberger et al., 2017a). Though promising in capabilities, these enhanced physics options have yet to be included in thorough validation studies to test their impact on *dB*/*dt* forecasts.

3.1.3. Extending MHD

The length and time scales of the solar wind–magnetosphere–ionosphere system span several orders of magnitude and cover a multitude of physical processes, both fluid and kinetic. This means that ideal MHD simply cannot hope to properly address all of the regions in a physically meaningful manner. Model developers have worked hard to overcome these deficiencies, both via expanded physics and through coupling with other, non-MHD models.

The most important improvement to global MHD for the Earth's magnetosphere has been two-way coupling with kinetic ring current models. Single fluid, single temperature MHD simply cannot accurately capture the energy dependent drifts that are critical in the inner magnetosphere (Heinemann and Wolf, 2001). Dedicated bounce-averaged models of the ring current excel at capturing these drifts and the build-up of energy in the inner magnetosphere (e.g., Wolf, 1983; Jordanova et al., 1999; Fok et al., 2001; Liemohn et al., 2001). However, these models require a plethora of inputs to run: particle flux about the outer boundary and electric and magnetic fields throughout their domain. A natural solution is to couple MHD to kinetic ring current models so that MHD provides all of the inputs to the ring current model, while the total plasma pressure from the ring current model is used to nudge the MHD pressure towards more realistic values.

This coupling was first achieved between BATS-R-US and the Rice Convection Model (RCM) (Harel et al., 1981; Sazykin and Stanislav, 2000; Toffoletto et al., 2003) through the SWMF (De Zeeuw et al., 2004). The initial conditions of the RCM model are provided by BATS-R-US: magnetic field flux tube volume and flux tube plasma content. Plasma is partitioned into RCM's energy grid by assuming a Maxwellian distribution about the MHD fluid temperature. Once the simulation begins, inputs to RCM are provided by either BATS-R-US (flux tube integral values about the RCM outer boundary) or by RIM (the convection electric field). Conversely, RCM calculates the total plasma pressure across its grid and returns this back to BATS-R-US. These values are used as a pressure source term in the MHD solver that gradually drives the original solution, P_{GM}, towards the RCM solution, P_{RCM}, converging over a time of $2\tau_{couple}$:

$$p'_{MHD} = p_{MHD} + \min\left(1, \frac{\Delta t}{\tau_{couple}}\right)\left(p_{RCM} - p_{MHD}\right). \quad (3.8)$$

Here, Δt is the MHD time step. The values are passed between all models at a user-defined frequency, typically $\sim$ 10 s. This approach leverages the SWMF to execute, synchronize, and couple all models. Following the success of the initial work, two-way couplings between other ring current models and other MHD codes have been completed (Pembroke et al., 2012; Glocer et al., 2013; Welling et al., 2015; Cramer et al., 2017).

Coupling between inner magnetosphere models and MHD codes stands as one of the great inter-model capabilities in space weather research. The initial goal of improving the thermal pressure solution within the MHD inner magnetosphere was reached immediately. This is shown in Figure 3.1, which shows the BATS-R-US MHD pressure without (left frame) and with (right frame) two-way coupling with the Comprehensive Ring Current Model (CRCM) (Fok et al., 2001). The pressure about the night-side ring current is increased by a factor of 6 or more. The increased inner magnetosphere pressure in the MHD solution drives strong pressure gradients and, therefore, stronger mid-latitude region-2 FACs. This impact is shown in Figure 3.2, which shows the ionospheric electrodynamics from a coupled LFM-RCM simulation. Without two way coupling (left frame), the low-latitude FACs are nearly non-existent (colored contours). With this coupling (right frame), the region-2 FACs are sharp and strong. The enhanced region-2 FACs also shields the convection electric field (line contours) from low-latitude regions. As will be shown later, the more realistic inner magnetosphere description plays a huge role in improving data-model comparisons.

Such coupling is not without challenges. The results are sensitive to choices of τ_{couple}. The location of the outer boundary of the ring current model is important, as it dictates where gradient-curvature drifts become important in the simulated plasma sheet. Further, in the real plasma sheet, adiabaticity can be broken for fast flows, meaning that neither bounce-averaged drift models nor MHD are valid. Despite these shortcomings, coupling still results in driving better, more operationally-relevant simulations.

Outside of ring current coupling, there have been many efforts to expand physics capabilities of global MHD. Multi-species (Ma et al., 2007; Welling and Ridley, 2010b) and multi-fluid (Winglee, 1998; Glocer et al., 2009; Wiltberger et al., 2010) approaches have been developed and employed to examine the role of composition throughout the system (Wiltberger, 2015). Anisotropic MHD has been employed on a global scale (Meng et al., 2012), extending the coupling capabilities to ring current models (Meng et al., 2013). Attempts to address limited reconnection physics by including resistive terms in Equation (3.4) have also been made. These studies have included both constant (e.g., Borovsky et al., 2008; Glocer et al., 2016a) and anomalous (Raeder et al., 2001a, 2010) resistivity values. Alternatively, Hall MHD, long used to simulate the reconnection region (e.g., Birn and Hesse, 2001), has been extended to global magnetosphere simulations (Tóth et al., 2008). More recently, the global effects of kinetic reconnection have been included in

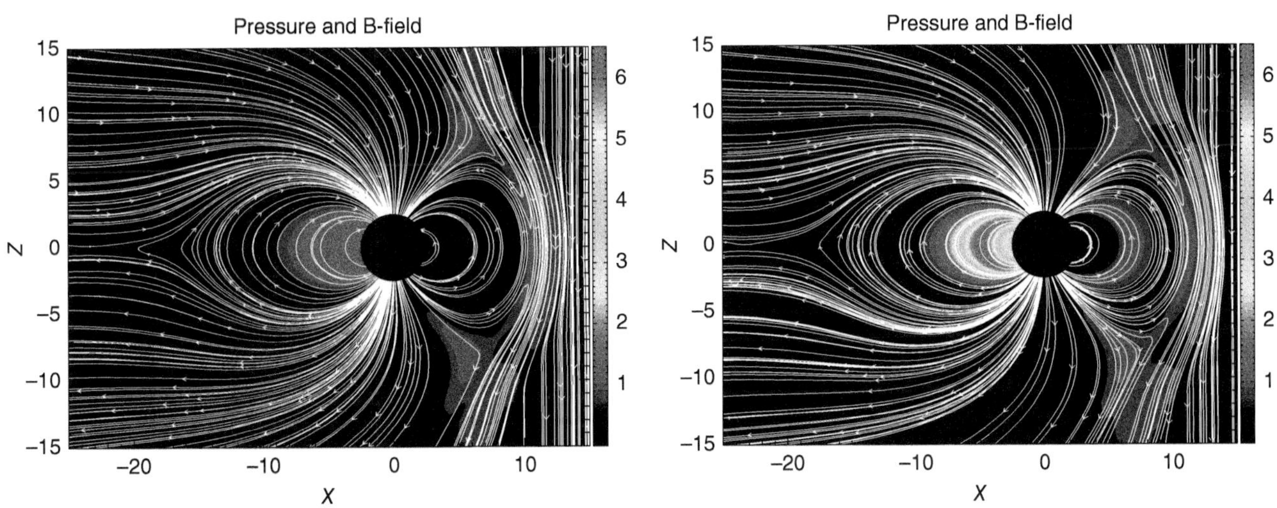

Figure 3.1 MHD pressure (contours) and magnetic field (white lines) from a BATS-R-US simulation with (left frame) and without (right frame) two-way coupling with the CRCM inner magnetosphere model. Source: Adapted from Glocer et al. (2013). Reproduced with permission of John Wiley and Sons. (*See electronic version for color representation of this figure.*)

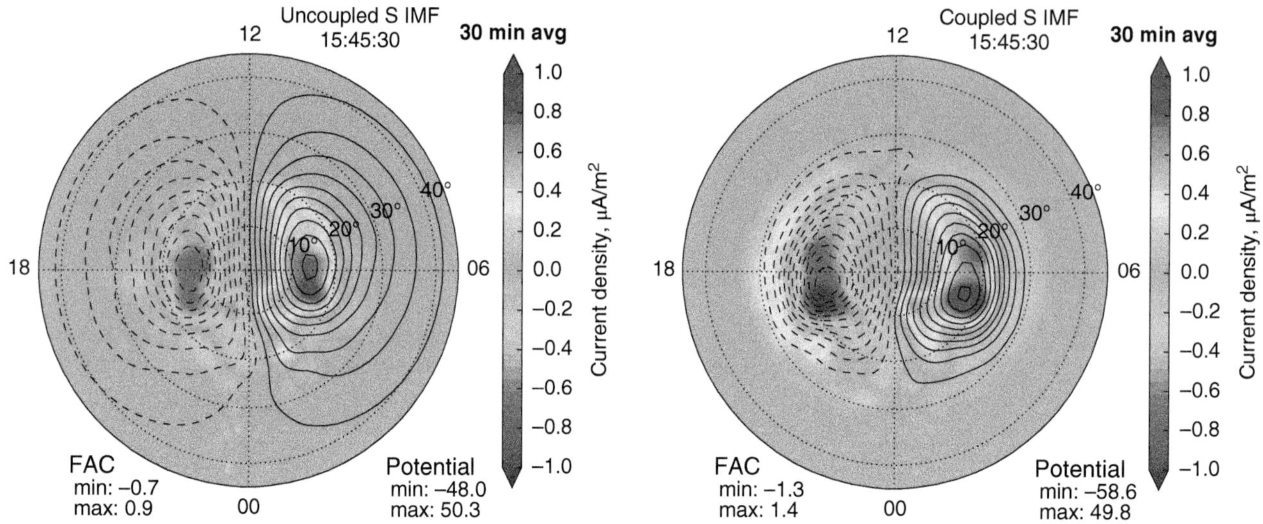

Figure 3.2 Ionospheric electrodynamics in terms of field-aligned currents (color contours) and electric potential (line contours) from an LFM MHD simulation without (left frame) and with (right frame) two way coupling to RCM. Source: Adapted from Pembroke et al. (2012). Reproduced with permission of John Wiley and Sons. (*See electronic version for color representation of this figure.*)

global MHD by embedding a fully kinetic particle-in-cell model to simulate the reconnection region (Chen et al., 2017; Tóth et al., 2017). While these expanded approaches have important if not niche applications in science, they have yet to be widely employed in GIC related research.

3.2. VALIDITY OF MHD AT EARTH

Before the question of MHDs applicability to GIC problems can be answered, the question about its validity at Earth must be explored. How successfully can MHD reproduce basic features of the magnetosphere? Can MHD accurately reproduce the spatial and temporal patterns of FACs? What about ionospheric electrodynamics, that are critical contributors to GICs? How does the validity depend on activity level? Further, is global MHD a viable alternative to empirical models that are far less computationally expensive to run? Addressing these concerns is a critical hurdle to operations.

Decades of studies have examined the question of the validity and applicability of MHD applied Earth's magnetosphere. Most validation of MHD occurs in small, event study qualitative comparisons. Such studies typically occur as part of a larger science investigation. A small fraction are quantitative in their comparisons; an even smaller minority represent in-depth, quantitative validation studies that span multiple events and a variety of activity. Reviewing the large scale, in-depth validation efforts reveals emergent patterns concerning MHD performance at Earth:

- MHD can successfully reproduce the shape of the magnetosphere, including the location of the magnetopause within a few R_E of the observed location.
- MHD models with a height-integrated ionosphere solver are able to successfully reproduce large-scale features of ionospheric electrodynamics, but results are sensitive to conductance.
- Tail dynamics, especially related to substorms, remain difficult to consistently reproduce.
- Inclusion of a two-way coupled ring current model drastically improves data-model comparisons.
- Results are sensitive to grid resolution, with higher resolutions typically yielding better results overall.

From a very early point, MHD has shown value in reproducing the shape of the magnetosphere. This is an important capability, as the shape of the magnetosphere dictates the location of currents within the system, the location of their closure into the ionosphere, and the size and shape of the polar caps. MHD's ability to capture the magnetosphere shape is now well established via many systematic, quantitative validation studies. Comparisons to magnetic field observations about geosynchronous orbit are a popular and powerful way to test MHD results. Systematic studies consistently show favorable data-model comparisons on a statistical level (Huang et al., 2006; Welling and Ridley, 2010a; Rastätter et al., 2011). Validation metrics, such as correlation coefficient and root-mean-squared error, tend to improve as the resolution of the models are increased and when the MHD is coupled to a dedicated ring current model (Huang et al., 2006; Pulkkinen et al., 2010; Rastätter et al., 2011; Ridley et al., 2016). Global MHD tends to perform as well as popular empirical models of the magnetic field (e.g., Tsyganenko, 1989, 2002), even out performing them during periods of strong driving (Huang et al., 2006; Rastätter et al., 2011). Similarly, predicting the location of the magnetopause can give strong indication of the MHDs ability to reproduce magnetospheric shape. Again, MHD shows a consistent ability to correctly predict magnetopause crossings, correctly predicting ~50–90% of crossings at geosynchronous orbit (Lopez et al., 2007; Welling and Ridley, 2010a). MHD models produce predictions of the magnetopause stand-off distance that are similar to popular empirical models (Gordeev et al., 2015), again showing that they are competitive with observational-based approaches.

These statistical successes are tempered by the amount of inter- and intra-model variation in forecast quality displayed in MHD results. This is well illustrated in comparisons of geosynchronous $|B|$ from Rastätter et al. (2011): when all of the predictions made by different MHD models and configurations are plotted together for a single event, the results vary by tens of nanotesla (i.e., up to ~50% of the observed magnitude). Rarely does a single simulation result outperfrom all others throughout a single event. Similarly, such variation is present with predictions of the magnetopause location. Samsonov et al. (2016) showed that the predictions of magnetopause locations given by various MHD and empirical models can span several R_E. Therefore, while it is safe to say that gross features are captured, the quality of any one MHD simulation can vary strongly. The extent to which dB/dt predictions are sensitive to errors in field geometry predictions is yet unknown.

Another common target of MHD validation studies is ionospheric electrodynamics. Ionospheric signatures, including the strength and pattern of FACs, the electric potential pattern, and cross polar cap potential (CPCP) are important indications that the model is correctly capturing energy transfer from the solar wind to the magnetosphere and that the magnetosphere response is realistic. Obtaining accurate ionospheric electrodynamics is also critically important to predicting dB/dt and therefore GICs. Because ionospheric currents contribute strongly to GICs, even a slight shift in the location of horizontal ionospheric currents can lead to a large discrepancy in the location of the predicted dB/dt peak. Therefore, accurate electrodynamics are essential to GIC forecasting.

As is the case with magnetic field, validation shows that MHD can reproduce the gross features of ionospheric electrodynamics. Comparisons of FAC patterns between MHD and reconstructions via observations from the Iridium satellite constellation are favorable across a range of activity (Korth et al., 2004, 2008): MHD was able to reproduce the polarity, rough shape, and approximate magnitude of the currents. Steady state MHD simulations for a variety of IMF field orientations are able to reproduce statistical Iridium FAC pictures and the integrated current amplitude within 10% (Korth et al., 2011). Two-dimensional correlation coefficents show moderate (~50% for northward IMF conditions) to strong (~80% for southward conditions) correlation with observations. Ability to capture FACs directly leads to reasonable reproduction of CPCP, as observed (Haiducek et al., 2017) and compared to leading empirical models (Zhang et al., 2011; Gordeev et al., 2015; Wiltberger et al., 2017b). This is illustrated in Figure 3.3, which shows the correlation between MHD models and the empirical Weimer 2005 (Weimer, 2005) model for CPCP and total upward FAC strength (left and right frames, respectively). For the most part, the model curves, where the colors indicate different global MHD codes, have a slope at or near 1, indicating good correspondence to the empirical relationship. These correspondences translate to reproduction of interesting magnetospheric phenomena, such as saturation of cross polar cap potential (Siscoe et al., 2002; Merkine et al., 2003; Ridley, 2007; Kivelson and Ridley, 2008).

Though global MHD can boast many successes with electrodynamics, there are many shortcomings that must be addressed in the future. Foremost are the models' sensitivity to ionospheric conductance, immediately illustrated in Figure 3.3. The darker gray lines (red lines in the electronic version) show the results for BATS-R-US with RIM via the SWMF: this combination appears to predict CPCP with very high skill (left frame) while under predicting the strength of the FACs. Conversely, the LFM model (dark blue lines in the electronic version) accurately captures FAC strength while consistently under predicting CPCP. Finding a validated and accurate conductance model remains an important outstanding task. Auroral dynamics remain challenging; MHD inconsistently reproduces auroral signatures via the Auroral Electrojet (AE) index (Davis and Sugiura, 1966), both on long time scales (Haiducek et al., 2017) and during substorms (Raeder et al., 2001a; Wiltberger et al., 2003). Comparisons to in-situ (Wang et al., 2008) and remote observations (Anderson et al., 2017) of FACs show a propensity of all MHD models to predict values at higher latitudes than the real system. This discrepancy tends to grow with increasingly strong southward IMF conditions (Korth et al., 2011). As is the case with magnetic field comparisons, high variance in the quality of the results is an issue with reliability.

Tail dynamics remain challenging for global MHD models. Without a dedicated ring current model, MHD has earned a reputation of producing a night side region that is too cold and too dense. This reputation has been borne out statistically via long-term comparisons between the LFM model and observations made by the Geotail satellite (Guild et al., 2008). Faster-than-observed flows in the MHD tail correspond to high convection electric field as indicated by high CPCP values. Plasma comparisons made with results from a coupled ring current model help improve tail dynamics

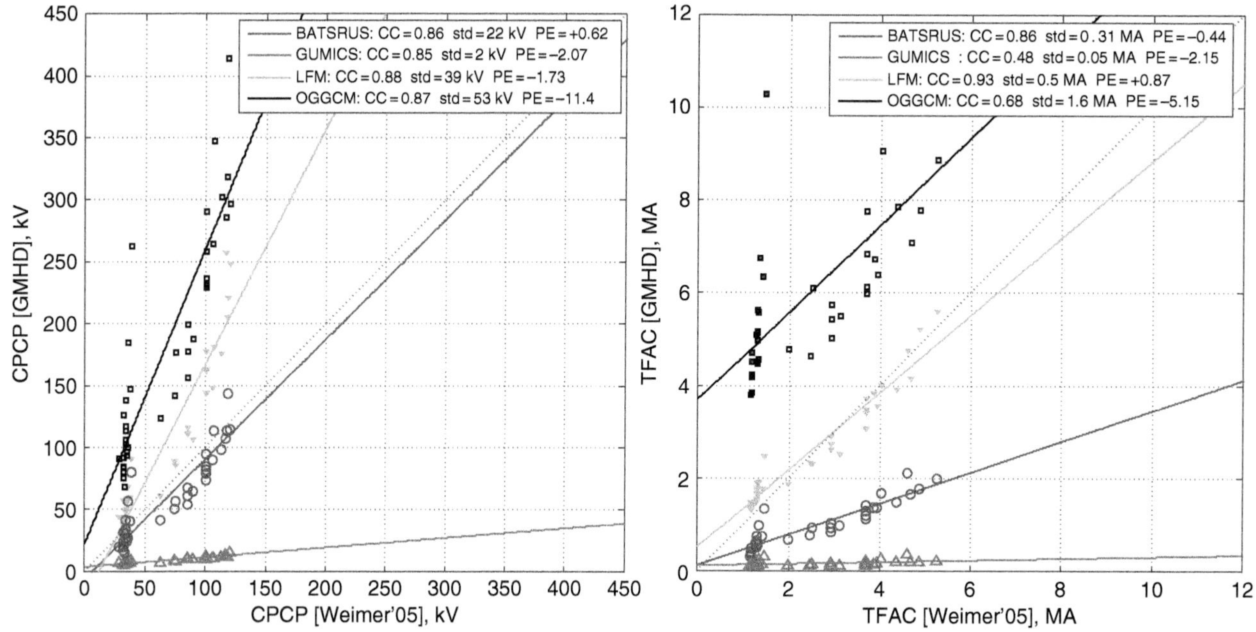

Figure 3.3 Correlation between MHD predictions (vertical axes) of CPCP (left frame) and total upward FAC (right frame) against empirically-based values from the Weimer2005 model. The color of the curve corresponds to different MHD models as shown in the figure legends. Source: Adapted from Gordeev et al. (2015). Reproduced with permission of John Wiley and Sons. (*See electronic version for color representation of this figure.*)

(Welling and Ridley, 2010a), but accurate plasma predictions remain elusive.

Challenges in tail modeling extend to substorm activity. Ideal and resistive MHD models can produce "substorms," at least in the sense that there is a tail stretching and plasmoid release that results in transient high-latitude currents (Raeder et al., 2001a; Welling and Ridley, 2010a). However, the timing of events compared to observations is inconsistent. The strength of modeled substorm events, as defined by the dipolarization and current flows, is typically much less than observed. When substorms are initiated via a north-to-south transistion of IMF B_Z, Gordeev et al. (2017) found critical differences between magnetic field convection rates in the tail during the growth and expansion phases. Of the three models investigated, the LFM model best reproduced tail convection patterns as understood via observations. Because substorms are important contributors to strong night side current flows, the inability to effectively capture these events represents a GIC vulnerability that cannot yet be reliably predicted.

This deficiency peaks when trying to reproduce global sawtooth oscillations. Sawtooth events are quasi-periodic, high-amplitude substorms that produce a response across a wide swath of magnetic local times (Borovsky, 2004; Henderson, 2004; Cai et al., 2006; Cai and Clauer, 2009; Morley et al., 2009; Morley and Henderson, 2010). These events get their name from the signature found in geosynchronous particle and field data (Henderson et al., 2006a,b; Ohtani et al., 2007) and are typically associated with periods of exceptionally strong driving (DeJong et al., 2009; Cai et al., 2011). Ideal MHD models fail at reproducing sawteeth oscillations (as noted by Goodrich et al. (2007); Kuznetsova et al. (2007) and Brambles et al. (2011)) unless some additional physics are added. Kuznetsova et al. (2007) applied a non-gyrotropic correction to the electric field within BATS-R-US MHD. Simulations using this correction were able to reproduce the behavior and periodicity of sawteeth. Brambles et al. (2011) included heavy ion outflow at the inner boundary of the multi-fluid LFM model. The outflow was specified via an empirical relationship tied to the MHD solution, yielding a causal source of oxygen. Sawtooth oscillations developed in the simulation with the period and amplitude showing a clear dependence on the scaling of the outflow relationship. While this work is promising, it remains an incomplete explanation of sawtooth development as observations show that high O^+ concentrations are a necessary but not sufficient condition for sawteeth to develop (Liao et al., 2014). Accounting for these important phenomena remains elusive in global MHD models.

Throughout validation studies, two themes emerge. The first is that absence of a coupled inner magnetosphere model is a clear limitation. Many studies note that inclusion of a ring current model would improve comparisons to FACs (Korth et al., 2008) and in-situ field comparisons (Rastätter et al., 2011). When ring current coupling is included, comparisons improve as predicted (Welling and Ridley, 2010a; Korth et al., 2011; Haiducek et al., 2017). The second theme is resolution: whenever comparisons are made between observations and results from different grid resolutions, the finer grid nearly always performs better (Guild et al., 2008; Haiducek et al., 2017; Wiltberger et al., 2017b). A dramatic example of this is shown in Figure 3.4. Each frame shows field aligned currents as identified via a machine learning algorithm and colored by type (upward and downward) and direction (regions 1 and 2). The left three frames show LFM FACs for three different resolutions: the coarse "single" resolution, "double" and "quad." The rightmost frame shows FACs as identified from the Weimer 2005 empirical model. Each grid refinement pushes the results much closer to the empirical patterns, demonstrating how strongly the results rely on resolution. The literature unambiguously demonstrates that both including ring current physics and properly resolving dynamics are vital to predicting realistic magnetosphere–ionosphere conditions.

Beyond resolution and coupling to the ring current, the myriad of settings on a typical MHD model can affect the output significantly. This was explored in detail by Ridley et al. (2010), which showed that both the MHD magnetosphere and the associated ionospheric dynamics depend on choice of solver, Boris factor, inner boundary density, and other settings. There is also broad variability in simulation quality between models as well as within model configurations (Ridley et al., 2016). Validation of global models is a big and ongoing task.

3.3. MHD AND GICs

The potential for MHD to be a valuable tool for forecasting and evaluating space weather events, including GICs, has long been acknowledged. Realizing this potential, however, required the development of *dB/dt* outputs and new validation efforts. The combined effort of the modeling community led to the adoption of MHD in operations, but also revealed new challenges and shortcomings to overcome.

3.3.1. Virtual Magnetometers

While previous studies were important for establishing MHDs relevance to space weather, it was the development of a new output tool, *virtual magnetometers*, that cemented MHD's importance in GIC forecasting. Virtual magnetometers calculate the perturbation of the magnetic field from background (ΔB) at a set location on the surface of the Earth. They do this by performing Biot–Savart integrals (Jackson, 1999; Griffiths, 2015) throughout the simulation domain:

$$\Delta \bar{B}(\bar{r}) = \frac{\mu_0}{4\pi} \int \frac{\bar{J}' \times (\bar{r} - \bar{r}')}{|\bar{r} - \bar{r}'|} d\bar{r}'^3. \tag{3.9}$$

This volume integral sums the perturbations at a fixed location, $\bar{r}$, due to infinitesimal current density elements, $\bar{J}'$, within volume element $d\bar{r}'^3$ centered at $\bar{r}'$. The associated *dB/dt* can be obtained by taking the time derivative of a series of Biot–Savart calculations.

Early use of virtual magnetometers, which included only a subset of the major current systems that contribute to

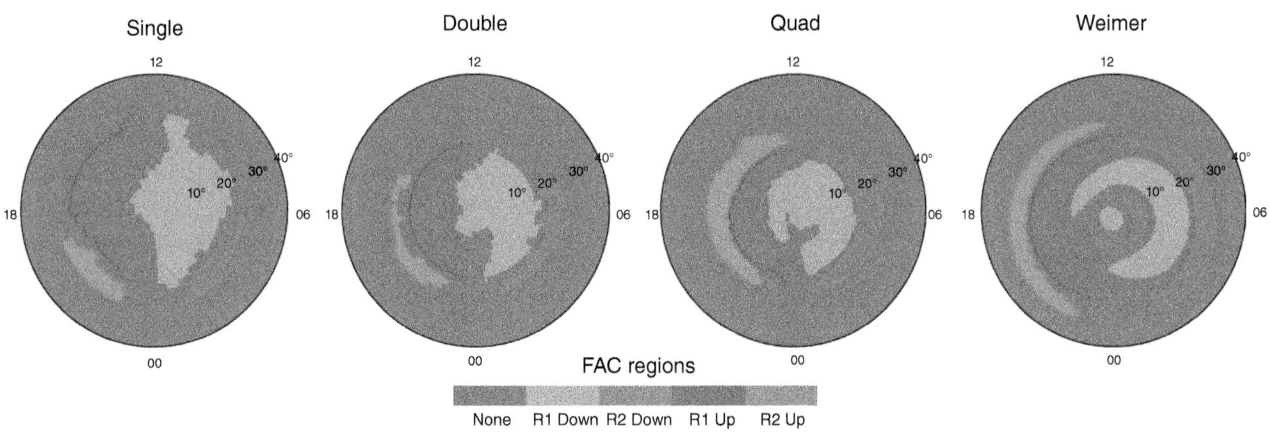

Figure 3.4 Machine identification of FACs from the LFM model using coarse (left frame) to most fine (second frame from the right) grid resolution. Rightmost frame shows FACs identified from Weimer 2005 empirical model. Colors show FAC region and direction. Source: Adapted from figure 6 of Wiltberger et al. (2017b). Reproduced with permission of John Springer Nature. *(See electronic version for color representation of this figure.)*

ΔB, demonstrated the promise of MHD for ΔB and *dB/dt* predictions. Initial studies leveraged Fukushima's Theorem (Fukushima, 1969, 1976) to simplify their calculation: for perfectly radial FACs, the perturbation from the FACs and the horizontal Pedersen currents will cancel each other, leaving the Hall currents as the dominant source of ΔB. OpenGGCM was employed to study Hall-related ΔB during the Bastille Day storm of July 2000 (Raeder et al., 2001b). Comparisons to observations were most favorable at high latitude; spectral information of *dB/dt* was best reproduced for lower frequencies (0 − 3 Hz). A similar study with OpenGGCM illustrated challenges with reproducing substorm-related ΔB (Raeder et al., 2001a), as discussed above. The LFM model showed similar behavior: substorm-related perturbations were under predicted by the model when only Hall currents (Shao et al., 2002) or Hall and Pedersen currents (Wiltberger et al., 2003) were used; comparisons were far more favorable on the day side than the night side. Yu and Ridley (2008) performed a thorough validation of modeled ΔB using only the ionospheric Hall currents from RIM (coupled to BATS-R-US) across 7 real world storm events and 150 magnetometer stations. Again, only horizontal ionospheric currents were considered in the integration. This study established that a full-system integration, not merely a regional one, was necessary to get the full perturbation. The comparisons with observations showed a propensity for under-prediction and a poor re-creation of ΔB at auroral latitudes. The authors also demonstrated that validation of *dB/dt* is extremely sensitive to the timing of spikes; validation over a temporal window instead of simultaneous comparisons proved to better characterize model performance. Pulkkinen et al. (2010) calculated ΔB using all ionospheric currents, but limited the range of contributions to a 1000 km radius about the location of each virtual magnetometer. Predictions from both the SWMF and OpenGGCM models were compared to observations across four real world storms. It was found that inclusion of a ring current model improved metrics markedly. Additionally, performance of virtual magnetometers appears strongly correlated to performance of the models to predict ΔB at geosynchronous locations. These initial studies would set the stage for the modern era of MHD *dB/dt* forecasting. Expanding on this work (Pulkkinen et al., 2011), performed the most comprehensive validation study up to this point, covering 6 models, 4 events, 12 ground observatories, and 5 different metrics. This study showed that assessment of model performance was highly sensitive to the chosen metric.

Virtual magnetometer capability made the largest stride forward because of the work of Yu et al. (2010). In an effort to address the shortcomings of the preceding studies (limited current inclusion, reliance on Fukushima's Theorem, etc.), the virtual magnetometer calculation was expanded to include all MHD and ionospheric currents. The Biot–Savart integration was now covering the entire MHD domain. Additionally, it was recognized that *gap-region* FACs were being neglected. These are FACs that flow between the inner boundary altitude of the MHD model (typically 1–2 R_E above the surface of the Earth) and the altitude of the ionosphere solver (typically around 100 km). While these currents are not explicitly modeled, they both exist and contribute to the total surface perturbation. To include these, field-aligned currents near the inner boundary are mapped along assumed dipole field lines to the ionosphere. This calculation is now an internal capability of the SWMF; Rastätter et al. (2014) generalized this methodology to create a cross-model post-processing tool for NASAs Community Coordinated Modeling Center (CCMC).

Figure 3.5 diagrams the methodology of Yu et al. (2010) [...], a full-color version of this figure is available in the electronic version. There are now four major current groups that contribute to the total ΔB. The first is J_{Mag}, or all currents flowing anywhere in the entire MHD domain. The top left frame of Figure 3.5 illustrates this, showing a slice through a subset of the BATS-R-US domain and enveloped by an orange border. The magnitude of current density is shown by the contours; magnetopause, field-aligned, ring and cross-tail currents are all observable. It should be noted that the MHD domain extends far past the limits shown in Figure 3.5. Next are the gap-region FACs, J_{FAC}. These are all located within the inner boundary of the MHD model (the gray region within the upper left frame of Figure 3.5) and are encircled by a red border. The last two groups are the horizontal currents from the ionosphere solver: J_{Hall} and J_{Peder}. The upper right panel of Figure 3.5 shows an example of Hall currents from RIM model when coupled to BATS-R-US. Under this calculation, there is no spatial limitation to what ionospheric currents are included in the integration; the entire model domain is included.

The bottom frame of 3.5 shows the resulting ΔB from the Yu et al. (2010) methodology in the magnetic north–south direction. The integration is performed at the Fürstenfeldbruck, Germany magnetometer observatory (FUR) at 48.4° geomagnetic latitude over the 14 December 2006 geomagnetic storm. The colored lines correspond to the perturbation from each major current group: orange for J_{Mag}, red for J_{FAC}, dark blue for J_{Hall}, and light blue for J_{Peder}. The total ΔB_N, shown in black, is the sum of the four colored lines. Of immediate interest is the dominant contribution from the magnetosphere (orange line, medium gray in print). At several points, strong contributions come from the gap-region FACs (red line; dark gray in print). These are not canceled out by the Pedersen currents (light blue line; light-gray in print), showing that Fukushima's theorem is easily violated at mid-latitudes. The Hall currents (dark

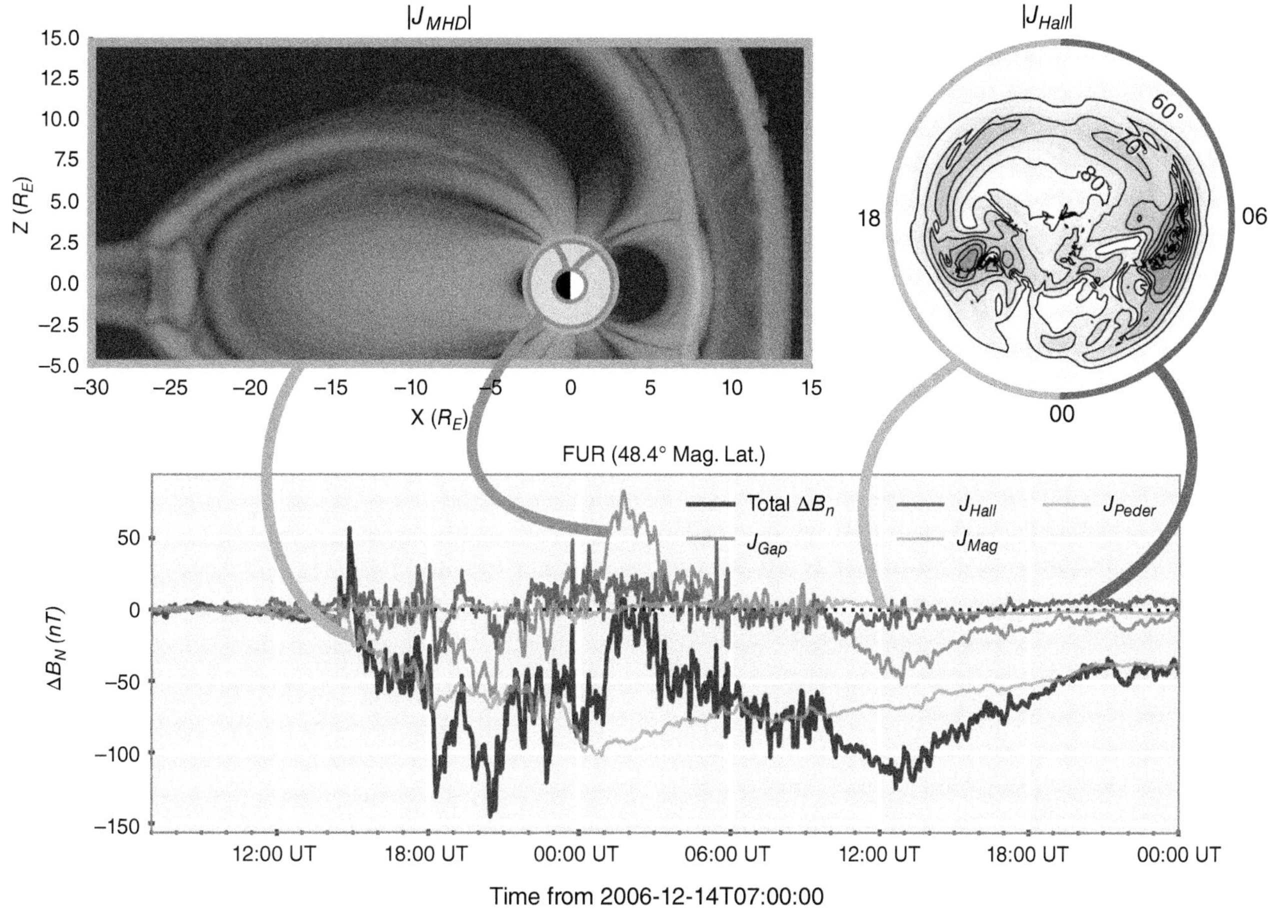

Figure 3.5 Contributions of different current density categories to the total virtual magnetometer ΔB for a mid-latitude station. (*See insert for color representation of the figure.*)

blue; darkest gray in print) are a minor contribution to the overall ΔB. These patterns could not have been captured by the more limited approaches.

While the algorithm of Yu et al. (2010) is more comprehensive, it is important to not overstate its advantages. Figure 3.6 shows the virtual magnetometer results corresponding to the Abisko, Sweden observatory (ABK) at 66.1° geomagnetic latitude. Again, the geomagnetic north–south component is shown for the 14 December 2006 event. The higher latitude result contrasts strongly with the example in Figure 3.5. The strongest contributor, by far, is the Hall current. Currents in the magnetosphere contribute the least to the total perturbation. The FAC and Pedersen currents nearly cancel, demonstrating that Fukushima's theorem is, for the most part, valid. While there are non-negligible contributions from the non-Hall currents, a less comprehensive ΔB prediction could have closely reproduced this figure. Data-model comparisons from 20 magnetometers and 9 storm events showed the strongest improvement at mid- and low-latitudes (Yu et al., 2010).

The most important impact of the establishment and validation of a uniform virtual magnetometer algorithm was enabling inter-model comparisons. Using the approach from Yu et al. (2010) and Rastätter et al. (2014) removes ambiguity concerning what is included in the final perturbation. This capability was an important step towards bringing MHD *dB*/*dt* predictions to operations.

3.3.2. Transition to Operations

The modern era of MHD in GIC research and operations began with what is colloquially known as the *CCMC-SWPC Validation Challenge*. With a desire to achieve *dB*/*dt* forecasting capabilities, NOAA's Space Weather Prediction Center (SWPC) connected with NASAs CCMC and representatives from the research community to test models that had demonstrated some *dB*/*dt* predictive capability. Three MHD models were included (LFM-MIX, OpenGGCM, and BATS-R-US via the SWMF), as well as two empirical models: the

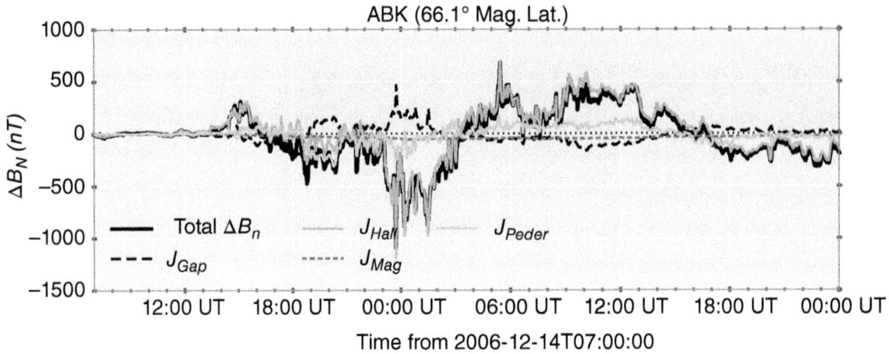

Figure 3.6 Similar to the bottom frame of Figure 3.5, but for a higher latitude station. (*See insert for color representation of the figure.*)

Weimer Empirical Ground Magnetic Field Prediction Model (Weimer et al., 2010; Weimer, 2013) and the Weigel Empirical Ground Magnetic Field Prediction Model (Weigel, 2007; Pulkkinen et al., 2013). All of these models can report ΔB (and therefore dB/dt) as a function of upstream solar drivers. The validation challenge tested predictions against observations of six different ground stations over six different real world events. The stations, events, and metrics of this effort have become the defacto standard for dB/dt forecast models today (Pulkkinen et al., 2013). The challenge was extended to local-K, a single-magnetometer index of geomagnetic activity (Glocer et al., 2016b).

Many unique aspects differentiate the CCMC-SWPC challenge from preceding validation efforts and modeling challenges. Foremost is *who* performed the validation: all simulations and metric calculations were performed by CCMC. Rather than allowing model developers to iterate simulation configurations to achieve the best comparisons, single configurations for each model were submitted to CCMC to be used across all events. To further prevent model calibration, two of the six events were revealed after model submission. The challenge was also unique in that it was done in the context of operations: the included models were required to run at real time speeds on only 24 processors. Because SWPCs interest was to transition one or more models into operations, the readiness of each model was also considered (ease of use, robustness, etc.). The methods, simulation configurations, results, and associated observations have been made publicly available for reanalysis (CCMC, 2013). All of these features contrast sharply with previous validations, typically led by members of model development teams.

Binary event analysis (Jolliffe and Stephenson, 2012), widely used in terrestrial weather forecasting, was employed to score model performance. An event is defined as a value of interest (i.e., dB_H/dt, or the magnitude of the horizontal components of $d\bar{B}/dt$) crossing of some set threshold (0.3, 0.7. 1.1, or 1.5 nT/s) within some non-overlapping time window (20 min). The top frame of Figure 3.7 shows a comparison between observations of dB_H/dt from the Yellowknife, Canada station (YKC, 68.9° geomagnetic latitude) and virtual magnetometer predictions from the OpenGGCM MHD model (light gray and black lines, respectively) during the 14 December 2006 storm. Horizontal dashed lines mark the four different event thresholds used in this study. For any given 20-min window, an event occurs if the observed level crosses the threshold. The interval is then categorized by the ability or failure of the model to correctly predict an event or non-event. The example in the top frame of Figure 3.7 displays examples of all four categorizations, including "hits," "misses," "true negatives," and "false positives." Metrics are calculated from these categorizations: Probability of Detection (the fraction of correctly forecast events), Probability of False Detection (the fraction of predicting an event when none was observed), and Heidke Skill Score (the fraction of correct predictions above what is expected based on random chance.) The models were then ranked based on these metrics. The bottom frame of Figure 3.7 shows an example of such a ranking based on Heidke Skill Score. The metrics were fully explored by Pulkkinen et al. (2013); a similar procedure was used to validate predictions of local K-index (Glocer et al., 2016b) with similar conclusions. Full reports on model performance were prepared by the joint CCMC-SWPC team CCMC (2013).

The results from the CCMC-SWPC validation challenge had an immediate impact on the community. The overarching conclusion was that all models demonstrated value for forecasters as potential operational tools. Surprisingly, it was the MHD models, not the empirical ones, that dominated the metric rankings overall. This ranking formalized the value of MHD models in GIC forecasting.

Of the three MHD models tested, the SWMF was selected by SWPC to be the first physics-based geospace model to be transitioned into operations. The model is

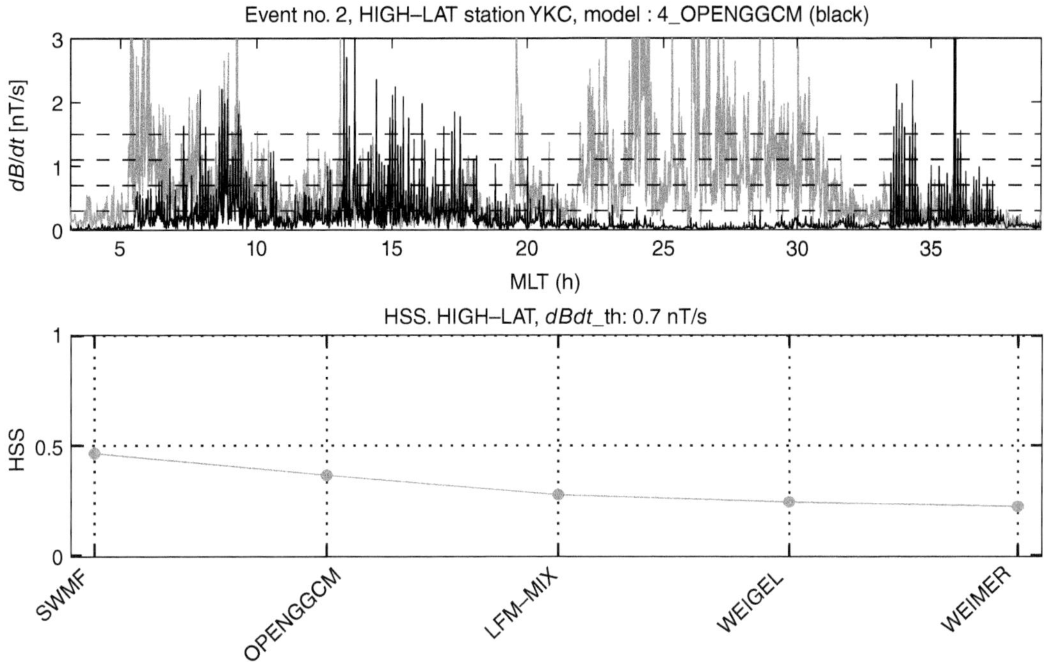

Figure 3.7 An illustration of the data-model comparisons leveraged in the SWPC-CCMC validation challenge (top frame). The light gray line shows observations, the black line shows model prediction. Horizontal dashed lines denote event threshold values. Total Heidke skill scores for high-latitude predictions at the 0.7 nT threshold are shown (bottom frame) for all five models. Source: Adapted from Pulkkinen et al. (2013). Reproduced with permission of John Wiley and Sons.

driven by real-time observations of the solar wind and IMF from either the ACE or DSCOVR satellites. Values are propagated to the upstream boundary of the model, which can provide ~15–30 min of lead time, depending on the solar wind velocity. The model went live on September 7th, 2016. Results, including predicted geomagnetic indices (D_{ST} and K_p), global and regional maps of ΔB, and visualizations of the MHD magnetosphere, are published to the web in real time.

3.3.3. Extreme Event Analysis

A distinct advantage of MHD and other first-principles-based approaches is their ability to explore the impact of hypothetical space weather conditions. This opens up the opportunity to examine conditions during historical extreme events and hypothetical worst-case-scenario events. This analysis is important for understanding the extent of GIC and other space weather threats.

The work of Ridley et al. (2006) illustrates the potential of hypothetical storm studies. In an attempt to better understand the conditions resulting from a Carrington-level storm, the extreme solar wind conditions predicted by (Manchester et al., 2006) were used to drive BATS-R-US via the SWMF. The simulation showed that the magnetopause would be pushed within 2 R_E and CPCP would exceed 1800 kV. The extreme shock displayed a unique interaction with the magnetosphere, "bouncing" off the cold, dense plasmasphere population. These results help illustrate new dynamics for extreme events, but have not been cast into terms relevant to GIC applications.

More contemporary studies leveraged virtual magnetometers to put extreme event simulations into GIC context. Ngwira et al. (2013) performed a "what-if" simulation using observations of the 23 July 2012 near-miss CME event from the STEREO-A satellite (Baker et al., 2013) and examined the impact on the Earth system. The SWMF was again used, this time with the RCM as well as BATS-R-US and RIM. The model reported *dB*/*dt* values of ~14 V/km at mid latitudes. Using a similar set up, Ngwira et al. (2014) took a second look at a hypothetical Carrington-class storm. Solar wind conditions were chosen to reproduce the observed ground perturbation of −1600 nT during the real Carrington event (Siscoe et al., 2006). Beyond extreme CPCP values and deep compression of the magnetopause, the authors examined δB, *dB*/*dt*, and geoelectric field at various magnetometer stations. Values of ~ −1600 nT, ~ 2 nT/s, and 26 V/km (respectively) are in line with estimates from other studies (Pulkkinen et al., 2012). In each of these cases, specialized model configuration was required to prevent non-physical results, especially towards the inner boundary.

3.3.4. Limitations

Despite its many successes, there are many limitations to MHD that must be considered, especially from the perspective of GIC modeling. The *dB/dt* values produced by MHD are still several steps removed from geoelectric field, which is the operationally relevant product. This will involve propagating the electromagnetic signal through the atmosphere and including the effects of the heterogeneous conducting crust and lithosphere of the Earth (e.g., Simpson, 2011; Bedrosian and Love, 2015). Initial work has taken place to translate MHD *dB/dt* into geoelectric field (Pulkkinen et al., 2007), but very little validation has taken place. The extent to which the assumption of an electrostatic ionosphere affects *dB/dt* predictions remains unknown. Electrostatic solvers neglect inductive effects, which can become non-negligible (Lotko, 2004; Song et al., 2009; Yoshikawa et al., 2010; Tu et al., 2011; Vanhamäki, 2011), especially during short time scale transients such as substorms and storm sudden commencements. Electrodynamic treatments of the ionosphere exist (Sokolov et al., 2013), but have yet to be integrated into the global MHD models.

These physical limitations build on performance shortcomings highlighted in the validation literature. The reconnection physics in the ideal MHD models are, of course, flawed. Despite the ability to reproduce the gross features of magnetosphere–ionosphere coupling, small structures still remain out of reach, especially on the night side and during substorms. The models cannot yet consistently predict the location of the auroral oval. These limitations will affect the ability to forecast localized *dB/dt* activity.

A reanalysis of the CCMC-SWPC validation results showed that the prediction error depends greatly on the activity level, revealing even more limitations (Welling et al., 2017). Normalized error, defined as the prediction minus the observation divided by the observation, was binned by both activity (as given by the SYM-H index) and driving (via the solar wind electric field, $V\ B_Z$) corresponding to the error value. Figure 3.8 summarizes the findings by plotting the median *dB/dt* error values for each bin as a function of $V\ B_Z$ (left frame) and SYM-H (right frame) for each of the MHD and empirical models tested in the CCMC-SWPC validation challenge. A perfect prediction corresponds to zero error; however, the lines are almost always below that. This indicates that all models, both MHD and empirical, have a predilection towards under prediction. The situation is worst during periods of strong driving and strong magnetospheric activity (i.e., the right side of the plots). Overall, the MHD models perform somewhat better than the empirical ones, but this is a concern when the models are employed for GIC and space weather applications.

Understanding how important these different issues are to *dB/dt* performance is challenging as validation efforts have not yet become granular enough to fully characterize these limitations. For example, the binary-event analysis employed by Pulkkinen et al. (2013) does not discriminate between strength of driving, type of activity, or local time of disturbance. It is not currently possible to understand if the models predict *dB/dt* better during storm sudden commencements or substorms, main storm phase or recovery. It is unknown how accurate reconnection physics translates into accurate geoelectric magnetic field predictions. These questions must be answered if MHD's role in GIC research will continue to expand.

3.4. THE FUTURE OF MHD AND GICs

Today, MHD provides value in both operations and GIC research. Short-term *dB/dt* forecasts from MHD are performed by NOAA SWPC. Research into how solar wind transients translate into *dB/dt* and geoelectric fields continue today. Exploration of extreme space weather conditions and their impact on the Earth system are made possible via MHD modeling.

Validation of MHD models is an ongoing process. Previous validation studies have made great strides in characterizing MHD performance, but many questions remain unanswered. The CCMC-SWPC validation challenge should be expanded to more events and more magnetometers in order to fully explore skill as a function of position (i.e., latitude and magnetic local time) and activity. A better understanding of what types of activity are best captured by global MHD is needed. For example, under what conditions can MHD best reproduce substorm events? Under what conditions will it regularly fail? What types of storms are best or least well reproduced by MHD? Answering these questions will provide more value to forecasters and a better understanding of MHD applicability during extreme event analysis.

Related to these questions is the impact of extended MHD on GIC applications. As discussed in Section 3.1.3, many advanced physics options exist but few have been tested in a systematic way. Many of these improvements can directly address the open questions above, for example, better reconnection physics may improve poor substorm modeling. Extended MHD versions must be exercised via the CCMC-SWPC validation challenge to understand the full benefit they may provide.

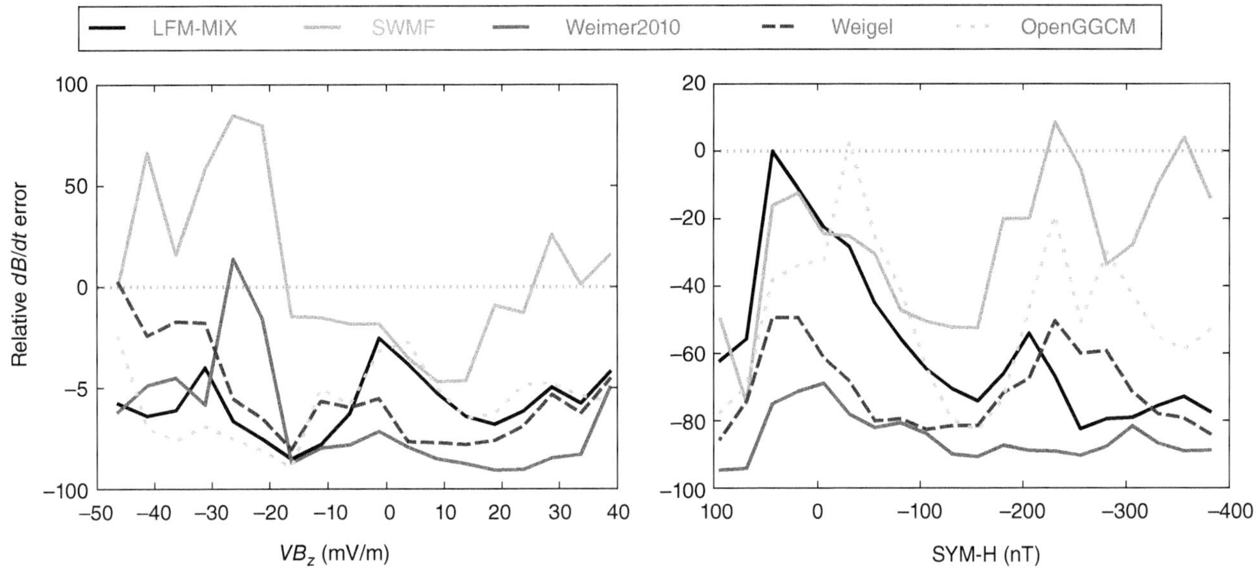

Figure 3.8 Median *dB*/*dt* prediction errors of different MHD and empirical models as a function of solar driving ($V\ B_z$, left frame) and geospace activity (SYM-H, right frame). In each frame, stronger space weather conditions lie on the right side of the *x*-axes. Source: Adapted from Welling et al. (2017). Reproduced with permission of John Wiley and Sons.

Finally, the goal of GIC forecasting is geoelectric field, not merely *dB*/*dt*. Pulkkinen et al. (2007) showed that such forecasts are possible, but this work is a first step in a long process. The geomagnetic response as calculated by MHD must be propagated through the atmosphere and down through the heterogeneous conducting Earth. Extensive effort is required to understand what methodologies and assumptions best balance predictive skill with computational efficiency. Achieving true geoelectric field forecast capabilities is the next big step for MHD models.

REFERENCES

O. Amm. Comment on "A three-dimensional iterative mapping procedure for the implementation of an ionosphere-magnetosphere anisotropic Ohm's law boundary condition in global magnetohydrodynamic simulations" by Michael L. Goodman. *Ann. Geophysicae*, *14*:773–774, 1996.

Brian J. Anderson, Haje Korth, Daniel T. Welling, Viacheslav G. Merkin, Michael J. Wiltberger, Joachim Raeder, Robin J. Barnes, Colin L. Waters, Antti A. Pulkkinen, and Lutz Rastaetter. Comparison of predictive estimates of high-latitude electrodynamics with observations of global-scale Birkeland currents. *Space Weather*, *15*(2):352–373, Feb 2017. ISSN 15427390. doi: 10.1002/2016SW001529. URL http://doi.wiley.com/10.1002/2016SW001529.

D. N. Baker, X. Li, A. Pulkkinen, C. M. Ngwira, M. L. Mays, A. B. Galvin, and K. D. C. Simunac. A major solar eruptive event in July 2012: Defining extreme space weather scenarios. *Space Weather*, *11*(10):585–591, Oct 2013. ISSN 15427390. doi: 10.1002/swe.20097. URL http://doi.wiley.com/10.1002/swe.20097.

Paul A. Bedrosian and Jeffrey J. Love. Mapping geoelectric fields during magnetic storms: Synthetic analysis of empirical United States impedances. *Geophysical Research Letters*, *42*(23):10160–10170, Dec 2015. ISSN 00948276. doi: 10.1002/2015GL066636. URL http://doi.wiley.com/10.1002/2015GL066636.

Joachim Birn and Michael Hesse. Geospace Environment Modeling (GEM) magnetic reconnection challenge: Resistive tearing, anisotropic pressure and Hall effects. *Journal of Geophysical Research*, *106*(A3):3737–3750, 2001. ISSN 0148-0227. doi: 10.1029/1999JA001001. URL http://doi.wiley.com/10.1029/1999JA001001.

Jay P. Boris. A physically motivated solution of the Alfven problem. Technical Report 2167, 1970. URL http://www.dtic.mil/docs/citations/AD0715774.

Joe Borovsky. Global sawtooth oscillations of the magnetosphere. *Eos, Transactions American Geophysical Union*, *85*(49):525–526, Dec 2004. ISSN 0096-3941. doi: 10.1029/2004EO490009 URL http://doi.wiley.com/10.1029/2004EO490009http://www.agu.org/pubs/crossref/2004/2004EO490009.shtml.

Joseph E. Borovsky, Michael Hesse, Joachim Birn, and Maria M. Kuznetsova. What determines the reconnection rate at the dayside magnetosphere? *Journal of Geophysical Research*, *113*(A7):A07210, Jul 2008. ISSN 0148-0227. doi: 10.1029/2007JA012645. URL http://doi.wiley.com/10.1029/2007JA012645.

T. J. M. Boyd and J. J. Sanderson. *The Physics of Plasmas*. Cambridge University Press, 2003. ISBN 0521459125.

O. J. Brambles, W. Lotko, B. Zhang, M. Wiltberger, J. Lyon, and R. J. Strangeway. Magnetosphere sawtooth oscillations induced by ionospheric outflow. *Science (New York, N.Y.)*, *332*(6034):1183–1186, Jun 2011. ISSN 1095-9203. doi: 10.1126/science.1202869.

Stephen H. Brecht. Global simulations using MHD codes: A few points to consider before you try one. *Space Science Reviews*, *42*(1–2):169–185, Oct 1985. ISSN 0038-6308. doi: 10.1007/BF00218231. URL http://link.springer.com/10.1007/BF00218231.

S. H. Brecht, J. Lyon, J. A. Fedder, and K. Hain. A simulation study of east-west IMF effects on the magnetosphere. *Geophysical Research Letters*, *8* (4):397–400, Apr 1981. ISSN 00948276. doi: 10.1029/GL008i004p00397. URL http://doi.wiley.com/10.1029/GL008i004p00397.

X. Cai and C. R. Clauer. Investigation of the period of sawtooth events. *Journal of Geophysical Research: Space Physics*, *114*(A6), Jun 2009. ISSN 01480227. doi: 10.1029/2008JA013764. URL http://doi.wiley.com/10.1029/2008JA013764.

X. Cai, C. R. Clauer, and A. J. Ridley. Statistical analysis of ionospheric potential patterns for isolated substorms and sawtooth events. *Annales Geophysicae*, *24*:1977–1991, 2006.

X. Cai, J.-C. Zhang, C. R. Clauer, and M. W. Liemohn. Relationship between sawtooth events and magnetic storms. *Journal of Geophysical Research*, *116* (A7):A07208, Jul 2011. ISSN 0148-0227. doi: 10.1029/2010JA016310.

NASA CCMC. Operational geospace model validation, 2013. URL https://ccmc.gsfc.nasa.gov/challenges/dBdt/.

Francis Chen. *Introduction to Plasma Physics and Controlled Fusion*. Springer International Publishing, Cham, 2016. ISBN 978-3-319-22308-7. doi: 10.1007/978-3-319-22309-4. URL http://link.springer.com/10.1007/978-3-319-22309-4.

Yuxi Chen, Gábor Tóth, Paul Cassak, Xianzhe Jia, Tamas I. Gombosi, James A. Slavin, Stefano Markidis, Ivy Bo Peng, Vania K. Jordanova, and Michael G. Henderson. Global three-dimensional simulation of Earth's dayside reconnection using a two-way coupled magnetohydrodynamics with embedded particle-in-cell model: Initial results. *Journal of Geophysical Research: Space Physics*, Sep 2017. ISSN 21699380. doi: 10.1002/2017JA024186. URL http://doi.wiley.com/10.1002/2017JA024186.

W. D. Cramer, J. Raeder, F. R. Toffoletto, M. Gilson, and B. Hu. Plasma sheet injections into the inner magnetosphere: Two-way coupled OpenGGCM-RCM model results. *Journal of Geophysical Research: Space Physics*, *122*(5):5077–5091, May 2017. ISSN 21699380. doi: 10.1002/2017JA024104. URL http://doi.wiley.com/10.1002/2017JA024104.

Ioannis A. Daglis, Richard M. Thorne, Wolfgang Baumjohann, and Stefano Orsini. The terrestrial ring current: Origin, formation, and decay. *Reviews of Geophysics*, *37*(4):407–438, Nov 1999. ISSN 87551209. doi: 10.1029/1999RG900009. URL http://doi.wiley.com/10.1029/1999RG900009http://onlinelibrary.wiley.com/store/10.1029/1999RG900009/asset/rog1545.pdf?v=1.

T. Neil Davis and Masahisa Sugiura. Auroral electrojet activity index AE and its universal time variations. *Journal of Geophysical Research*, *71*(3): 785–801, Feb 1966. ISSN 01480227. doi: 10.1029/JZ071i003p00785. URL http://doi.wiley.com/10.1029/JZ071i003p00785.

a. D. DeJong, a. J. Ridley, X. Cai, and C. R. Clauer. A statistical study of BRIs (SMCs), isolated substorms, and individual sawtooth injections. *Journal of Geophysical Research*, *114*(A8):A08215, Aug 2009. ISSN 0148-0227. doi: 10.1029/2008JA013870.

D. L. De Zeeuw, T. I. Gombosi, C. P. T. Groth, K. G. Powell, and Q. F. Stout. An adaptive MHD method for global space weather simulations. *IEEE Transactions on Plasma Science*, *28*(6):1956–1965, 2000. ISSN 00933813. doi: 10.1109/27.902224. URL http://ieeexplore.ieee.org/lpdocs/epic03/wrapper.htm?arnumber=902224.

Darren L. De Zeeuw, Stanislav Sazykin, Richard A. Wolf, Tamas I. Gombosi, Aaron J. Ridley, and Gabor Tóth. Coupling of a global MHD code and an inner magnetospheric model: Initial results. *Journal of Geophysical Research: Space Physics*, *109*(A12):A12219, 2004. ISSN 21699402. doi: 10.1029/2003JA010366. URL http://doi.wiley.com/10.1029/2003JA010366.

John C. Dorelli. A new look at driven magnetic reconnection at the terrestrial subsolar magnetopause. *Journal of Geophysical Research*, *109*(A12):A12216, 2004. ISSN 0148-0227. doi: 10.1029/2004JA010458. URL http://doi.wiley.com/10.1029/2004JA010458.

Charles R. Evans and John F. Hawley. Simulation of magnetohydrodynamic flows – A constrained transport method. *The Astrophysical Journal*, *332*: 659, Sep 1988. ISSN 0004-637X. doi: 10.1086/166684. URL http://adsabs.harvard.edu/doi/10.1086/166684.

J. A. Fedder and J. G. Lyon. Region 1 Birkeland currents from a global MHD simulation of the magnetosphere, Sep 1984. URL http://oai.dtic.mil/oai/oai?verb=getRecord&metadataPrefix=html&identifier=ADA147435.

J. A. Fedder and J. G. Lyon. The Earth's magnetosphere is 165 R E long: Self-consistent currents, convection, magnetospheric structure, and processes for northward interplanetary magnetic field. *Journal of Geophysical Research: Space Physics*, *100*(A3):3623–3635, Mar 1995. ISSN 01480227. doi: 10.1029/94JA02633. URL http://doi.wiley.com/10.1029/94JA02633.

J. A. Fedder, S. P. Slinker, J. G. Lyon, and R. D. Elphinstone. Global numerical simulation of the growth phase and the expansion onset for a substorm observed by Viking. *Journal of Geophysical Research*, *100*(A10): 19083, Oct 1995. ISSN 0148-0227. doi: 10.1029/95JA01524. URL http://doi.wiley.com/10.1029/95JA01524.

M.-C. Fok, R. A. Wolf, R. W. Spiro, and T. E. Moore. Comprehensive computational model of Earth's ring current. *Journal of Geophysical Research: Space Physics*, *106*(A5):8417–8424, May 2001. ISSN 01480227. doi: 10.1029/2000JA000235. URL http://doi.wiley.com/10.1029/2000JA000235.

R. A. Frahm, J. D. Winningham, J. R. Sharber, R. Link, G. Crowley, E. E. Gaines, D. L. Chenette, B. J. Anderson, and T. A. Potemra. The diffuse aurora: A significant source of ionization in the middle atmosphere. *Journal of Geophysical Research: Atmospheres*, *102*(D23):28203–28214, Dec 1997.

ISSN 01480227. doi: 10.1029/97JD02430. URL http://doi.wiley.com/10.1029/97JD02430.

N. Fukushima. Equivalence in ground geomagnetic effect of chapman–vestine's and birkeland–alfven's electric current-systems for polar magnetic storms. *Report of Ionosphere and Space Research in Japan*, *23*:219–227, Jan 1969. URL https://www.osti.gov/scitech/biblio/4755464.

N. Fukushima. Generalized theorem for no ground magnetic effect of vertical currents connected with Pedersen currents in the uniform-conductivity ionosphere. *Report of Ionosphere and Space Research in Japan*, *30*(1):35–40, 1976. URL http://ci.nii.ac.jp/naid/10030183891/.

A. Glocer, G. Tóth, Y. Ma, T. Gombosi, J.-C. Zhang, and L. M. Kistler. Multifluid block-adaptive-tree solar wind roe-type upwind scheme: Magnetospheric composition and dynamics during geomagnetic stormsInitial results. *Journal of Geophysical Research*, *114*(A12):A12203, Dec 2009. ISSN 0148-0227. doi: 10.1029/2009JA014418.

A. Glocer, M. Fok, X. Meng, G. Toth, N. Buzulukova, S. Chen, and K. Lin. CRCM + BATS-R-US two-way coupling. *Journal of Geophysical Research: Space Physics*, *118*(4):1635–1650, Apr 2013. ISSN 21699380. doi: 10.1002/jgra.50221. URL http://doi.wiley.com/10.1002/jgra.50221.

A. Glocer, J. Dorelli, G. Toth, C. M. Komar, and P. A. Cassak. Separator reconnection at the magnetopause for predominantly northward and southward IMF: Techniques and results. *Journal of Geophysical Research: Space Physics*, *121*(1):140–156, Jan 2016a. ISSN 21699380. doi: 10.1002/2015JA021417. URL http://doi.wiley.com/10.1002/2015JA021417.

A. Glocer, L. Rastätter, M. Kuznetsova, A. Pulkkinen, H. J. Singer, C. Balch, D. Weimer, D. Welling, M. Wiltberger, J. Raeder, R. S. Weigel, J. McCollough, and S. Wing. Community-wide validation of geospace model local K-index predictions to support model transition to operations. *Space Weather*, *14*(7):469–480, Jul 2016b. ISSN 15427390. doi: 10.1002/2016SW001387. URL http://doi.wiley.com/10.1002/2016SW001387.

T. I. Gombosi, D. L. Dezeeuw, C. P. T. Groth, K. G. Powell, and P. Song. The length of the magnetotail for northward IMF : Results of global 3D MHD simulations. *Physics of Space Plasmas*, *15*:121–128, 1998.

Tamas I. Gombosi, Gábor Tóth, Darren L. De Zeeuw, Kenneth C. Hansen, Konstantin Kabin, and Kenneth G. Powell. Semirelativistic magnetohydrodynamics and physics-based convergence acceleration. *Journal of Computational Physics*, *177*(1):176–205, Mar 2002. ISSN 00219991. doi: 10.1006/jcph.2002.7009. URL http://linkinghub.elsevier.com/retrieve/pii/S0021999102970099.

M. L. Goodman. A three-dimensional, iterative mapping procedure for the implementation of an ionosphere-magnetosphere anisotropic Ohm's law boundary condition in global magnetohydrodynamic simulations. *Annales Geophysicae*, *13*(8):843–853, 1995. ISSN 1432-0576. doi: 10.1007/s00585-995-0843-z. URL http://www.ann-geophys.net/13/843/1995/.

C. C. Goodrich, T. I. Pulkkinen, J. G. Lyon, and V. G. Merkin. Magnetospheric convection during intermediate driving: Sawtooth events and steady convection intervals as seen in Lyon-Fedder-Mobarry global MHD simulations. *Journal of Geophysical Research*, *112*(A8):A08201, Aug 2007. ISSN 0148-0227. doi: 10.1029/2006JA012155.

E. Gordeev, V. Sergeev, I. Honkonen, M. Kuznetsova, L. Rastätter, M. Palmroth, P. Janhunen, G. Tóth, J. Lyon, and M. Wiltberger. Assessing the performance of community-available global MHD models using key system parameters and empirical relationships. *Space Weather*, *13*(12):868–884, Dec 2015. ISSN 15427390. doi: 10.1002/2015SW001307. URL http://doi.wiley.com/10.1002/2015SW001307.

E. Gordeev, V. Sergeev, N. Tsyganenko, M. Kuznetsova, L. Rastïetter, J. Raeder, G. Tïth, J. Lyon, V. Merkin, and M. Wiltberger. The substorm cycle as reproduced by global MHD models. *Space Weather*, *15* (1):131–149, Jan 2017. ISSN 15427390. doi: 10.1002/2016SW001495. URL http://doi.wiley.com/10.1002/2016SW001495.

David J. Griffiths. *Introduction to Electrodynamics*. Prentice Hall, Upper Saddle River, New Jersey, 4th edition, 2015. ISBN 8120347765.

Clinton P. T. Groth, Darren L. De Zeeuw, Tamas I. Gombosi, and Kenneth G. Powell. Global three-dimensional MHD simulation of a space weather event: CME formation, interplanetary propagation, and interaction with the magnetosphere. *Journal of Geophysical Research*, *105*(A11):25053, 2000. ISSN 0148-0227. doi: 10.1029/2000JA900093. URL http://doi.wiley.com/10.1029/2000JA900093.

Timothy B. Guild, Harlan E. Spence, E. Larry Kepko, Viacheslav Merkin, John G. Lyon, Michael Wiltberger, and Charles C. Goodrich. Geotail and LFM comparisons of plasma sheet climatology: 1. Average values. *Journal of Geophysical Research: Space Physics*, *113*(A4), Apr 2008. ISSN 01480227. doi: 10.1029/2007JA012611. URL http://doi.wiley.com/10.1029/2007JA012611.

John D. Haiducek, Daniel T. Welling, Natalia Y. Ganushkina, Steven K. Morley, and Dogacan Su Ozturk. SWMF global magnetosphere simulations of January 2005: Geomagnetic indices and cross-polar cap potential. *Space Weather*, *15*(12):1567–1587, Dec 2017. ISSN 15427390. doi: 10.1002/2017SW001695. URL http://doi.wiley.com/10.1002/2017SW001695.

M. Harel, R. A. Wolf, P. H. Reiff, R. W. Spiro, W. J. Burke, F. J. Rich, and M. Smiddy. Quantitative simulation of a magnetospheric substorm 1. Model logic and overview. *Journal of Geophysical Research*, *86*(A4): 2217, 1981. ISSN 0148-0227. doi: 10.1029/JA086iA04p02217. URL http://adsabs.harvard.edu/abs/1981JGR....86.2217H.

M. Heinemann and R. A. Wolf. Relationships of models of the inner magnetosphere to the Rice Convection Model. *Journal of Geophysical Research (Space Physics)*, *106554*(1):545–515, Aug 2001. ISSN 01480227. doi: 10.1029/2000JA000389. URL http://doi.wiley.com/10.1029/2000JA000389.

M. G. Henderson. The May 23, 1986 CDAW-9C interval: A sawtooth event. *Geophysical Research Letters*, *31*(11):L11804, 2004. ISSN 0094-8276. doi: 10.1029/2004GL019941. URL http://doi.wiley.com/10.1029/2004GL019941.

M. G. Henderson, G. D. Reeves, R. Skoug, M. F. Thomsen, M. H. Denton, S. B. Mende, T. J. Immel, P. C. Brandt, and H. J. Singer. Magnetospheric and auroral activity during the 18 April 2002 sawtooth event. *Journal of Geophysical Research*, *111*(A1):A01S90, 2006a. ISSN 0148-0227. doi: 10.1029/2005JA011111.

M. G. Henderson, R. Skoug, E. Donovan, M. F. Thomsen, G. D. Reeves, M. H. Denton, H. J. Singer, R. L. McPherron, S. B. Mende, T. J. Immel, J. B. Sigwarth, and L. a. Frank. Substorms during the 1011 August 2000 sawtooth event. *Journal of Geophysical Research*, *111*(A6):A06206, 2006b. ISSN 0148-0227. doi: 10.1029/2005JA011366.

Chia-Lin Huang, Harlan E. Spence, John G. Lyon, Frank R. Toffoletto, Howard J. Singer, and Stanislav Sazykin. Storm-time configuration of the inner magnetosphere: Lyon-Fedder-Mobarry MHD code, Tsyganenko model, and GOES observations. *Journal of Geophysical Research*, *111*(A11):A11S16, Nov 2006. ISSN 0148-0227. doi: 10.1029/2006JA011626. URL http://doi.wiley.com/10.1029/2006JA011626.

John David Jackson. *Classical Electrodynamics*. Wiley, 1999. ISBN 047130932X.

P. Janhunen, M. Palmroth, T. Laitinen, I. Honkonen, L. Juusola, G. Facskó, and T.I. Pulkkinen. The GUMICS-4 global MHD magnetosphereionosphere coupling simulation. *Journal of Atmospheric and Solar-Terrestrial Physics*, *80*: 48–59, May 2012. ISSN 13646826. doi: 10.1016/j.jastp.2012.03.006. URL http://www.sciencedirect.com/science/article/pii/S1364682612000909 http://linkinghub.elsevier.com/retrieve/pii/S1364682612000909.

Ian T. Jolliffe and David B. Stephenson. *Forecast Verification: A Practitioner's Guide in Atmospheric Science*. Wiley 2012. ISBN 1119961076. URL https://books.google.com/books?hl=en&lr=&id=DCxsKQeaBH8C&pgis=1.

V. K. Jordanova, R. B. Torbert, R. M. Thorne, H. L. Collin, J. L. Roeder, and J. C. Foster. Ring current activity during the early B z < 0 phase of the January 1997 magnetic cloud. *Journal of Geophysical Research*, *104* (A11):24895, 1999. ISSN 0148-0227. doi: 10.1029/1999JA900339. URL http://doi.wiley.com/10.1029/1999JA900339.

S. R. Kaeppler, D. L. Hampton, M. J. Nicolls, A. Strømme, S. C. Solomon, J. H. Hecht, and M. G. Conde. An investigation comparing ground-based techniques that quantify auroral electron flux and conductance. *Journal of Geophysical Research: Space Physics*, *120*(10):9038–9056, Oct 2015. ISSN 21699380. doi: 10.1002/2015JA021396. URL http://doi.wiley.com/10.1002/2015JA021396.

Margaret G. Kivelson and Aaron J. Ridley. Saturation of the polar cap potential: Inference from Alfvén wing arguments. *Journal of Geophysical Research: Space Physics*, *113*(A5), May 2008. ISSN 01480227. doi: 10.1029/2007JA012302. URL http://doi.wiley.com/10.1029/2007JA012302.

Stephen Knight. Parallel electric fields. *Planetary and Space Science*, *21*(5): 741–750, May 1973. ISSN 00320633. doi: 10.1016/0032-0633(73)90093-7. URL http://linkinghub.elsevier.com/retrieve/pii/0032063373900937.

B Barry Koren. A robust upwind discretization method for advection, diffusion and source terms. *Numerical Methods for Advection-Diffusion Problems*, *45*: 117, 1993. URL https://www.narcis.nl/publication/RecordID/oai:library.tue.nl:737323.

H. Korth, Brian J. Anderson, Michael J. Wiltberger, John G. Lyon, and P. C. Anderson. Intercomparison of ionospheric electrodynamics from the Iridium constellation with global MHD simulations. *Journal of Geophysical Research*, *109*(A7):A07307, Jul 2004. ISSN 0148-0227. doi: 10.1029/2004JA010428. URL http://doi.wiley.com/10.1029/2004JA010428.

H. Korth, B. J. Anderson, J. G. Lyon, and M. Wiltberger. Comparison of Birkeland current observations during two magnetic cloud events with MHD simulations. *Annales Geophysicae*, *26*(3):499–516, Mar 2008. ISSN 1432-0576. doi: 10.5194/angeo-26-499-2008. URL http://www.ann-geophys.net/26/499/2008/.

H. Korth, L. Rastätter, B. J. Anderson, and A. J. Ridley. Comparison of the observed dependence of large-scale Birkeland currents on solar wind parameters with that obtained from global simulations. *Annales Geophysicae*, *29*(10):1809–1826, Oct 2011. ISSN 1432-0576. doi: 10.5194/angeo-29-1809-2011. URL http://www.ann-geophys.net/29/1809/2011/angeo-29-1809-2011. htmlhttp://www.ann-geophys.net/29/1809/2011/.

M. M. Kuznetsova, M. Hesse, L. Rastätter, A. Taktakishvili, G. Toth, D. L. De Zeeuw, A. Ridley, and T. I. Gombosi. Multiscale modeling of magnetospheric reconnection. *Journal of Geophysical Research*, *112*(A10): A10210, Oct 2007. ISSN 0148-0227. doi: 10.1029/2007JA012316.

J. N. Leboeuf, T. Tajima, C. F. Kennel, and J. M. Dawson. Global simulations of the three-dimensional magnetosphere. *Geophysical Research Letters*, *8*(3): 257–260, Mar 1981. ISSN 00948276. doi: 10.1029/GL008i003p00257. URL http://doi.wiley.com/10.1029/GL008i003p00257.

J. Liao, X. Cai, L. M. Kistler, C. R. Clauer, C. G. Mouikis, B. Klecker, and I. Dandouras. The relationship between sawtooth events and O+in the plasma sheet. *Journal of Geophysical Research: Space Physics*, *119*(3): 1572–1586, Mar 2014. ISSN 21699380. doi: 10.1002/2013JA019084. URL http://doi.wiley.com/10.1002/2013JA019084.

M. W. Liemohn, J. U. Kozyra, C. R. Clauer, and A. J. Ridley. Computational analysis of the near-Earth magnetospheric current system during two-phase decay storms. *Journal of Geophysical Research*, *106*(A12):29531, 2001. ISSN 0148-0227. doi: 10.1029/2001JA000045. URL http://doi.wiley.com/10.1029/2001JA000045.

Jing Liu, Wenbin Wang, Meers Oppenheim, Yakov Dimant, Michael Wiltberger, and Slava Merkin. Anomalous electron heating effects on the E region ionosphere in TIEGCM. *Geophysical Research Letters*, *43*(6):2351–2358, Mar 2016. ISSN 00948276. doi: 10.1002/2016GL068010. URL http://doi.wiley.com/10.1002/2016GL068010.

R. E. Lopez, S. Hernandez, M. Wiltberger, C. L. Huang, E. L. Kepko, H. Spence, C. C. Goodrich, and J. G. Lyon. Predicting magnetopause crossings at geosynchronous orbit during the Halloween storms. *Space Weather*, *5*(1), Jan 2007. ISSN 15427390. doi: 10.1029/2006SW000222. URL http://doi.wiley.com/10.1029/2006SW000222.

W. Lotko. Inductive magnetosphereionosphere coupling. *Journal of Atmospheric and Solar-Terrestrial Physics*, *66*(15–16):1443–1456, Oct 2004. ISSN 1364-6826. doi: 10.1016/J.JASTP.2004.03.027. URL https://www.sciencedirect.com/science/article/pii/S1364682604001518?via%3Dihub.

J. G. Lyon, J. A. Fedder, and C. M. Mobarry. The Lyon-Fedder-Mobarry (LFM) global MHD magnetospheric simulation code. *Journal of Atmospheric and Solar-Terrestrial Physics*, *66*(15–16):1333–1350, Oct 2004. ISSN 13646826. doi:

10.1016/j.jastp.2004.03.020. URL http://adsabs.harvard.edu/abs/2004JASTP..66.1333L.

Ying-Juan Ma, Andrew F. Nagy, Gabor Toth, Thomas E. Cravens, Christopher T. Russell, Tamas I. Gombosi, Jan-Erik Wahlund, Frank J. Crary, Andrew J. Coates, César L. Bertucci, and Fritz M. Neubauer. 3D global multi-species Hall-MHD simulation of the Cassini T9 flyby. *Geophysical Research Letters*, *34*(24):L24S10, Dec 2007. ISSN 0094-8276. doi: 10.1029/2007GL031627. URL http://doi.wiley.com/10.1029/2007GL031627.

W. B. Manchester, A. J. Ridley, T. I. Gombosi, and D. L. DeZeeuw. Modeling the Sun-to-Earth propagation of a very fast CME. *Advances in Space Research*, *38*(2):253–262, Jan 2006. ISSN 02731177. doi: 10. 1016/j.asr.2005.09.044. URL http://linkinghub.elsevier.com/retrieve/pii/S027311770501210X.

Xing Meng, Gábor Tóth, Igor V. Sokolov, and Tamas I. Gombosi. Classical and semirelativistic magnetohydrodynamics with anisotropic ion pressure. *Journal of Computational Physics*, *231*(9):3610–3622, May 2012. ISSN 00219991. doi: 10.1016/j.jcp.2011.12.042. URL http://www.sciencedirect.com/science/article/pii/S0021999112000113.

X. Meng, G. Tóth, A. Glocer, M.-C. Fok, and T. I. Gombosi. Pressure anisotropy in global magnetospheric simulations: Coupling with ring current models. *Journal of Geophysical Research: Space Physics*, *118*(9):5639–5658, Sep 2013. ISSN 21699380. doi: 10.1002/jgra.50539. URL http://doi.wiley.com/10.1002/jgra.50539.

V. G. Merkin and J. G. Lyon. Effects of the low-latitude ionospheric boundary condition on the global magnetosphere. *Journal of Geophysical Research: Space Physics*, *115*(A10), Oct 2010. ISSN 01480227. doi: 10.1029/2010JA015461. URL http://doi.wiley.com/10.1029/2010JA015461.

V. G. Merkin, G. Milikh, K. Papadopoulos, J. Lyon, Y. S. Dimant, A. S. Sharma, C. Goodrich, and M. Wiltberger. Effect of anomalous electron heating on the transpolar potential in the LFM global MHD model. *Geophysical Research Letters*, *32*(22). Nov 2005a. ISSN 00948276. doi: 10.1029/2005GL023315. URL http://doi.wiley.com/10.1029/2005GL023315.

V. G. Merkin, A. S. Sharma, K. Papadopoulos, G. Milikh, J. Lyon, and C. Goodrich. Relationship between the ionospheric conductance, field aligned current, and magnetopause geometry: Global MHD simulations. *Planetary and Space Science*, *53*(9):873–879, 2005b. ISSN 00320633. doi: 10.1016/j.pss. 2005.04.001.

V. G. Merkine, K. Papadopoulos, G. Milikh, A. S. Sharma, X. Shao, J. Lyon, and C. Goodrich. Effects of the solar wind electric field and ionospheric conductance on the cross polar cap potential: Results of global MHD modeling. *Geophysical Research Letters*, *30*(23), Dec 2003. ISSN 00948276. doi: 10.1029/2003GL017903. URL http://doi.wiley.com/10.1029/2003GL017903.

C. M. Mobarry, J. A. Fedder, and J. G. Lyon. Equatorial plasma convection from global simulations of the Earth's magnetosphere. *Journal of Geophysical Research: Space Physics*, *101*(A4):7859–7874, Apr 1996. ISSN 01480227. doi: 10.1029/95JA02301. URL http://doi.wiley.com/10.1029/95JA02301.

Jøran Moen and Asgeir Brekke. The solar flux influence on quiet time conductances in the auroral ionosphere. *Geophysical Research Letters*, *20* (10):971–974, May 1993. ISSN 00948276. doi: 10.1029/92GL02109. URL http://doi.wiley.com/10.1029/92GL02109.

S. K. Morley, A. P. Rouillard, and M. P. Freeman. Recurrent substorm activity during the passage of a corotating interaction region. *Journal of Atmospheric and Solar-Terrestrial Physics*, *71*(10–11):1073–1081, Jul 2009. ISSN 1364-6826. doi: 10.1016/J.JASTP.2008.11.009. URL https://www.sciencedirect.com/science/article/pii/S1364682608003611.

Steven K. Morley and Michael G. Henderson. Comment on investigation of the period of sawtooth events by X. Cai and C. R. Clauer. *Journal of Geophysical Research*, *115*(A2):A02216, Feb 2010. ISSN 0148-0227. doi: 10.1029/2009JA014721. URL http://doi.wiley.com/10.1029/2009JA014721.

P. T. Newell, T. Sotirelis, K. Liou, C.-I. Meng, and F. J. Rich. A nearly universal solar wind-magnetosphere coupling function inferred from 10 magnetospheric state variables. *Journal of Geophysical Research*, *112*(A1): A01206, Jan 2007. ISSN 0148-0227. doi: 10.1029/2006JA012015. URL http://doi.wiley.com/10.1029/2006JA012015.

P. T. Newell, T. Sotirelis, and S. Wing. Diffuse, monoenergetic, and broadband aurora: The global precipitation budget. *Journal of Geophysical Research: Space Physics*, *114*(A9), Sep 2009. ISSN 01480227. doi: 10.1029/2009JA014326. URL http://doi.wiley.com/10.1029/2009JA014326.

Chigomezyo M. Ngwira, Antti Pulkkinen, M. Leila Mays, Maria M. Kuznetsova, A. B. Galvin, Kristin Simunac, Daniel N. Baker, Xinlin Li, Yihua Zheng, and Alex Glocer. Simulation of the 23 July 2012 extreme space weather event: What if this extremely rare CME was Earth directed? *Space Weather*, *11*(12):671–679, Dec 2013. ISSN 15427390. doi: 10.1002/2013SW000990. URL http://doi.wiley.com/10.1002/2013SW000990.

Chigomezyo M. Ngwira, Antti Pulkkinen, Maria M. Kuznetsova, and Alex Glocer. Modeling extreme Carrington-type space weather events using three-dimensional global MHD simulations. *Journal of Geophysical Research: Space Physics*, *119*(6):4456–4474, Jun 2014. ISSN 21699380. doi: 10.1002/2013JA019661. URL http://doi.wiley.com/10.1002/2013JA019661.

Tatsuki Ogino and Raymond J. Walker. A magnetohydrodynamic simulation of the bifurcation of tail lobes during intervals with a northward interplanetary magnetic field. *Geophysical Research Letters*, *11*(10):1018–1021, Oct 1984. ISSN 00948276. doi: 10.1029/GL011i010p01018. URL http://doi.wiley.com/10.1029/GL011i010p01018.

S. Ohtani, H. Korth, P. C. Brandt, L. G. Blomberg, H. J. Singer, M. G. Henderson, E. a. Lucek, H. U. Frey, Q. Zong, J. M. Weygand, Y. Zheng, and a. T. Y. Lui. Cluster observations in the inner magnetosphere during the 18 April 2002 sawtooth event: Dipolarization and injection at r = 4.6 R E. *Journal of Geophysical Research*, *112*(A8):A08213, Aug 2007. ISSN 0148-0227. doi: 10.1029/2007JA012357. URL http://doi.wiley.com/10.1029/2007JA012357.

J. E. Ouellette, O. J. Brambles, J. G. Lyon, W. Lotko, and B. N. Rogers. Properties of outflow-driven sawtooth substorms. *Journal of Geophysical Research: Space Physics*, *118*(6):3223–3232, Jun 2013. ISSN 21699402. doi: 10.1002/jgra.50309. URL http://doi.wiley.com/10.1002/jgra.50309.

J. E. Ouellette, J. G. Lyon, and B. N. Rogers. A study of asymmetric reconnection scaling in the Lyon-Fedder-Mobarry code. *Journal of Geophysical Research: Space Physics*, *119*(3):1673–1682, 2014. ISSN 21699402. doi: 10.1002/2013JA019366.

M. Palmroth, P. Janhunen, T. I. Pulkkinen, and H. E. J. Koskinen. Ionospheric energy input as a function of solar wind parameters: Global MHD simulation results. *Annales Geophysicae*, *22*(2):549–566, Jan 2004. ISSN 1432-0576. doi: 10.5194/angeo-22-549-2004. URL https://hal.archives-ouvertes.fr/hal-00317237/http://www.ann-geophys.net/22/549/2004/.

Asher Pembroke, Frank Toffoletto, Stanislav Sazykin, Michael Wiltberger, John Lyon, Viacheslav Merkin, and Peter Schmitt. Initial results from a dynamic coupled magnetosphere-ionosphere-ring current model. *Journal of Geophysical Research*, *117*(A2):A02211, Feb 2012. ISSN 0148-0227. doi: 10.1029/2011JA016979. URL http://doi.wiley.com/10.1029/2011JA016979.

N. J. Perlongo, A. J. Ridley, M. W. Liemohn, and R. M. Katus. The effect of ring current electron scattering rates on magnetosphere-ionosphere coupling. *Journal of Geophysical Research: Space Physics*, *122*(4):4168–4189, Apr 2017. ISSN 21699380. doi: 10.1002/2016JA023679. URL http://doi.wiley.com/10.1002/2016JA023679.

K. G. Powell, P. L. Roe, and T. J. Linde. A solution-adaptive upwind scheme for ideal magnetohydrodynamics. *Journal of Computational Physics*, *154*(2):284–309, Sep 1999. ISSN 00219991. doi: 10.1006/jcph.1999.6299. URL http://linkinghub.elsevier.com/retrieve/pii/S002199919996299X http://www.sciencedirect.com/science/article/pii/S002199919996299X.

A. Pulkkinen, M. Hesse, M. Kuznetsova, and L. Rastätter. First-principles modeling of geomagnetically induced electromagnetic fields and currents from upstream solar wind to the surface of the Earth. *Annales Geophysicae*, *25*(4): 881–893, May 2007. ISSN 1432-0576. doi: 10.5194/angeo-25-881-2007. URL http://www.ann-geophys.net/25/881/2007/.

A. Pulkkinen, L. Rastätter, M. Kuznetsova, M. Hesse, A. Ridley, J. Raeder, H. J. Singer, and A. Chulaki. Systematic evaluation of ground and geostationary magnetic field predictions generated by global magnetohydrodynamic models. *Journal of Geophysical Research: Space Physics*, *115*(A3), Mar 2010. ISSN 01480227. doi: 10.1029/2009JA014537. URL http://doi.wiley.com/10.1029/2009JA014537.

A. Pulkkinen, M. Kuznetsova, A. Ridley, J. Raeder, A. Vapirev, D. Weimer, R. S. Weigel, M. Wiltberger, G. Millward, L. Rastätter, M. Hesse, H. J. Singer, and A. Chulaki. Geospace Environment Modeling 2008-2009 challenge: Ground magnetic field perturbations. *Space Weather*, *9*(2), Feb 2011. ISSN 15427390. doi: 10.1029/2010SW000600. URL http://doi.wiley.com/10.1029/2010SW000600.

A. Pulkkinen, E. Bernabeu, J. Eichner, C. Beggan, and A. W. P. Thomson. Generation of 100-year geomagnetically induced current scenarios. *Space Weather*, *10*(4), Apr 2012. ISSN 15427390. doi: 10.1029/2011SW000750. URL http://doi.wiley.com/10.1029/2011SW000750.

A. Pulkkinen, L. Rastätter, M. Kuznetsova, H. Singer, C. Balch, D. Weimer, G. Toth, A. Ridley, T. Gombosi, M. Wiltberger, J. Raeder, and R. Weigel. Community-wide validation of geospace model ground magnetic field perturbation predictions to support model transition to operations. *Space Weather*, *11*(6):369–385, Jun 2013. ISSN 15427390. doi: 10.1002/swe.20056. URL http://doi.wiley.com/10.1002/swe.20056.

Joachim Raeder. Modeling the magnetosphere for northward interplanetary magnetic field: Effects of electrical resistivity. *Journal of Geophysical Research*, *104*(A8):17357–17367, Aug 1999. ISSN 0148-0227. doi: 10.1029/1999ja900159. URL http://doi.wiley.com/10.1029/1999JA900159.

Joachim Raeder. Global magnetohydrodynamics a tutorial. In *Space Plasma Simulation*, pages 212–246. Springer Berlin Heidelberg, Berlin, Heidelberg, 2003. doi: 10.1007/3-540-36530-3_11. URL http://link.springer.com/10.1007/3-540-36530-3_11.

J. Raeder, R. J. Walker, and M. Ashour-Abdalla. The structure of the distant geomagnetic tail during long periods of northward IMF. *Geophysical Research Letters*, *22*(4):349–352, Feb 1995. ISSN 00948276. doi: 10.1029/94GL03380. URL http://doi.wiley.com/10.1029/94GL03380.

J. Raeder, J. Berchem, and M. Ashour-Abdalla. The Geospace Environment Modeling grand challenge: Results from a Global Geospace Circulation Model. *J. Geophys. Res.*, *103*(A7):14787–14797, 1998. ISSN 0148-0227. doi: 10.1029/98JA00014. URL http://dx.doi.org/10.1029/98JA00014 http://doi.wiley.com/10.1029/98JA00014.

J. Raeder, R. L. McPherron, L. A. Frank, S. Kokubun, G. Lu, T. Mukai, W. R. Paterson, J. B. Sigwarth, H. J. Singer, and J. A. Slavin. Global simulation of the Geospace Environment Modeling substorm challenge event. *Journal of Geophysical Research*, *106*(A1):381, 2001a. ISSN 0148-0227. doi: 10.1029/2000JA000605. URL http://doi.wiley.com/10.1029/2000JA000605.

J. Raeder, Y. L. Wang, T. J. Fuller-Rowell, and H. J. Singer. Global simulation of magnetospheric space weather effects of the bastille day storm. *Solar Physics*, *204*(1/2):323–337, 2001b. ISSN 00380938. doi: 10.1023/A:1014228230714. URL http://link.springer.com/10.1023/A:1014228230714.

Joachim Raeder, Douglas Larson, Wenhui Li, Emil L. Kepko, and Timothy Fuller-Rowell. OpenGGCM simulations for the THEMIS mission. *Space Science Reviews*, *141* (1–4):535–555, Sep 2008. ISSN 0038-6308. doi: 10.1007/s11214-008-9421-5. URL http://link.springer.com/10.1007/978-0-387-89820-9_22 http://link.springer.com/10.1007/s11214-008-9421-5.

J. Raeder, P. Zhu, Y. Ge, and G. Siscoe. Open Geospace General Circulation Model simulation of a substorm: Axial tail instability and ballooning mode preceding substorm onset. *Journal of Geophysical Research*, *115*:A00I16, Dec 2010. ISSN 0148-0227. doi: 10.1029/2010JA015876. URL http://doi.wiley.com/10.1029/2010JA015876.

L. Rastätter, M. M. Kuznetsova, A. Vapirev, A. Ridley, M. Wiltberger, A. Pulkkinen, M. Hesse, and H. J. Singer. Geospace environment modeling 2008–2009 challenge: Geosynchronous magnetic field. *Space Weather*, *9*(4): S04005, Apr 2011. ISSN 1542-7390. doi: 10.1029/2010SW000617. URL http://onlinelibrary.wiley.com/doi/10.1029/2010SW000617/full.

Lutz Rastätter, Gábor Tóth, Maria M. Kuznetsova, and Antti A. Pulkkinen. CalcDeltaB: An efficient postprocessing tool to calculate ground-level magnetic perturbations from global magnetosphere simulations. *Space Weather*, *12*(9):553–565, Sep 2014. ISSN 15427390. doi: 10.1002/2014SW001083. URL http://doi.wiley.com/10.1002/2014SW001083.

A. D. Richmond and Y. Kamide. Mapping electrodynamic features of the high-latitude ionosphere from localized observations: Technique. *Journal of Geophysical Research*, *93*(A6):5741, 1988. ISSN 0148-0227. doi: 10.1029/JA093iA06p05741. URL http://doi.wiley.com/10.1029/JA093iA06p05741.

A. J. Ridley. Alfven wings at Earth's magnetosphere under strong interplanetary magnetic fields. *Annales Geophysicae*, *25*:533–542, 2007. ISSN 09927689. URL www.ann-geophys.net/25/533/2007/.

A. J. Ridley, D. L. De Zeeuw, T. I. Gombosi, and K. G. Powell. Using steady state MHD results to predict the global state of the magnetosphere-ionosphere system. *Journal of Geophysical Research*, *106*(A12):30067, 2001. ISSN 0148-0227. doi: 10.1029/2000JA002233. URL http://doi.wiley.com/10.1029/2000JA002233.

A. J. Ridley, T. I. Gombosi, and D. L. DeZeeuw. Ionospheric control of the magnetosphere: Conductance. *Annales Geophysicae*, *22*(2):567–584, Jan 2004. ISSN 1432-0576. doi: 10.5194/angeo-22-567-2004. URL http://www.ann-geophys.net/22/567/2004/.

A. J. Ridley, D. L. De Zeeuw, W. B. Manchester, and K. C. Hansen. The magnetospheric and ionospheric response to a very strong interplanetary shock and coronal mass ejection. *Advances in Space Research*, *38*(2): 263–272, Jan 2006. ISSN 02731177. doi: 10.1016/j.asr.2006.06.010. URL http://linkinghub.elsevier.com/retrieve/pii/S0273117706003966.

A. J. Ridley, T. I. Gombosi, I. V. Sokolov, G. Tóth, and D. T. Welling. Numerical considerations in simulating the global magnetosphere. *Annales Geophysicae*, *28*(8):1589–1614, Aug 2010. ISSN 1432-0576. doi: 10.5194/angeo-28-1589-2010. URL http://www.ann-geophys.net/28/1589/2010/angeo-28-1589-2010.html http://www.ann-geophys.net/28/1589/2010/.

A. J. Ridley, D. L. De Zeeuw, and L. Rastätter. Rating global magnetosphere model simulations through statistical data-model comparisons. *Space Weather*, *14*(10):819–834, Oct 2016. ISSN 15427390. doi: 10.1002/2016SW001465. URL http://doi.wiley.com/10.1002/2016SW001465.

R. M. Robinson, R. R. Vondrak, K. Miller, T. Dabbs, and D. Hardy. On calculating ionospheric conductances from the flux and energy of precipitating electrons. *Journal of Geophysical Research*, *92*(A3):2565, 1987. ISSN 0148-0227. doi: 10.1029/JA092iA03p02565. URL http://doi.wiley.com/10.1029/JA092iA03p02565.

A. A. Samsonov, E. Gordeev, N. A. Tsyganenko, J. Šafránková, Z. Němeček, J. Šimnek, D. G. Sibeck, G. Tóth, V. G. Merkin, and J. Raeder. Do we know the actual magnetopause position for typical solar wind conditions? *Journal of Geophysical Research A: Space Physics*, *121*(7):6493–6508, Jul 2016. ISSN 21699402. doi: 10.1002/2016JA022471. URL http://doi.wiley.com/10.1002/2016JA022471.

Stanislav Sazykin. Theoretical studies of penetration of magnetospheric electric fields to the ionosphere. *Thesis (PhD). UTAH STATE UNIVERSITY, Source DAI-B 61/12, p. 6521, Jun 2001, 278 pages*, 2000.

X. Shao, P. N. Guzdar, G. M. Milikh, K. Papadopoulos, C. C. Goodrich, A. Sharma, M. J. Wiltberger, and J. G. Lyon. Comparing ground magnetic field perturbations from global MHD simulations with magnetometer data for the 10 January 1997 magnetic storm event. *Journal of Geophysical Research: Space Physics*, *107*(A8):SMP 11–1–SMP 11–10, Aug 2002. ISSN 21699402. doi: 10.1029/2000JA000445. URL http://doi.wiley.com/10.1029/2000JA000445.

Jamesina J. Simpson. On the possibility of high-level transient coronal mass ejection-induced ionospheric current coupling to electric power grids. *Journal of Geophysical Research: Space Physics*, *116*(A11), Nov 2011. ISSN 01480227. doi: 10.1029/2011JA016830. URL http://doi.wiley.com/10.1029/2011JA016830.

G. L. Siscoe. Transpolar potential saturation: Roles of region 1 current system and solar wind ram pressure. *Journal of Geophysical Research*, *107*(A10):1321, 2002. ISSN 0148-0227. doi: 10.1029/2001JA009176. URL http://doi.wiley.com/10.1029/2001JA009176.

G. L. Siscoe, G. M. Erickson, B. U. Ö. Sonnerup, N. C. Maynard, J. A. Schoendorf, K. D. Siebert, D. R. Weimer, W. W. White, and G. R. Wilson. Hill model of transpolar potential saturation: Comparisons with MHD simulations. *Journal of Geophysical Research*, *107*(A6):1075, Jun 2002. ISSN 0148-0227. doi: 10.1029/2001JA000109. URL http://doi.wiley.com/10.1029/2001JA000109.

G. Siscoe, N. U. Crooker, and C. R. clauer. Dst of the Carrington storm of 1859. *Advances in Space Research*, *38*(2):173–179, Jan 2006. ISSN 02731177. doi: 10.1016/j.asr.2005.02.102. URL http://www.sciencedirect.com.proxy.lib.umich.edu/science/article/pii/S0273117705012081http://linkinghub.elsevier.com/retrieve/pii/S0273117705012081.

S. P. Slinker, J. A. Fedder, B. A. Emery, K. B. Baker, D. Lummerzheim, J. G. Lyon, and F. J. Rich. Comparison of global MHD simulations with AMIE simulations for the events of May 19-20, 1996. *Journal of Geophysical Research: Space Physics*, *104*(A12):28379–28395, Dec 1999. ISSN 01480227. doi: 10.1029/1999JA900403. URL http://doi.wiley.com/10.1029/1999JA900403.

Igor V. Sokolov, Tamas I. Gombosi, and Aaron J. Ridley. Non-potential Electric Field Model of Magnetosphere-Ionosphere Coupling. In *Inner Magnetosphere Interactions: New Perspectives from Imaging*, volume *159*, pages 141–152. American Geophysical Union, 2013. ISBN 9781118666128. doi: 10.1029/159GM10. URL http://www.agu.org/books/gm/v159/159GM10/159GM10.shtml.

P. Song, V. M. Vasylinas, and X.-Z. Zhou. Magnetosphere-ionosphere/thermosphere coupling: Self-consistent solutions for a one-dimensional stratified ionosphere in three-fluid theory. *Journal of Geophysical Research: Space Physics*, *114*(A8), Aug 2009. ISSN 01480227. doi: 10.1029/2008JA013629. URL http://doi.wiley.com/10.1029/2008JA013629.

Frank Toffoletto, Stanislav Sazykin, Robert Spiro, and Richard Wolf. Inner magnetospheric modeling with the Rice Convection Model. *Space Science Reviews*, (*1*):175–196, 2003. doi: 10.1023/A:1025532008047. URL http://adsabs.harvard.edu/abs/2003SSRv..107..175T.

E. F. Toro. *Riemann Solvers and Numerical Methods for Fluid Dynamics : A Practical Introduction*. Springer, 2006. ISBN 9783540252023. doi: 10.1007/978-3-540-49834-6.

Gábor Tóth. The $\nabla \dot{B} = 0$ Constraint in shock-capturing magnetohydrodynamics codes. *Journal of Computational Physics*, *161*(2):605–652, Jul 2000. ISSN 00219991. doi: 10.1006/jcph.2000.6519. URL http://www.sciencedirect.com/science/

article/pii/S0021999100965197 http://linkinghub.elsevier.com/retrieve/pii/S0021999100965197.

Gábor Tóth, Igor V. Sokolov, Tamas I. Gombosi, David R. Chesney, C. Robert Clauer, Darren L. De Zeeuw, Kenneth C. Hansen, Kevin J. Kane, Ward B. Manchester, Robert C. Oehmke, Kenneth G. Powell, Aaron J. Ridley, Ilia I. Roussev, Quentin F. Stout, Ovsei Volberg, Richard a. Wolf, Stanislav Sazykin, Anthony Chan, Bin Yu, and József Kóta. Space Weather Modeling Framework: A new tool for the space science community. *Journal of Geophysical Research*, *110*(A12):A12226, 2005. ISSN 0148-0227. doi: 10.1029/2005JA011126. URL http://www.agu.org/pubs/crossref/2005/2005JA011126.shtml.

Gábor Tóth, Darren L. De Zeeuw, Tamas I. Gombosi, Ward B. Manchester, Aaron J. Ridley, Igor V. Sokolov, and Ilia I. Roussev. Sun-to-thermosphere simulation of the 28-30 October 2003 storm with the Space Weather Modeling Framework. *Space Weather*, *5*(6), Jun 2007. ISSN 15427390. doi: 10.1029/2006SW000272. URL http://doi.wiley.com/10.1029/2006SW000272.

Gábor Tóth, Yingjuan Ma, and Tamas I. Gombosi. Hall magnetohydrodynamics on block-adaptive grids. *Journal of Computational Physics*, *227*(14):6967–6984, Jul 2008. ISSN 00219991. doi: 10.1016/j.jcp.2008.04.010. URL http://dl.acm.org/citation.cfm?id=1379915.1380233.

Gábor Tóth, Bart van der Holst, Igor V. Sokolov, Darren L. De Zeeuw, Tamas I. Gombosi, Fang Fang, Ward B. Manchester, Xing Meng, Dalal Najib, Kenneth G. Powell, Quentin F. Stout, Alex Glocer, Ying-Juan Ma, and Merav Opher. Adaptive numerical algorithms in space weather modeling. *Journal of Computational Physics*, *231*(3):870–903, Feb 2012. ISSN 00219991. doi: 10.1016/j.jcp.2011.02.006. URL http://www.sciencedirect.com/science/article/pii/S002199911100088X.

Gábor Tóth, Yuxi Chen, Tamas I. Gombosi, Paul Cassak, Stefano Markidis, and Ivy Bo Peng. Scaling the ion inertial length and its implications for modeling reconnection in global simulations. *Journal of Geophysical Research: Space Physics*, Sep 2017. ISSN 21699380. doi: 10.1002/2017JA024189. URL http://doi.wiley.com/10.1002/2017JA024189.

N.A. Tsyganenko. A magnetospheric magnetic field model with a warped tail current sheet. *Planetary and Space Science*, *37*(1):5–20, Jan 1989. ISSN 00320633. doi: 10.1016/0032-0633(89)90066-4. URL http://linkinghub.elsevier.com/retrieve/pii/0032063389900664.

N. A. Tsyganenko. A model of the near magnetosphere with a dawn-dusk asymmetry 2. Parameterization and fitting to observations. *Journal of Geophysical Research: Space Physics*, *107*(A8):SMP 10–1–SMP 10–17, Aug 2002. ISSN 01480227. doi: 10.1029/2001JA000220. URL http://doi.wiley.com/10.1029/2001JA000220.

Jiannan Tu, Paul Song, and Vytenis M. Vasylinas. Ionosphere/thermosphere heating determined from dynamic magnetosphere-ionosphere/thermosphere coupling. *Journal of Geophysical Research*, *116*:A09311, Sep 2011. ISSN 21699402. URL http://doi.wiley.com/10.1029/2011JA016620.

H. Vanhamäki. Inductive ionospheric solver for magnetospheric MHD simulations. *Annales Geophysicae*, *29*(1):97–108, Jan 2011. ISSN 09927689. doi: 10.5194/angeo-29-97-2011. URL http://www.ann-geophys.net/29/97/2011/.

James F. Vickrey, Richard R. Vondrak, and Stephen J. Matthews. The diurnal and latitudinal variation of auroral zone ionospheric conductivity. *Journal of Geophysical Research*, *86*(80):65, 1981. ISSN 0148-0227. doi: 10.1029/JA086iA01p00065.

H. Wang, H. Lühr, a. Ridley, P. Ritter, and Y. Yu. Storm time dynamics of auroral electrojets: CHAMP observation and the Space Weather Modeling Framework comparison. *Annales Geophysicae*, *26*:555–570, 2008. ISSN 1432-0576. doi: 10.5194/angeo-26-555-2008.

R. S. Weigel. Solar wind time history contribution to the day-of-year variation in geomagnetic activity. *Journal of Geophysical Research: Space Physics*, *112*(A10), Oct 2007. ISSN 01480227. doi: 10.1029/2007JA012324. URL http://doi.wiley.com/10.1029/2007JA012324.

D. R. Weimer. Improved ionospheric electrodynamic models and application to calculating Joule heating rates. *Journal of Geophysical Research*, *110* (A5):A05306, 2005. ISSN 0148-0227. doi: 10.1029/2004JA010884. URL http://doi.wiley.com/10.1029/2004JA010884.

D. R. Weimer. An empirical model of ground-level geomagnetic perturbations. *Space Weather*, *11*(3):107–120, Mar 2013. ISSN 15427390. doi: 10.1002/swe. 20030. URL http://doi.wiley.com/10.1002/swe.20030.

D. R. Weimer, C. R. Clauer, M. J. Engebretson, T. L. Hansen, H. Gleisner, I. Mann, and K. Yumoto. Statistical maps of geomagnetic perturbations as a function of the interplanetary magnetic field. *Journal of Geophysical Research*, *115*(A10):A10320, 2010. ISSN 0148-0227. doi: 10.1029/2010JA015540. URL http://doi.wiley.com/10.1029/2010JA015540.

D. T. Welling and M. W. Liemohn. Outflow in global magnetohydrodynamics as a function of a passive inner boundary source. *Journal of Geophysical Research: Space Physics*, *119*(4):2691–2705, Apr 2014. ISSN 21699380. doi: 10.1002/2013JA019374. URL http://doi.wiley.com/10.1002/2013JA019374.

D. T. Welling and A. J. Ridley. Validation of SWMF magnetic field and plasma. *Space Weather*, *8*(3), Mar 2010a. ISSN 15427390. doi: 10.1029/2009SW000494. URL http://doi.wiley.com/10.1029/2009SW000494.

D. T. Welling and A. J. Ridley. Exploring sources of magnetospheric plasma using multispecies MHD. *Journal of Geophysical Research*, *115*(A4):A04201, Apr 2010b. ISSN 0148-0227. doi: 10.1029/2009JA014596. URL http://doi.wiley.com/10.1029/2009JA014596.

D. T. Welling, V. K. Jordanova, A. Glocer, G. Toth, M. W. Liemohn, and D. R. Weimer. The two-way relationship between ionospheric outflow and the ring current. *Journal of Geophysical Research: Space Physics*, *120*(6): 4338–4353, Jun 2015. ISSN 21699380. doi: 10.1002/2015JA021231. URL http://doi.wiley.com/10.1002/2015JA021231.

Daniel T. Welling, Brian J. Anderson, Geoff Crowley, Antti A. Pulkkinen, and L. Rastätter. Exploring predictive performance: A reanalysis of the geospace model transition challenge. *Space Weather*, *15*(1):192–203, Jan 2017. ISSN 15427390. doi: 10.1002/2016SW001505. URL http://doi.wiley.com/10.1002/2016SW001505.

Michael Wiltberger. Review of global simulation studies of effect of ionospheric outflow on magnetosphere-ionosphere system dynamics. In *Magnetotails in the Solar System*, chapter 22, pages 373–392. Wiley, Hoboken, NJ, 2015. doi: 10.1002/9781118842324.

M. Wiltberger, J. G. Lyon, and C. C. Goodrich. Results from the Lyon-Fedder-Mobarry global magnetospheric model for the electrojet challenge. *Journal of Atmospheric and Solar-Terrestrial Physics*, *65*(11–13): 1213–1222, Jul 2003. ISSN 13646826. doi: 10.1016/j.jastp.2003.08.003. URL http://www.sciencedirect.com/science/article/pii/S1364682603001718.

M. Wiltberger, W. Wang, A. G. Burns, S. C. Solomon, J. G. Lyon, and C. C. Goodrich. Initial results from the coupled magnetosphere ionosphere thermosphere model: Magnetospheric and ionospheric responses. *Journal of Atmospheric and Solar-Terrestrial Physics*, *66*(15–16 special issue):1411–1423, Oct 2004. ISSN 13646826. doi: 10.1016/j.jastp.2004.03.026. URL http://www.sciencedirect.com/science/article/pii/S136468260400149X#bbib22.

M. Wiltberger, R. S. Weigel, W. Lotko, and J. a. Fedder. Modeling seasonal variations of auroral particle precipitation in a global-scale magnetosphere-ionosphere simulation. *Journal of Geophysical Research*, *114* (A1):A01204, 2009. ISSN 0148-0227. doi: 10.1029/2008JA013108. URL http://doi.wiley.com/10.1029/2008JA013108.

M. Wiltberger, W. Lotko, J. G. Lyon, P. Damiano, and V. Merkin. Influence of cusp O + outflow on magnetotail dynamics in a multifluid MHD model of the magnetosphere. *Journal of Geophysical Research*, *115*(June):A00J05, Oct 2010. ISSN 0148-0227. doi: 10.1029/2010JA015579. URL http://doi.wiley.com/10.1029/2010JA015579.

M. Wiltberger, V. Merkin, B. Zhang, F. Toffoletto, M. Oppenheim, W. Wang, J. G. Lyon, J. Liu, Y. Dimant, M. I. Sitnov, and G. K. Stephens. Effects of electrojet turbulence on a magnetosphere-ionosphere simulation of a geomagnetic storm. *Journal of Geophysical Research: Space Physics*, *122*(5):5008–5027, May 2017a. ISSN 21699380. doi: 10.1002/2016JA023700. URL http://doi.wiley.com/10.1002/2016JA023700.

M. Wiltberger, E. J. Rigler, V Merkin, and J. G. Lyon. Structure of high latitude currents in Magnetosphere-Ionosphere Models. *Space Science Reviews*, *206* (1–4):575–598, Mar 2017b. ISSN 0038-6308. doi: 10.1007/s11214-016-0271-2. URL http://download.springer.com/static/pdf/72/art%253A10.1007%252Fs11214-016-0271-2.

R. M. Winglee. Multi-fluid simulations of the magnetosphere: The identification of the geopause and its variation with IMF. *Geophysical Research Letters*, *25*(24):4441–4444, Dec 1998. ISSN 00948276. doi: 10.1029/1998GL900217. URL http://doi.wiley.com/10.1029/1998GL900217.

R. M. Winglee. Mapping of ionospheric outflows into the magnetosphere for varying IMF conditions. *Journal of Atmospheric and Solar-Terrestrial Physics*, *62*:527–540, 2000. URL http://www.sciencedirect.com/science/article/pii/S1364682600000158.

R. M. Winglee. Circulation of ionospheric and solar wind particle populations during extended southward interplanetary magnetic field. *Journal of Geophysical Research*, *108*(A10):1385, 2003. ISSN 0148-0227. doi: 10.1029/2002JA009819. URL http://doi.wiley.com/10.1029/2002JA009819.

R. A. Wolf. *The Quasi-Static (Slow-Flow) Region of the Magnetosphere*, pages 303–368. Springer Netherlands, 1983. doi: 10.1007/978-94-009-7194-3_14. URL http://www.springerlink.com/index/10.1007/978-94-009-7194-3_14.

C. C. Wu, R. J. Walker, and J. M. Dawson. A three dimensional MHD model of the Earth's magnetosphere. *Geophysical Research Letters*, *8*(5): 523–526, May 1981. ISSN 00948276. doi: 10.1029/GL008i005p00523. URL http://doi.wiley.com/10.1029/GL008i005p00523.

Akimasa Yoshikawa, Hiroyuki Nakata, Aoi Nakamizo, Teiji Uozumi, Masahiro Itonaga, Shigeru Fujita, Kiyohumi Yumoto, and Takashi Tanaka. Alfvenic-coupling algorithm for global and dynamical magnetosphere-ionosphere coupled system. *Journal of Geophysical Research*, *115*(A4):1–7, Apr 2010. ISSN 01480227. doi: 10.1029/2009JA014924. URL http://doi.wiley.com/10.1029/2009JA014924.

Yiqun Yu and Aaron J. Ridley. Validation of the space weather modeling framework using ground-based magnetometers. *Space Weather*, *6*(5):S05002, May 2008. ISSN 1542–7390. doi: 10.1029/2007SW000345. URL http://www.agu.org/pubs/crossref/2008/2007SW000345.shtml.

Yiqun Yu, Aaron J. Ridley, D. T. Dan T. Welling, and Gabor Tóth. Including gap region field-aligned currents and magnetospheric currents in the MHD calculation of ground-based magnetic field perturbations. *Journal of Geophysical Research*, *115*(A8):A08207, Aug 2010. ISSN 0148-0227. doi: 10.1029/2009JA014869. URL http://doi.wiley.com/10.1029/2009JA014869.

Yiqun Yu, Vania K. Jordanova, Aaron J. Ridley, Gabor Toth, and Roderick Heelis. Effects of electric field methods on modeling the midlatitude ionospheric electrodynamics and inner magnetosphere dynamics. *Journal of Geophysical Research: Space Physics*, *122*(5):5321–5338, May 2017. ISSN 21699380. doi: 10.1002/2016JA023850. URL http://doi.wiley.com/10.1002/2016JA023850.

B. Zhang, W. Lotko, M. J. Wiltberger, O. J. Brambles, and P. A. Damiano. A statistical study of magnetosphereionosphere coupling in the Lyon-Fedder-Mobarry global MHD model. *Journal of Atmospheric and Solar-Terrestrial Physics*, *73*(5–6):686–702, Apr 2011. ISSN 13646826. doi: 10.1016/j.jastp.2010.09.027. URL http://www.sciencedirect.com/science/article/pii/S1364682610002877 http://linkinghub.elsevier.com/retrieve/pii/S1364682610002877.

B. Zhang, W. Lotko, O. Brambles, M. Wiltberger, and J. Lyon. Electron precipitation models in global magnetosphere simulations. *Journal of Geophysical Research: Space Physics*, *120*(2):1035–1056, Feb 2015. ISSN 21699402. doi: 10.1002/2014JA020615. URL http://doi.wiley.com/10.1002/2014JA020615.

4

Empirical Modeling of the Geomagnetic Field for GIC Predictions

D. R. Weimer

ABSTRACT

Empirical models can be used to predict the geomagnetic perturbations that cause Geomagnetic Induced Currents (GIC). Using examples from a model developed at Virginia Tech, the advantages and weaknesses of empirical models are discussed. The primary advantages are low computational cost and high speeds, which makes such models useful for predictions of when and where troublesome GIC may occur. In comparison to other model types, the Virginia Tech model excels in some types of metrics and poorly in others. While the low-frequency variations are predicted well, the chaotic, high-frequency fluctuations and substorm perturbations are not. To illustrate these strengths and weaknesses, examples from three events are shown. Two strategies are discussed for improving model performance. One method is to superpose high-frequency fluctuations that have spectral characteristics appropriate for the given IMF conditions, based on an analysis of the data. Another technique is adding evolving, substorm patterns at appropriate times. While these strategies would improve model skills on indices and metric evaluations, adding the high-frequency components might not be as useful for exact calculations of GIC within a specific network of conductors. The coherency of rapid field variations over long distances needs to be known.

Key Points

- Empirical models can be useful for predicting the geomagnetic perturbations that cause GIC
- Strengths and weaknesses of the Virginia Tech empirical model are discussed
- Adding higher-frequency components would improve some types of forecasts

4.1. INTRODUCTION

The solar wind and its embedded, Interplanetary Magnetic Field (IMF) interact with the Earth's magnetosphere to produce strong currents in the polar ionosphere. These auroral currents are the source of magnetic perturbations on the ground that cause Geomagnetically Induced Currents (GIC). These GIC can flow within conducting structures such as power transmission lines, pipelines, and communication cables that span long distances (Boteler, 2001; Gummow, 2002; Lanzerotti, 2001). Due to the undesired effects of GICs, described in more detail elsewhere in this Monograph, there is a need for prediction of geomagnetic variations.

One technique for making these predictions is through the use of an empirical model of geomagnetic perturbations on the ground, which is fed with real-time measurements of the solar wind and IMF. One such model was developed at Virginia Tech with support from the NSF National Space Weather Program. This model is based on data collected from 149 different magnetometer stations in the Northern Hemisphere. These measurements span a nearly 8-year time period, from February 1998 to December 2005. The starting date is when NASA's

Bradley Department of Electrical and Computer Engineering and Center for Space Science and Engineering Research, Virginia Tech, Blacksburg, VA, USA

National Institute of Aerospace, Hampton, Virginia, USA

Geomagnetically Induced Currents from the Sun to the Power Grid, Geophysical Monograph 244, First Edition.
Edited by Jennifer L. Gannon, Andrei Swidinsky, and Zhonghua Xu.

Advanced Composition Explorer (ACE) began scientific operations, as measurements of the interplanetary space environment were required for the model development. The Magnetic Field Instrument (Smith et al., 1998) and the Solar Wind Electron, Proton, and Alpha Monitor (McComas et al., 1998) instruments provided the data required for the model development, and later tests and operation.

The empirical model is based on Spherical Cap Harmonic Analysis (SCHA) (Haines, 1985). As the "polar cap" that is used in the model extends to the magnetic equator, the half-angle of the spherical cap is 90°, so Legendre polynomials having integer degrees are used in the SCHA formulas, which simplifies the calculations. At smaller half-angles Associated Legendre functions are required, that use real (fractional) degrees.

Because electrodynamic parameters in the ionosphere, such as electric fields and currents, are well organized in geomagnetic coordinates but not in geographic coordinates, the empirical model uses "Modified Apex" coordinates (Emmert et al., 2010; Richmond, 1995; VanZandt et al., 1972). Apex coordinates are nearly identical to the Altitude Adjusted Corrected Geomagnetic system (Baker and Wing, 1989) at high latitudes. The modified apex coordinates are valid at all latitudes, which is the main advantage over the corrected coordinates, that cannot be computed for some locations near the magnetic equator. Magnetic longitudes are also converted to magnetic local time (MLT), which varies with the time of day. The magnetic fields that are calculated in this magnetic coordinate system need to be translated and rotated for the output of the perturbation values in geographic coordinates.

Details about the model construction are provided by Weimer (2013). Specific information about the processing of the extensive magnetometer data and initial tests in model development are described by Weimer et al. (2010). During these initial tests, it was found that the ground-level magnetometer data have the best correlation with IMF measurements when 15–20-min averages of the IMF are used for the comparison, which apparently is the time that it takes for the entire magnetosphere–ionosphere system to reconfigure to a change in the IMF. Therefore, the model has been driven by using IMF that has been processed with a running average. This does have the effect of limiting the predicted magnetic field changes to frequencies of no more than one complete cycle in 30 min ($5.6 \cdot 10^{-4}$ Hz), if driven with 15-min averages.

Examples of model operation using real-time solar wind and IMF data have been shown on the web site at http://mist.nianet.org/weimerGeomag.html, running nearly continuously since 2011. The data from the ACE spacecraft had been used for input up until the end of July 2016, when the Deep Space Climate Observatory (DSCOVR) spacecraft became operational, replacing ACE for the source of the real-time data stream provided by the National Oceanic and Atmospheric Administration's (NOAA) Space Weather Prediction Center (SWPC). In November of 2016, the program for the real-time demonstration was modified to use these data from the Plasma-Magnetometer (PlasMag) instrument on DSCOVR.

As ACE and DSCOVR are located approximately 248 times the Earth's Radius (R_E) ahead of Earth in the "upstream" solar wind, it takes approximately 1 h, depending on velocity, for the solar wind and IMF to reach the Earth's magnetosphere after the radioed data, traveling at light speed, reach the Earth. Therefore, predictions of the geomagnetic perturbations can be obtained about 1 h ahead of their impact.

The magnetic field vectors in the solar wind are known to have a complex structure, which can cause variations in the arrival timing of field transitions at the Earth, as documented by Weimer et al. (2002). To compensate for variability in the orientation of the IMF "phase fronts," the technique described by Weimer and King (2008) was used to calculate the time delays for the model development, and for real-time operation.

This calculation of phase front orientation uses a combination of the vector cross product of the magnetic fields at different times, and the minimum variance of the magnetic field over some time interval, with the constraint that the average field along the phase front's normal vector is zero (Haaland et al., 2004; Sönnerup et al., 2004). For brevity, the abbreviation CPVM is used to refer to this combination of the cross product and minimum variance methods.

4.2. EMPIRICAL MODEL STRENGTHS AND WEAKNESSES

Returning to the topic of empirical models, their major strength is the very low computational cost and high speed, even when used on very small computers. For example, the web server that displays results from the Virginia Tech model uses an old, Mac "mini" desktop model, yet it takes only a few seconds for each real-time update to complete; in each update the data download and graphics output use the majority of this computing time. On the other hand, first-principles-based numerical models require a supercomputer in order to keep up with real-time data input and forecasts, at much greater cost. Another advantage of the empirical model is that the effects of the variations in the ionospheric conductivity are implicitly included in both the measured and modeled magnetic fields. In the numerical models, the values of the ionospheric conductivity tend to be a quantity that has the least precision, and approximations are often used.

Since the empirical model is based on measured magnetic field values that include the total field from all sources, it has another advantage in that it inherently incorporates the contributions to the perturbation magnetic field that result from underground, induced currents (Tanskanen et al., 2001). These induced currents tend to amplify the horizontal component of the magnetic field, while reducing the vertical component. The magnetic effects of the internal currents are not usually included in the global numerical simulations, and typically some compensation is done through adjustments in the ionospheric conductivities.

The Community Coordinated Modeling Center (CCMC) at NASA's Goddard Space Flight Center (GSFC) has conducted three studies of the performances of several geomagnetic prediction models. At the time of the first test (Pulkkinen et al., 2011) the model described here was still under development, and an early version was used. Nevertheless, in a comparison with magnetometer measurements during selected time periods, this empirical model had the best score for "prediction efficiency" and second-best score for "root-mean-square difference." These metric scores are defined in detail by Pulkkinen et al. (2011). Other comparisons in the same study used a metric for "log-spectral distance." As these tests were based on the frequency spectrums of magnetometer data having 1 min sample rates, the spectrums were compared for periods in the range of 2–20 min, or frequencies up to $8.3 \cdot 10^{-3}$ Hz. Not surprisingly, since the empirical model's variations do not have periods of less than 30 min, the numerical models did better than the empirical models on this metric.

The second study by the CCMC evaluated only the magnetic field's rate of change over time, looking at how the magnitude of $|dB/dt|$ in the model compared with the measured values over some time intervals (Pulkkinen et al., 2013). Only $|dB/dt|$ values were compared, regardless of the sign and magnitude of the actual magnetic field. In this study, the Virginia Tech model placed in the middle to low end of the scores for the selected events. Section 4.4 will discuss strategies for making model improvements that would result in better scores on such metric evaluations, and the merits of such changes.

In the third study by the CCMC, the objective was to test model performance on an improvised version of the planetary K index (K_p), that is intended to be localized rather than global. For test purposes, this "local K-index" was calculated for a specific set of magnetometer locations that were selected to represent different latitudes or regions (Glocer et al., 2016). This index was calculated from the maximum range of $\Delta\mathbf{B}$ in the two horizontal directions:

$$\text{Range} = \max\left[\left(\Delta B_{x,\max} - \Delta B_{x,\min}\right), \left(\Delta B_{y,\max} - \Delta B_{y,\min}\right)\right] \tag{4.1}$$

This calculation is done within a 3 h time window, like the K_p index, using the maximum and minimum values in the window of the horizontal north and east components of the magnetic field, in geomagnetic dipole coordinates. This range is then converted to a numerical index in the range of 0–9 by application of scaling factors and a lookup table. In order to provide this index with a better temporal resolution, the 3 h window is moved ahead at 15 min increments. These indices were calculated separately for each magnetometer site location that were part of the test suite. At each site, the range that was calculated from (4.1) used only the magnetic field values at that particular location.

The performance of each model was derived by comparing the predicted and observed K values at the selected magnetometer locations, as described in more detail by Glocer et al. (2016). The metric scores were based on threshold crossings at each index level, basically counting how often the index derived from each model exceeded the given level at the same times that the corresponding index from the measurements exceeded the same threshold. The metric score computations in this study included penalties for over predictions, or false positives, so that if a model consistently predicted K indices higher than the measured value, then their score would be reduced. On the other hand, since (4.1) only looked at the difference between the maximum and minimum values, it was possible to obtain good metric scores with predicted $\Delta\mathbf{B}$ values that had a lot of variability but did not actually match the magnitudes or signs of the measured values.

In the first metric evaluation that was done by the CCMC Pulkkinen et al. (2011), model outputs provided by each modeling group were used, and afterwards the CCMC switched to running the model calculations at their own facility. During the course of the K index evaluations it was discovered that the program used by the CCMC to run the Virginia Tech model for the prior study (Pulkkinen et al., 2013) had generated output in geographic coordinates rather than geomagnetic coordinates and, due to a typographical error, the Y component of the IMF input to this model was always set to zero. After a correction was made, the result was that "The 6_WEIMER statistical model performed exceptionally well for midlatitudes for a threshold of K=6, the top performer in this category" (Glocer et al., 2016), meaning that for $\Delta\mathbf{B}$ in the range of 120–200 nT this model had better prediction efficiencies than the other models. At the higher threshold of $K = 8$, with $\Delta\mathbf{B}$ in the range of 330–500 nT, "the model performance decreased but was still strong," while at higher latitudes the score for $K = 8$ dropped below the majority of the numerical models.

4.3. EXAMPLES FROM THREE EVENTS

Examples of the Virginia Tech empirical model's predictions are shown in Figures 4.1–4.3 for three recent

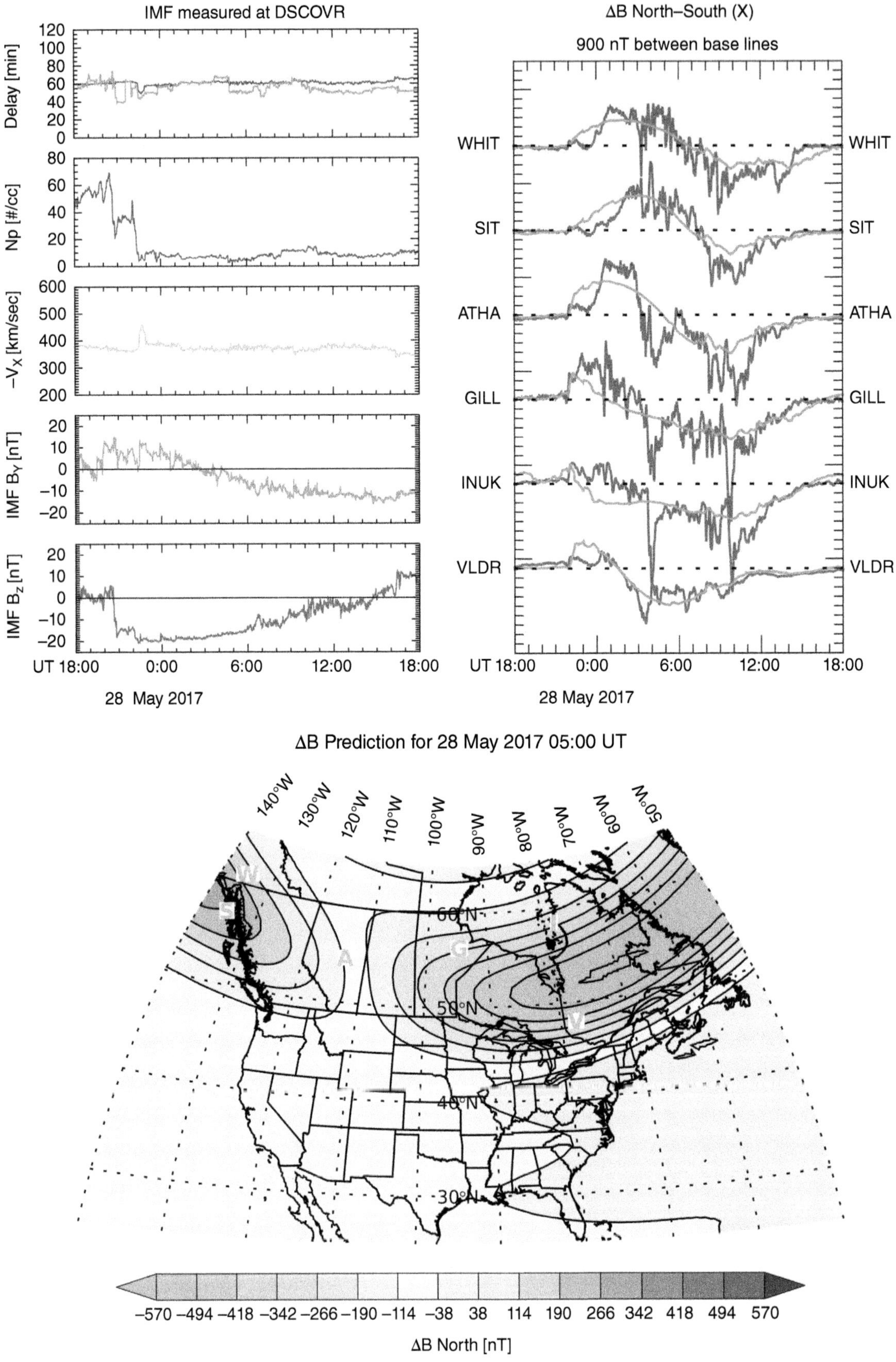

Figure 4.1 Empirical model prediction for 27–28 May 2017. The solar wind and IMF measured at the DSCOVR spacecraft are graphed in the upper left corner, predicted (blue) and measured (red) magnetic field perturbations in the North–South direction are shown in the upper right, and a map of these perturbations at 05:00 UT is shown at the bottom. The locations of the magnetometer stations are indicated with their first letter. (*See insert for color representation of the figure.*)

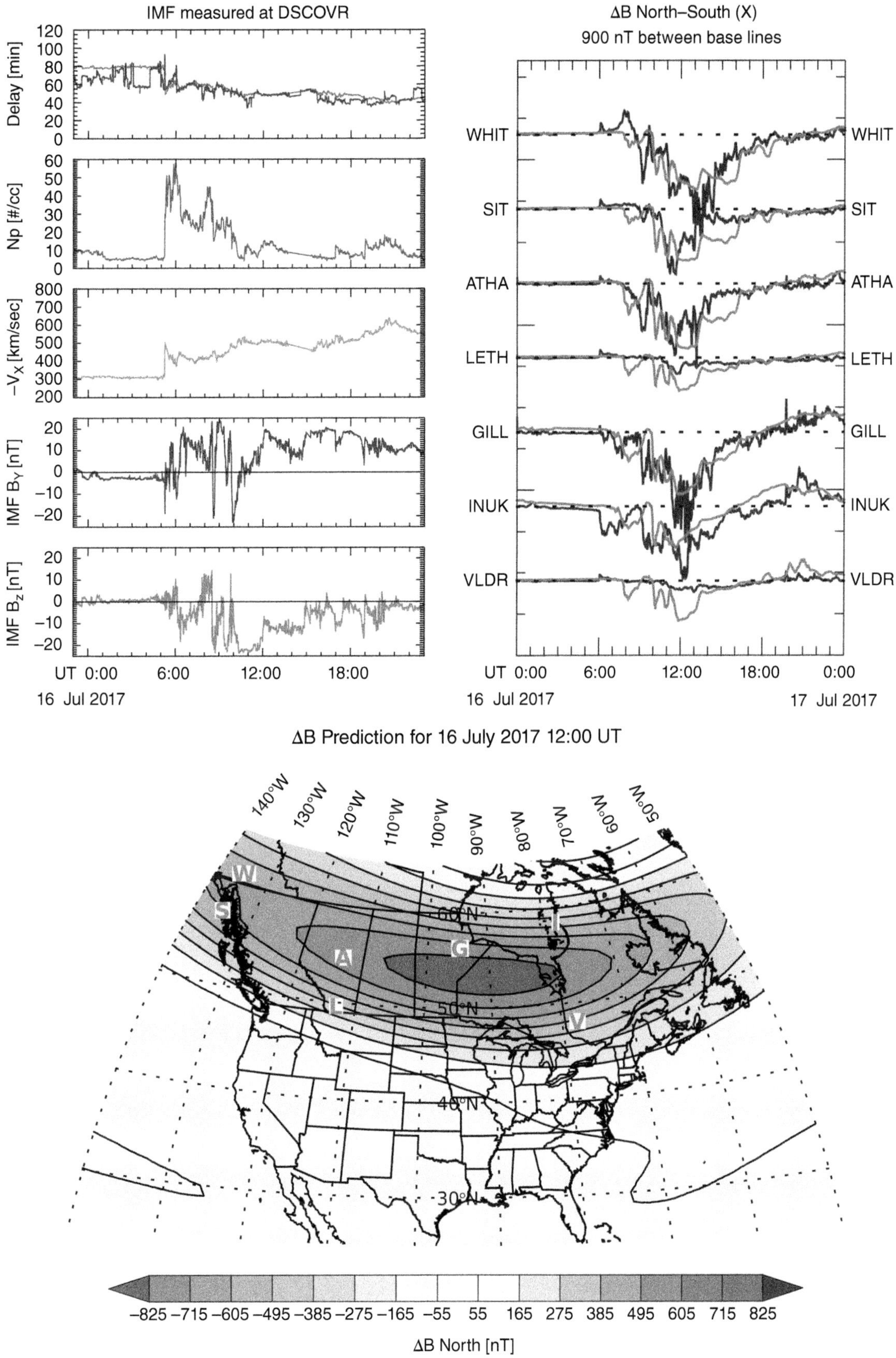

Figure 4.2 Empirical model prediction for 16 July 2017. The solar wind and IMF measured at the DSCOVR spacecraft are graphed in the upper left corner, predicted (blue) and measured (red) magnetic field perturbations in the North–South direction are shown in the upper right, and a map of these perturbations at 12:00 UT is shown at the bottom. The locations of the magnetometer stations are indicated with their first letter. (*See insert for color representation of the figure.*)

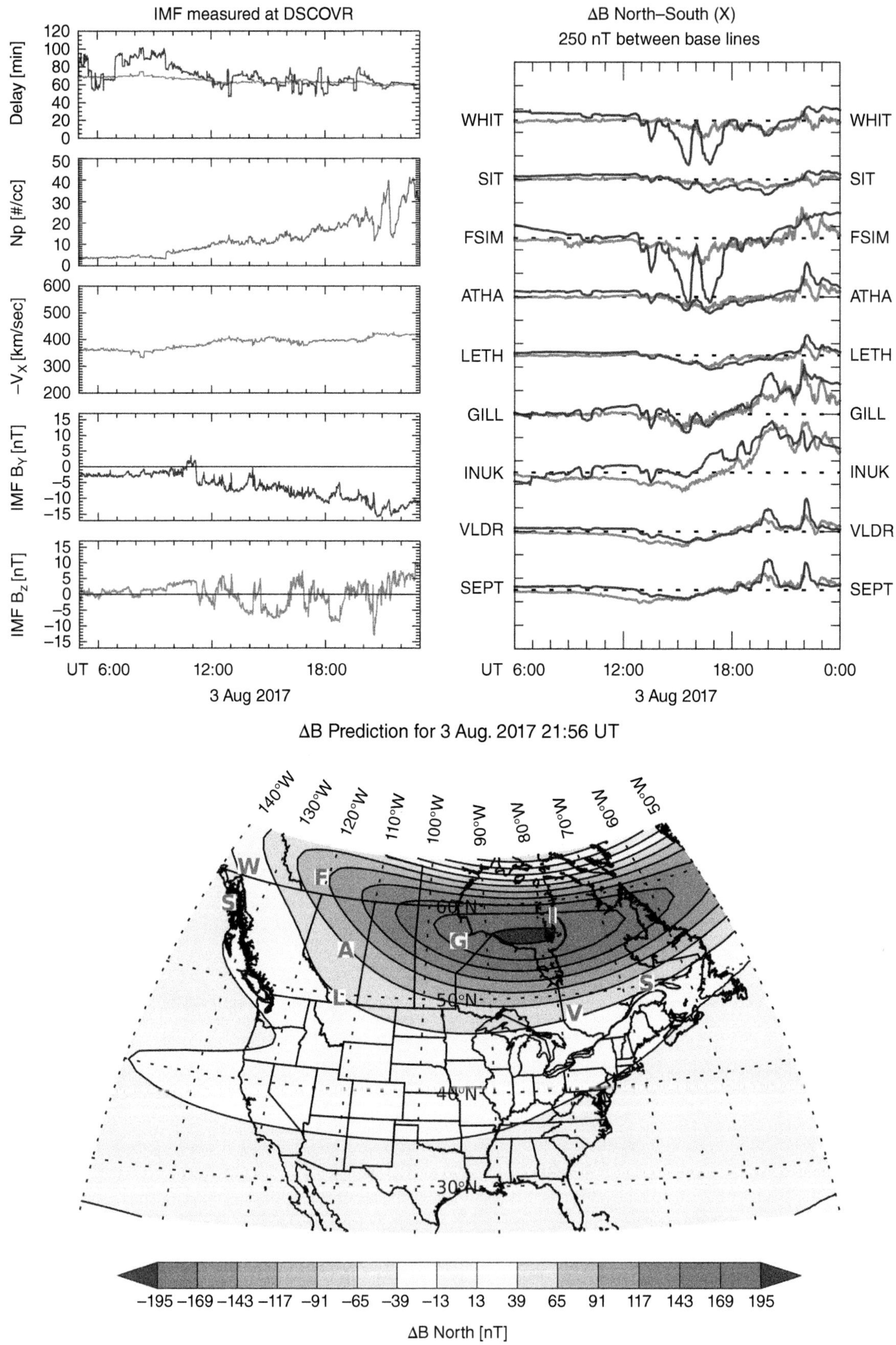

Figure 4.3 Empirical model prediction for 3 August 2017. The solar wind and IMF measured at the DSCOVR spacecraft are graphed in the upper left corner, predicted (blue) and measured (red) magnetic field perturbations in the North–South direction are shown in the upper right, and a map of these perturbations at 21:56 UT is shown at the bottom. The locations of the magnetometer stations are indicated with their first letter. (*See insert for color representation of the figure.*)

time periods, using solar wind and IMF measurements from the DSCOVR spacecraft as input, obtained from https://www.ngdc.noaa.gov/dscovr. In all of these figures the solar wind and IMF measured at the DSCOVR spacecraft are graphed in the upper left corner. The panels from bottom to top show the Z and Y components of the IMF in Geocentric Solar Magnetic (GSM) coordinates, the component of the solar wind velocity flowing toward the Earth, the proton number density, and the propagation time delay to reach the Earth's bow shock. The green line indicates the delay obtained using "flat plane" approximation, and the blue line shows the delay calculated with the varying phase front angles, using the CPMV method. The abscissa indicates the time of measurements at the location of DSCOVR.

The upper right graphs in these figures contain the ground magnetic field perturbations as a function of time, with model predictions plotted with the blue lines, and measured magnetic fields shown with the red lines, showing the North–South component in geomagnetic coordinates. These data are from stations that are associated with the Time History of Events and Macroscale Interactions During Substorms (THEMIS) mission, and available at http://themis.ssl.berkeley.edu/gmag. The three to four-letter station identifiers used by the THEMIS project are indicated alongside each plot line. The full station names and locations are shown in Table 4.1. In two of the events not all stations are shown due to the lack of data availability for that particular time period. For the comparisons with the magnetic perturbations that are due to ionospheric sources, baseline levels of the magnetic field, due to Earth's internal dynamo, were subtracted from the measured values. These offsets were derived at a preceding time when the activity was low.

The low parts of the three figures contain contour maps of the North–South component of the magnetic field perturbations over North America in geographic coordinates, at the specific time indicated above the map. The locations of the stations are indicated by their first letter, with the "S" that represents Sitka being the one farthest to the West, and Sept Iles is in the East. Generally, the order of the stations from top to bottom on these graphs are at locations that progress from North to South and West to East on the map.

Figure 4.1 shows these results for a long-duration magnetic cloud on 27–28 May 2017, during which the Z component of the IMF reached −20 nT and the Y component varied between −10 and +10 nT. The map at the bottom is for 0500 UT. During the passage of this magnetic cloud the division between the eastward and westward electrojets passed over the magnetometers in North America, as seen in the contour map of the North–South component of the magnetic field. While the low-frequency variation in the magnetic field (periods of 30 min or longer) is predicted very well by the model, the characteristic variations at higher frequencies are not included. These fluctuations, at periods on the order of a couple of minutes, are due to auroral currents overhead. These currents are usually caused by the activity in the magnetotail, where there are turbulent flows. Being chaotic, these currents are not exactly predictable by any model, empirical or numeric. These currents are sometimes very localized and the corresponding magnetic perturbations are not coherent over large distances, which is particularly evident in comparing the measurements between Whitehorse (WHIT) and Sitka (SIT) in the west and Inukjauk (INUK) and Val-d'Or (VLDR) in the east. Also during this time period in May 2017 there were two substorms, which are evident in the data from four of the magnetometers. As the substorm variations span larger distances they may have a greater effect on GIC than the 2-min fluctuations. The inclusion of substorm variations in a future version of the prediction model is discussed later.

An event in which the Z component reached −20 nT and lower, but for a shorter duration, is illustrated in Figure 4.2, for 16 July 2017. The map at the bottom

Table 4.1 Magnetometer Station Names and Locations.

		Geographic		Magnetic	
Station location	Code name	Latitude	Longitude	Latitude	Longitude
Whitehorse, YT	WHIT	61.01	224.78	63.49	279.42
Sitka, AK	SIT	57.06	224.67	59.48	281.32
Fort Simpson, NT	FSIM	61.76	238.77	67.13	294.14
Athabasca, AB	ATHA	54.71	246.69	61.78	307.03
Lethbridge, AB	LETH	49.64	247.13	56.80	309.06
Gillam, MB	GILL	56.35	265.34	65.98	332.97
Inukjuak, QC	INUK	58.47	281.92	68.02	359.12
Val-d'Or, QC	VLDR	48.05	282.22	58.00	358.74
Sept Iles, QC	SEPT	50.22	293.73	58.51	15.17

illustrates magnetic perturbations at 1200 UT. Similar to the prior event, the model does a decent job at predicting the overall envelope of the lower-frequency component of the geomagnetic fluctuations during this event, and over-predicts the magnetic perturbation at some stations at the center of this event. The best match is observed at Gillam (GILL), which is just north of where the model predicts a negative (southward) perturbation exceeding 825 nT.

The last time period shown in Figure 4.3 is not as extreme in magnitude, and it had an IMF in which the Y component, at about −15 nT, which was stronger than the Z component. This event on 3 August 2017 is very recent, occurring while this manuscript was in preparation. The map at the bottom shows the prediction for 2156 UT, close to when the stations at Gillamm (GILL) and Inukjauk (INUK) recorded their largest variations. While this event was in progress, the prediction from the numerical model running in real time at the NOAA SWPC was examined. Figure 4.4 shows the magnitude of **ΔB** at the same time as illustrated at the bottom of Figure 4.3, obtained from the web site at http://www.swpc.noaa.gov/products/geospace-ground-magnetic-perturbation-maps. Curiously, the numerical model had predicted no **ΔB** of significance at Gillamm and Inukjauk, and very little happening under 65° latitude. Archived plots or model output from past time periods are not presently available at the SWPC web site, so it is difficult to make anything other than cursory comparisons with the ground station data.

4.4. IMPROVING PREDICTION OF HIGHER FREQUENCY VARIATIONS

In Section 4.2, the results of three different studies performed by the CCMC were discussed. In the first model comparison (Pulkkinen et al., 2011), the difference between the models' and measured **ΔB** was compared directly with the models' output, with the nearest match producing the highest metric scores. In this study, the Virginia Tech empirical model often had the best score for "prediction efficiency" and second-best score for "root-mean-square difference," while performing much worse on "log-spectral distance." The empirical model's scores were even worse in the second CCMC study, in which only $|dB/dt|$ values were compared (Pulkkinen et al., 2013), although a few years later it was found that the model was handicapped by a programming error at CCMC. This problem was corrected during the third study (Glocer et al., 2016) that compared the range of variability, through Equation (4.1), and the empirical model performed better for some conditions. These

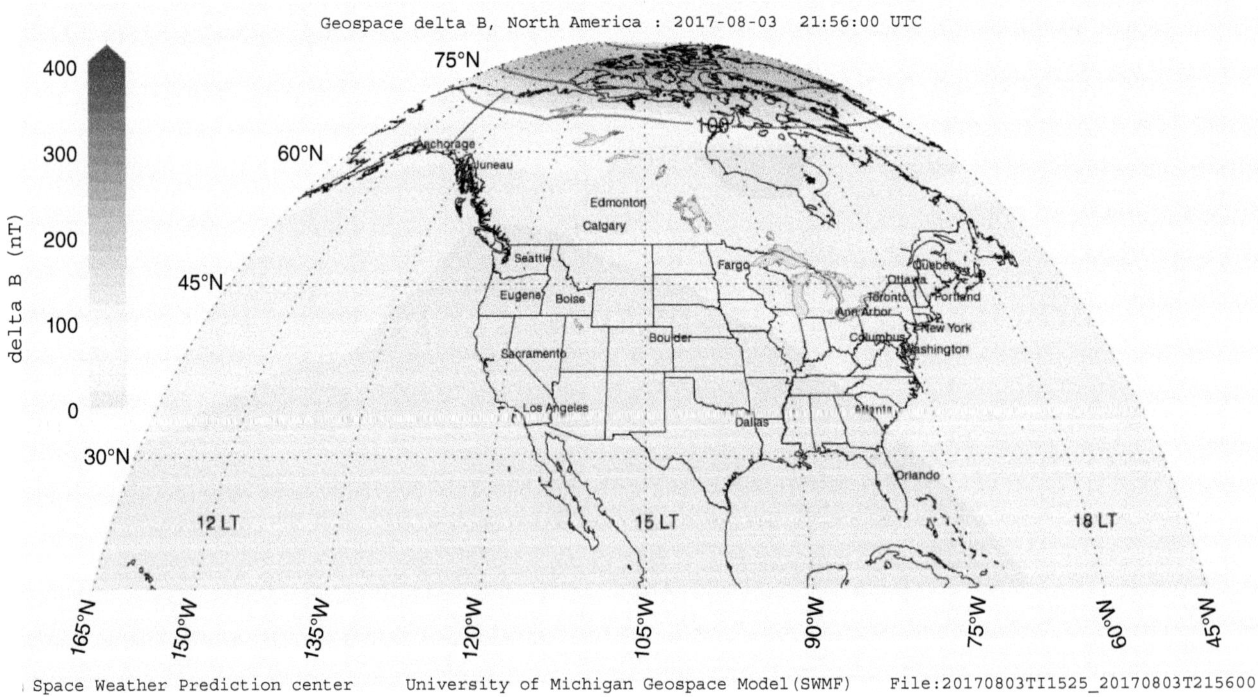

Figure 4.4 The SWPC/SWMF numerical model prediction for 3 August 2017, 21:56 UT. This map is from the NOAA Space Weather Prediction Center's web site that shows the predictions of the SWMF numerical model, using the real-time data stream from the DSCOVR spacecraft, at http://www.swpc.noaa.gov/products/geospace-ground-magnetic-perturbation-maps. (*See electronic version for color representation of this figure.*)

results demonstrate the high value that was placed on the spectral content and variability of $\Delta\mathbf{B}$, for predicting when an event may produce high GIC levels, while not necessarily predicting the signs and exact levels very well.

As shown in the examples in Figures 4.1–4.3, while the empirical model generally matches observations of the slowly-varying component of $\Delta\mathbf{B}$, the higher frequency variations are not reproduced well. As mentioned earlier in this paper and previously by Weimer (2013), the rapid variations may be beyond the ability of any model to predict with exact precision, as the locations of their sources in the magnetosphere are random.

Even so, the largest of these rapid variations tend to occur when and where the magnitude of the predicted $\Delta\mathbf{B}$ is high, as shown the illustrations by Weimer (2013). While exact predictions cannot be made, the results from the empirical model can indicate where large dB/dt are expected to occur, as these high-frequency fluctuations tend to scale in magnitude with the level of low-frequency variations that are better predicted.

To do better on dB/dt modeling and metric scores it would be necessary to account for the chaotic fluctuations, that are likely due to structure and motion of auroral currents in the ionosphere. Some MHD simulations have shown turbulent flows in the magnetosphere (Ge et al., 2012; White et al., 2001), and such turbulence is very likely the cause of the rapid variations. Along this same train of thought, Toth et al. (2014) had developed an empirical, power law relationship between B_H and dB_H/dt, having a correlation coefficient near 0.8. Knowing the spectral characteristics of the $\Delta\mathbf{B}$ is also important. For example, De Michelis et al. (2015) had looked at the latitudinal dependence of short timescale fluctuations; they used an "empirical mode decomposition'" method to separate magnetic field data into a set of "intrinsic mode functions," each having a characteristic frequency that represents a local oscillatory mode.

In order to improve the empirical prediction model it would be possible to address the issue of short period fluctuations using either Fourier or wavelet analysis (Xu et al., 2008, 2013). By decomposing geomagnetic data into time varying components with different frequency bands, a new statistical model of spectral amplitudes could be constructed, for all three orthogonal components. Such a model might complement the existing empirical model. Amplitudes at several frequency bands might be specified as a function of geomagnetic location and the solar wind/IMF parameters. The output from such a model could be used to reconstruct the spectral content for a specified location, provided with an input of the real-time IMF. Alternatively, a random time series of $\Delta\mathbf{B}$ having similar frequency components could be superposed with the values of $\Delta\mathbf{B}$ from the existing model. This would only resemble, but not exactly match, the perturbations from the random auroral motions. The result could have improved metric scores for dB/dt, spectral characteristics, and local K index.

A hybrid model combination like this might enable better regional estimates of magnetic perturbations, and thus better induced electric fields that are useful for indices and alerts. On the other hand, predictions that contain the high-frequency components may not be as useful for exact calculations of GIC within a specific network of conductors, as the variations are not likely to be coherent over long distances. For example, the dB_H/dt at one end of a long conductor might have a sign that is opposite to that on the other end, and these opposing signs of dB_H/dt could possibly cancel each other in the total induced voltage. Further studies are required to determine the coherency of the magnetic variations as a function of latitudinal and longitudinal length scales as well as frequency.

4.5. IMPROVING PREDICTIONS OF SUBSTORM PERTURBATIONS

Section 4.3 had examples of predictions from three events, and in the first event in Figure 4.1 there were two substorms that were not predicted/captured by the empirical model. These substorms are indicated where more than one station had substantial and simultaneous, negative perturbations in the North–South component of the magnetic field and lasting more than an hour. As the map at the bottom of Figure 4.1 shows, at this time the stations are near the division between the eastward and westward electrojets near local midnight, where substorms are most common.

With further development work it would be possible to improve the prediction efficiency of the empirical model by adding a substorm component. Preliminary work toward this goal was reported by Pothier et al. (2015), using the same database as was used in the development of the empirical model Weimer (2013). These data were searched for substorm signatures, and a list was created that included the time of onset and the most negative value of all perturbations within our database. This value is a pseudo-AL index that is referred to as AL^*, very much like the way that similar indices from the SuperMAG project are referred to as "SAL" (Newell and Gjerloev, 2011). These indices use many more stations than the official AL index that traditionally relies on a fixed set of magnetometers. After the subtraction of baseline levels obtained at substorm onset, the residual magnetic fields at all stations were fit using SCHA at 5 min time increments beyond onset. The result was an ensemble average of the substorm perturbation as a function of time, for all three vector components. The data used in the fits were sorted for different seasons and IMF orientations. One example of the results from Pothier et al.

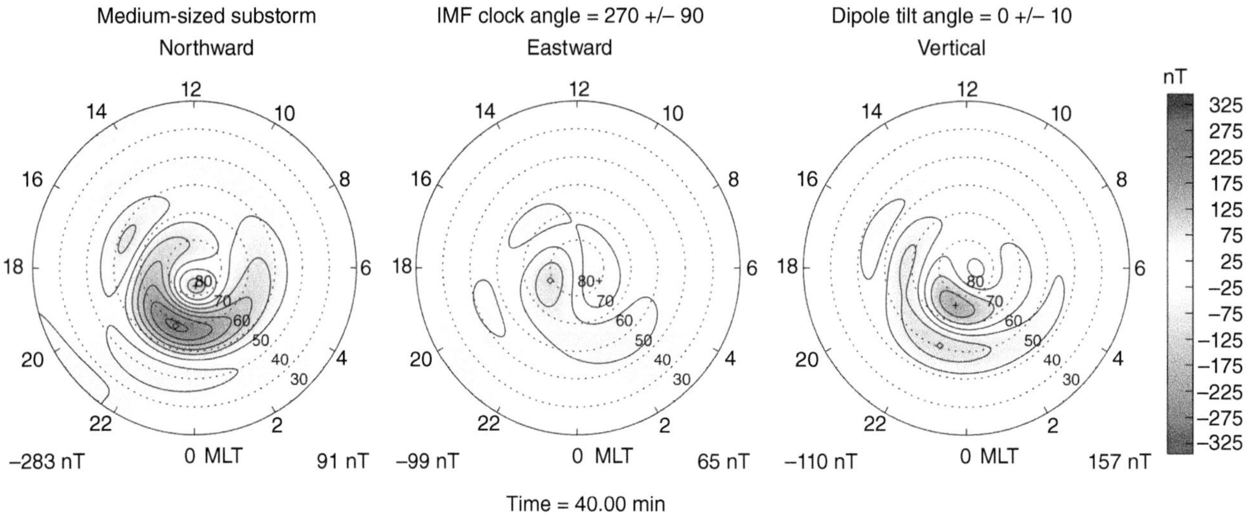

Figure 4.5 Maps of geomagnetic perturbations for a clock angle of 270° ± 90° ($-B_Y$ IMF) and dipole tilt angle of 0° ± 10° (near equinox). The measurements are from substorms having their greatest $|AL^*|$ index in the range of 300–500 nT and taken at 40 min after onset of the expansion phase. Each of the three vector components is shown. These plots show the results of least error fits using spherical cap harmonics, with the baseline (time = 0 min) removed. (*See insert for color representation of the figure.*)

(2015) is shown in Figure 4.5, for substorms that had their maximum AL^* in the range of 300–500 nT.

Substorms can be added to the prediction model by superposing their perturbation patterns as a function of time, using the substorm pattern that is appropriate for the solar wind and IMF conditions at that time, as well as their prior history. Of course, the trick is to know when to activate each substorm, as their timings and locations in local time are related to that of magnetic reconnections, which are also notoriously difficult to predict.

A number of substorm characteristics are known that can help. One fact is that most often substorms occur after moderate to strong solar wind driving, leading to the growth phase (Borovsky and Yakymenko, 2017; Liou et al., 2013; Newell and Liou, 2011), and related to polar cap (region of open field lines) expansion and contraction (Sotirelis et al., 2017). It is also known that substorms tend to occur with a periodicity of approximately 3 h, which seems to be the time of the substorm loading and unloading cycle (Borovsky et al., 1993). It has been confirmed by Borovsky and Yakymenko (2017) that the occurrence of periodic substorm is associated with strong driving of the magnetosphere by the solar wind, and that very strong activity leads to a shorter recurrence period.

Knowing these facts, it likely will be possible for an algorithm to determine when to activate a superposition of the substorm time sequence to the empirical prediction model. The use of neural networks (Weigel et al., 1999) for this purpose is feasible. Even if the substorm predictions are not precise in their timing, the result will improve the accuracy of "local K-index" and GIC alert forecasts.

4.6. SUMMARY

The primary advantages of using empirical models for the prediction of geomagnetic perturbations are computational cost and speed, which makes them useful for computing GIC in a variety of situations. One such model was developed at Virginia Tech with prior support from the National Space Weather Program (Weimer, 2013; Weimer et al., 2010). In a comparison to other types of prediction models done by the NASA CCMC, this model excelled in some types of metrics and did poorly in others. In the cases where the model did not do well on metric skill scores it was due to the inability to include chaotic, high-frequency fluctuations and substorm variations. Examples from three events have been shown, illustrating the empirical model's strengths and weaknesses. Two strategies have been discussed for the possibility of improving performances of future models. The addition of high-frequency fluctuations having the appropriate spectral characteristics for the given IMF conditions and location would improve some metric scores and indices used for issuing alerts. The forecasting ability of this model would also be improved by superposing substorm disturbances at the appropriate times.

ACKNOWLEDGMENTS, SAMPLES, AND DATA

The development of the empirical model was supported by NSF grant ATM-0817751 to Virginia Tech, from March 2009 to January 2013. Preparation of this manuscript was supported by NSF grant AGS-1638270. Measurements from the DSCOVR spacecraft were obtained from the National Centers for Environmental Information at https://www.ngdc.noaa.gov/dscovr. Figure 4.4 was obtained from the NOAA Space Weather Prediction Centerwebsiteathttp://www.swpc.noaa.gov/products/geospace-ground-magnetic-perturbation-maps. The magnetometer data shown in Figures 4.1–4.3 were distributed by the THEMIS mission at http://themis.ssl.berkeley.edu/gmag. We acknowledge NASA contract NAS5-02099 and V. Angelopoulos for use of data from the THEMIS Mission web site. The following individuals and grants are acknowledged for providing magnetometer data through the THEMIS Mission: S. Mende, E. Donovan and the Canadian Space Agency for logistical support in fielding and data retrieval from the Ground Based Observatory stations, and NSF for support through grant AGS-1004736. C. T. Russell and NSF support through grant AGS-1004814. I. R. Mann, D. K. Milling and the rest of the Canadian Array for Realtime Investigations of Magnetic Activity (CARISMA) team; CARISMA is operated by the University of Alberta, funded by the Canadian Space Agency. Martin Connors and the team at the University of Athabasca. The USGS Geomagnetism Program (http://geomag.usgs.gov).

REFERENCES

Baker, K. B., & Wing, S. (1989), A new magnetic coordinate system for conjugate studies at high latitudes, *J. Geophys. Res.*, *94*, 9139.

Borovsky, J. E., & Yakymenko, K. (2017), Substorm occurrence rates, substorm recurrence times, and solar wind structure, *J. Geophys. Res. Space Physics*, *122*, 2973–2998, https://doi.org/10.1002/2016JA023625

Borovsky, J. E., Nemzek, R. J., & Belian, R. D. (1993), The occurrence rate of magnetospheric-substorm onsets: Random and periodic substorms, *J. Geophys. Res. Space Physics*, *98*, 3807–3813, https://doi.org/10.1029/92JA02556

Boteler, D. H. (2001), Space weather effects on power systems, in *Space Weather*, Geophysical Monograph Series, vol. *125*, edited by P. Song, H. J. Singer, & G. L. Siscoe, pp. 347–352, American Geophysical Union, Washington, DC.

De Michelis, P., Consolini, G., & Tozzi, R. (2015), Latitudinal dependence of short timescale fluctuations during intense geomagnetic storms: A permutation entropy approach, *J. Geophys. Res.*, *120*, 5633–5644, https://doi.org/10.1002/2015JA021279

Emmert, J. T., Richmond, A. D., & Drob, D. P. (2010), A computationally compact representation of Magnetic Apex and Quasi Dipole coordinates with smooth base vectors, *J. Geophys. Res.*, *115*, A08322, https://doi.org/10.1029/2010JA015326

Ge, Y. S., Zhou, X.-Z., Liang, J., Raeder, J., Gilson, M. L., Donovan, E., Angelopoulos, V., & Runov, A. (2012), Dipolarization fronts and associated auroral activities: 1. Conjugate observations and perspectives from global MHD simulations, *J. Geophys. Res.*, *117*, A10226, https://doi.org/10.1029/2012JA017676

Glocer, A., Rastätter, L., Kuznetsova, M., Pulkkinen, A., Singer, H. J., Balch, C., Weimer, D., Welling, D., Wiltberger, M., Raeder, J., Weigel, R. S., McCollough, J., & Wing, S. (2016), Community-wide validation of geospace model local k-index predictions to support model transition to operations, *Space Weather*, *14*, 469–480.

Gummow, R. A. (2002), GIC effects on pipeline corrosion control systems, *J. Atmos. Sol. Terr. Phys.*, *64*, 1755.

Haaland, S., Sönnerup, B. U. O., Dunlop, M. W., Balogh, A., Georgescu, E., Hasegawa, H., Klecker, B., Paschmann, G., Puhl-Quinn, P., Rème, H., Vaith, H., & Vaivads, A. (2004), Four-spacecraft determination of magnetopause orientation, motion and thickness: Comparison with results from single-spacecraft methods, *Ann. Geophys.*, *22*, 1347–1365.

Haines, G. V. (1985), Spherical cap harmonic analysis, *J. Geophys. Res.*, *90*(B3), 2583, https://doi.org/10.1029/JB090iB03p02583

Lanzerotti, L. J. (2001), Space weather effects on technologies, in *Space Weather*, Geophysical Monograph Series, vol. *125*, edited by P. Song, H. J. Singer, & G. L. Siscoe, pp. 11–22, American Geophysical Union, Washington, DC.

Liou, K., Newell, P. T., Zhang, Y.-L., & Paxton, L. J. (2013), Statistical comparison of isolated and non-isolated auroral substorms, *J. Geophys. Res. Space Physics*, *118*(5), 2466–2477, https://doi.org/10.1002/jgra.50218

McComas, D. J., Bame, S. J., Barber, P., Feldman, W. C., Phillips, J. L., & Riley, P. (1998), Solar wind electron, proton, and alpha monitor (SWEPAM) on the Advanced Composition Explorer, *Space Sci. Rev.*, *86*, 563.

Newell, P. T., & Gjerloev, J. W. (2011), Evaluation of SuperMag auroral electrojet indices as indicators of substorms and auroral power, *J. Geophys. Res. Space Physics*, *116*(A12), a12211, https://doi.org/10.1029/2011JA016779.

Newell, P. T., & Liou, K. (2011), Solar wind driving and substorm triggering, *J. Geophys. Res. Space Physics*, *116*(A3), 229, https://doi.org/10.1029/2010JA016139

Pothier, N. M., Weimer, D. R., & Moore, W. (2015), Quantitative maps of geomagnetic perturbation vectors during substorm onset and recovery, *J. Geophys. Res.*, *120*, 1197–1214, https://doi.org/10.1002/2014JA020602

Pulkkinen, A., Kuznetsova, M., Ridley, A., Raeder, J., Vapirev, A., Weimer, D., Weigel, R. S., Wiltberger, M., Millward, G., Rastätter, L., Hesse, M., Singer, H. J., & Chulaki, A. (2011), Geospace environment modeling 2008–2009 challenge: Ground magnetic field perturbations, *Space Weather*, *9*, S02004, https://doi.org/10.1029/2010SW000600

Pulkkinen, A., Rastätter, L., Kuznetsova, M., Singer, H., Balch, C., Weimer, D., Toth, G., Ridley, A., Gombosi, T., Wiltberger, M., Raeder, J. & Weigel, R., (2013), Community-wide geospace model validation to support model transition to operations, *Space Weather*, *11*, 369–385, https://doi.org/10.1002/swe.20056

Richmond, A. D. (1995), Ionospheric electrodynamics using magnetic apex coordinates, *J. Geomagn. Geoelectr.*, *47*, 191, https://doi.org/10.5636/jgg.47.191

Smith, C. W., L'Heureux, J., Ness, N. F., Acuna, M. H., Burlaga, L. F., & Scheifele, J. (1998), The ACE magnetic field experiment, *Space Sci. Rev.*, *86*, 613.

Sönnerup, B. U. O., Haaland, S., Paschmann, G., Lavraud, B., Dunlop, M. W., Rème, H., & Balogh, A. (2004), Orientation and motion of a discontinuity from single-spacecraft measurements of plasma velocity and density: Minimum mass flux residue, *J. Geophys. Res.*, *109*, A03221, https://doi.org/10.1029/2003JA010230

Sotirelis, T., Keller, M. R., Liou, K., Smith, D., Barnes, R. J., Talaat, E., & Baker, J. B. H. (2017), Testing the expanding-contracting polar cap paradigm, *J. Geophys. Res. Space Physics*, *122*(7), 7077–7086, https://doi.org/10.1002/2017JA024238

Tanskanen, E. I., Viljanen, A., Pulkkinen, T. I., Pirjola, R., Häkkinen, L., Pulkkinen, A., & Amm, O. (2001), At substorm onset, 40% of AL comes from underground, *J. Geophys. Res.*, *106*(A7), 13119, https://doi.org/10.1029/2000JA900135

Toth, G., Meng, X., Gombosi, T. I., & Rastätter, L. (2014), Predicting the time derivative of local magnetic perturbations, *J. Geophys. Res.*, *119*, 310–321, https://doi.org/10.1002/2013JA019456

VanZandt, T. E., Clark, W. L., & Warnock, J. M. (1972), Magnetic apex coordinates: A magnetic coordinate system for the ionospheric f_2 layer, *J. Geophys. Res.*, *77*, 2406, https://doi.org/10.1029/JA077i013p02406

Weigel, R. S., Horton. W., Tajima, T., & Detman, T. (1999), Forecasting auroral electrojet activity from solar wind input with neural networks, *Geophys. Res. Lett.*, *26*(10), 1353–1356, https://doi.org/10.1029/1999GL900280

Weimer, D. R. (2013), An empirical model of ground-level geomagnetic perturbations, *Space Weather*, *11*, 107–120, https://doi.org/10.1002/swe.20030

Weimer, D. R., & King, J. H. (2008), Improved calculations of interplanetary magnetic field phase front angles and propagation time delays, *J. Geophys. Res.*, *113*, A01105, https://doi.org/10.1029/2007JA012452

Weimer, D. R., Ober, D. M., Maynard, N. C., Burke, W. J., Collier, M. R., McComas, D. J., Nagai, T., & Smith, C. W. (2002), Variable time delays in the propagation of the interplanetary magnetic field, *J. Geophys. Res.*, 107, A8, https://doi.org/10.1029/2001JA009102

Weimer, D. R., Clauer, C. R., Engebretson, M. J., Hansen, T. L., Gleisner, H., Mann, I., & Yumoto, K. (2010), Statistical maps of geomagnetic variations as a function of the interplanetary magnetic field, *J. Geophys. Res.*, *115*, A10320, https://doi.org/10.1029/2010JA015540

White, W., Schoendorf, J., Siebert, K., Maynard, N., Weimer, D., Wilson, G., Sönnerup, B., Siscoe, G., & Erickson, G. (2001), MHD simulation of magnetospheric transport at the mesoscale, in *Space Weather*, Geophysical Monograph Series, vol. *125*, edited by P. Song, H. J. Singer, & G. L. Siscoe, pp. 229–240, American Geophysical Union, Washington, DC.

Xu, Z., Zhu, L., Sojka, J., Kokoszka, P., & Jach, A. (2008), An assessment study of the wavelet-based index of magnetic storm activity (WISA) and its comparison to the Dst index, *J. Atmos. Sol. Terr. Phys.*, *70*(11–12), 1579–1588.

Xu, Z., Gannon, J. L., & Rigler, E. (2013), Report of geomagnetic pulsation indices for space weather applications, *Open-File Report 2013–1166*, U.S. Geological Survey.

5

Geoelectric Field Generation by Field-Aligned Currents

J. R. Woodroffe

ABSTRACT

In this paper, we describe a model of the geoelectric fields produced by field-aligned currents (FACs) that are incident on the Earth's ionosphere. Using this model, we demonstrate how FACs produce ground-level magnetic perturbations through the Hall effect and show how a finite ground conductivity results in nonzero geoelectric fields. We apply our model to determine the spatial structure of geoelectric fields that result from representative FAC structures and compare the results that are obtained from the standard combination of Biot–Savart integration and a one-dimensional magnetotelluric method ("the Biot–Savart approach"). We find that the accuracy of the Biot–Savart approach is frequency- and conductivity-dependent, and that it is a particularly poor approximation at frequencies above 30 mHz. Our findings indicate that the Biot–Savart approach is generally sufficient for estimating the impacts of long-period, large-scale disturbances, but that it is likely to produce significant inaccuracies when applied to localized and/or temporally variable structures such as those that might be associated with substorms. Finally, we discuss some scenarios that could result in the production of spatially localized geoelectric fields, revealing how particular configurations of FAC are associated with localized and enhanced geoelectric fields.

Key Points

- Finite ground conductivity is required for nonzero geoelectric fields.
- Biot–Savart integration of ionospheric currents may give inaccurate results.
- Geoelectric field localization is likely due to local closure of ionospheric currents.

5.1. INTRODUCTION

Electromagnetic signatures that originate in the magnetosphere–ionosphere system that are observable on the ground have a variety of different names, depending on the field of interest. Space physics has typically referred to these as magnetic pulsations whereas hazards and applications studies refer to them as geo(electro) magnetic disturbances (GMDs); the present study will adopt this latter terminology. Whatever name is used to refer to them, it has lately been recognized that GMDs are among the most significant space weather threats to our technological society (see e.g., Knipp, 2015, and works therein).

The dynamics responsible for producing these signatures are largely describable using the formalism of magnetohydrodynamics (MHD). Small-scale effects are typically important only within a limited region of space near the magnetospheric disturbance, and as discussed in example, Lysak and Yoshikawa (2006), disturbances whose scale lengths approach the collisional depth in the ionosphere are potentially subject to strong attenuation. It is a fundamental result of MHD theory that time-varying field aligned currents (FACs) can be interpreted as a superposition of Alfvén waves, and so analysis of these currents

Space Science and Applications, Los Alamos National Laboratory, Los Alamos, NM, USA

Geomagnetically Induced Currents from the Sun to the Power Grid, Geophysical Monograph 244, First Edition.
Edited by Jennifer L. Gannon, Andrei Swidinsky, and Zhonghua Xu.

and their subsequent production of ground-observable signatures can leverage considerable theoretical work that has already been done in this area of ultra-low frequency (ULF) wave propagation and ionospheric transmission.

For the purposes of understanding and modeling geoelectric fields, much of the existing ULF wave theory is limited by the common assumption of infinite ground conductivity. For example, a thorough analysis of trans-ionospheric ULF wave propagation was presented in Sciffer and Waters (2002), their assumption of infinite ground conductivity requires the geoelectric field to vanish at the Earth's surface. Other work has considered this problem in a more general sense, most notably Alperovich and Fedorov (2007), but even that work did not pay any particular attention to the geoelectric field.

The goal of the present work is to develop a physically representative and analytically tractable model of ground-level electric field (geoelectric) fields associated with FACs that are incident upon the Earth's ionosphere. Because such a model is simplest in Fourier space but most useful in physical space, we will use illustrative examples of GMDs resulting from characteristic FAC patterns to demonstrate how various parameters affect GMD properties. We will then close by discussing the implications of our results for standard methods of geoelectric field estimation and understanding the recently-identified phenomena of localized geoelectric fields (Ngwira et al., 2015).

5.2. A MODEL OF GEOELECTRIC FIELD GENERATION

5.2.1. Model Description

For simplicity, we assume the background magnetic field to uniformly directed along the $\hat{z}$ direction, $\mathbf{B}_0 = B_0\hat{z}$. The magnetosphere is taken to be a uniform plasma for all $z > 0$. The ionosphere is an infinitely thin conducting layer at $z = 0$, with its electrodynamic response described by an anisotropic conductance; the fields above and below this layer are related by "jump conditions" (Fujita and Tamao, 1988; Lysak and Yoshikawa, 2006),

$$[\hat{z} \cdot \mathbf{B}] = 0 \tag{5.1}$$

$$[\hat{z} \times \mathbf{B}] = \mu_0 \left(\Sigma_P \mathbf{E} + \Sigma_H \left(\hat{z} \times \mathbf{E} \right) \right) \tag{5.2}$$

where [...] indicates the difference of a quantity above and below the ionosphere and $\Sigma_{H,P}$ are the Hall and Pedersen conductances.

Below the ionosphere is an atmospheric layer of width H and below that is the solid ground, occupying all $z < -H$. We assume a two-layer model of the ground, with a finite-width layer of conductivity σ_1 and width Δ on top of a half-infinite region of conductivity σ_2. All regions are assumed to be horizontally uniform. The physical domain is shown schematically in Figure 5.1.

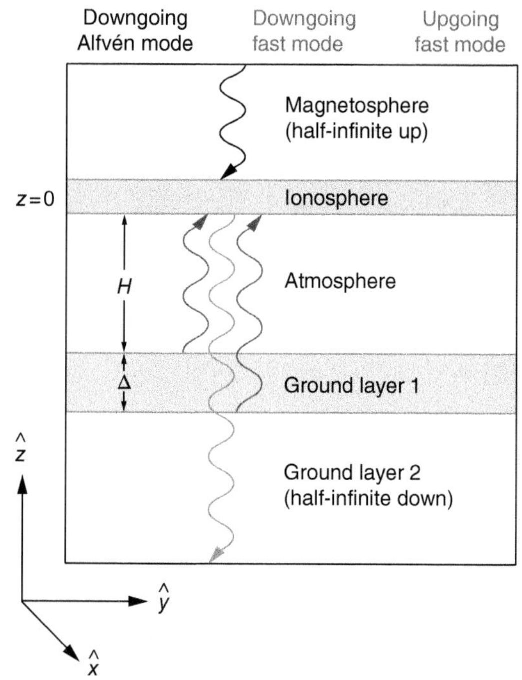

Figure 5.1 Cartoon illustration of the domain and coordinate system of the model described in this paper. (*See electronic version for color representation of this figure.*)

Although the entire domain is subdivided into multiple regions of finite vertical extent, each of these regions is itself vertically uniform. Combined with the horizontal uniformity, this allows us to express an arbitrary quantity as a the superposition of plane wave components via Fourier analysis. Allowing for the possibility of reflection from boundaries, we can write the plane wave solution for a quantity, ψ, in any given region as

$$\psi = \left(\psi_i e^{-ik_z z} + \psi_r e^{ik_z z} \right) e^{i(\mathbf{k}_\perp \cdot \mathbf{x}_\perp - \omega t)} \tag{5.3}$$

where k_z is the component of the wave vector in the direction of the background magnetic field, $\mathbf{k}_\perp = k_x\hat{x} + k_y\hat{y}$ is the component of the wavevector that perpendicular to the background magnetic field, and the subscripts $_{i,r}$ indicate the incident (downgoing) and reflected (upgoing) components of the wave.

5.2.2. Characteristic Variables

We begin with Maxwell's equations, including polarization current in Ampere's law. This gives us the following set of vector differential equations,

$$\frac{\partial \mathbf{B}}{\partial t} = -\nabla \times \mathbf{E} \tag{5.4}$$

$$\frac{\epsilon}{c^2}\frac{\partial \mathbf{E}}{\partial t} = \nabla \times \mathbf{B} - \mu_0 \mathbf{J}. \tag{5.5}$$

Because these equations are derived from ideal MHD, it is also implicit within these equations that $\mathbf{E}\cdot\mathbf{B}=0$ or, equivalently, $\mathbf{E}\cdot\hat{z}=0$.

The material response is contained within the dielectric constant for each medium, which is given in the magnetosphere by $\epsilon = 1 + c^2/V_A^2$ and in the ground by $\epsilon = 1 + i\sigma_l/\epsilon_0\omega$, where V_A is the Alfvén speed and ω is the wave (angular) frequency.

For our analysis, it is convenient to introduce four "characteristic variables,"

$$\zeta = \nabla \cdot \mathbf{E} \tag{5.6}$$

$$J = (\nabla \times \mathbf{B})\cdot\hat{z} \tag{5.7}$$

$$\Gamma = (\nabla \times \mathbf{E})\cdot\hat{z} \tag{5.8}$$

$$B = \mathbf{B}\cdot\hat{z}. \tag{5.9}$$

Aside from minor differences in notation and constant scaling factors, these are the same quantities used in a study of trans-ionospheric wave propagation by Lysak (1997). Using these characteristic variables, Equations (5.4) and (5.5) can be written as

$$\frac{\epsilon}{c^2}\frac{\partial \zeta}{\partial t} = -\frac{\partial J}{\partial z} \tag{5.10}$$

$$\frac{\partial J}{\partial t} = -\frac{\partial \zeta}{\partial z} \tag{5.11}$$

$$\frac{\epsilon}{c^2}\frac{\partial \Gamma}{\partial t} = -\nabla^2 B \tag{5.12}$$

$$\frac{\partial B}{\partial t} = -\Gamma. \tag{5.13}$$

From these expressions, we can derive two uncoupled wave equations,

$$\frac{\epsilon}{c^2}\frac{\partial^2 \zeta}{\partial t^2} = \frac{\partial^2 \zeta}{\partial z^2} \tag{5.14}$$

$$\frac{\epsilon}{c^2}\frac{\partial^2 \Gamma}{\partial t^2} = \nabla^2 \Gamma. \tag{5.15}$$

The first of these equations describes the propagation of signals entirely in the field-aligned direction and the second describes an isotropically-propagating signal; these correspond to the Alfvén and Fast modes, respectively. Using the plane wave ansatz from before, we can use Equations (5.14) and (5.15) to derive the following well-known dispersion relations in a uniform medium:

$$\omega = k_{\parallel} c_A \tag{5.16}$$

$$\omega = \sqrt{k_\perp^2 + k_\parallel^2}\, c_A \tag{5.17}$$

where in this case $k_\perp^2 = k_x^2 + k_y^2$ is the squared magnitude of the horizontal wave vector and $c_A = c/\sqrt{\epsilon}$ is the Alfvén speed with displacement current correction. In terms of our characteristic variables, the boundary conditions between any two regions are given by

$$[B] = 0 \tag{5.18}$$

$$\left[\frac{\partial B}{\partial z}\right] = -\mu_0(\Sigma_P \Gamma + \Sigma_H \zeta) \tag{5.19}$$

$$[J] = -\mu_0(\Sigma_P \zeta - \Sigma_H \Gamma) \tag{5.20}$$

where the latter two equations were derived by taking the parallel curl and divergence of Equation (5.2). It is important to note that these equations can be applied across any infinitesimal (i.e., small compared to wavelength) transition – allowing us to include, for example, a finite jump in magnetospheric Alfvén speed – but the RHS at each point is equal to zero except across the ionosphere, as this is the only location where the there is finite conductance in an "infinitesimally thin" region at $z=0$.

It will be useful going forward to express the jump conditions in terms of the electric field quantities, ζ and Γ, only. Using the definitions provided above and assuming harmonic time dependence, we have

$$[\Gamma] = 0 \tag{5.21}$$

$$\left[\frac{\partial \Gamma}{\partial z}\right] = -i\mu_0\omega(\Sigma_P \Gamma + \Sigma_H \zeta) \tag{5.22}$$

$$\left[\frac{\partial \zeta}{\partial z}\right] = i\mu_0\omega(\Sigma_P \zeta - \Sigma_H \Gamma). \tag{5.23}$$

5.2.3. The Solution Below the Ionosphere

The atmosphere is, to a very good level of approximation, a perfectly insulating medium. As a consequence, we can take $\epsilon = 1$ in this region and we can ignore the flow of vertical currents, thus implying $J=0$ and $\zeta=0$. This latter condition provides an extreme simplification of the problem below the atmosphere, allowing us to focus entirely on the Fast mode.

At the interface between the conducting layers within the ground, we must enforce continuity, since $[\Gamma]=0$ and $[\partial\Gamma/\partial z]=0$. Because the bottom conducting layer is half-infinite, we do not expect there to be any reflected components, and the magnitude of the upgoing component in this region can be set to zero. Evaluating the jump conditions at $z=-D$ (where $D=H+\Delta$) leads us to the following set of equations:

$$\Gamma_2 e^{ik_2 D} = \Gamma_1\left(e^{ik_1 D} + R_{12} e^{-ik_1 D}\right) \tag{5.24}$$

$$-ik_2\Gamma_2 e^{ik_2 D} = -ik_1\Gamma_1\left(e^{ik_1 D} - R_{12}e^{-ik_1 D}\right) \quad (5.25)$$

here k_i is the vertical wavenumber in region i and $R_{12}=\Gamma_{1,r}/\Gamma_{1,i}$ is the reflection coefficient, defined as the amplitude ratio of the incoming to outgoing components of the wave.

Combining Equations (5.24) and (5.25), we obtain

$$R_{12} = \frac{k_1 - k_2}{k_1 + k_2} e^{2ik_1 D}. \quad (5.26)$$

It is straightforward to show that

$$k_i = \sqrt{\frac{\omega^2}{c^2}\left(1 + i\frac{\sigma_i}{\epsilon_0\omega}\right) - k_\perp^2}. \quad (5.27)$$

It is generally the case that low frequencies are most important for storm-time geoelectric fields (see e.g., Winter et al., 2017), so assuming frequencies of ~ 1 Hz or less we have $\sigma_i \gg \epsilon_0\omega$ and, consequently,

$$k_i = \frac{1+i}{\delta_i}\sqrt{1 + i\frac{k_\perp^2\delta_i^2}{2}} \quad (5.28)$$

where $\delta_i = \sqrt{2/\mu_0\sigma_i\omega}$ is the conductor skin depth in the corresponding region. From this expression, it follows that horizontal variations only affect the characteristics of the ground solution when they correspond to wavelengths comparable to or smaller than the skin depth.

At the atmosphere–ground interface, we again have to enforce continuity of both Γ and $\partial\Gamma/\partial z$, but now there are upgoing and downgoing components in both regions:

$$\Gamma_1\left(e^{ik_1 H} + R_{12}e^{-ik_1 H}\right) = \Gamma_{atm}\left(e^{ik_{atm}H} + R_g e^{-ik_{atm}H}\right) \quad (5.29)$$

$$-ik_1\Gamma_1\left(e^{ik_1 H} - R_{12}e^{-ik_1 H}\right) = -ik_{atm}\Gamma_{atm}\left(e^{ik_{atm}H} - R_g e^{-ik_{atm}H}\right). \quad (5.30)$$

We now have $k_{atm} = \sqrt{\omega^2/c^2 - k_\perp^2}$ and $R_g=\Gamma_{atm,r}/\Gamma_{atm,i}$ is the coefficient of reflection of the signal from the compound ground. For frequencies and wavelengths of interest it is always the case that $k_\perp \gg \omega/c$, so we have $k_{atm} = ik_\perp$; thus, the atmospheric wave is evanescent. Using the jump equations applied to the atmosphere–ground interface, we can solve for the ground reflection coefficient R_g, finding

$$R_g = \frac{k_{atm} - k_g}{k_{atm} + k_g} e^{2ik_{atm}H} \quad (5.31)$$

where k_g is an effective wavenumber in the structured ground that is given by

$$k_g = k_1\left(\frac{1 - R_{12}e^{-2ik_1 H}}{1 + R_{12}e^{-2ik_1 H}}\right) = k_1\left(\frac{k_2 - ik_1\tan(k_1\Delta)}{k_1 - ik_2\tan(k_1\Delta)}\right). \quad (5.32)$$

Consequently, the total wave in the atmosphere is given by

$$\Gamma_{atm} = \Gamma_i\left(e^{ik_{atm}z} + R_g e^{-ik_{atm}z}\right). \quad (5.33)$$

At $z=0$, the jump condition tells us $\Gamma_{atm}=\Gamma_{mag}$, where Γ_{mag} is the total field above the ionosphere. This allows us to determine Γ_i,

$$\Gamma_i = \frac{\Gamma_{mag}}{1 + R_g} \quad (5.34)$$

the Appendix describes how this method that can be applied to an arbitrary number of layers, demonstrating that it is actually a generalization of the standard one-dimensional magnetotelluric method. The atmospheric field is thus given in terms of the magnetospheric field by

$$\Gamma_{atm} = \Gamma_{mag}\frac{e^{-ik_{atm}z} + R_g e^{ik_{atm}z}}{1 + R_g}. \quad (5.35)$$

In the limit of $\sigma \to \infty$, $R_g \to -\exp(-2ik_{atm}H)$ and Equation (5.35) becomes

$$\Gamma_{atm} = \Gamma_{mag}\frac{\sinh\left(k_{atm}(z+H)\right)}{\sinh\left(k_{atm}H\right)}, \quad (5.36)$$

which is equivalent to equation 36 of *Lysak and Yoshikawa* (2006). Evaluated at the ground ($z=-H$), Equation (5.35) yields

$$\Gamma_{gnd} = \frac{k_{atm}\Gamma_{mag}}{k_{atm}\cos\left(k_{atm}H\right) + ik_g\sin\left(k_{atm}H\right)} \quad (5.37)$$

which demonstrates that the ground signal is determined by a complex but well-defined mixture of the atmospheric input and ground response (the compressional magnetic field can be obtained by simply substituting $\Gamma \to B$).

Finally, we will later find it useful to know the vertical derivative of this field evaluated below the ionosphere,

$$\left(\frac{\partial\Gamma}{\partial z}\right)_{atm} = -ik_{atm}\Gamma_{mag}\frac{1 - R_g}{1 + R_g}. \quad (5.38)$$

We must now continue above the ionosphere in order to complete our solution.

5.2.4. The Solution Above the Ionosphere

The solution above the ionosphere is much more complex than the solution below the atmosphere because both the Alfvén and Fast modes are present in this region, and they are coupled by the anisotropically-conducting ionosphere.

Let us suppose we are dealing with an incident wave with horizontal wavevector $\mathbf{k}_\perp$ above the ionosphere. Due to the horizontal uniformity of the media, we may treat $k_\perp$ as a constant throughout the model domain. Since both $k_\perp$ and ω are constant, it follows from the dispersion relationships that the Alfvén and Fast modes have different vertical wavenumbers that are given by

$$k_{z,A} = \frac{\omega}{c_A} \equiv k_A \qquad k_{z,F} = \sqrt{\frac{\omega^2}{c_A^2} - k_\perp^2} \equiv k_F. \quad (5.39)$$

As previously discussed, we can express each wave mode as a superposition of upgoing and downgoing plane waves; for a single $(\omega,\mathbf{k}_\perp)$ pair we have

$$\zeta = \left(\zeta_i e^{-ik_A z} + \zeta_r e^{ik_A z}\right) e^{i(\mathbf{k}_\perp \cdot \mathbf{x}_\perp - \omega t)} \quad (5.40)$$

$$\Gamma = \left(\Gamma_i e^{-ik_F z} + \Gamma_r e^{ik_F z}\right) e^{i(\mathbf{k}_\perp \cdot \mathbf{x}_\perp - \omega t)}. \quad (5.41)$$

Evaluated at $z=0$, these equations give us $\zeta=\zeta_i+\zeta_r$ and $\Gamma=\Gamma_i+\Gamma_r$, as should be expected. Using Equations (5.40) and (5.41), we can write the vertical derivatives of ζ and γ (evaluated at $z=0$) as

$$\frac{\partial \zeta}{\partial z} = -ik_A\left(\zeta_i - \zeta_r\right) \quad (5.42)$$

$$\frac{\partial \Gamma}{\partial z} = -ik_F\left(\Gamma_i - \Gamma_r\right). \quad (5.43)$$

We can then write the jump conditions in terms of incident and reflected components of ζ and Γ, finding

$$\Sigma_A\left(\zeta_i - \zeta_r\right) = \Sigma_P\left(\zeta_i + \zeta_r\right) - \Sigma_H\left(\Gamma_i + \Gamma_r\right) \quad (5.44)$$

$$\Sigma_F\left(\Gamma_i - \Gamma_r\right) - \Sigma_{atm}\left(\Gamma_i + \Gamma_r\right) = \Sigma_P\left(\Gamma_i + \Gamma_r\right) + \Sigma_H\left(\zeta_i + \zeta_r\right) \quad (5.45)$$

where we have introduced the "wave conductances"

$$\Sigma_{A,F} = \frac{k_{A,F}}{\mu_0 \omega} \quad (5.46)$$

and the "atmospheric conductance,"

$$\Sigma_{atm} = \frac{k_{atm}}{\mu_0 \omega}\left(\frac{1-R_g}{1+R_g}\right). \quad (5.47)$$

Equations (5.44) and (5.45) can be combined into a single matrix equation that can be solved to obtain the reflected components as a function of the incident components. Doing this, we find

$$\begin{pmatrix} \zeta_r \\ \gamma_r \end{pmatrix} = \begin{pmatrix} R_{A\to A} & R_{F\to A} \\ R_{A\to F} & R_{F\to F} \end{pmatrix} \cdot \begin{pmatrix} \zeta_i \\ \gamma_i \end{pmatrix} \quad (5.48)$$

where the individual components of the "reflection matrix" are

$$R_{A\to A} = \frac{\Sigma_A - \Sigma_P - \Sigma_{FA}}{\Sigma_A + \Sigma_P + \Sigma_{FA}} \quad (5.49)$$

$$R_{F\to F} = \frac{\Sigma_F - \Sigma_P - \Sigma_{AF}}{\Sigma_F + \Sigma_P + \Sigma_{AF}} \quad (5.50)$$

$$R_{A\to F} = -\frac{\Sigma_{FA}}{\Sigma_H}\left(1 + R_{A\to A}\right) \quad (5.51)$$

$$R_{F\to A} = \frac{\Sigma_H}{\Sigma_P + \Sigma_A}\left(1 + R_{F\to F}\right) \quad (5.52)$$

and the two auxiliary conductances, Σ_{FA} and Σ_{AF}, are defined as

$$\Sigma_{FA} = \frac{\Sigma_H^2}{\Sigma_F + \Sigma_P + \Sigma_{atm}} \quad (5.53)$$

$$\Sigma_{AF} = \Sigma_{atm} + \frac{\Sigma_H^2}{\Sigma_A + \Sigma_P}. \quad (5.54)$$

These expressions are formally identical to those of Lysak and Yoshikawa (2006), but by using our definitions of the various quantities, they now allow for a nonzero geoelectric field.

5.2.5. The Geoelectric Field

We showed earlier than $\Gamma(z=-H)$ is given by Equation (5.37). It is instructive to examine the behavior of this function for $k_{atm} \to i\,|k_\perp|$, as this approximation should frequently be valid. Using the identities $\cos(ix)=\cosh(x)$ and $\sin(ix)=i\sinh(x)$, we have

$$\Gamma_{gnd} \approx \frac{\Gamma_{mag}}{\cosh\left(|k_\perp|H\right)} \frac{1}{1-\left(k_g/|k_\perp|\right)\tanh\left(|k_\perp|H\right)}. \quad (5.55)$$

If $|k_\perp|H \gg 1$, then $1/\cosh\left(|k_\perp|H\right) \sim \exp\left(-|k_\perp|H\right)$. This indicates that the field is exponentially attenuated with a vertical scale length proportional to the horizontal scale size of the disturbance – smaller magnetospheric scale sizes are more strongly attenuated than larger scale sizes. It is also important to note that the ground-level value of Γ (or equivalently B) is linearly proportional to the magnetospheric value.

Assuming that the disturbance is originally due to an ionospherically-incident field-aligned current, we can write Equation (5.37) as

$$\Gamma_{gnd} = -\frac{c_A k_{atm}}{k_{atm}\cos\left(k_{atm}H\right) + ik_g \sin\left(k_{atm}H\right)} R_{A\to F} J_i \quad (5.56)$$

where we have used $\Gamma_{mag} = R_{A\to F}\zeta_i$, and used Equation (5.10) to replace ζ_i by J_i while assuming that the incident FAC is expressible as a superposition of downgoing plane waves such that $\partial J/\partial z \to -ik_A J$.

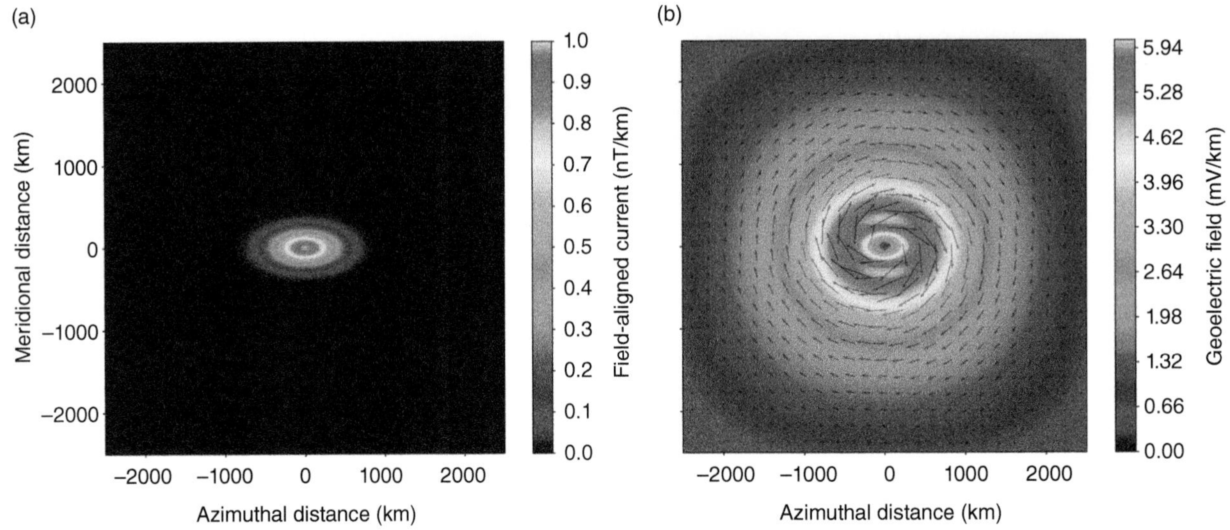

Figure 5.2 Two-dimensional profile of (a) the ionospherically-incident localized Gaussian FAC and (b) the resultant geoelectric fields. In (b), arrows show the direction and relative magnitude of the geoelectric fields. (*See insert for color representation of the figure.*)

In order to relate this result to the geoelectric field, we can combine Equations (5.6) and (5.8) with the plane wave representations of the fields to obtain a matrix equation for the electric fields,

$$\begin{pmatrix} E_x \\ E_y \end{pmatrix} = \frac{1}{k_\perp^2} \begin{pmatrix} k_x & -k_y \\ k_y & k_x \end{pmatrix} \cdot \begin{pmatrix} \zeta \\ \Gamma \end{pmatrix}. \tag{5.57}$$

Since $\zeta = 0$ below the ionosphere, it follows that the geoelectric fields are related to the characteristic variable Γ by

$$E_x = -\frac{k_y}{k_\perp^2}\Gamma \qquad E_y = \frac{k_x}{k_\perp^2}\Gamma. \tag{5.58}$$

5.3. GROUND SIGNATURE OF A LOCALIZED FAC

Let us suppose an incident field-aligned current given by

$$J_i = J_0 \exp\left(-\frac{x^2}{\alpha^2} - \frac{y^2}{\beta^2}\right) e^{-i(k_A z + \omega t)} \tag{5.59}$$

where α and β are the scale sizes in the $\hat{x}$- and $\hat{y}$-directions. In order to convert our characteristic variable J to more familiar units, it may be useful to note that 1 nT/km is equivalent to $10/4\pi$ (≈ 0.8) μA/m^2.

Although it is possible to analytically Fourier transform Equation (5.59), the inversion necessary in order to obtain the geoelectric fields from the resultant expression is sufficiently complex that our only viable recourse is to use numerical methods, that is, Fast Fourier Transforms (FFTs).

Combining Equations (5.58) and (5.56), we can obtain a two-dimensional spatial profile of the geoelectric fields resulting from an ionospherically localized FAC; such a profile, obtained for a source with $\alpha = 200$ km and $\beta = 400$ km, is shown in Figure 5.2.

There are a number of interesting features to the geoelectric field profile. First, the field is at a minimum directly beneath the center of the FAC. Second, the field maximizes beyond the scale length of the FAC profile, with a minimum of geoelectric field within a few scale lengths of the footprint and peak geoelectric fields occurring in an oblate annular region having a width of a few scale lengths beyond that. The magnitude of the electric field peaks most strongly in the direction of the narrowest dimension of the driver, but the component of the electric field which is strongest is in the direction of largest dimension – in this case, it peaks most strongly along the $\hat{y}$-axis with the largest electric field in the $\hat{x}$-direction.

As a second example, let us consider a "ribbon" of field-aligned current that is a cartoon model of an auroral form. The ionospheric footprint of this feature, shown in Figure 5.3a, has a 1000 km wavelength in the $\hat{x}$-direction and a 400 km amplitude of oscillation in the $\hat{y}$-direction, with the width of the structure being defined by a 100 km Gaussian envelope.

The ground signature of our model auroral form is shown in Figure 5.3. We can clearly see that, as before, there is a minimum of geoelectric field magnitude directly underneath the FAC structure. However, unlike the previous case, we now see is that the field magnitude maximizes near the kinks in the ribbon rather than at some finite distance from the center (although such a ring-like enhancement is still noticeable in the background). The direction of the field is relatively uniform, circulating around the entirety of the

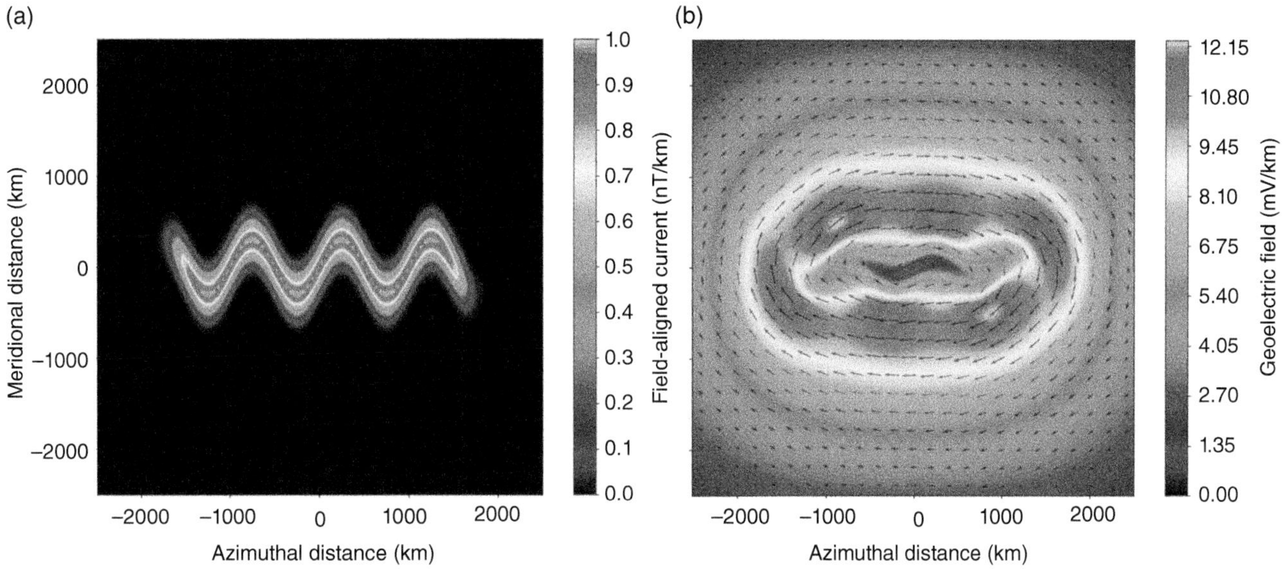

Figure 5.3 Two-dimensional profile of (a) the ionospherically-incident localized ribbon FAC and (b) the resultant geoelectric fields. In (b), arrows show the direction and relative magnitude of the geoelectric fields. (*See insert for color representation of the figure.*)

current structure – it is only directly underneath that the ribbon-like shape is recognizable. Thus, only the "envelope" of the FAC system is visible at locations beyond the direct footprint of the FAC.

5.4. RELATION TO PREVIOUS MODELS

Although previous models of ionospheric transmission typically did not allow for geoelectric fields due to their assumption of infinite ground conductivity, they did provide accurate estimates of the horizontal ground-level magnetic fields (Woodroffe and Lysak, 2012; Mann et al., 2014). In order to compare the results of our model with previous works, we can obtain the horizontal magnetic field components by noting that, since $\nabla \cdot \mathbf{B} = 0$ and $\nabla \times \mathbf{B} = 0$ in the atmosphere and at the ground, it follows that the magnetic field may be derived from a scalar potential, $\mathbf{B} = \nabla \Psi$. This means that our characteristic variable is related to the scalar potential by $B = \partial\Psi/\partial z$, and so

$$\Gamma_{atm} = -i\omega \frac{\partial \Psi}{\partial z} = \Gamma_{mag} \frac{e^{-ik_{atm}z} + R_g e^{ik_{atm}z}}{1 + R_g} \tag{5.60}$$

where the horizontal and time dependences have been suppressed. Solving Equation (5.60) for $\partial\Psi/\partial z$ and then integrating to obtain Ψ, it follows (again using $k_{atm} \approx ik_\perp$) that

$$\Psi = \Gamma_{mag} \frac{\omega}{k_\perp} \frac{e^{k_\perp z} - R_g e^{-k_\perp z}}{1 + R_g} e^{i(\mathbf{k}_\perp \cdot \mathbf{x}_\perp - \omega t)}. \tag{5.61}$$

The horizontal components of the magnetic field are thus given by $B_{x,y} = ik_{x,y}\Psi$, which when evaluated at the ground gives us

$$B_{x,y} = iB_{mag} \frac{k_{x,y}}{k_\perp} \frac{e^{-k_\perp H} - R_g e^{k_\perp H}}{1 + R_g} e^{i(\mathbf{k}_\perp \cdot \mathbf{x}_\perp - \omega t)}. \tag{5.62}$$

In the limit of a perfectly conducting ground, we obtain the following expression for the horizontal magnetic perturbations

$$B_{x,y} = i \frac{k_{x,y}}{k_\perp} \frac{B_{mag}}{\sinh(k_\perp H)} e^{i(\mathbf{k}_\perp \cdot \mathbf{x}_\perp - \omega t)}. \tag{5.63}$$

This is identical to equation 11 of Lysak and Yoshikawa (2006) under similar assumptions, demonstrating the asymptotic equivalence of the two formulations.

5.5. COMPARISON TO THE BIOT–SAVART APPROACH

In the global MHD modeling context, the typical method for calculating geoelectric fields is to apply Biot–Savart integration to global current systems in order to obtain the ground-level magnetic fields followed by the application of a magnetotelluric transfer function to obtain the corresponding electric field. Our self-consistent calculation framework provides us with a means to evaluate the validity and potential limitations of the non-self-consistent Biot–Savart approach.

5.5.1. Biot–Savart Integration

Fundamentally, the Biot–Savart approach requires us to evaluate a three-dimensional integral for every point of interest,

$$\mathbf{B}(\mathbf{x}) = \frac{\mu_0}{4\pi}\iiint \frac{\mathbf{J}(\mathbf{x}')\times(\mathbf{x}-\mathbf{x}')}{\|\mathbf{x}-\mathbf{x}'\|^3}dx'dy'dz' \quad (5.64)$$

where the primed coordinates correspond to the source currents and the unprimed coordinates correspond to the observation point. For the problem we are considering, there are three current systems of interest: FACs, Pedersen currents, and Hall currents. However, because we are dealing with a vertically incident magnetic field at the conducting ionosphere, we can simplify the problem by invoking Fukushima's theorem, which states that under these conditions, the ground magnetic field due to the FACs is exactly canceled by the ionospheric Pedersen currents (*Fukushima*, 1976). Thus, we need only integrate over the Hall currents in order to obtain the ground fields.

The Hall current density for an infinitesimally thin ionosphere is given by

$$\mathbf{J}_H = \Sigma_H \hat{z}\times\mathbf{E}_\perp \delta(z) \quad (5.65)$$

where $\delta(z)$ represents the fact that the current is strictly confined to the $z=0$ plane. In order to proceed with our calculation, we must now determine the ionospheric electric field that is associated due to a specified FAC source.

5.5.1.1. The Ionospheric Electric Field

Under the assumption of current continuity, $\nabla_\perp \mathbf{J}_\perp + \partial J_\parallel/\partial z = 0$, it can be shown that the ionospheric electric field is given by

$$\nabla_\perp\cdot(\Sigma_P\mathbf{E}_\perp + \Sigma_H\hat{z}\times\mathbf{E}_\perp) = -J_\parallel \quad (5.66)$$

where it has been assumed that $\mathbf{J}_\perp = \sigma_P\mathbf{E}_\perp + \sigma_H\hat{z}\times\mathbf{E}_\perp$ and the vertical variation of the electric fields is negligible across the height of the ionosphere so that integration with respect to height affects only the conductivities.

If the conductances are horizontally uniform, then the Hall currents drop out of this equation and we have

$$\Sigma_P\nabla_\perp\cdot\mathbf{E}_\perp = -J_\parallel. \quad (5.67)$$

Assuming a quasi-static situation such that the electric field can be derived from a scalar potential, $\mathbf{E}_\perp = -\nabla_\perp\phi$, we obtain a Poisson-type equation,

$$\nabla_\perp^2\phi = \frac{J_\parallel}{\Sigma_P}. \quad (5.68)$$

Because we are assuming horizontal uniformity, we can readily solve Equation (5.68) using Fourier methods, finding

$$\tilde{\phi} = -\frac{\tilde{J}_\parallel}{k_\perp^2\Sigma_P} \quad (5.69)$$

where a $\tilde{\cdot}$ indicates the Fourier transform of a quantity. We can then obtain the electric field by using $\tilde{\mathbf{E}}_\perp = i\mathbf{k}_\perp\tilde{\phi}$. For an arbitrary two-dimensional source function, we can again use the FFT algorithm to perform these transforms numerically, and upon obtaining the Fourier-space electric field we can apply the inverse transform to get our ionospheric electric fields and thus, using Equation (5.65), the source term for our ground magnetic fields.

5.5.2. Ground Fields

Inserting Equation (5.65) into Equation (5.64) and applying the triple product identity, we obtain a convenient decomposition of terms:

$$\mathbf{B}(\mathbf{x}) = \frac{\mu_0\Sigma_H}{4\pi}\iiint \frac{\hat{z}((\mathbf{x}-\mathbf{x}')\cdot\mathbf{E}_\perp) - (\hat{z}\cdot(\mathbf{x}-\mathbf{x}'))\mathbf{E}_\perp}{\|\mathbf{x}-\mathbf{x}'\|^3}\delta(z')dx'dy'dz'. \quad (5.70)$$

In order to determine the horizontal fields at the surface, we need only consider the second half of the numerator in Equation (5.70) which, following integration in the vertical direction, gives us

$$\mathbf{B}_\perp(x,y) = \frac{\mu_0\Sigma_H H}{4\pi}\iint \frac{\mathbf{E}_\perp}{\|\mathbf{x}-\mathbf{x}'\|^3}dx'dy' \quad (5.71)$$

where it is understood that $z - z' = -H$ in the evaluation of the source-to-observation distance, since $z'=0$ due to the delta function and $z=-H$ due to our specific interest in the fields at ground level.

Given the ground magnetic field from Equation (5.71), we can then apply the magnetotelluric method (Boteler, 2015) to obtain the geoelectric fields. In order to do this, we first double the magnitude of the ground magnetic field, consistent with the (generally good) assumption that the ground is a very good conductor. We then calculate a frequency-dependent impedance factor $Z(\omega)$ that provides us with the electric fields through a simple matrix relationship,

$$\begin{pmatrix} E_x \\ E_y \end{pmatrix} = \begin{pmatrix} 0 & Z \\ -Z & 0 \end{pmatrix}\cdot\begin{pmatrix} B_x \\ B_y \end{pmatrix}. \quad (5.72)$$

Using the same input fields as described above, we can obtain results that are suitable for direct comparison. Although there are many possible questions to ask an answer about how the results of these two methods might differ, we will focus on just two – what are the differences in their spatial distributions, and what are the differences in their peak geoelectric field magnitudes?

Figure 5.4 shows model ionospheric drivers (a and b) along with the geoelectric fields from the Biot–Savart

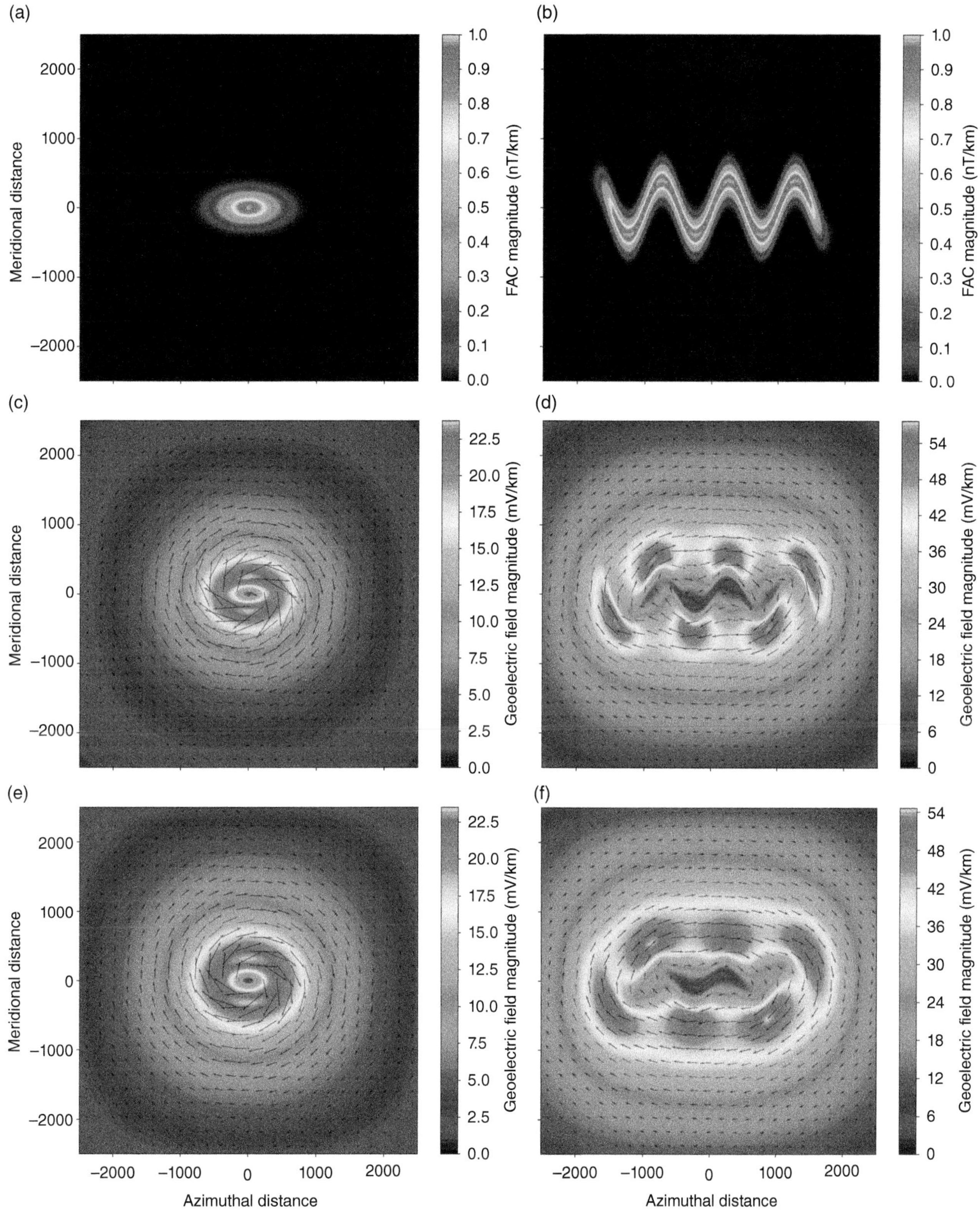

Figure 5.4 Spatial distribution of the ionospheric FACs and the resultant geoelectric fields. (a) Gaussian FAC structure; (b) Ribbon FAC structure; (c and d) Geoelectric fields from the Biot–Savart approach assuming an 18 mHz frequency; (e and f) Geoelectric fields from the wave method assuming a 22 mHz frequency. In (c–f), arrows indicate the local strength and direction of the geoelectric field. (*See insert for color representation of the figure.*)

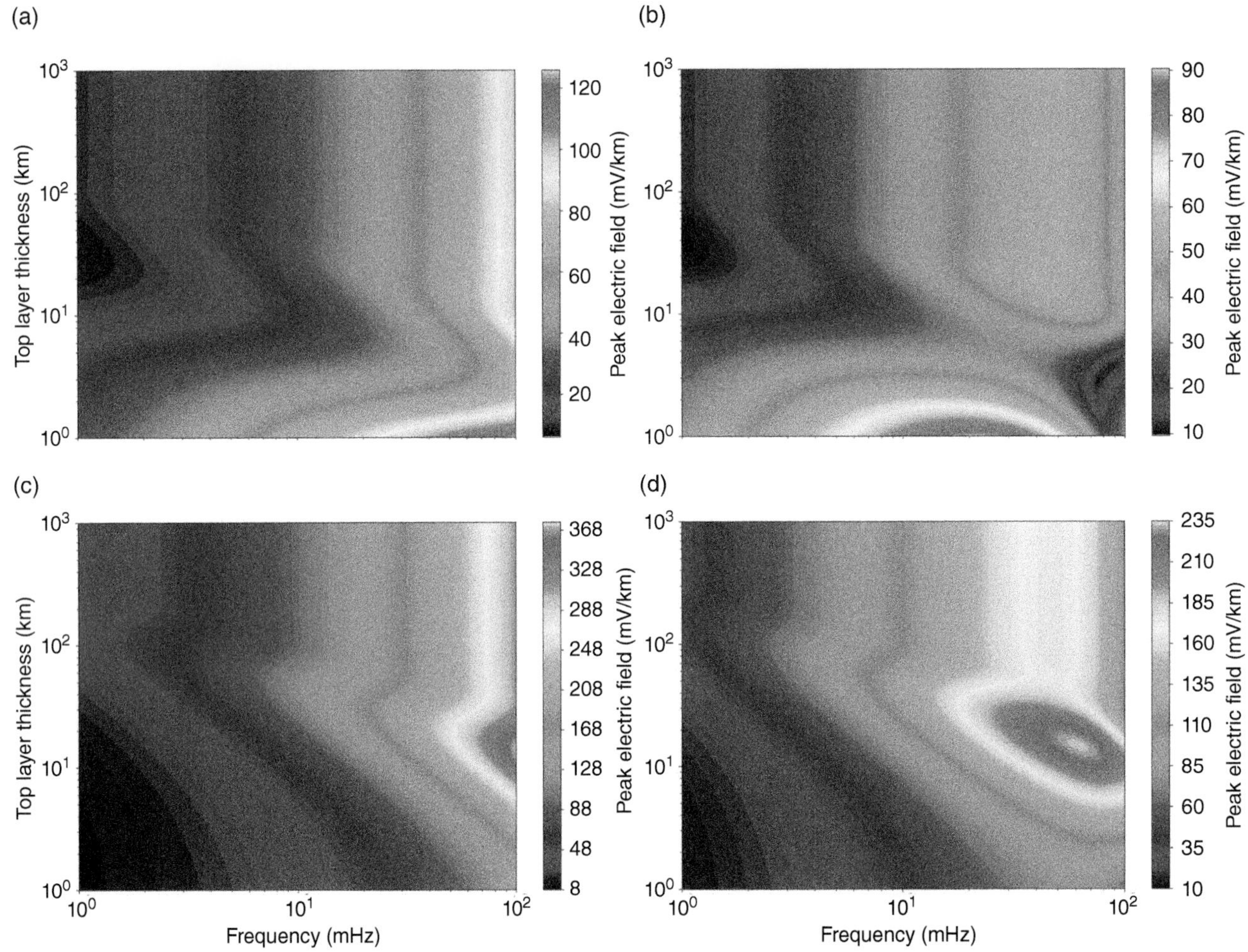

Figure 5.5 Comparison of peak geoelectric field magnitudes as a function of frequency and top-layer depth for (a) Biot–Savart approach with $\sigma_1 = 10^{-1}$ S/m and $\sigma_2 = 10^{-2}$ S/m; (b) Wave method with $\sigma_1 = 10^{-1}$ S/m and $\sigma_2 = 10^{-2}$ S/m; (c) Biot–Savart approach with $\sigma_1 = 10^{-2}$ S/m and $\sigma_2 = 10^{-1}$ S/m; (d) Wave method with $\sigma_1 = 10^{-2}$ S/m and $\sigma_2 = 10^{-1}$ S/m. (*See insert for color representation of the figure.*)

approach (c and d) and wave method (e and f) assuming $\sigma_1 = 10^{-1}$ S/m, $\sigma_2 = 10^{-2}$ S/m, $\Delta = 100$ km, and a frequency of 10 mHz. Generally speaking, the Biot–Savart approach produces more sharply localized signatures – this can be seen in both the peaks (particularly those associated with the ribbon) and with the central minimum; the wave method produces a spatial distribution that has more significant spatial averaging, owing to the non-uniform reflection of components with different perpendicular scale sizes. This suggests that the Biot–Savart approach can overestimate the spatial localization of geoelectric fields and that, in reality, we should expect geoelectric fields to be uniform over a relatively larger region.

Figure 5.5 shows the peak geoelectric field generated anywhere in the 2D domain by the FAC ribbon as a function of frequency and top-layer depth for the cases of a conductive top layer and resistive bottom layer (a and b) as well as in the opposite case of a resistive top layer and a conductive bottom layer (c and d). The results are for the most part indistinguishable for frequencies below 30 mHz, but display significant variation at higher frequencies. The reason for this difference is that the method described in this paper accounts for horizontal as well as vertical structure of the incident signal, and this changes the reflection and transmission properties of the conducting ground.

In order to more clearly illustrate the regions of agreement and disagreement between the more accurate wave method and the Biot–Savart approach, Figure 5.6 shows the dependence of the peak geoelectric field anywhere within the spatial domain as a function of frequency. The two methods are clearly in agreement for frequencies below 30 mHz, but the wave method predicts a decrease in field magnitudes at higher frequencies whereas the Biot–Savart approach predicts a continued increase. As frequencies approach 100 mHz, there is

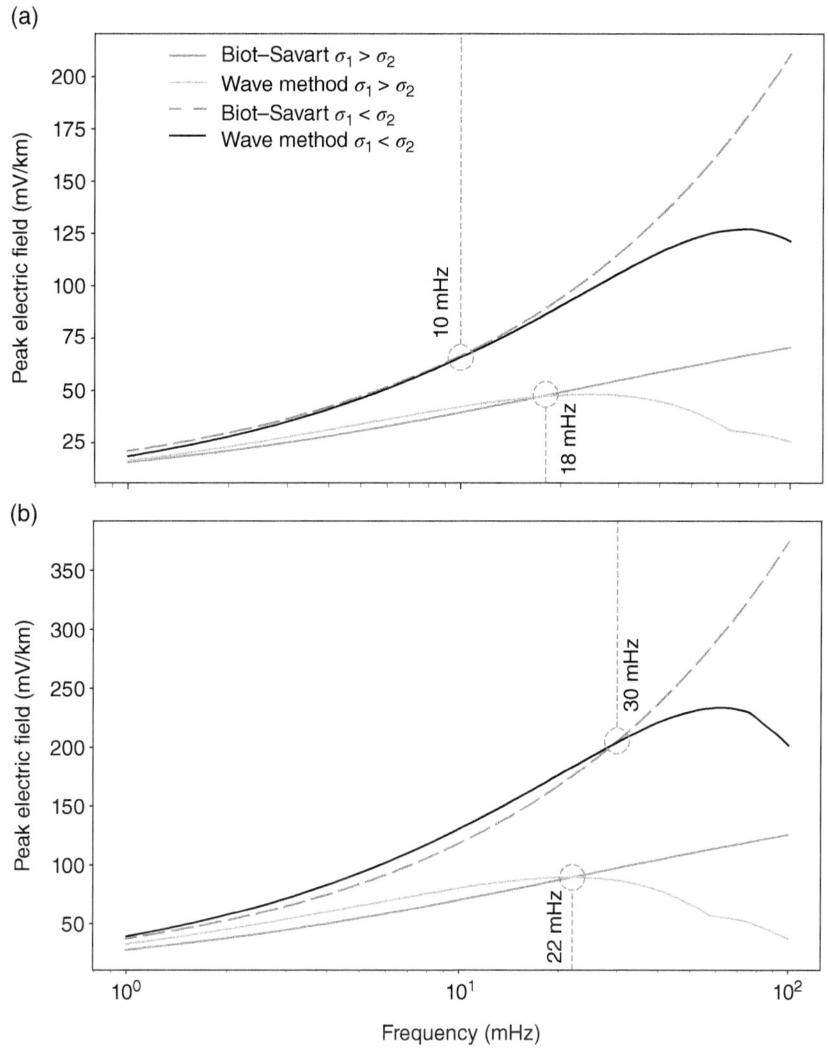

Figure 5.6 Frequency dependence of largest geoelectric field for all top-layer depths in Figure 5.5. Vertical dashed lines and circles indicate the points at which the two methods provide identical results; beyond these crossover points, the accuracy of the Biot–Savart approach decreases rapidly. Top-layer depth = 5 km (a) and 10 km (b).

nearly a factor of two discrepancy between these two models, and the applicability of the Biot–Savart approach is questionable.

5.5.3. Localized Geoelectric Fields

Before concluding our brief investigation into the factors that affect the characteristics of geoelectric fields, it is worth looking into whether there are particular circumstances in which localized signatures are likely to occur. One such case is with higher-frequency waves and a conductive top layer. This combination of factors produces smaller skin depths in the upper layer which in turn reduces the difference between the two methods' representation of the ground fields by, for example, rendering the $k_\perp^2\delta_i^2$ terms in Equation (5.28) negligible.

Another element that should be considered is that ionospheric currents are rarely as simple as those considered above – our earlier analysis of current structures did not include closure currents. In real circumstances, a downgoing current system will be coupled to an upgoing current system elsewhere by Pedersen currents. The closure current might be imagined as an opposite-polarity copy of the incident FAC, in which case the fields can be inferred by a combination of spatial shifting, direction inversion, and field superposition. Consequently, the Hall currents between the incident FAC and the closure will be enhanced, while the currents beyond them will be reduced. This is one means of localizing the signatures, as fields will be strongest between the two FACs. An example of this is shown in Figure 5.7, where the fields resulting from an incident Gaussian FAC and its corresponding closure currents are shown.

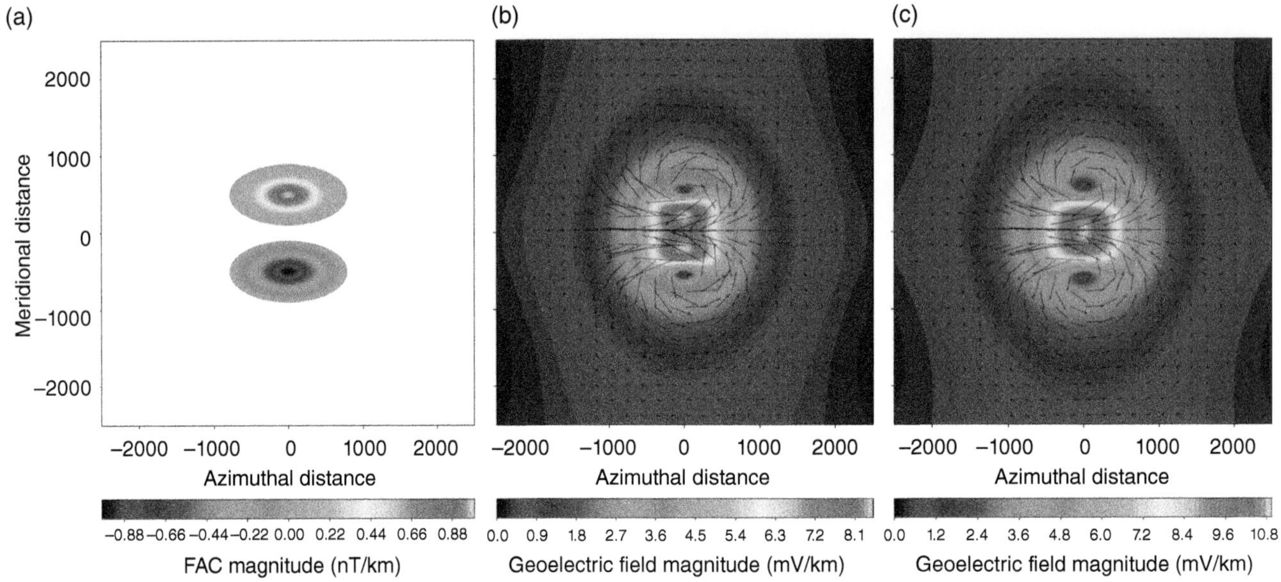

Figure 5.7 Spatial distribution of geoelectric fields for FACs that include a closure current. (a) Closed Gaussian FACs; (b) Geoelectric field calculated using the Biot–Savart approach; (c) Geoelectric field calculated using the wave method. In (a), the region of near-zero values beyond the FAC footprint is not plotted in order to highlight the geometry of the FACs; in (b and c), arrows indicate the direction and magnitude of the geoelectric field. (*See insert for color representation of the figure.*)

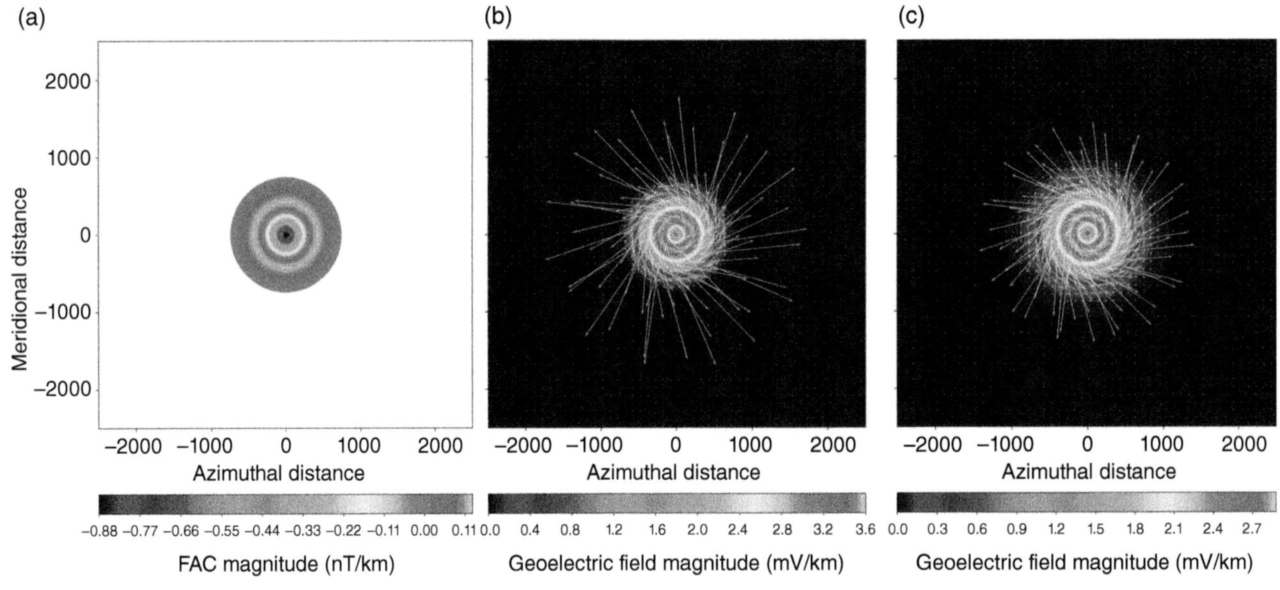

Figure 5.8 Geoelectric fields associated with a self-closing coaxial FAC. (a) Field aligned current; (b) Geoelectric field calculated using Biot–Savart approach; (c) Geoelectric field calculated using wave method. In (a), the region of near-zero values beyond the FAC footprint is not plotted in order to highlight the geometry of the FACs; in (b and c), arrows indicate the direction and magnitude of the geoelectric field. (*See insert for color representation of the figure.*)

Because the magnitude of the fields typically drops with distance from the source, the largest geoelectric fields are realized for the smallest separation distances, so long as the current systems are not placed so close together as to cancel each other out. This also results in further localization of the ground signatures, as the size of the region of enhanced fields is decreased even as the magnitude of the fields are increased. In this case, the region of the enhanced geoelectric field is physically smaller and has larger amplitude than when calculated using the wave method. This suggests that in the case of finely-structured currents, the Biot–Savart approach may underestimate the magnitude of geoelectric fields while also producing an incorrect spatial distribution.

It is also possible for a current system to be intrinsically self-localizing. Perhaps the simplest example of such a system is that which results from the imposition of a Gaussian electric potential,

$$\phi = \phi_0 e^{-\frac{x^2+y^2}{\alpha^2}}. \tag{5.73}$$

Inserting the potential from Equation (5.73) into Equation (5.68), we find that the FAC associated with this potential structure is

$$J_{\|} = \frac{4}{\alpha^2}\frac{\phi_0}{\Sigma_P}\left(\frac{x^2}{\alpha^2}+\frac{y^2}{\alpha^2}-1\right)e^{-\frac{x^2}{\alpha^2}-\frac{y^2}{\alpha^2}}. \tag{5.74}$$

This current produces a downward FAC within the circular region defined by $x^2 + y^2 \leq \alpha^2$, and an upward FAC beyond – in essence, the return current encircles the incident current, and as can be seen in Figure 5.8, this "coaxial current" system produces negligible perturbations beyond its immediate footprint – the effects are intrinsically localized to the characteristic scale size of the potential structure.

5.6. CONCLUSIONS

In this paper, we have presented a new analytical model of geoelectric field generation by FACs that are incident upon the ionosphere. Our new method is a generalization of previous wave transmission models that considers for the first time the effect of a layered conducting ground and examines the effects of varying the configuration of these layers for two-dimensional horizontally-structured FACs. Application of our model and comparison of its results to the Biot–Savart approach, consisting of a static integration of ionospheric currents and a one-dimensional magnetotelluric method, demonstrates that the Biot–Savart approach is accurate at low frequencies (<30 mHz) – which are dominant during geomagnetic storms – but should be used only with extreme caution at higher frequencies such as are associated with substorm phenomena. Additionally, we have found that ionospheric structure is likely less well-resolved on the ground than would be predicted by the Biot–Savart approach, owing to the interaction and attenuation of components with different wavenumbers in the more accurate wave model.

It is important to note that our model does possess limitations. First and foremost, we have assumed a vertically incident magnetic field. This ignores the potential for direct production of geoelectric fields by FACs as described in Tamao (1986) (equivalent to a reduction of the Fukushima shielding effect, with the Pedersen currents no longer compensating for the FACs). This effect would be most important during extremely large geomagnetic storms when FACs are pushed down to lower latitudes. Moreover, an oblique magnetic field yields a meaningful difference between the shear mode polarizations at the ionosphere, with the toroidal and poloidal shear modes being differently reflected and yielding different couplings to the fast mode (Sciffer and Waters, 2002). A second limitation of our model is that it only allows for spatially uniform conductances and conductivities, whereas in the real case we expect that meaningful gradients will exist in both the ionosphere and in the ground. However, in both cases, we would not expect further complications to produce better agreement between the models.

Overall, we suggest in order to most effectively and accurately model geoelectric field production, our results indicate that a generalization of the wave model such as finite difference time domain (FDTD) simulation will produce the best and most accurate estimates of the geoelectric field, and we encourage the development and incorporation of the FDTD approach as a means of improving geoelectric field estimation in space weather models.

APPENDIX: REFLECTION FROM A VERTICALLY-STRUCTURED GROUND PROFILE

Let us assume that the ground is stratified into N layers with the interface between layers n and $n-1$ located at a depth H_{N-1}. The bottom layer is "half-infinite" in the sense that there is no further vertical structure below H_{N-1}. In order to determine the overall effect of this stratified ground on incident signals, we can generalize the formalism described in the main text.

In every region, we suppose that the characteristic field Γ can be represented by a superposition of downgoing (incident) and upgoing (reflected) plane waves:

$$\Gamma_n = \Gamma_{n,i}\left(e^{-ik_n z} + R_n e^{ik_n z}\right) \tag{A.1}$$

where we have defined the reflection coefficient as the ratio of the upgoing to downgoing wave amplitudes, $R_n = \Gamma_{n,r}/\Gamma_{n,i}$, and k_n is given by Equation (5.28).

We begin our consideration just below the interface at H_{N-1}. Because there is no vertical structure below, there is no reflected component of the wave in this region; consequently, $R_N = 0$.

In every layer but the bottom layer, we define an "effective wavenumber" for the previous region,

$$\kappa_n = k_n \frac{1 - R_n e^{-2ik_n H_n}}{1 + R_n e^{-2ik_n H_n}}. \tag{A.2}$$

In layer $N-1$, the absence of reflections in the previous layer ($R_N=0$) implies that $\kappa_N=k_N$. It can be shown that the reflection coefficient at the interface between layers n and $n-1$ (for any n) is given by

$$R_{n-1} = \frac{k_{n-1}-\kappa_n}{k_{n-1}+\kappa_n} e^{2ik_{n-1}H_{n-1}}. \tag{A.3}$$

The overall reflection coefficient at the top layer, which we have previously labeled R_g, is determined by recursively applying the effects of reflections at lower layers. This result was derived explicitly for a two-layer model in the main text, but the method need not be limited to such a profile. It is worthwhile to note that this is analogous to the one-dimensional magnetotelluric method of Cagniard (1953), but it comes at the problem from a somewhat different direction, formulating it as a natural extension of the magnetospheric transmission problem.

ACKNOWLEDGMENTS

This research was supported by the LANL/Laboratory Directed Research and Development (LDRD) Program funded by the U.S. Department of Energy.

REFERENCES

Alperovich, L. S. and E. N. Fedorov (2007), Hydromagnetic waves in the magnetosphere and the ionosphere, *Astrophys. Space Sci. Libr.*, *353*, ISBN:978-1-4020-6637-5.

Boteler, D. H. (2015), The evolution of Québec earth models used to model geomagnetically induced currents, *IEEE Trans. Pow. Deliv.*, *30*(5), doi:10.1109/TPWRD.2014.2379260.

Cagniard, L. (1953), Basic theory of the magneto-telluric methods of geophysical prospecting, *Geophysics*, *18*(3): 605–635.

Fujita, S., and T. Tamao (1988), Duct propagation of hydromagnetic waves in the upper ionosphere, 1, Electromagnetic field disturbances in high latitudes associated with localized incidence of a shear Alfvén wave, *J. Geophys. Res.*, *93*(A12):14665–14673, doi:10.1029/JA093iA12p14665.

Fukushima, N. (1976), Generalized theorem for no ground magnetic effect of vertical currents connected with Pedersen currents in the uniform-conductivity ionosphere, *Rep. Ionos. Space Res. Jap.*, *30*:35–50.

Knipp, D. J. (2015), Forward to space weather collection on geomagnetically induced currents: Commentary and research, *Space Weath.*, *13*:742–746, doi:10.1002/2015SW001318.

Lysak, R. L. (1997), The propagation of Alfvén waves through the ionosphere, *Phys. Chem. Earth*, *22*(7–8):757–766.

Lysak, R. L. and A. Yoshikawa (2006), Resonant cavities and waveguides in the ionosphere and atmosphere, in *Magnetospheric ULF Waves: Synthesis and New Directions* (eds K. Takahashi, P. J. Chi, R. E. Denton and R. L. Lysak), American Geophysical Union, Washington, DC, doi: 10.1029/169GM19.

Mann, I. R., M. E. Usanova, K. Murphy, M. T. Robertson, D. K. Milling, A. Kale, C. Kletzing, J. Wygant, S. Thaller, and T. Raita (2014), Spatial localization and ducting of EMIC waves: Van Allen Probes and ground?based observations, *Geophys. Res. Lett.*, *41*:785–792, doi:10.1002/2013GL058581.

Ngwira, C. M., A. A. Pulkkinen, E. Bernabeu, J. Eichner, A. Viljanen, and G. Crowley (2015), Characteristics of extreme geoelectric fields and their possible causes: Localized peak enhancements, *Geophys. Res. Lett.*, *42*:6916–6921, doi:10.1002/2015GL065061.

Sciffer, M. D., and C. L. Waters, Propagation of ULF waves through the ionosphere: Analytic solutions for oblique magnetic fields, *J. Geophys. Res.*, *107*(A10):1297, doi:10.1029/2001JA000184.

Tamao, T. (1986), Direct contribution of oblique field-aligned currents to ground magnetic fields, *J. Geophys. Res.*, *91*(A1):183–189.

Winter, L. M., J. Gannon, R. Pernak, S. Huston, R. Quinn, E. Pope, A. Ruffenach, P. Bernardara, and N. Crocker (2017), Spectral scaling technique to determine extreme Carrington-level geomagnetically induced currents effects, *Space Weath.*, *15*:713–725, doi:10.1002/2016SW001586.

Woodroffe, J. R., and R. L. Lysak (2012), Ultra-low frequency wave coupling in the ionospheric Alfvén resonator: Characteristics and implications for the interpretation of ground magnetic fields, *J. Geophys. Res.*, *117*, A03223, doi:10.1029/2011JA017057.

Part II
Geomagnetic Induction

6
Empirical Estimation of Natural Geoelectric Hazards

Jeffrey J. Love[1], Paul A. Bedrosian[2], Anna Kelbert[1], and Greg M. Lucas[1]

ABSTRACT

Geoelectric field time series can be estimated by convolving estimates of Earth-surface impedance, such as those obtained from magnetotelluric survey measurements, with historical records of geomagnetic variation obtained at magnetic observatories. This straightforward procedure permits the mapping of geoelectric field variation during magnetic storms. Statistical analysis of the time series allows extrapolation to extreme-value amplitudes, such as might be realized during an intense magnetic storm in the future. The development of these products is illustrated for the Mid-Atlantic United States, using impedances obtained from EarthScope survey data and geomagnetic variation records obtained at the Fredericksburg observatory operated by the U.S. Geological Survey. For this region, 100-year geoelectric exceedance amplitudes have a range of almost three orders of magnitude (from 0.04 V/km at a site in southern Pennsylvania to 24.29 V/km at a site in central Virginia), and they have significant geographic granularity, which is due to site-to-site differences in surface impedance (and subsurface electrical conductivity structure). Maps of 100-year exceedance amplitudes resemble those of geoelectric amplitudes for the March 1989 magnetic storm, and, in that sense, the March 1989 storm resembles what might be loosely called a "100-year" event.

Key Points

- Storm-time geoelectric amplitudes and polarizations can show significant geographic granularity.
- This geographic granularity is a function of both geomagnetic activity and Earth-surface impedance.
- In terms of geoelectric amplitudes, the March 1989 storm resembles a one-hundred-year storm.

6.1. INTRODUCTION

Geoelectric fields are induced in the Earth's electrically conducting crust, mantle, and ocean by natural time-dependent geomagnetic field variation that is generated by dynamic processes in the space environment surrounding the Earth. This induction occurs all the time, during both calm and stormy space weather conditions, but intense magnetic storms can induce geoelectric fields of sufficient strength to interfere with the operation of bulk electric-power grids, sometimes causing blackouts and, even, damaging transformers (e.g., Molinski, 2002; Piccinelli and Krausmann, 2014). A dramatic demonstration of this natural hazard occurred in March 1989 (e.g., Allen et al., 1989), when an intense magnetic storm caused the collapse of the entire Hydro-Québec electric-power grid in Canada (Bolduc, 2002; Béland and Small, 2005). Historically, the Carrington event of 1859 (Tsurutani et al., 2003; Siscoe et al., 2006; Cliver and Dietrich, 2013) was, by some measures, the most intense magnetic storm ever recorded; it disrupted telegraph systems around the world (e.g., Boteler, 2006). Another event, the Halloween magnetic storm of October 2003,

[1]*Geomagnetism Program, United States Geological Survey, Denver, CO, USA*

[2]*Geology, Geophysics, and Geochemistry Science Center, United States Geological Survey, Denver, CO, USA*

Geomagnetically Induced Currents from the Sun to the Power Grid, Geophysical Monograph 244, First Edition.
Edited by Jennifer L. Gannon, Andrei Swidinsky, and Zhonghua Xu.

caused operational failures in Sweden's grid systems (Pulkkinen et al., 2005). Some researchers have suggested that the future arrival of a rare magnetic superstorm (e.g., Cliver and Dietrich, 2013; Lakhina and Tsurutani, 2018) could cause long-lasting interruption of electric-power transmission (e.g., Odenwald and Green, 2006; Kappenman, 2012). Should such an event occur, it would carry significant economic and societal cost (Barnes and Van Dyke, 1990; Baker et al., 2008; Lloyd's of London, 2013; Eastwood et al., 2017; Worman et al., 2018).

In the context of these historical events and the possibility that an extreme-event magnetic storm might occur in the future, the United States (U.S.) Federal Energy Regulatory Commission (2013, Order No. 779) directed American utility companies to assess the vulnerability of U.S. domestic power-grid systems to storm-time magnetic disturbance. This, in turn, motivated a study to estimate threshold geoelectric amplitudes that might only be exceeded once every 100 years (on average) across the continental United States (North American Electric Reliability Corporation, 2014a). Separately, the U.S. National Science and Technology Council (2015, Goal 1.1) (NSTC), has coordinated the development of 100-year geoelectric "benchmarks" (as well as benchmarks for other space-weather phenomena). The subject is rapidly developing (e.g., Pulkkinen et al., 2017), and much has undoubtedly been learned since the deleterious events of March 1989.

In this brief article, we review and illustrate the basic physical principles of electromagnetic induction and empirical estimation of geoelectric hazards, with analysis and results given for the Mid-Atlantic United States (Love et al., 2018a). This geographic region encompasses the megalopolis of Philadelphia–Baltimore–Washington, with critically important power-grid infrastructure that experienced operational stress during the 1989 magnetic storm. The Mid-Atlantic region is both geologically complicated and occasionally subjected to intense storm-time geomagnetic activity. Sufficient geomagnetic monitoring data and magnetotelluric survey data are also available from this region to enable realistic estimates of geoelectric hazards.

6.2. DEFINING THE HAZARD

A natural "hazard" can be defined as a phenomenon or event that might have a negative impact on people, infrastructure, or the environment. Depending on the nature of the hazard, the word is often further specialized to mean the probability that a damaging event will occur within a specified window of time and within a given geographic area [Varnes, 1984, p. 10; Panel on Seismic Hazard Analysis, 1988, p. 94]; this is, essentially, how the word is used in the insurance industry (e.g., Smolka, 2006). One can, furthermore, define "risk" as the product of hazard probability, times the "exposure" of an asset to the hazard, times "vulnerability" of the asset should the hazard occur (e.g., Smith and Petley, 2009, chapter 1). From this, one can understand that risk can be relatively high, even for the rare occurrence of a hazard, if the exposure and vulnerability of the asset are high. For an electric-power grid system, geoelectric fields induced in the Earth's conducting interior during magnetic storms are the relevant hazard (e.g., Boteler, 2001), and with their estimation, geomagnetically induced currents (GIC) in power-grid distribution lines can be calculated (e.g., Horton et al., 2012). The operational interference and infrastructure damage that these GICs can wreak on a power-grid system are issues of exposure and vulnerability that can be assessed as part of engineering studies (e.g., North American Electric Reliability Corporation, 2013, 2014b).

In principle, the geoelectric field could be usefully monitored over a given geographic region from a network of fixed-site stations, and, again, in principle, statistical analysis of long geoelectric time series from such stations could be done to estimate geoelectric hazards. In practice, however, long-term geoelectric monitoring is performed at only a very few stations around the world, for example, at the Kakioka observatory in Japan (Fujii et al., 2015) and at the Boulder observatory in the United States (Blum et al., 2017). Data from these stations are useful for validating modeling and prediction methods. However, due to the localized geological complexity (and attendant complexity in Earth conductivity structure), storm-time geoelectric fields measured at one site can differ significantly from geoelectric fields measured at another just a hundred or so kilometers away (e.g., McKay and Whaler, 2006; Bedrosian and Love, 2015; Bonner and Schultz, 2017). This, together with the effort required to maintain geoelectric monitoring systems (e.g., Ferguson, 2012), make it impractical to maintain a geoelectric monitoring network of sufficient density to evaluate geoelectric hazards on a continental scale.

6.3. INDUCTION IN THE CONDUCTING EARTH

The most practical way to estimate geoelectric fields across a given geographic region is through a parameterization of their storm-time induction (e.g., Love et al., 2014). At the Earth's surface, geomagnetic field variation $\mathbf{B}(t, x, y)$ and geoelectric field variation $\mathbf{E}(t, x, y)$ are both functions of time t and geographic location (x, y). For the frequency range of interest to GIC studies, electromagnetic variation within the Earth can be approximated without displacement currents in terms of the classical laws of "quasi-static" electromagnetism for a conducting medium (e.g., Stratton, 1941, chapter 5).

Additional simplification can be obtained by assuming that geomagnetic field variation, over a given region, can be approximated as a plane-wave (Wait, 1982, chapter VI). With this, the relationship between horizontal-component (north $\hat{\mathbf{x}}$, east $\hat{\mathbf{y}}$), geomagnetic $\mathbf{B}_h = [B_x, B_y]$, and geoelectric $\mathbf{E}_h = [E_x, E_y]$ field variation at a given geographic location can be described (approximately) in terms of a linear equation,

$$\mathbf{E}_h(t, x, y) = \frac{1}{\mu}(\mathbf{Z} * \mathbf{B}_h)(t, x, y) \tag{6.1}$$

where μ is magnetic permeability, and where impedance is usually expressed as a function of the Fourier frequency f (for sinusoidal period $T = 1/f$)

$$\mathbf{Z}(f, x, y \mid \sigma(\mathbf{r})) = \begin{bmatrix} Z_{xx} & Z_{xy} \\ Z_{yx} & Z_{yy} \end{bmatrix} (f, x, y \mid \sigma(\mathbf{r})). \tag{6.2}$$

This 2×2 tensor is complex; it has real and imaginary parts (e.g., Simpson and Bahr, 2005; Unsworth, 2007; Weidelt and Chave, 2012). The response function $\mathbf{Z}/\mu$ has units of (V/km)/nT, and the impedance tensor $\mathbf{Z}(f, x, y)$ has units of ohms (Ω). Physically, the Earth's surface impedance is a nonlinear function of subsurface conductivity structure, $\sigma(\mathbf{r})$ (or, equivalently, subsurface resistivity structure $\rho(\mathbf{r}) = 1/\sigma$), where $\mathbf{r}$ is position within the Earth. Rock conductivity depends on numerous ancillary properties, including minerology, melt and solid phases, water and clay content, porosity, cracks, and grain boundaries (e.g., Yoshino, 2011; Evans, 2012); the spatial distribution of these properties is reflective of the Earth's geology; ocean water is electrically conductive, though much more uniformly than the solid Earth. Qualitatively, field variation is attenuated within the Earth as a diffusive "skin effect," conventionally defined for a frequency dependent "apparent" resistivity (the inverse of "apparent" conductivity) and corresponding characteristic length scale

$$\frac{1}{\sigma^A(f)} = \rho^A(f) \simeq \frac{|\mathbf{Z}(f)|^2}{2\pi\mu f} \quad \text{and} \quad \delta^A(f) \simeq \sqrt{\frac{\rho^A(f)}{\pi\mu f}}, \tag{6.3}$$

for a tensor norm $|\cdot|$ (e.g., Berdichevsky and Dmitriev, 2008; Weidelt and Chave, 2012). From this, it can be understood that resistive (conductive) Earth structure has a long (short) diffusive length scale. In more detail, one can also discuss apparent resistivities for individual impedance elements, for example, $\rho^A_{xy}(f)$ for Z_{xy}, etc.

Magnetotellurics (e.g., Simpson and Bahr, 2005; Unsworth, 2007; Chave and Jones, 2012) is a geophysical method for obtaining model estimates of the Earth's interior electrical conductivity structure. Passive simultaneous measurement (e.g., Ferguson, 2012) of geomagnetic $\mathbf{B}_h(t, x, y)$ and geoelectric $\mathbf{E}_h(t, x, y)$ field variation is made at various geographic locations as part of a coordinated survey of a given geographic area. From these measurements and using Equation (6.1), impedance tensors $\mathbf{Z}(f, x, y)$ are constructed (e.g., Egbert, 2007a; Chave, 2012); these tensors are then used, together with computer-based algorithm packages (e.g., Kelbert et al., 2014), in an inverse-modeling procedure (e.g. Egbert, 2007b; Rodi and Mackie, 2012) to obtain estimates of Earth conductivity structure $\sigma(\mathbf{r})$ (e.g., Murphy and Egbert, 2017). But Equation (6.1) also shows that it is possible to exploit the work of magnetotelluric scientists to estimate geoelectric hazards. The combination of impedance tensors $\mathbf{Z}(f, x, y)$ and geomagnetic monitoring time series $\mathbf{B}_h(t, x, y)$ can be used to construct geoelectric time series $\mathbf{E}_h(t, x, y)$ (e.g., Pirjola, 2002; Kelbert et al., 2017; Weigel, 2017) at magnetotelluric survey sites. This construction can be done either retrospectively, using historical geomagnetic time series to construct historical geoelectric time series, or t might even be done for situational monitoring using real-time geomagnetic data feeds (e.g., Love et al., 2018b). It is the former approach that will be used to illustrate our discussion of geoelectric hazards.

6.4. A LOOK AT SOME DATA

Insight into the nature of geomagnetic storms and geoelectric fields induced in the electrically conducting Earth can be obtained by simply looking at some data. In Figure 6.1a, we plot 1-min resolution, horizontal-component $\mathbf{B}_h(t)$ magnetometer time series recorded at the U.S. Geological Survey (USGS) Fredericksburg, Virginia magnetic observatory (Love and Finn, 2011) during the March 1989 magnetic storm. The geomagnetic disturbance seen in Figure 6.1a is the result of a superposition of geomagnetic signals generated by external electric currents in the magnetosphere and ionosphere, as well as a smaller contribution coming from electric currents induced within the Earth beneath the observatory by external geomagnetic disturbance. Briefly: The magnetic storm commenced suddenly with the arrival 01:28 UT on March 13 at the Earth of a shock-wave in the solar wind driven by a coronal-mass ejection from the Sun. This was followed by main-phase intensification of the magnetospheric ring-current system, the strength of which is measured by the storm-time disturbance index *Dst*. The March 1989 storm reached its maximum ring-current intensity with $-Dst = 589$ nT at 01:30 UT on March 14 (the highest value since the 1957 start of the *Dst* time series); the storm also attained a mid-latitude disturbance level of $K_p = 9$ (or G5 or "extreme" on a scale developed by the National Oceanic and Atmospheric Administration). Over the course of this storm, and especially leading up to

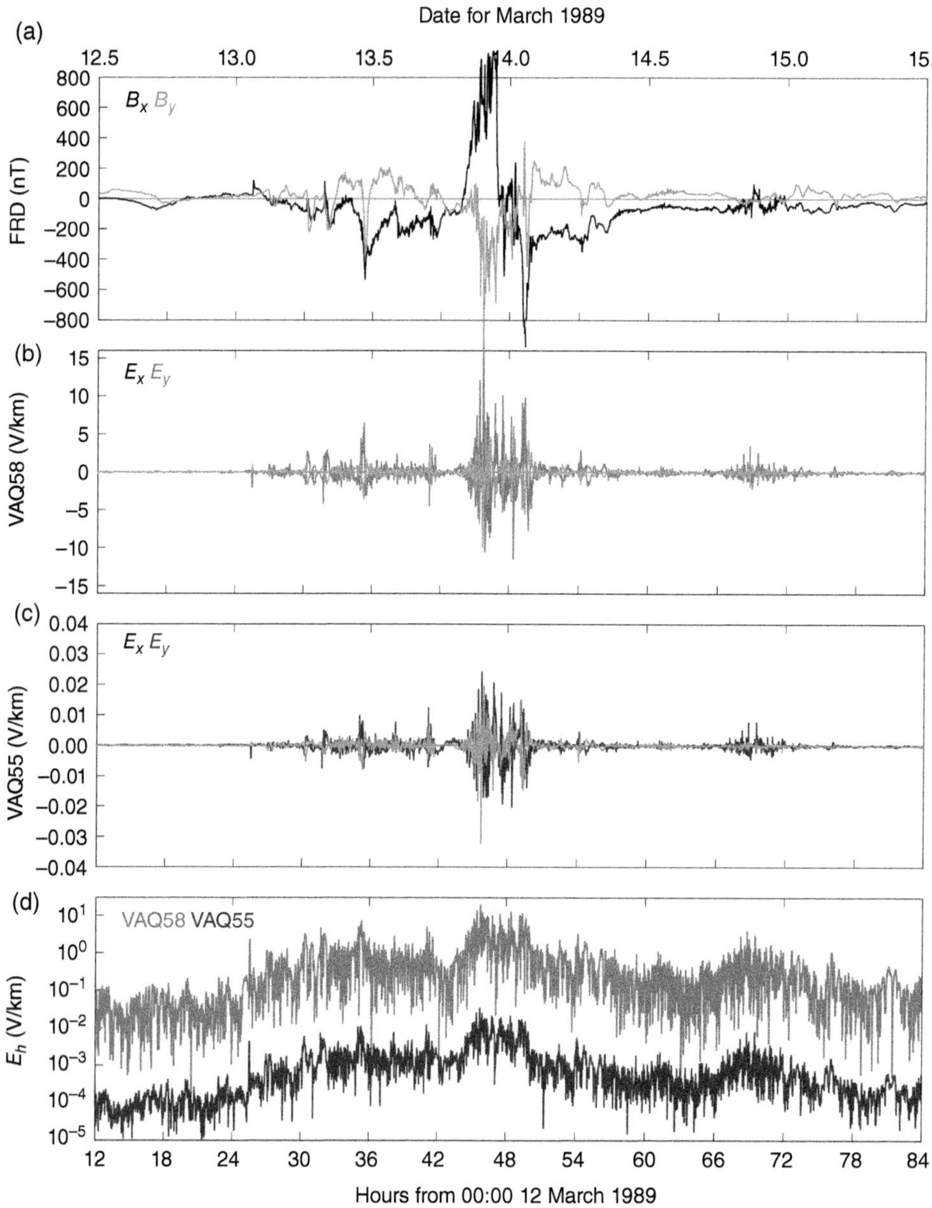

Figure 6.1 Time series of 1-min resolution (a) north $B_x(t)$ (black) and east $B_y(t)$ (gray) component variation recorded at the USGS Fredericksburg, Virginia (FRD), observatory during the magnetic storm of March 1989; (b and c) corresponding geoelectric north $E_x(t)$ and east $E_y(t)$ component variation calculated for EarthScope magnetotelluric survey sites VAQ58 (gray, red) central Virginia (VA), and VAQ55 (gray, blue), western Virginia (VA); and, (d) on a semilog plot, geoelectric amplitude $E_h(t)$ for VAQ58 (red) and VAQ55 (blue). (*See electronic version for color representation of this figure.*)

this maximum − *Dst*, magnetic field lines in the magnetosphere, and possibly in the magnetotail, underwent multiple abrupt rearrangements; these drove substorm diversions of electric currents into and out of the ionosphere. Some of this is indicated in the local *K* index for Fredericksburg, which reached a maximum defined value of 9 from 09:00–12:00 UT, March 13 and from 18:00 UT, March 13 to 03:00 UT, March 14.

In using data from a given ground-based magnetic observatory, it is important to consider the size of the geographic area over which they might be considered to be representative of geomagnetic field variation. Since storm-time disturbance is generated by electric currents at heights greater than about 100 km, one can reasonably assume that, by purely geometric attenuation, the horizontal length scale characterizing ground-level geomagnetic variation will exceed 100 km. More specifically, analysis of 1-min, horizontal-component geomagnetic variation (at high latitudes that are more active than the mid-latitude Mid-Atlantic region of the United States) shows reasonably good coherence in signals recorded by magnetometers separated in latitude by $\sim 2^{\circ}$ ($\sim$200 km) (Watermann et al., 2006). We push the plane-wave approximation a bit and assume that 1-min, horizontal-component variation measured at the Fredericksburg observatory is representative of geomagnetic variation across a $\sim 6° \times 6°$ ($\sim 600 \times 600$ km) latitude–longitude square region centered on the location of the observatory: (38.20°N, 77.37°W).

We use magnetotelluric impedance tensors $\mathbf{Z}(f, x, y)$ (Schultz et al., 2006–2018; Kelbert et al., 2011) for estimating geoelectric fields across the Mid-Atlantic region. These tensors are constructed from data time series obtained at different geographic sites (x, y) during surveys (Schultz, 2010) performed from 2015 to 2016 as part of the National Science Foundation's EarthScope project (Williams et al., 2010). The tensors correspond to survey sites having a 70-km nominal spacing; they are well-defined within a period band of 10–10,000 s (a frequency band of 10^{-1}–10^{-4} Hz). In Figure 6.2, we show apparent resistivities $\rho^A_{xx}(T)$, $\rho^A_{xy}(T)$, etc. across this period band for two survey sites in Virginia (VA). Site VAQ58 (37.38°N, 77.58°W) is located on a geological formation known as the Petersburg batholith, a large igneous intrusion within the Piedmont of the Appalachian mountain range (e.g., Gates and Glover III, 1989). For geomagnetic variation with a period of 2 min (Nyquist for the $\mathbf{B}(t)$ data analyzed, here), and using Equation (6.3), the apparent resistivity at this site is $\rho^A \simeq 8 \times 10^4$ Ω-m, with induction occurring throughout the crustal column and deep into the mantle $\delta^A \simeq 1500$ km. In more detail, Figure 6.2a shows considerable asymmetry; from 10 to 10,000 s, ρ^A_{xy} does not nearly equal ρ^A_{yx}, and, over periods of 10^3–10^4 s, ρ^A_{xx} is comparable to ρ^A_{xy}; over periods of 1000–10, 000 s, ρ^A_{yy} is comparable to ρ^A_{yx}. These observations indicate that Earth-conductivity structure beneath and surrounding the site is complex and three-dimensional. In comparison, we can see in Figure 6.2b that resistivities are much smaller at site VAQ55 (37.88°N, 79.81°W), which is located on sedimentary rocks of the fold and thrust fault belt that is the Appalachian Valley and Ridge (e.g., Virginia Division of Mineral Resources, 1993); for geomagnetic variation with a period of 2 min, the apparent resistivity at this site is $\rho^A \simeq 1.00$ Ω-m and induction is focused at relatively shallow depths, $\delta^A \simeq 5$ km. Furthermore, from 10 to 10,000 s, ρ^A_{xy} approximately equals ρ^A_{yx}, and both ρ^A_{xx} and ρ^A_{yy} are small. This near symmetry indicates that Earth conductivity structure beneath and surrounding the site is nearly a one-dimensional function of depth.

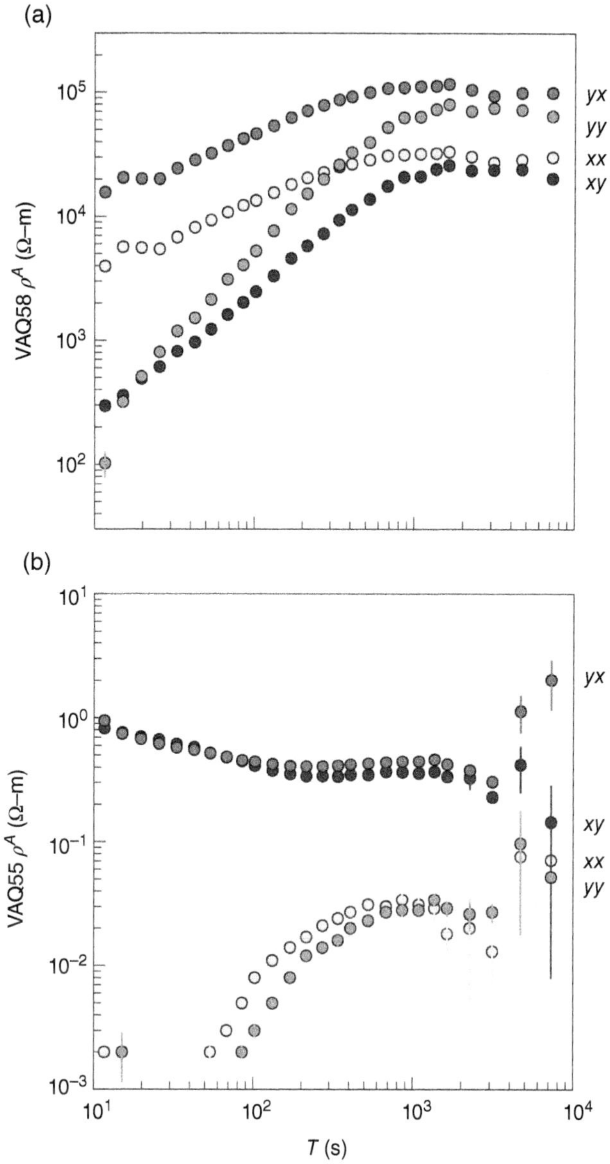

Figure 6.2 Apparent resistivities, $\rho^A_{xx}(T)$ (yellow), $\rho^A_{xy}(T)$ (blue), $\rho^A_{yx}(T)$ (red), $\rho^A_{yy}(T)$ (green), each plotted as a function of variational period T, for EarthScope survey sites (a) VAQ58 and (b) VAQ55. One-standard-deviation error bars are also shown. *(See electronic version for color representation of this figure.)*

6.5. GEOELECTRIC TIME SERIES

Using Equation (6.1), we calculate 1-min resolution, horizontal-component geoelectric time series $\mathbf{E}_h(t, x, y)$ at each EarthScope magnetotelluric survey site across the Mid-Atlantic. Before continuing, however, it is important to consider the accuracy of these calculations. On 8 May 2016, a magnetic storm occurred with maximum intensity $-Dst = 88$ nT, NS $K_p = 7$ (G3 or "strong"). During this storm, the magnetotelluric survey of the Mid-Atlantic United States was in progress – in particular, magnetometer and geoelectric time series were being acquired at site WVO64 (39.04°N, 80.07°W) located 252 km from the Fredericksburg observatory. The survey data provide us with an opportunity to estimate the errors in our (retrospective) calculation of geoelectric fields. In Figure 6.3a and c, we plot the geoelectric field $\mathbf{E}_h(t_i)$ measured at WVO64 during the storm; in the same figure, we plot estimated geoelectric fields calculated using the WVO64 impedance tensor and the geomagnetic field variation measured at WVO64. We also plot estimated geoelectric fields calculated using the WVO64 impedance tensor and the geomagnetic field variation measured at Fredericksburg. Here, we see that the three time series generally track each other, but, to permit closer scrutiny, in Figure 6.3b and d, we plot the residual differences ΔE_x and ΔE_y between the calculated geoelectric fields and the measured fields. For geoelectric fields calculated using the locally measured geomagnetic field, the root-mean-

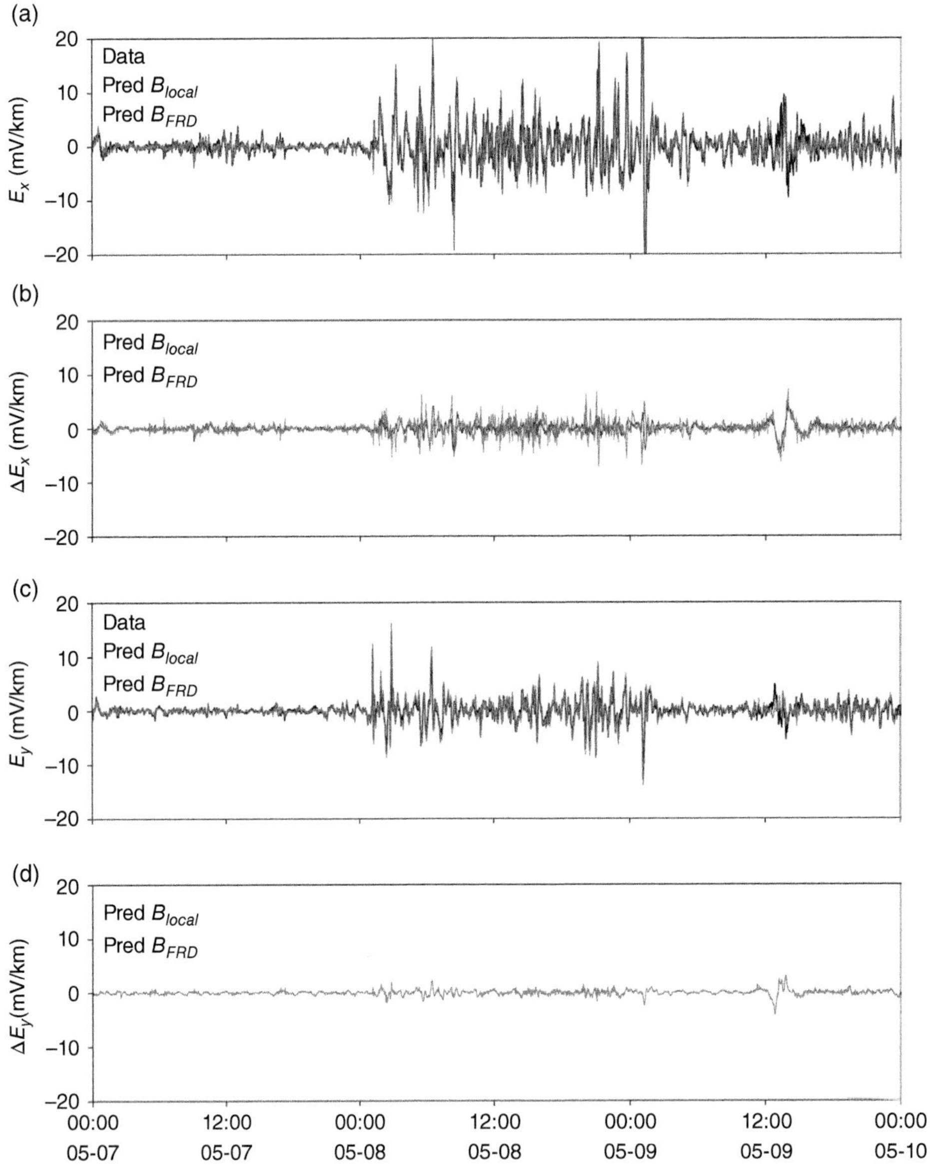

Figure 6.3 Time series of 1-min resolution (a and c) north $E_x(t)$ and east $E_y(t)$ geoelectric fields as (black) measured at the the EarthScope magnetotelluric survey site during a magnetic storm that occurred in May 2016; also shown dark gray are estimated geoelectric fields calculated using the WVO64 impedance tensor and the geomagnetic field variation measured at WVO64, and light gray estimated geoelectric field calculated using the WVO64 impedance tensor and the geomagnetic field variation measured at Fredericksburg; in (b and d), the residual differences ΔE_x and ΔE_y between the calculated geoelectric fields and the measured fields. (*See electronic version for color representation of this figure.*)

square (RMS) of the residual differences divided by the RMS of the measured variation for E_x and E_y are, respectively, 0.23 and 0.20; for geoelectric fields calculated using geomagnetic field variation measured at Fredericksburg, the ratios are 0.38 and 0.25. From results like these, we estimate that the errors in our calculated geoelectric fields, obtained using geomagnetic field variation measured at Fredericksburg, are about 30%. Since the discrepancies shown in Figure 6.3b and d, for both locally measured geomagnetic field variation and that measured at Fredericksburg, closely track each other, we can infer that errors in our calculation of geoelectric field variation has less to do with regional differences in geomagnetic field variation, and more to do with inadequacies in the parameterization of induction given by Equation (6.1).

We now consider the geoelectric time series for the March 1989 storm. In Figure 6.1b, we plot horizontal components for survey site VAQ58, and in Figure 6.1c we plot components for survey site VAQ55. These components are each calculated using geomagnetic field variation recorded at Fredericksburg; in Figure 6.1d, we plot geoelectric field amplitude, $E_h(t,x,y)=\sqrt{E_x^2+E_y^2}$. Generally speaking, at both sites, geoelectric disturbance levels are high when geomagnetic disturbance levels are high (and low when they are low), though it is noteworthy that geoelectric amplitude is much higher at VAQ58 than it is at VAQ55. This result is to be expected, given the apparent resistivities at each site (e.g., Cuttler et al., 2018); from Figure 6.2 the impedance at VAQ58 is more than two orders of magnitude higher than it is at VAQ55; this range dwarfs the 30% error we estimate for geoelectric fields calculated at each site. The maximum geoelectric amplitude during the 1989 storm and at site VAQ58 was attained at 21:45 UT on 13 March 1989, with $E_h^m = 19.97$ V/km. This extreme amplitude was almost simultaneous with the tripping of compensators that led to the blackout of the Hydro-Québec power-grid system in Canada (Bolduc, 2002). Over the subsequent several hours of the magnetic storm, numerous operational anomalies were reported by electric-power companies in the Mid-Atlantic United States, including for the Allegheny Power System, the Pennsylvania–New Jersey–Maryland Interconnection, and Virginia Power (North American Electric Reliability Corporation, 1990).

In Figure 6.4, we show maps of horizontal-component geoelectric vectors $\mathbf{E}_h(t,x,y)$ for the various magnetotelluric survey sites across the Mid-Atlantic, each for three different instances in time t during the March 1989 magnetic storm. Notably, the geoelectric vectors changed quickly over time, and, at each instance, geoelectric polarization and amplitude differ significantly from one survey site to another. Across the Mid-Atlantic, at any particular moment in time, the geoelectric vector at one site points at odd angles relative to

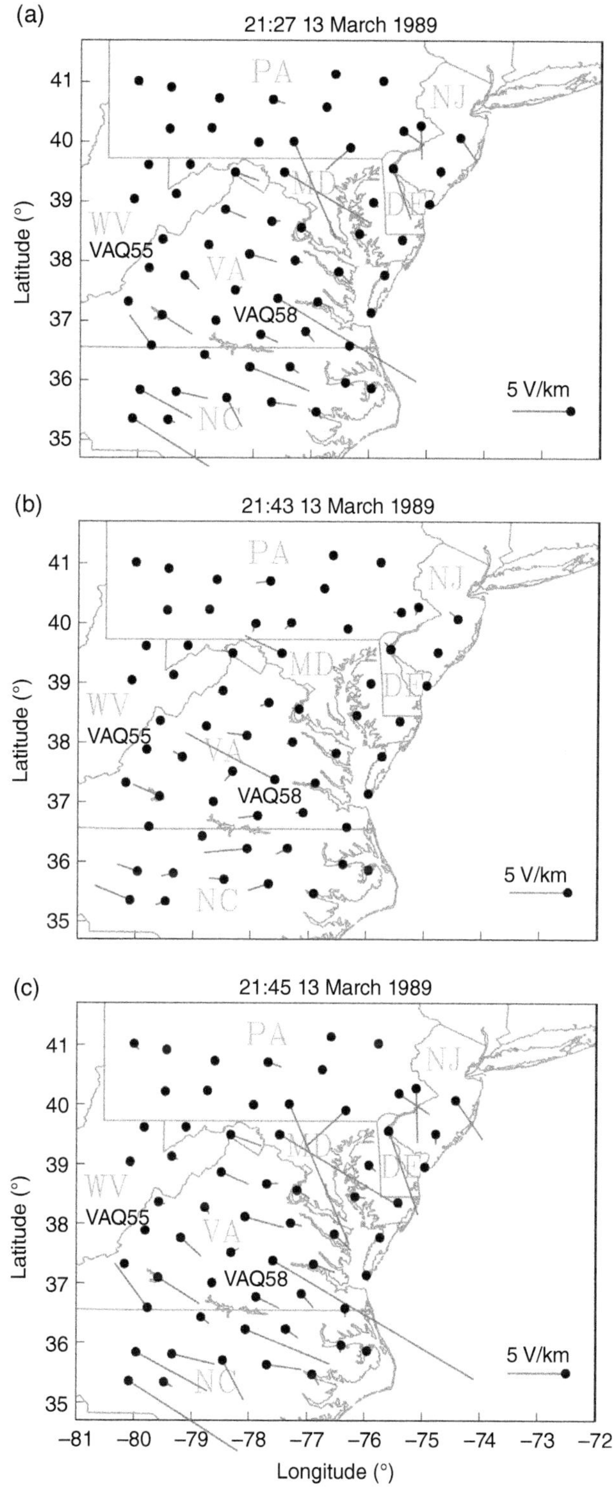

Figure 6.4 Maps of horizontal-component geoelectric field vectors $\mathbf{E}_h(t, x, y)$ for the various magnetotelluric survey sites across the Mid-Atlantic, each for three different instances in time t: (a) 21:27 UT, (b) 21:43 UT, and (c) 21:45 UT on 13 March 1989.

vectors at other sites, and the range in amplitude spans more than two orders of magnitude. Since all of the calculations we used to develop these maps of geoelectric vectors assume a uniform geomagnetic field, the spatial complexity of the geoelectric field seen in Figure 6.4 is entirely the result of localized complexity in the surface impedance of the Earth, that is, itself, due to a complex subsurface conductivity structure (e.g., McKay and Whaler, 2006; Bedrosian and Love, 2015; Bonner and Schultz, 2017).

6.6. 100-YEAR AND 1989 HAZARD MAPS

Extreme-value, storm-maximum geoelectric amplitudes E_h^m can be estimated for the Mid-Atlantic by extrapolating statistical models of geoelectric time series constructed by convolving long, multi-decade historical Fredericksburg geomagnetic time series with magnetotelluric impedance tensors from across the region (Love et al., 2018a). In Figure 6.5a, a hazard color-coded map is given for the Mid-Atlantic of geoelectric amplitudes that are only exceeded, on average, once per 100 years. Among the various sites across the Mid-Atlantic, 100-year amplitudes exceed 10 V/km at 4 of the 61 sites and exceed 3 V/km at 15 of the 61 sites. In more detail, exceedance amplitude is highest at survey site VAQ58 at 24.29 V/km; for year-2000 geomagnetic latitude and physiographic regions of this site, this is more than an order of magnitude higher than the 100-year benchmark amplitude (2.25 V/km) developed by NERC (North American Electric Reliability Corporation, 2014a) using one-dimensional Earth models. Exceedance amplitude is lowest at the nearby site VAQ55 at 0.05 V/km (while for NERC, it is much higher at 0.66 V/km). All of this reminds us of the importance of using realistic estimates of Earth-surface impedance when estimating geoelectric hazards.

The results presented in Figure 6.5a need to be properly interpreted – the map depicts point-wise statistical estimates of thresholds in geoelectric amplitude that are only exceeded, on average, once per century; the map does not show geoelectric amplitudes occurring during a single intense storm. Indeed, when a 100-year geoelectric amplitude is exceeded for a particular storm and at a particular site, it might not be attained for another site during the same storm. For comparison, in Figure 6.5b, we plot maximum E_h^m amplitudes for the various survey sites across the Mid-Atlantic during the March 1989 magnetic storm (this is not a statistical extrapolation). It is noteworthy that only small differences are seen between Figure 6.5a and b; the maximum E_h^m that occurred at VAQ58, for example, during the 1989 storm was 19.98 V/km. Furthermore, during the 1989 storm, E_h^m exceeded 10 V/km at 4 of the 61 sites considered; it exceeded 3 V/km at 12 of the 61 sites. Apparently, geoelectric amplitudes attained during the 1989 storm were only slightly less than our estimated 100-year exceedances. In this respect, the 1989 storm was nearly a "100-year event" for the Mid-Atlantic.

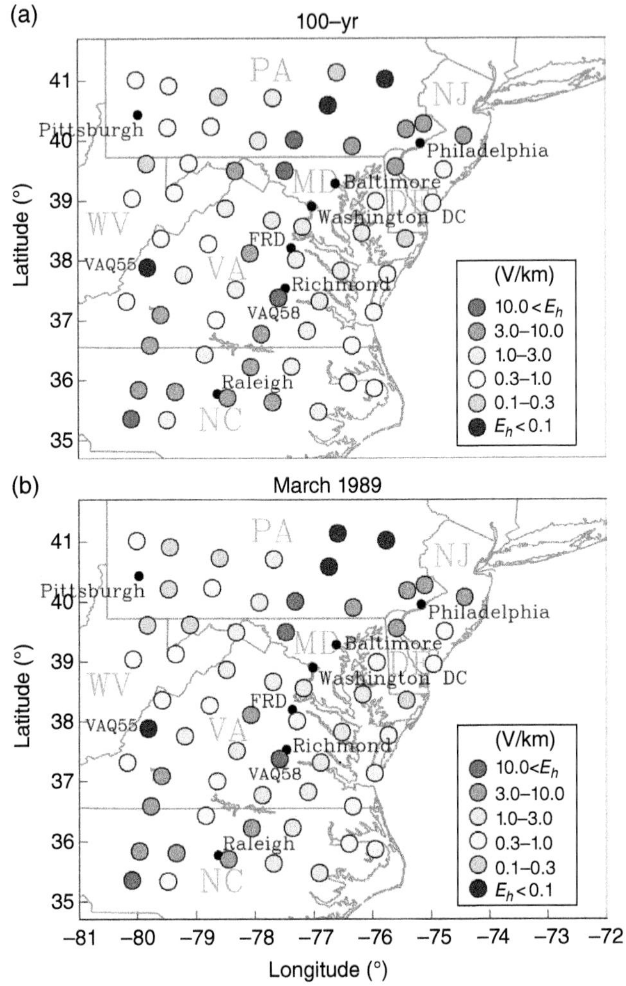

Figure 6.5 Maps showing (a) 100-year geoelectric exceedance amplitudes at the various EarthScope magnetotelluric survey sites across the Mid-Atlantic region of the United States, (b) maximum E_h^m that occurred during March 1989 magnetic storm. (*See insert for color representation of the figure.*)

The geographic granularity of geoelectric hazards seen in Figure 6.5 is correlated with geologic structure. The Mid-Atlantic region has, over more than a billion years, been shaped by plate-tectonic continental collisions and rifting events (e.g., Whitmeyer et al., 2007), creating southwest to northeast trending mountain belts and basement uplifts (e.g., King et al., 1974; Rast, 1989; Hatcher, 2005). Generally, low geoelectric hazards are seen surrounding Pittsburgh, a part of the Appalachian Plateau with electrically conductive sedimentary basins. Similarly, low geoelectric hazards are seen along the Mid-Atlantic Coastal Plain, a region covered by sediments that eroded from the mountains to the west. High geoelectric hazards are seen in the electrically resistive "crystalline Appalachians" that run southwest to northeast, from around Raleigh up to west of Philadelphia and which include the Blue Ridge Mountains and the Piedmont, which, respectively, consist of metamorphosed basement rocks, and igneous rocks and a metamorphosed assembly of island-arc and micro-continental terranes.

It is noteworthy that the geoelectric hazard maps do not show a clear "coast effect" – a localized distortion (and possible relative enhancement) of the geoelectric field caused by the conductivity contrast between continental crust and seawater (e.g., Lilley, 2007; Pirjola, 2013). We speculate that a coast effect might be obscured in our maps of geoelectric hazard by the complex shape of the Mid-Atlantic coast and the relatively shallow depth of the near-coast ocean. The Chesapeake Bay, which forms the eastern coast of Virginia and the inland bay of Maryland, is only about 6 m deep on average, but the diffusive skin depth for a half space of seawater at 2 min is $\delta_A \simeq 3$ km – in other words, the near-coast ocean is possibly too shallow to have an appreciable effect on the geoelectric fields analyzed here. Other factors might further obscure a coast effect, including the relatively conductive coastal-plane sediments that continue offshore and over the continental shelf (under the ocean) for over 100 km, and the spatial complexity of the Mid-Atlantic's solid-Earth conductivity structure. It is also possible that the geographic spacing of the EarthScope survey sites is too sparse to resolve a coast effect. Although further investigation of a possible Mid-Atlantic coast effect might be worthwhile, empirical results, such as those presented here, remain valid regardless of physical explanations.

6.7. DISCUSSION

This summary illustrates the importance of both long-term geomagnetic monitoring and modern magnetotelluric surveys for making realistic assessments of geoelectric hazards of concern to the electric-power grid industry. With respect to the effects on electric-power grid systems, the voltage on a grid line is obtained by integration

$$V = \int_G \mathbf{E}_h \cdot d\mathbf{l} \tag{6.4}$$

over the path of the power-grid line denoted as G. Since this path can have characteristic dimensions of hundreds of kilometers, the geomagnetically induced voltage on power grid lines can, very easily, attain hundreds of volts (e.g., Lucas et al., 2018). Under Ohm's law, and assuming line resistance of (say) 1 Ω, hundreds of GIC amps can be realized during an intense storm; such currents are sufficient to damage high-voltage transformers (e.g., Girgis and Vedante, 2002). Looking beyond the retrospective results presented here, a related follow-on project would be the development of a real-time mapping service, though the challenges for such a project are considerable (e.g., Love et al., 2018b). Additional geomagnetic monitoring stations and denser magnetotelluric surveying, both in the United States and in southern Canada, would lead to improved geoelectric hazard mapping needed to assess the vulnerability of the greater integrated North-American electric-power grid network.

ACKNOWLEDGMENTS

We thank C. A. Finn, J. McCarthy, E. J. Rigler, and J. L. Slate for reviewing a draft manuscript. We thank C. C. Balch and A. Pulkkinen for useful conversations. This work was supported by the USGS Geomagnetism Program; G. M. Lucas is supported by a USGS Mendenhall postdoctoral fellowship. Magnetic observatory data used in this analysis are available from the USGS Geomagnetism Program (www.usgs.gov; 1983-2014); some of them can be obtained from INTERMAGNET (www.intermagnet.org; 1991–2014). EarthScope impedance tensors can be obtained from the Data Management Center of the Incorporated Research Institutions for Seismology (ds.iris.edu/ds/products/emtf) (Kelbert et al., 2011).

REFERENCES

Allen, J., L. Frank, H. Sauer, and P. Reiff, Effects of the March 1989 solar activity, *Eos Trans. Am. Geophys. Union*, *70*(46), 1479, 1486–1488, doi:10. 1029/89EO00409, 1989.

Baker, D. N., et al., *Severe Space Weather Events – Understanding Societal and Economic Impacts*, pp. 1–144, The National Academy Press, Washington, DC, doi:10.17226/12507, 2008.

Barnes, P. R., and J. W. Van Dyke, Economic consequences of geomagnetic storms (a summary), *IEEE Power Eng. Rev.*, *10*(11), 3–4, doi:10.1109/39. 60450, 1990.

Bedrosian, P. A., and J. J. Love, Mapping geoelectric fields during magnetic storms: Synthetic analysis of empirical United States impedances, *Geophys. Res. Lett.*, *42*(23), 10160–10170, doi:10.1002/2015GL066636, 2015.

Béland, J., and K. Small, Space weather effects on power transmission systems: The cases of Hydro-Québec and Transpower New Zealand Ltd, in *Effects of Space Weather on Technology Infrastructure*, edited by I. A. Daglis, pp. 287–299, Springer, Dordrecht, The Netherlands, 2005.

Berdichevsky, M. N., and V. I. Dmitriev, *Models and Methods of Magnetotellurics*, pp. 1–563, Springer-Verlag, Berlin, Germany, 2008.

Blum, C., T. C. White, E. A. Sauter, D. C. Stewart, P. A. Bedrosian, and J. J. Love, Geoelectric monitoring at the boulder magnetic observatory, *Geosci. Instrum. Method. Data Syst.*, *6*, 447–452, doi:10.5194/gi-2017-27, 2017.

Bolduc, L., GIC observations and studies in the Hydro-Québec power system, *J. Atmos. Solar-Terr. Phys.*, *64*(16), 1793–1802, doi:10.1016/S1364-6826(02) 00128-1, 2002.

Bonner, L. R., and A. Schultz, Rapid prediction of electric fields associated with geomagnetically induced currents in the presence of three-dimensional ground structure: Projection of remote magnetic observatory data through magnetotelluric impedance tensors, *Space Weather*, *15*(1), 204–227, doi:10.1002/2016SW001535, 2017.

Boteler, D. H., Assessment of geomagnetic hazard to power systems in Canada, *Nat. Hazards*, *23*(2–3), 101–120, 2001.

Boteler, D. H., The super storms of August/September 1859 and their effects on the telegraph system, *Adv. Space Res.*, *38*(2), 159–172, 2006.

Chave, A. D., Estimation of the magnetotelluric response function, in *The Magnetotelluric Method*, edited by A. D. Chave and A. G. Jones, pp. 165–218, Cambridge University Press, Cambridge, UK, 2012.

Chave, A. D., and A. G. Jones (Eds.), *The Magnetotelluric Method*, pp. 1–552, Cambridge University Press, Cambridge, UK, 2012.

Cliver, E. W., and W. F. Dietrich, The 1859 space weather event revisited: Limits of extreme activity, *J. Space Weather Space Clim.*, *3*, A31, doi:10. 1051/swsc/2013053, 2013.

Cuttler, S. W., J. J. Love, and A. Swidinsky, Geoelectric hazard assessment: The differences of geoelectric responses during magnetic storms within common physiographic zones, *Earth Planets Space*, *70*, doi:10.1186/s40623-018-0807-7, 2018.

Eastwood, J. P., et al., The economic impact of space weather: Where do we stand? *Risk Anal.*, *37*(2), 206–218, doi:10.1111/risa.12765, 2017.

Egbert, G. D., Robust electromagnetic transfer functions estimates, in *Encyclopedia of Geomagnetism and Paleomagnetism*, edited by D. Gubbins and E. Herrero-Bervera, pp. 866–870, Springer, Dordrecht, The Netherlands, 2007a.

Egbert, G. D., EM modeling, inverse, in *Encyclopedia of Geomagnetism and Paleomagnetism*, edited by D. Gubbins and E. Herrero-Bervera, pp. 219–223, Springer, Dordrecht, The Netherlands, 2007b.

Evans, R. L., Earth's electromagnetic environment: 3A. Conductivity of Earth materials, in *The Magnetotelluric Method*, edited by A. D. Chave and A. G. Jones, pp. 50–95, Cambridge University Press, Cambridge, UK, 2012.

Federal Energy Regulatory Commission, Reliability standards for geomagnetic disturbances, *Fed. Regist.*, *78*(100), 30747–30762, 2013.

Ferguson, I. J., Instrumentation and field procedures, in *The Magnetotelluric Method*, edited by A. D. Chave and A. G. Jones, pp. 421–479, Cambridge University Press, Cambridge, UK, 2012.

Fujii, I., T. Ookawa, S. Nagamachi, and T. Owada, The characteristics of geoelectric fields at Kakioka, Kanoya, and Memambetsu inferred from voltage measurements during 2000 to 2011, *Earth Planets Space*, *67*, 62, doi:10.1186/s40623-015-0241-z, 2015.

Gates, A. E., and L. Glover III, Alleghanian tectono-thermal evolution of the dextral transcurrent Hylas zone, Virginia Piedmont, *U.S.A., J. Struct. Geol.*, *11*(4), 407–419, doi:10.1016/0191-8141(89)90018-7, 1989.

Girgis, R., and K. Vedante, Effects of GIC on power transformers and power systems, *IEEE Trans. Power Delivery*, *17*(4), 1002–1008, doi:10.1109/TPWRD.2002.803710, 2002.

Hatcher, R. D., North America: Southern and Central Appalachians, in *Encyclopedia of Geology*, edited by R. C. Selley, L. Cocks, M. Robin, and I. R. Plimer, pp. 72–81, Elsevier, Oxford, UK, 2005.

Horton, R., D. Boteler, T. J. Overbye, R. Pirjola, and R. C. Dugan, A test case for the calculation of geomagnetically induced currents, *IEEE Trans. Power Delivery*, *27*(4), 2368–2373, doi:10.1109/TPWRD.2012.2206407, 2012.

Kappenman, J., A perfect storm of planetary proportions, *IEEE Spectrum*, *49*, 26–31, doi:10.1109/MSPEC.2012.6139230, 2012.

Kelbert, A., S. Erofeeva, C. Trabant, R. Karstens, M. Van Fossen, G. D. Egbert, and A. Schultz, IRIS DMC Data services products: EMTF, the magnetotelluric transfer functions, U.S. Geological Survey, doi:10.17611/DP/EMTF, 2011.

Kelbert, A., N. Meqbel, G. D. Egbert, and K. Tandon, ModEM: A modular system for inversion of electromagnetic geophysical data, *Comput. Geosci.*, *66*, 40–53, doi:10.1016/j.cageo.2014.01.010, 2014.

Kelbert, A., C. C. Balch, A. Pulkkinen, G. D. Egbert, J. J. Love, E. J. Rigler, and I. Fujii, Methodology for time-domain estimation of storm time geoelectric fields using the 3-D magnetotelluric response tensors, *Space Weather*, *15*(7), 874–894, doi:10.1002/2017SW001594, 2017.

King, P. B., H. M. Beikman, and G. J. Edmonston, *Geologic Map of the United States (Exclusive of Alaska and Hawaii)*, U.S. Geological Survey Report, 1974.

Lakhina, G. S., and B. T. Tsurutani, Super geomagnetic storms: Past, present and future, in *Extreme Space Weather: Origins, Predictability, and Consequences*, edited by N. Buzulukova, chap. 7, pp. 157–185, Elsevier, Amsterdam, The Netherlands, 2018.

Lilley, T., Coast effect of induced currents, in *Encyclopedia of Geomagnetism and Paleomagnetism*, edited by D. Gubbins and E. Herrero-Bervera, pp. 61–65, Springer, Dordrecht, The Netherlands, 2007.

Lloyd's of London, *Emerging Risk Report: Solar Storm Risk to the North American Electric Grid*, pp. 1–22, London, UK, 2013.

Love, J. J., and C. A. Finn, The USGS Geomagnetism Program and its role in space weather monitoring, *Space Weather*, *9*(7), S07001, doi:10.1029/2011SW000684, 2011.

Love, J. J., E. J. Rigler, A. Pulkkinen, and C. C. Balch, Magnetic storms and induction hazards, *Eos Trans. Am. Geophys. Union*, *95*(48), 445–446, doi:10.1002/2014EO480001, 2014.

Love, J. J., G. M. Lucas, A. Kelbert, and P. A. Bedrosian, Geoelectric hazard maps for the Mid-Atlantic United States: 100-year extreme values and the 1989 magnetic storm, *Geophys. Res. Lett.*, *45*(1), 5–15, doi:10.1002/2017GL076042, 2018a.

Love, J. J., E. J. Rigler, A. Kelbert, C. A. Finn, P. A. Bedrosian, and C. C. Balch, On the feasibility of real-time mapping of the geoelectric fields across North America, *USGS Open-File Report, 2018-1043*, pp. 1–16, doi:10.3133/ofr20181043, 2018b.

Lucas, G. M., J. J. Love, and A. Kelbert, Calculation of voltages in electric power transmission lines during historic geomagnetic storms: An investigation using realistic Earth impedances, *Space Weather*, *16*(2), 185–195, doi:10. 1002/2017SW001779, 2018.

McKay, A. J., and K. A. Whaler, The electric field in northern England and southern Scotland: Implications for geomagnetically induced currents, *Geophys. J. Int.*, *167*(2), 613–625, doi:10.1111/j.1365-246X.2006.03128.x, 2006.

Molinski, T. S., Why utilities respect geomagnetically induced currents, *J. Atmos. Solar-Terr. Phys.*, *64*(16), 1765–1778, doi:10.1016/S1364-6826(02) 00126-8, 2002.

Murphy, B. S., and G. D. Egbert, Electrical conductivity structure of Southeastern North America: Implications for lithospheric architecture and Appalachian topographic rejuvenation, *Earth Planet. Sci. Lett.*, *462*, 66–75, doi:10.1016/j.epsl.2017.01.009, 2017.

National Science and Technology Council, *National Space Weather Action Plan*, pp. 1–38, Executive Office, Washington, DC, 2015.

North American Electric Reliability Corporation, March 13, 1989 geomagnetic disturbance, in *1989 System Disturbances*, pp. 36–60, Princeton, NJ, 1990.

North American Electric Reliability Corporation, *Application Guide: Computing Geomagnetically-Induced Current in the Bulk-Power System*, pp. 1–39, Atlanta, GA, 2013.

North American Electric Reliability Corporation, *Benchmark Geomagnetic Disturbance Event Description*, pp. 1–26, Atlanta, GA, 2014a.

North American Electric Reliability Corporation, *Transformer Thermal Impact Assessment: Project 2013-03 (Geomagnetic Disturbance Mitigation)*, pp. 1–16, Atlanta, GA, 2014b.

Odenwald, S. F., and J. L. Green, Bracing for a solar superstorm, *Sci. Am.*, *299*(8), 80–87, doi:10.1038/scientificamerican080852980, 2006.

Panel on Seismic Hazard Analysis, *Probabilistic Seismic Hazard Analysis*, pp. 1–97, National Academy Press, Washington, DC, 1988.

Piccinelli, R., and E. Krausmann, *Space Weather and Power Grids – A Vulnerability Assessment*, pp. 1–53, European Union, Luxembourg, 2014.

Pirjola, R., Review on the calculation of surface electric and magnetic fields and of geomagnetically induced currents in ground-based technological systems, *Surv. Geophys.*, *23*(1), 71–90, doi:10.1023/A:10148160, 2002.

Pirjola, R., Practical model applicable to investigating the coast effect on the geoelectric field in connection with studies of geomagnetically induced currents, *Adv. App. Phys.*, *1*(1), 9–28, 2013.

Pulkkinen, A., S. Lindahl, A. Viljanen, and R. Pirjola, Geomagnetic storm of 29-31 October 2003: Geomagnetically induced currents and their relation to problems in the Swedish high-voltage power transmission system, *Space Weather*, *3*(8), S08C03, doi:10.1029/2004SW000,123, 2005.

Pulkkinen, A., et al., Geomagnetically induced currents: Science, engineering and applications readiness, *Space Weather*, *15*(7), 828–856, doi:10.1002/2016SW001501, 2017.

Rast, N., The evolution of the Appalachian chain, in *The Geology of North America: An Overview*, vol. *A*, edited by A. W. Bally and A. R. Palmer, pp. 323–348, Geological Society of America, Boulder, CO, doi:10.1130/DNAG-GNA-A.323, 1989.

Rodi, W. L., and R. L. Mackie, The inverse problem, in *The Magnetotelluric Method*, edited by A. D. Chave and A. G. Jones, pp. 347–420, Cambridge University Press, Cambridge, UK, 2012.

Schultz, A., A continental scale magnetotelluric observatory and data discovery resource, *Data Sci. J.*, *8*, IGY6–IGY20, 2010.

Schultz, A., G. D. Egbert, A. Kelbert, T. Peery, V. Clote, B. Fry, S. Erofeeva, and staff of the National Geoelectromagnetic Facility and their contractors, USArray TA magnetotelluric transfer functions, doi:10.17611/DP/EMTF/USARRAY/TA, 2006–2018.

Simpson, F., and K. Bahr, *Practical Magnetotellurics*, pp. 1–254, Cambridge University Press, Cambridge, UK, 2005.

Siscoe, G. L., N. U. Crooker, and C. R. Clauer, *Dst* of the Carrington storm of 1859, *Adv. Space Res.*, *38*(2), 173–179, 2006.

Smith, K., and D. N. Petley, *Environmental Hazards: Assessing Risk and Reducing Disaster*, pp. 1–382, Routledge, New York, NY, 2009.

Smolka, A., Natural disasters and the challenge of extreme events: Risk management from an insurance perspective, *Phil. Trans. R. Soc. Lond. Ser. A*, *364*, 2147–2165, 2006.

Stratton, J. A., *Electromagnetic Theory*, pp. 1–615, McGraw-Hill Book Company, New York, NY, 1941.

Tsurutani, B. T., W. D. Gonzalez, G. S. Lakhina, and S. Alex, The extreme magnetic storm of 1-2 September 1859, *J. Geophys. Res.*, *108*(A7), 1268, doi:10.1029/2002JA009504, 2003.

Unsworth, M., Magnetotellurics, in *Encyclopedia of Geomagnetism and Paleomagnetism*, edited by D. Gubbins and E. Herrero-Bervera, pp. 670–673, Springer, Dordrecht, The Netherlands, 2007.

Varnes, D. J., *Landslide Hazard Zonation: A Review of Principles and Practices*, pp. 1–63, United Nations Educational, Scientific and Cultural Organization, Paris, France, 1984.

Virginia Division of Mineral Resources, *Geologic Map of Virginia*, Charlottesville, VA, 1993.

Wait, J. R., *Geo-electromagnetism*, pp. 1–268, Academic Press, New York, NY, 1982.

Watermann, J., O. Rasmussen, P. Stauning, and H. Gleisner, Temporal versus spatial geomagnetic variations along the west coast of Greenland, *Adv. Space Res.*, *37*(6), 1163–1168, doi:10.1016/j.asr.2005.08.019, 2006.

Weidelt, P., and A. D. Chave, The magnetotelluric response function, in *The Magnetotelluric Method*, edited by A. D. Chave and A. G. Jones, pp. 122–164, Cambridge University Press, Cambridge, UK, 2012.

Weigel, R. S., A comparison of methods for estimating the geoelectric field, *Space Weather*, *15*(2), 430–440, doi:10.1002/2016SW001504, 2017.

Whitmeyer, S. J., L. S. Fichter, and E. J. Pyle, New directions in Wilson Cycle concepts: Supercontinent and tectonic rock cycles, *Geosphere*, *3*(6), 511–526, doi:10.1130/GES00091.1, 2007.

Williams, M. L., K. M. Fischer, J. T. Freymueller, B. Tikoff, A. M. Tréhu, and others, *Unlocking the Secrets of the North American Continent: An EarthScope Science Plan for 2010-2020*, pp. 1–78, EarthScope, 2010.

Worman, S., S. Taylor, T. Onsager, J. Adkins, D. N. Baker, and K. F. Forbes, The social and economic impacts of moderate and sever space weather, in *Extreme Space Weather: Origins, Predictability, and Consequences*, edited by N. Buzulukova, chap. 29, pp. 701–710, Elsevier, Amsterdam, The Netherlands, 2018.

Yoshino, T., Electrical properties of rocks, in *Encyclopedia of Solid Earth Geophysics*, edited by H. K. Gupta, pp. 270–276, Springer-Verlag, Dordrecht, The Netherlands, 2011.

7

The Magnetotelluric Method and Its Application to Understanding Geomagnetically Induced Currents

Esteban Bowles-Martinez and Adam Schultz

ABSTRACT

The electric power transmission infrastructure is exposed to hazards from geomagnetically induced currents (GICs) in the Earth. We review the theory of geomagnetic induction and how electromagnetic (EM) signals measured on the surface are used to create magnetotelluric (MT) impedance tensor data, which describe how the relationship between the electric and magnetic fields measured on the Earth's surface depends on the Earth's internal electrical conductivity. MT data are usually used to create subsurface resistivity models from which geologic interpretations are made, but they can also be used to predict ground electric fields induced by disturbances to the geomagnetic field, such as those produced by a solar coronal mass ejection. This chapter describes the complete MT workflow with attention to GICs, including the basic theory and physics of MT, instrumentation, data acquisition, data processing, inversion, and geologic interpretation of resistivity models.

Key Points

• Earth's magnetic field induces electric fields that can be used to determine the three-dimensional structure of the Earth's interior and predict geomagnetic disturbances.
• Magnetotelluric data acquisition must be done with instruments sensitive to the frequency range of interest, which usually depends on the depth to subsurface targets of interest.
• It is highly advantageous to model Earth's conductivity in three dimensions.

7.1. INTRODUCTION

The relationship between Earth's electric and magnetic fields can be used to determine the electrical properties of the subsurface (Tikhonov, 1950; Cagniard, 1953). This chapter introduces the magnetotelluric (MT) geophysical method, which is used to create electrical resistivity models of the Earth's interior as well as for modeling electric fields related to geomagnetically induced currents (GICs). Thorough reviews of magnetotellurics are provided by Ward and Hohmann (1988), Vozoff (1991), Simpson and Bahr (2005), Chave and Jones (2012), and Zhdanov (2017). MT employs the principle of electromagnetic induction where the Earth's time-varying magnetic field (H) produces an electric field (E) within the conductive materials of Earth's interior. These signals are analyzed in the frequency domain over a spectrum that can range from tens of kHz to sub-mHz, with lower-frequency signals corresponding to greater depth of penetration into the Earth's subsurface. This broad frequency range allows the crust and upper mantle to be investigated. This chapter provides an introduction to MT by covering the following topics: the physics of geoelectromagnetic induction, how electromagnetic (EM) signals are used in MT, data acquisition, data processing and inversion, and a brief overview of resistivity model interpretation.

College of Earth, Ocean, and Atmospheric Sciences, Oregon State University, Corvallis, OR, USA

Geomagnetically Induced Currents from the Sun to the Power Grid, Geophysical Monograph 244, First Edition.
Edited by Jennifer L. Gannon, Andrei Swidinsky, and Zhonghua Xu.

Estimating ground-level electric fields for GIC prediction and subsequent vulnerability assessment and mitigation requires measuring electromagnetic (EM) signals and converting the electric and magnetic field data to MT data in the frequency domain. MT data are typically represented as the impedance tensor **Z**, a frequency dependent 2×2 matrix that relates electric and magnetic fields through the equation $\mathbf{E} = \mathbf{ZH}$. The mapping between inducing magnetic fields and induced electric fields depends on the volumetric variations in subsurface electrical conductivity (see Section 7.3.3 for more information on the impedance tensor). Data acquisition must be on a time-scale – and therefore within a frequency band – of relevance to the natural signal content of space weather events that are known to impact the power grid. Broadly speaking, to monitor the response to geomagnetic disturbances associated with GICs, current practice suggests that MT impedances are required over a frequency band spanning $\sim 10^{-4}$–10^{-1} Hz. Such a band yields information of adequate spatial resolution on electrical conductivity structure from mid-crustal depths (~10 km below ground surface) to the upper mantle (~300 km depth), depending on the vertically integrated conductivity (conductance) of the overburden. While such data have reduced resolving power at shallow crustal depths, they remain sensitive to the conductance from the surface to the upper mantle, as well as to unresolvably fine-scale, near-surface galvanic scattering or distortion effects that impact the electric field. Obtaining MT data in the aforementioned frequency band is the province of long-period MT, typically, although not exclusively, obtained by combining observatory-quality triaxial fluxgate magnetometer measurements of the magnetic field with horizontal electric dipole measurements, typically at a sample rate of 1–10 samples per second.

Through the impedance tensor, GICs can be predicted from magnetic field measurements. Space weather events that cause GMDs strong enough to produce dangerous GICs are a rare occurrence, but occur frequently enough and with high enough potential for infrastructure damage that it is important to predict their effects using MT data. The earliest and strongest GMD known to affect infrastructure occurred in 1859 causing damage to telegraph systems (Carrington, 1859; Boteler, 2006; Shea & Smart, 2006), with other significant events in 1989, 1972, 1958, 1940, 1921, and 1872 (Boteler, 2001; Silverman, 2006). There is also current interest in the problem of intense, but geographically more localized geomagnetic disturbances and related GICs induced by exo-atmospheric detonation of nuclear devices, also called electromagnetic pulse (EMP) weapons. The signal produced by an EMP weapon has three stages (Foster et al., 2004). The first, or E1 stage, occurs as a high-amplitude impulsive EM energy burst within the first nanoseconds after detonation and is capable of damaging electronic devices. The intermediate E2 stage follows within the first second, with lower amplitude energy and an effect similar to a weak but widespread lightning strike affecting an area thousands of kilometers wide (Foster et al., 2008). The final stage, E3, lasts for a minute or more and rather than propagating as an electromagnetic wave, it results from distortions to the Earth's magnetic field caused by the nuclear detonation, inducing currents that can damage infrastructure in an area about half as wide as the E2 stage (Foster et al., 2008). The frequency range of the E3 stage of these disturbances requires the use of EM signals that extend to higher frequencies than those created by GMDs, generally in the band of 10^{-3}–10^{3} Hz, and that involves electromagnetic inductive responses in the upper ~10 km of the Earth's crust as well as in deeper crust and upper mantle sections. By using instruments sensitive to this frequency range, the same methods used for estimating ground electric fields from GMDs can be applied to these artificially induced signals.

7.2. THE MAGNETOTELLURIC METHOD

7.2.1. EM Induction in the Earth

Electric and magnetic fields within the Earth, like elsewhere, are governed by Maxwell's equations. In MT, the range of frequencies used and assumptions about the plausible range of the electrical conductivity of geologic materials under crust and upper mantle conditions allow us to simplify the relationships between the fields. Variables and units are defined in Table 7.1. The reduced governing equations for MT are therefore Faraday's Law:

$$\nabla \times E = -\mu_0 \frac{\partial H}{\partial t}. \tag{7.1}$$

Table 7.1 Units Used in Electromagnetism.

Variable	Description	Units
E	Electric field	V/m
H	Magnetic field	A/m
B	Magnetic flux density	Tesla
t	Time	s
T	Period	s
ω	Angular frequency	s^{-1}
μ_0	Magnetic permeability of free space	$4\pi \times 10^{-7}$ V s A^{-1} m^{-1}
ε	Electrical permittivity	8.85×10^{-12} F/m
$\boldsymbol{\beta}$	Charge density	C/m^3
$\mathbf{J}$	Current density	A/m^2
σ	Conductivity	S/m
ρ	Resistivity	Ωm

and Ampere's Law:

$$\nabla \times H = \sigma E + \varepsilon_0 \frac{\partial E}{\partial t}. \tag{7.2}$$

In this text, we will refer to the magnetic field H with units of V/m rather than magnetic flux density B, which is equal to $\mu_0 H$ and has units of Teslas. We will also neglect small differences in permeability μ, which varies only by parts per thousand for most materials, whereas conductivity varies by orders of magnitude. We therefore assume free space permeability ($\mu = 4\pi \cdot 10^{-7}$ H/m) for all cases for the remainder of this chapter.

We will also use Gauss's law to relate electric field divergence with charge density β:

$$\nabla \cdot E = \frac{\beta}{\varepsilon_0}. \tag{7.3}$$

Faraday's Law (Equation (7.1)) shows that a time-varying magnetic field produces a circulating electric field in a plane orthogonal to the magnetic field. Similarly, Ampere's law shows that a circulating magnetic field is produced by an electric current and/or a time-varying electric field.

Taking the curl of both sides of Equation (7.1), we have:

$$\nabla \times (\nabla \times E) = -\mu \frac{\partial}{\partial t}(\nabla \times H). \tag{7.4}$$

Applying the curl of curl vector identity to the left-hand side gives:

$$\nabla \times (\nabla \times E) = -\nabla^2 E + \nabla(\nabla \cdot E). \tag{7.5}$$

Using Equation (7.3) and assuming charge density β is uniform, the final term of Equation (7.5) becomes zero. The derivation shown in Eqs. (7.4) through (7.7) describes a homogeneous medium, where charge density is uniform, though this is not necessarily true in realistic environments that may have gradients in the conductivity of materials. Returning to Equation (7.4) and substituting on the right-hand side, we now have:

$$\nabla^2 E = \mu\sigma \frac{\partial E}{\partial t} + \mu\varepsilon_0 \frac{\partial^2 E}{\partial t^2}, \tag{7.6}$$

which is the electromagnetic wave equation. Before continuing, two assumptions about MT must be specified. The first is that the EM signals used in MT are at low enough frequencies that the fields change at rates slow enough that they can be considered static fields. This quasi-static assumption along with the very small values of the constants μ_0 and ε_0, makes the second derivative term in Equation (7.6) negligible. The second assumption is that the Earth is sufficiently conductive that the first term on the right-hand side of Equation (7.6) is much larger than the second term. While there is wide variability in conductivity of individual crystals and fluids in the Earth between about 10^{10} and 10^{-10} S/m, the most resistive rocks on a scale large enough to be measured by MT have bulk conductivities above about 10^{-5} S/m, with typical rocks ranging from 10^{1} to 10^{-4} S/m (Tyburczy & Fisler, 1995; Yoshino, 2010; Pommier & Le-Trong, 2011). Therefore, the second term becomes negligible relative to the first. Discarding it results in the diffusion equation for electric fields:

$$\nabla^2 E - \mu\sigma \frac{\partial E}{\partial t} = 0. \tag{7.7}$$

A similar procedure can be followed to give the equivalent equation for magnetic fields. Thus, in the quasi-static approximation, in both time- and frequency-domains, it is common practice to approximate the MT governing equations as diffusion rather than wave equations. It should be noted, however, that a transformation exists into the *fictitious wave domain* (Mittet, 2015) in which the low frequency electromagnetic induction problem can be expressed to a close approximation as a wave propagation rather than diffusion phenomenon.

7.2.2. 2D and 3D MT

The charge-free assumption is not valid when current can travel across boundaries between materials of contrasting conductivity, as is the case in 2D and 3D environments. In 2D MT, where MT measurements are taken along a line of stations traversing a target of interest, it is common to separate the orthogonal E and H signals into two sets of data. The first set involves current that travels across conductivity boundaries (across geoelectric strike) with magnetic fields that are measured parallel to strike, called the "transverse magnetic" or TM mode. The second set of signals is the "transverse electric" or TE mode, whose magnetic field measurements are taken across strike and electric fields are measured parallel to strike. Because the TM mode involves current that crosses a conductivity boundary, and current is continuous, the Ohm's Law relationship $\sigma E = J$ (where J is current density) requires charge to exist along conductivity boundaries because E must be discontinuous if σ is discontinuous. Because TM mode accounts for boundary charges, complex structures are usually resolved better using the TM mode (Wannamaker et al., 1984), though it can also be very strongly affected by small-scale near-surface features that alter the observed conductivity at all frequencies (see Section 7.3.5), which do not affect the TE mode as strongly. Additionally, the TE mode may be more useful in resolving structures that are resistive rather than conductive (Berdichevsky et al., 1998). It is common to solve both modes simultaneously to obtain a robust 2D solution.

The presence of surface charge in 2D and 3D MT requires that Equation (7.4) (containing the analytically cumbersome $\nabla \times \nabla \times E$ term) is solved. This is usually done with the integral form of Maxwell's equations and/or in the frequency domain for mathematical simplicity. A common way of solving these problems is with a finite difference approach. E and H can be approximated in two or three dimensions by subdividing the Earth's interior into a grid of rectangular blocks (rectangles or prisms for 2D and 3D, respectively) and integrating E and H along simple rectangular loops through the block faces (or edges). While this is not as accurate as solving Maxwell's equations at every point, this simple geometry eases the computational burden sufficiently that a large enough number of blocks can be used to obtain a satisfactorily accurate solution within a reasonable time (e.g., Yee, 1966; Mackie et al., 1994; Smith, 1996; Siripunvaraporn et al., 2005; Egbert & Kelbert, 2012). Additional information about dimensionality in MT will be given in Section 7.3.4.

7.3. MT FUNDAMENTAL PRINCIPLES

7.3.1. Signal Source

The source of MT data is ideally modeled as a time-varying magnetic field produced at a sufficiently great distance above the Earth's surface (e.g., at ionospheric heights of 90 km or greater) that it is assumed to be a plane wave when it reaches the surface, inducing an electric field in the same horizontal plane as the incoming magnetic field. Conventional wisdom holds that this assumption is valid at mid-geomagnetic latitudes distant from the auroral current systems and equatorial electrojet. However, studies in Antarctica have shown that high-quality data is obtainable despite close proximity to geomagnetic poles for periods between 0.01 and 4000 s (Beblo & Liebig, 1990; Wannamaker et al., 1996, 2004; Peacock & Selway, 2016). Nonplanar source effects at polar latitudes are present only intermittently and can be identified and removed through time-frequency analysis (Chant & Hastie, 1992; Peacock & Selway, 2016).

7.3.2. Skin Depth

Because MT is most typically implemented as a frequency domain method, E and H are considered periodic signals with an $e^{i\omega t}$ term. Whereas previous sections of this paper used time domain expressions, this section will use the frequency domain. Equation (7.7) is then restated as:

$$\nabla^2 E - i\omega\mu\sigma E = 0. \tag{7.8}$$

To consider how this frequency-domain expression maps into the spatial domain as the electric and magnetic fields penetrate into the Earth, we must consider it in the wavenumber domain as well as the frequency domain. The $i\omega\mu\sigma$ term in Equation (7.8) is the square of the complex wavenumber $k^2 = i\omega\mu\sigma$, or $k = \sqrt{i\omega\mu\sigma}$, which is a value that describes EM field attenuation in a conductor and has units of m^{-1}, where ω is the radian frequency ($\omega = 2\pi f$), μ is the magnetic permeability of the geologic material (usually taken to have the free space value $\mu_0 = 1.257 \times 10^{-6}$ H/m), and σ is the electrical conductivity (S/m), with its reciprocal being the electrical resistivity $\rho = \sigma^{-1}$ (Ωm). It should also be noted that some geophysics texts use the convention that $k^2 = -i\omega\mu\sigma$ (e.g., Ward & Hohmann, 1988). This is the reduced, or quasi-static form of the wavenumber. The full form needed for higher frequencies beyond those typically used in MT (generally above the audio band $f > 20$ kHz) is $k^2 = i\omega\mu\sigma - \mu\omega^2\varepsilon$.

The reciprocal of the real part of this expression gives the skin depth $\delta(\omega,\sigma)$, the depth to which an inducing magnetic field of a given frequency ω at the surface of a conducting body decays in amplitude to $1/e$ of its surface amplitude. The real part of the expression for k is as follows:

$$k = \sqrt{i\omega\mu\sigma} = \sqrt{i}\sqrt{\omega\mu\sigma} = \frac{1+i}{\sqrt{2}}\sqrt{\omega\mu\sigma} = \sqrt{\omega\mu\sigma/2} + i\sqrt{\omega\mu\sigma/2}, \tag{7.9}$$

with the reciprocal of the real part of Equation (7.9) being $\sqrt{2/\omega\mu\sigma}$. Rather than using units of conductivity, its reciprocal – resistivity – is preferred in modern studies. Furthermore, the EM signals produced by GICs vary over such long time scales that it is appropriate to work with period rather than frequency. Restating the previous expression in terms of resistivity and period gives:

$$\delta = \sqrt{2/\left(8\pi^2 \times 10^{-7}\right)}\sqrt{\rho T} \approx 503.29\sqrt{\rho T}. \tag{7.10}$$

The skin depth is often used as a rough estimate of the depth of sensitivity for data at a given period for an assumed half-space resistivity.

7.3.3. Impedance Tensor

The magnetotelluric impedance tensor is a complex 2×2 matrix that represents the relationship between the electric and magnetic fields at a specified frequency. This relationship depends on the resistivity of the subsurface, and the ratio E/H at a given frequency defines resistivity in a half-space. More complex subsurface resistivity structures will be discussed in the next section. Generally, a rank-2 tensor relates two vectors in every combination of the basis vectors in which they exist. However, the assumptions used in MT eliminate terms in the impedance tensor involving the vertical direction, leaving only the x and y basis vectors, which represent the north and east directions, respectively. The complete 3×3 impedance tensor using x, y, and z directions is shown in Figure 7.1a. Because electric fields measured at the surface have no vertical component, the bottom row is eliminated. And because the MT method uses only the electric fields polarized in the x and y directions and a vertical magnetic field would create circulating electric fields, the right column is eliminated. This results in the simplified 2×2 impedance tensor that is used in MT, shown in Figure 7.1b.

Even though the vertical magnetic field is not used in MT for GIC estimation, it is worth mentioning that it is helpful in resistivity modeling to use the vertical magnetic transfer function (VTF), also called the tipper, which relates the vertical and horizontal magnetic fields. VTFs are complementary to MT data by helping to indicate dimensionality and for creating frequency-dependent induction vectors that point toward (or away from, depending on the direction convention used) conductive bodies (e.g., Parkinson, 1962; Gregori & Lanzerotti, 1980; Weaver & Agarwal, 1991).

In the following discussion, variables that represent matrices are identified by bold font. The impedance tensor, **Z**, is used as a transfer function that relates the electric and magnetic fields according to Equation (7.11), where **Z** is simply **E**/**H**. Since **Z** and **H** are multiplied in the frequency domain, **Z** can be thought of as the transfer function that describes a filter producing the output **E** from the input **H**. Equation (7.12) gives the impedance tensor and its relation to the **E** and **H** fields when the independent x and y polarizations are both used. As illustrated in Figure 7.1, the off-diagonal Z_{xy} and Z_{yx} components come from orthogonal E and H fields. When E and H fields have a parallel component, the diagonal Z_{xx} and Z_{yy} impedance tensor elements are nonzero. This holds where the geoelectrical structure is three-dimensional, which is generally the case in real data. Dimensionality will be discussed in more detail in the next section.

$$\mathbf{E} = \mathbf{Z}\mathbf{H}. \tag{7.11}$$

$$\begin{bmatrix} E_x \\ E_y \end{bmatrix} = \begin{bmatrix} Z_{xx} & Z_{xy} \\ Z_{yx} & Z_{yy} \end{bmatrix} \begin{bmatrix} H_x \\ H_y \end{bmatrix}. \tag{7.12}$$

Impedance is closely related to apparent resistivity ρ_a, which is the resistivity of a uniform half-space for data at a given frequency. Impedance can be converted to apparent resistivity through the equation:

$$\rho_a = \frac{1}{\omega\mu}|Z|^2. \tag{7.13}$$

where each component of **Z** is calculated independently. The phase of the impedance tensor is also calculated from **Z**, referring to the phase delay of the magnetic field relative to the electric field. Phase is calculated as follows:

$$\Phi = \tan^{-1}\frac{\text{imag}(\mathbf{Z})}{\text{real}(\mathbf{Z})}. \tag{7.14}$$

The phase above a half-space of uniform resistivity is 45° at all frequencies. This comes from Equation (7.9)'s $\sqrt{i}$ term, which has a phase angle of 45°. Phase angle increases when the signal penetrates into a more conductive material and decreases when a more resistive material is reached.

7.3.4. Dimensionality

An important part of an MT study is analyzing the dimensionality of the subsurface electrical structure. One-dimensional (1D), two-dimensional (2D), and three-dimensional (3D) structures have characteristic

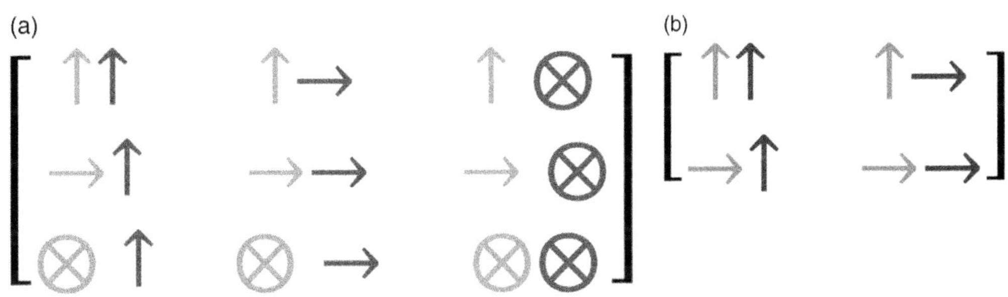

Figure 7.1 (a) Vector combinations used in the impedance tensor when vertical fields are considered. Light gray arrows represent electric field directions and dark gray arrows represent magnetic field directions. x is positive north, y is positive east, and z is positive down. (b) Eliminating tensor components that are not used in MT gives the simplified 2×2 impedance tensor, which has only horizontal electric and magnetic fields.

properties in the impedance tensor and allow for different assumptions to be made that can greatly simplify modeling and analysis. However, it is never appropriate to use lower dimensionality analysis methods on data that shows strong higher dimensionality characteristics, such as 1D methods on data with a 2D geoelectrical structure. While the savings in processing time may be tempting, one must be aware that the resulting interpretation is inevitably inaccurate and possibly misleading.

Other than a uniform half-space, the simplest resistivity model is that of a 1D earth. In this model, only horizontal layers of infinite lateral extent exist. The ratio E/H is the same in every direction and the resulting impedance tensor has equal amplitudes for the two components whose E and H fields are orthogonal (Z_{xy} and Z_{yx}), and their phases are 180° apart, resulting in:

$$\mathbf{Z} = \begin{bmatrix} 0 & Z_{1D} \\ -Z_{1D} & 0 \end{bmatrix}. \tag{7.15}$$

For a 1D Earth, the horizontal electric and magnetic fields are always orthogonal to each other, so any rotation in the polarization of the inducing magnetic fields leads to an identical relative rotation in the polarization of the induced electric field. Furthermore, the intensity of the induced electric field does not change under such a rotation of the magnetic field, if the intensity of the magnetic field does not change. This simple relationship between inducing magnetic and induced electric field is unique to the 1D induction problem, and does not hold for higher dimensional cases.

It is useful to mention that present day power utility industry practice is to apply regional-scale 1D conductivity models to the problem of estimating the predicted intensity of GICs due to geomagnetic disturbances. The impact of the forced orthogonality between electric and magnetic fields, as well as the azimuthally independent response of such a 1D structure to inducing magnetic fields of different polarizations on estimates of GICs in a real world where 3D resistivity structures are ubiquitous is an area of considerable research.

A 2D geoelectrical structure arises when the resistivity varies with depth and horizontally in one direction, with structures extending infinitely in the other horizontal direction. A geologic analog is a fault, where rock properties change going across the fault but do not change parallel to the fault over distances that MT data are sensitive to. In this setting, the E and H fields orthogonal to the fault are independent of those measured parallel to the fault, and $Z_{xy} \neq Z_{yx}$. With resistivity structure varying in only one horizontal dimension, there is still no way for a magnetic field to induce a parallel electric field, so the diagonal elements of Z remain zero and the impedance tensor reduces to:

$$\mathbf{Z} = \begin{bmatrix} 0 & Z_{xy} \\ Z_{yx} & 0 \end{bmatrix}. \tag{7.16}$$

A 3D geoelectrical structure is one where resistivity can vary with depth and with position in both x and y directions. All geologic settings exhibit this behavior to some extent, but it is often ignored if the 3D effect is small enough to justify 2D or even 1D analysis. 3D MT data exhibits magnetic fields that can induce electric fields in any direction, including parallel to the inducing magnetic field. This can happen when, for example, a magnetic field induces current in a conductor running diagonally to the magnetic field vector. This can occur when the conductor is arcuate or amorphous, or if the data are not rotated to align the x polarization with geoelectric strike. The full impedance tensor is required to describe these situations:

$$\mathbf{Z} = \begin{bmatrix} Z_{xx} & Z_{xy} \\ Z_{yx} & Z_{yy} \end{bmatrix}. \tag{7.17}$$

It is important to note that for the general 3D induction case, even a small rotation of the inducing magnetic field can lead to large rotations in the induced electric field as well as large changes in the electric field intensity, even if the intensity of the inducing magnetic field is constant during the change in its polarization. This has profound consequences when calculating the GICs induced in electric transmission lines by GMDs since the coupling of induced ground electric fields into transmission lines depends on the instantaneous polarization of the ground electric field and the degree to which it is aligned with the direction of each transmission line segment. Ground electric fields resulting from 3D conductivity structures may have entirely different field polarizations at a given moment during a GMD than those from a 1D structure. Whereas the peak GIC for a 1D scenario always occurs during the peak of a GMD, this is not the case in 3D, since the instantaneous polarization of the electric field, as well as its intensity, determines the intensity of the GIC (e.g., Bonner & Schultz, 2017).

7.3.5. Distortion

The assumption that charge density is uniform (Equation (7.3)) is only true in a half-space and in 1D models. Higher dimensional models allow current to cross an interface between materials of contrasting resistivity. Because current density is the product of conductivity and electric field, conservation of current requires an electric field discontinuity at the interface. This is accommodated by accumulation of charge on the interface to counter the electric field on the conductive side and strengthen it on the resistive side of the interface. 2D and 3D MT implements the complete

$\nabla \times \nabla \times E$ solution without the charge-free simplifications described in Section 7.2. However, local heterogeneities in the conductivity structure can be too small to be accurately measured or modeled. Accumulated charge along local resistivity boundaries with lateral dimensions much smaller than a skin depth deflects currents in ways that the data cannot account for, known as galvanic distortion. This non-inductive component to the data distorts it in several ways, though some of the effect is recoverable (Bahr, 1988; Jiracek, 1990). One of these effects is amplification of the electric fields in one direction, changing the impedance in the affected direction so that the apparent resistivity curve is shifted to appear higher or lower by a constant amount for every frequency, an effect known as static shift. Another effect is an apparent rotation or twist to the data when shallow current flow is diverted to follow a curved path near the measurement location. The effect of distortion is evaluated in terms of a 2×2 distortion tensor **D** that operates on the true impedance along with a rotation matrix **R** to create the measured impedance as follows (Groom & Bailey, 1989):

$$\mathbf{Z}_{\mathbf{distorted}} = \mathbf{RDZ}_{\mathbf{true}}\mathbf{R}^{T}. \tag{7.18}$$

Determining the contents of **D** and **R** was the great challenge of the 1D and 2D eras of MT, while the application of distortion removal methods in 3D MT is an area of considerable discussion, with some practitioners explicitly removing distortion terms prior to inversion (see Section 7.5.2 for inversion information) of the impedance tensor (Ledo et al., 1998; Utada & Munekane, 2000; Sasaki, 2004; Zhdanov et al., 2011) and others adding shallow scattering layers into their 3D model to accommodate the effects of unresolvably small electric field distorting structures (e.g., Meqbel et al., 2014).

It is not possible to fully determine **D** and **R**, but distortion can be decomposed into components that isolate the determinable parts from the indeterminable parts (Groom & Bailey, 1989). This allows as much of the unknown near-surface behavior to be removed as possible, helping to remove apparent 3D effects from data that should be 2D. The distortion tensor **D** is real, so it leaves the phase of **Z** unaffected (Bahr, 1988). Therefore, MT analysis methods that seek to fit the impedance phase as a discrete step in the modeling process, such as phase tensor analysis (Caldwell et al., 2004; Booker, 2014) and phase tensor inversion (Patro et al., 2013; Tietze et al., 2015), are also helpful in solving distortion problems. In practice, 3D inverse modeling is able to account for distortion if shallow parts of the model are finely discretized and allowed to change as necessary to satisfy the distortion, while the user must be aware that the shallow features thereby created do not represent true geologic structures.

Caution must be used when using impedance tensors for GIC prediction if their distortion components were removed for inverse modeling. If the distortion components have been stripped from the impedance tensor, then the tensor no longer fully represents the electrical properties of the site where it was measured and predictions will produce undistorted electric fields that do not match the observed, distorted electric fields (Bonner & Schultz, 2017). This could be a problem if electric fields are predicted using methods that use a modeled impedance tensor rather than an observed impedance tensor in areas where the subsurface materials have strong anisotropy.

7.4. DATA ACQUISITION

7.4.1. Array Design

The process of acquiring MT data begins with designing an MT station array of appropriate aperture, inter-station spacing and frequency bandwidth suitable for discriminating the location and properties of geoelectric target features in the crust and, when appropriate, the mantle below. The frequency bandwidth is determined by the desired range of depths of penetration beneath ground level required to delineate the survey targets. This determines the period of time a given MT station must operate before removal and relocation to another location on the survey grid, since lower frequency data will require a longer period of data acquisition than higher frequency data. In order to obtain MT impedance tensor data of sufficient quality (for instance, nominally 5% confidence limits in apparent resistivity and less than 3° uncertainty in phase) over a frequency band of 10^{-4}–10^{-1} Hz, as is characteristic of data from the EarthScope MT Transportable Array (Schultz et al., 2006–2018), typically requires anywhere from 5 to 28 days of continuous recording at each site, depending on the level of GMD activity during the recording period and the level of local electromagnetic noise sources. This further determines the selection of MT data acquisition hardware to use, since different sensor and data acquisition technologies are optimized for different frequency ranges. Thus, this first stage of array design is foundational to the data acquisition process.

The goal of most MT investigations is to address hypotheses that have been posed in relation to potential targets, such as the economic viability of specific geothermal energy resource (e.g., Fitterman, 1988; Wannamaker et al., 2004; Newman et al., 2008; Peacock et al., 2012), groundwater (e.g., Bernard et al., 1990; Nichols et al., 1994), mineral or hydrocarbon resources (e.g., Stevens & McNeice, 1998; Tuncer et al., 2006); or the relationship of deep-seated crustal and upper mantle hydrous and magmatic fluids to regional seismicity or surface volcanic features (e.g., Meqbel et al., 2014; McGary et al., 2014; Peacock et al., 2016; Bedrosian et al., 2018). To-date, there has been a paucity of MT data acquisition efforts specifically designed to address the concerns of the electric utilities in determining ground electric fields associated with GMDs. The use of MT data for this application has been *a posteriori* and opportunistic, and involves data from MT array designs optimized to address geologic targets that might be somewhat different than array configurations optimized for power grid resilience purposes.

Increasingly, MT surveys are components of integrated, multidisciplinary data acquisition efforts, where MT data are combined with seismic, potential fields and other observations in order to better constrain the geological process of interest (e.g., Pritchard et al., 2018). Up until the mid-2000s such investigations had been limited to 2D interpretations and survey designs that restricted data acquisition to a small number of transects along lines believed to be parallel or orthogonal to regional geologic and geoelectric strike directions. There are many cases at present where 2D survey designs persist, in large measure because of the greater cost of acquiring 3D data on a grid relative to acquiring 2D data on specific profiles along a line.

Such 2D survey arrays have increasingly given way over the past 15 years to 3D surveys, where MT stations are deployed in quasi-regular latitude-longitude grids. The transition from 2D to 3D surveys has been made possible in part by the availability of 3D MT inverse modeling codes designed to produce images of resistivity structure from arrays of impedance tensors (Siripunvaraporn et al., 2005; Egbert & Kelbert, 2012; Kelbert et al., 2014). The development of large MT facilities capable of deploying many tens of MT instruments simultaneously has made possible the collection of the large array data sets required for 3D interpretation. The long-running National Science Foundation (NSF) EarthScope Magnetotelluric Program (Schultz et al., 2006–2018) and the subsequent SinoProbe (Dong et al., 2013) and AusLAMP (Robertson et al., 2016; Thiel et al., 2016) MT Programs were the first to carry out continental-scale 3D MT reconnaissance mapping with large arrays, further promoting the adoption of 3D MT survey designs. In contrast to previous, more targeted 2D surveys, these large array efforts can provide broad-scale identification of targets of follow-on interest in otherwise "greenfield" areas.

The ability to resolve structural details in electrical resistivity is determined by the aperture of the MT array both inline and broadside to potential subsurface targets, the spacing between MT stations, and the frequency bandwidth of the recorded MT data. There is a theoretical resolving power limit beyond which no additional improvement can be made by adjusting these parameters, but in practice this is rarely reached. In practical terms, the main limitations in optimal array design are economic, with the cost of operations being the key factor.

While it is possible to install permanent or quasi-permanent MT instrumentation (a deployment mode we refer to as a "Level 0" survey), to enable continuous monitoring of the electric and magnetic fields over periods of months and years, this is rarely done. There have been a limited number of such installations at a handful of widely-spaced magnetic observatories (coordinated through the International Association of Geomagnetism and Aeronomy's INTERMAGNET program), and also within the NSF EarthScope program, seven "Backbone" MT stations of this type were operated for a number of years, following earlier ultra-long period MT work in deep lakebottoms (Schultz et al., 1987, 1993). The many advantages of using long-term, continuous monitoring of the ground electric and magnetic fields include the following: to study the spatial variability of Earth–ionospheric coupling; to obtain constraints on deep mantle electrical resistivity structure; to study temporal changes in resistivity structure associated with transient events; and to obtain a set of baseline observations for monitoring ground electric fields for GIC studies. However, the challenge of maintaining stable long-period electric field measurements – as well as the cost of maintaining such stations – inhibits their general use. Rather than permanent installation of MT equipment, the general deployment mode is for the temporary installation of a relocatable array of MT instruments.

We refer to broad-scale 3D reconnaissance mapping, typically using temporary installations of long-period MT instruments to create baseline 3D geological maps and potentially to identify targets for subsequent higher-resolution studies as a "Level 1" survey. In order to cover a large territory efficiently, Level 1 survey designs trade-off potential spatial resolving power against the cost and efficiency of operations by using as large an inter-station spacing as possible, up to the point where aliasing of potential resistivity target signatures becomes a dominating factor. For example, the EarthScope MT Transportable Array station spacing of 70 km has proven adequate for delineation of major structural features in the mid-crust to upper mantle. With somewhat greater interest in shallower mineral resource questions, the AusLAMP array employs ~55 km (1/2°) station spacing. Serendipitously, Level 1 MT data sets have proven extremely important to those researching space weather effects on the power grid (e.g., Love et al., 2016, 2018; Bonner & Schultz, 2017; Lucas et al., 2018).

We refer to long-period MT arrays with finer station spacing, also using long-period MT instruments designed to identify mid-crustal to upper mantle targets, but to discriminate finer-scale features closer to the potential long-period resolving power limit of the MT technique, as

"Level 2" surveys. Such surveys might employ inter-station spacing as fine as ~10–20 km. There is some evidence that data from Level 2 MT surveys may improve the accuracy of estimates of ground electric fields used for calculating GICs on the power grid (Bedrosian & Love, 2015; Bonner & Schultz, 2017).

We refer to wideband MT arrays, potentially with even finer inter-station spacing (<1–20 km) as "Level 3" surveys. Level 3 surveys can be designed to discriminate features in the upper crust, including the near-surface and even in the critical zone. While no studies exist to show if Level 3 MT survey data provides meaningful improvements to the accuracy of GIC estimates, physical scaling relations suggest that Level 3 MT survey data would provide improved GIC estimates for artificially induced GICs caused by exo-atmospheric detonation of nuclear devices, that is electromagnetic pulse or "EMP" weapons, since the signature of the artificially induced GMDs from these events includes intense signals at higher frequencies than for natural GMDs, thus also generating electromagnetic induction effects at finer scales within upper crustal resistivity structures.

7.4.2. MT Sensors and Data Acquisition Systems

The MT signal band can be divided into the long-period band associated with space weather modulations of ionospheric and magnetospheric electric fields (typically below ~1–10 Hz) and wideband variations associated with ionospheric waveguide propagation of signals due to global lightning activity (above 10 Hz). The natural background power spectral density for geomagnetic field variations follows a $1/f$ law at long periods, but in the transition between these two bands, known as the "long-period MT dead band," the signal level drops off sharply, but rises again at about 1 Hz and above (Vozoff, 1991). Available technologies for measuring vector magnetic fields also follow similar low and high frequency bandwidth divisions, and while a variety of different magnetic field sensor technologies have been employed in the past, contemporary surveys generally use one or both of induction coil and/or fluxgate magnetometer technologies to cover one or both bands.

Induction coils (Tumanski, 2007; Pronenko & Korepanov, 2009) employ fine solenoidal windings surrounding highly magnetically permeable core material in a linear configuration, so that each induction coil is sensitive to time-variations ($\partial B/\partial t$) of the magnetic field in the direction of the axis of the solenoid. In order to measure the full vector magnetic field variation, three coils are employed and typically deployed with each pointing in one of the three cardinal directions of the compass. Since induction coils are directly sensitive to time rate of changes in the magnetic field rather than to the field directly, their sensitivity drops off at lower frequencies, although this is compensated for by conditioning circuitry that amplifies and integrates the signal. The induction coils can be tuned for sensitivity over a given frequency bandwidth by modifying various design parameters, such as the length, configuration and mass of the permeable core material and surrounding windings, and the number, gauge, and resistance of the windings (the latter factor impacting the Johnson or thermal noise limit of the sensors). All other things being equal, longer induction coils tend to have improved long-period sensitivity than shorter coils, whereas the reverse is also true; shorter coils tend to have improved high-frequency sensitivity than longer coils.

Induction coils are suitable for Level 1 and Level 2 MT surveys that are used for power grid GIC applications, but in order to cover the 10^{-4} –10^{-1} Hz frequency bandwidth needed for such studies, wideband coils must be employed. The current state-of-the-art for such wideband coils are designs that can exceed 1 m in length for each coil. Noise levels as low as 30–50 fT/√Hz at 1 Hz are achievable with wideband induction coils, which provides a sensor noise ceiling that is below the noise floor of natural MT signals across the required frequency band. Wideband induction coils can also be used for Level 3 MT surveys, with usable sensor response up to ~10 kHz or above, depending on the sensor design.

The major disadvantages of induction coils are their bulk and weight, their cost, and to a lesser extent, their power consumption.

The fluxgate (Ripka, 2003) is the other predominant magnetic field sensor technology used in contemporary MT, which generally comes in two configurations; linear core and ring core. Fluxgates are directly sensitive to the magnetic field rather than its time-derivative. A highly permeable magnetic core is surrounded by a winding (either solenoidal for linear cores, or helical for ring cores). An alternating current is imposed in the winding that generates a magnetic field that drives the core into and out of magnetic saturation. The Earth's magnetic field influences the time during each AC cycle at which the core is saturated, leading to a detectible hysteresis that is picked up by a secondary winding surrounding the core. The phase (time lead or lag relative to zero external magnetic field saturation) of the signal scales with the intensity of the external magnetic field. In order to boost sensitivity, some of the highest quality fluxgate sensors are approximately nulling sensors, where most of the main part of the external magnetic field is bucked out by an artificially imposed magnetic field generated by Helmholtz coils that surround the fluxgate, and the residual hysteresis signal is then digitized.

A vector fluxgate sensor consists of three sets of cores and windings, with each set sensitive to one vector component of the magnetic field. Fluxgate sensors can be extremely low power, with <10 mW power consumption for a precision three-component vector fluxgate instrument representing the current state of the art. Fluxgates are also compact, with a three-component sensor fitting within a cube of less than 7 cm on a side. These attributes, as well as relatively low cost (roughly half to one-third the cost of a wideband set of three induction coils) make fluxgates a popular choice for Level 1 and Level 2 long-period MT surveys, although a complicating factor in recent years has been the difficulty in securing high-quality permeable core material suitable for construction of the highest quality fluxgate sensors. Whereas induction coils can be configured for sensitivity to the natural MT signal at frequencies extending into the kHz band, the current state-of-the-art for precision fluxgates limits them to sampling frequencies below ~10 Hz or lower. The best available fluxgate sensors currently on the market have a noise figure of ~7–10 pT/$\sqrt{\text{Hz}}$ at 1 Hz, which is roughly two orders-of-magnitude higher than the best induction coils.

The MT method requires measurement of the horizontal vector components of the electric field at the ground surface. Most typically, pairs of carefully matched non-polarizing electrodes (Petiau, 2000) are carefully grounded to the soil to make galvanic/electrolytic contact of low impedance, separated by a certain distance (100 m being a commonly used separation). With a sufficiently large amplifier input impedance, the flow of electric currents between the electrode pairs can be inhibited, therefore maintaining their stability by preventing the electrodes from becoming polarized by electrolytic deposition of metal salt compounds from one electrode serving as an anode to the other's cathode. Two electrode pairs are usually used, with one pair orthogonal to the other, or otherwise with a large angle separating the pairs so the data from them may subsequently rotated and projected to orthogonal coordinates. Continuous recordings are made of the voltages between the matched electrode pairs, which – divided by the electrode separation distance – provide a very close approximation to the electric field at the mid-point of the cable connecting the electrodes.

The signal-to-noise level of the electric field measurements is directly impacted by the distance between electrode pairs, with larger separations producing larger signal levels of the electric field (units: V m^{-1}). Obtaining stable electric field measurements is progressively more challenging at lower frequencies, since a variety of environmental factors can produce spurious potentials at the electrodes, including thermal drift effects, streaming potentials and changes in the electropotential of topsoil related to flowing surface water, etc. These effects manifest as time-variable self-potentials (SP) that are superimposed on the induced electric fields of interest. SP effects can be mitigated to an extent by careful attention to detail, such as preparing shallow holes for electrode installation with mud made from native soils, introducing an intermediate medium such as water-saturated (hydrophilic) bentonite or kaolinite clay, which can optionally be mixed with a salt solution and then buried and thermally insulated. In areas that are extremely arid or frozen, it can be difficult to achieve good, low-impedance electrolytic coupling with ground, and alternative approaches can be taken, such as the use of high impedance buffer amplifiers (Wannamaker et al., 1996), or capacitively coupled electrodes (Hibbs et al., 2012), but generally at a cost of poor performance at frequencies below about 1 Hz. Given this consideration, for Level 0–3 MT surveys appropriate for GIC studies, which require good low frequency coupling, conventional galvanically coupled electrodes are preferred.

Variations in the MT installation design include use of non-orthogonal electric dipole sensors, or magnetic field sensors installed in coordinates rotated from the cardinal directions, or electrodes and magnetic field sensors placed on the surface of the ground rather than buried, with special measures taken to achieve stable measurements and to rotate the resulting MT data into cardinal directions. Such sub-aerial installations can simplify permitting requirements for sites where ground disturbances are to be avoided. Two versions of a typical MT station installation are shown in Figure 7.2. Both versions have the same electric dipole lengths, but with the center points in different locations to accommodate different field conditions where one configuration might be preferred for convenience or accessibility in difficult terrain.

The electric and magnetic field sensor data are conditioned and digitized using a variety of commercially available, or custom-designed data acquisition systems. In order to achieve stable measurements at very low frequencies ($f \leq 10^{-4}$ Hz), special attention may be required to ensure that the data acquisition system analog and linear ground planes are either directly grounded to the center point of the electric dipole array, or the ground state potential is directly measured, otherwise long-term drift may result. In recent years there has been a transition from MT recording systems capable of digitizing data with 24-bits precision, to a number of different 32-bit systems capable of achieving digital sampling rates exceeding 4096–24,000 samples-per-second, with up to 27 noise-free bits at sample speeds of ~1–16 samples-per-second. Generally the faster the sample rate, the greater the number of noise bits in each sample.

In modern data acquisition systems, system timing is provided by a time stamp provided by GPS/GNSS satellite navigation systems, which has timing jitter well below the threshold required to achieve less than half a degree phase uncertainty at 4 kHz. Data are generally stored locally in removable flash memory devices, and in

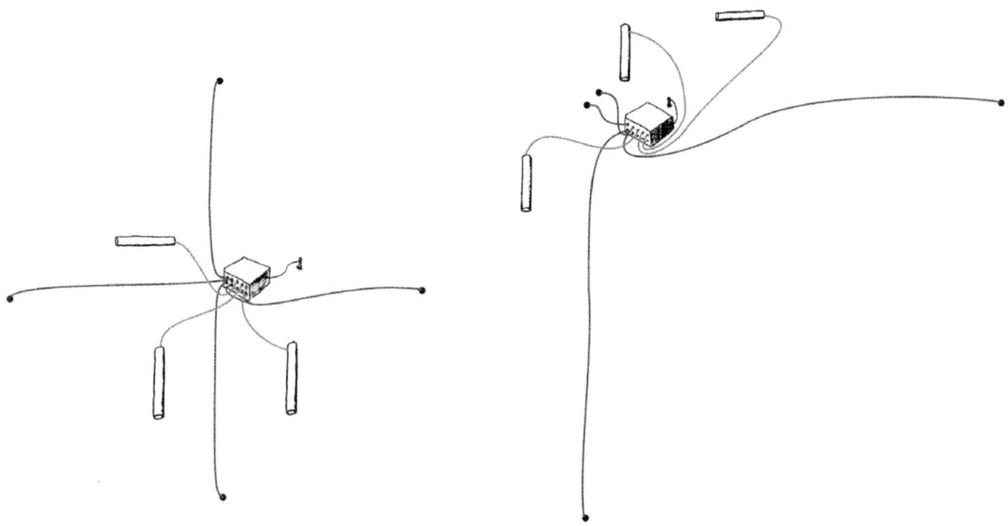

Figure 7.2 Schematic of typical MT station installation geometry. (left) Two sets of matched electrode pairs are aligned in cardinal coordinates (north–south and east–west) each forming an electric dipole sensor ~100 m in length. A fifth electrode is shown, which is a grounding rod for the recording system ground plane. Three magnetic field sensors (induction coils, as shown, or a unitary tri-axial fluxgate sensor (not shown)) are buried in shallow trenches or holes. The magnetic field sensors are aligned to cardinal coordinates typically with an accuracy of ±1°, and the electrodes are positioned with an accuracy of 0.5 m over a distance of 100 m. (right) As in figure at left, but rather than a cross-shaped electric dipole array, an "L" shaped array is used instead.

some cases on-demand or real-time telemetry of data, or of system state-of-health information is provided.

7.5. DATA PROCESSING AND INVERSION

Converting time-series EM field data to impedance is not as simple as dividing E by H in the frequency domain. Resulting from complex natural phenomena, MT source signal strength is highly variable and out of the user's control, and noise is always present. Noise can be natural, such as swaying trees or groundwater flow, or manmade, such as electrical infrastructure, and its effects can be either highly localized or contaminate data over 100 km away from the noise source (Egbert & Livelybrooks, 1996; Egbert et al., 2000; Kappler et al., 2010). To minimize noise contamination, uncorrelated noise is rejected and an ensemble averaging approach is used to estimate impedance tensors and their variances. An example of MT time-series data recorded during a two-day period of moderately strong geomagnetic activity is given in Figure 7.3.

7.5.1. Processing Methods

Impedances are estimated through the following steps, although there are several variations in approaches taken by different practitioners:

1. Divide the EM time-series data into short segments of equal length, detrend each segment and apply other prewhitening methods, then use a windowing function to taper the start and end of each segment in order to reduce spectral leakage. Although the entire time-series can have many millions of samples, each segment usually has only hundreds-to-tens of thousands of samples.

2. Fourier transform each segment, producing an ensemble of Fourier coefficients for the range of frequencies covered by the segment length. Data calibration is usually done in this step to correct for sensor and amplifier responses.

3. For each segment, calculate the impedance tensor and variance by solving Equation (7.12) for the four components of **Z**. With ideal, noise-free data, the solution is given by:

$$\mathbf{Z} = \langle \mathbf{EH}^* \rangle \langle \mathbf{HH}^* \rangle^{-1}, \tag{7.19}$$

where $\langle EH^* \rangle$ is an outer product resulting in a 2×2 matrix of cross powers. Likewise, $\langle HH^* \rangle$ gives auto powers. The brackets denote averages over the time windows. Therefore, the impedance comes from the coherence of **E** and **H**. The ideal impedance estimate is the one that minimizes uncorrelated noise. The equation that must be minimized for each frequency and **Z** component is the squared residuals r calculated from observed E-fields and the E-fields predicted by **H** through the transfer function **Z**.

Cross- and auto powers of frequencies within a narrow band can be binned together in this step, such that each decade of the spectrum is represented by no more than a small number (typically ~8–10 frequencies). There

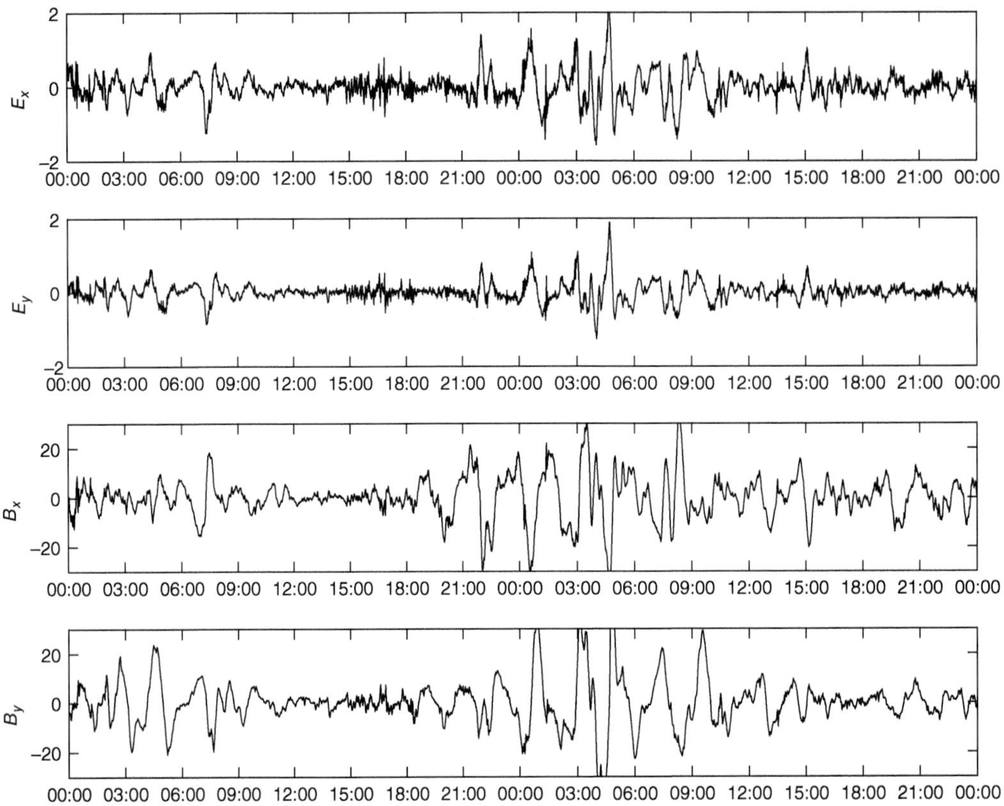

Figure 7.3 A two-day recording of raw, uncalibrated MT time series data from a site in southwestern Idaho, United States, during a time of strong geomagnetic activity. Data were recorded in a coordinate system where positive *x* is north and positive *y* is east.

is always a tradeoff between the number of discrete frequencies over which MT impedances are calculated, and the variance of those estimates, so by binning a larger number of auto- and cross powers over adjoining frequencies over a wider frequency band, a smaller confidence limit can be placed on the resulting impedance estimate but at the cost of the ability to resolve finer scale variations versus frequency. Using more frequencies allows for more subtle features to be seen, but when site spacing is wide and data coverage is sparse, which is usually the case, the diffusive nature of EM fields limits how small of a feature and how sharp of a boundary can be sensed, thus the marginal utility of using more frequencies diminishes quickly. It is also sometimes necessary to use fewer frequencies per decade with very noisy data, allowing a broader band of Fourier coefficients to be averaged in each bin. It can also be useful to discard certain frequencies that are noisy in order to avoid having them contaminate the frequency band that they would be binned into.

4. Longer time-series segments are used to estimate impedances of longer periods. Processing time is reduced by using cascade decimation, in which the time-series is lowpass filtered and downsampled, and the previous steps are then repeated.

Additional steps can be followed to improve the quality of the impedance estimates. In this method of MT data processing, we are assuming that the data are drawn from a normal distribution, but this assumption is usually violated. Outliers are considered to be data drawn from a different population than the main population, and ordinary least-squares fitting can give outliers a strong influence over the final impedance. Outliers must therefore be downweighted using a robust regression method to reduce their influence. One approach to this is Huber weighting, in which data whose residuals exceed some threshold are downweighted based on the size of their residuals. Once weights are determined, the impedance is estimated again using the weighted residuals. This process is repeated iteratively until the residuals do not change significantly (Egbert & Booker, 1986; Chave & Thomson, 1989). Additional weighting or deselection of individual auto- and cross powers can be made on the basis of the coherence between the *E* and *H* fields associated with those estimates, or other factors such as violations of the causality between *E* and *H* fields that are required for 1D and 2D Earth models (Sutarno, 2005). Robust processing of MT data provides much higher quality impedance estimates (Jones et al., 1989). MT data are usually plotted as apparent resistivity and phase as functions of period or frequency using logarithmic axes (Fig. 7.4).

The effects of noise in MT data can be reduced by using a magnetic field data from a second site. Noise at one MT site will not be correlated with the signal at a second MT site if the second site is sufficiently far away from the first. If the magnetic field signal H is the sum of the true regionally broadscale coherent magnetic field signal H_{true} and a local noise component H_{noise}, then $H = H_{true} + H_{noise}$. During times of weak geomagnetic activity when the signal to noise ratio is small, the H_{noise} component can make the overall power of H significantly greater. This results in impedances that suggest a more conductive subsurface in the noisy frequencies. The solution to this is remote reference processing (Gamble et al., 1979). The magnetic field signal does not change significantly over distances that are greater than the scale of most noise sources, therefore two sites will have nearly the same H_{true} but different H_{noise}. Remote referencing keeps only the coherent part of the two magnetic sites, removing incoherent noise from H. Remote reference processing is very important in noisy settings and in low-noise settings during times of weak source signal and should be done whenever possible. It is therefore advantageous to conduct an MT survey with more than one site in operation at all times or with a dedicated remote reference site operating through the duration of the survey. If this cannot be done, it may be possible to use data from a permanent geomagnetic observatory for a remote reference, such as the observatories operated by the Geomagnetism Program of the United States Geological Survey in the United States or the global INTERMAGNET Program.

Remote referencing is done by modifying Equation (7.19) to include the cross powers of the local and remote reference sites as shown in Equation (7.20):

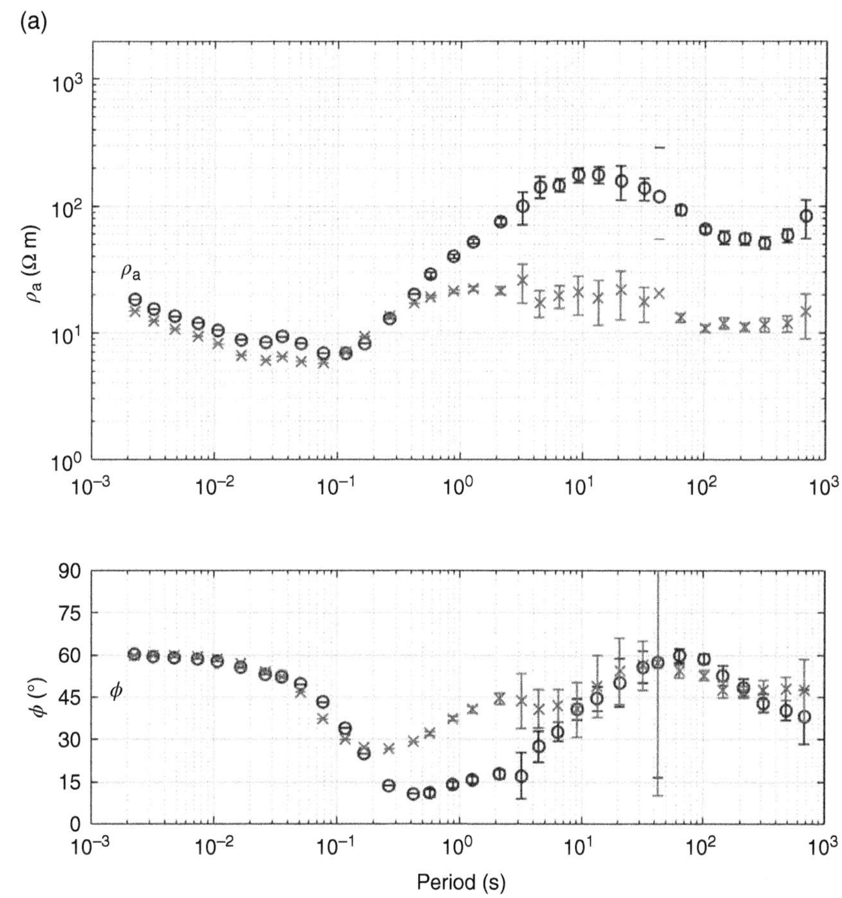

Figure 7.4 (a) Example of MT data with error bars. Only the Z_{xy} (light gray) and Z_{yx} (dark gray) components of the impedance tensor are shown, converted to apparent resistivity (top) and phase (bottom). The slight offset in apparent resistivity at the shortest periods illustrates static shift resulting from distortion. Remote referencing has been used to correct the effects of noise. The uncorrected data are shown in (b). Without remote referencing, the data at periods between about $10^{-1}$and 10^2 s reflect noise in the magnetic field data rather than the internal structure of the Earth. After corrections, data quality at periods greater than about 3 s is significantly improved. Error bars that extend beyond the axis limits are not plotted. These plots were made by processing ~48 h of time-series data. Source: from dataset of Schultz et al. (2006–2018). (*See electronic version for color representation of this figure.*)

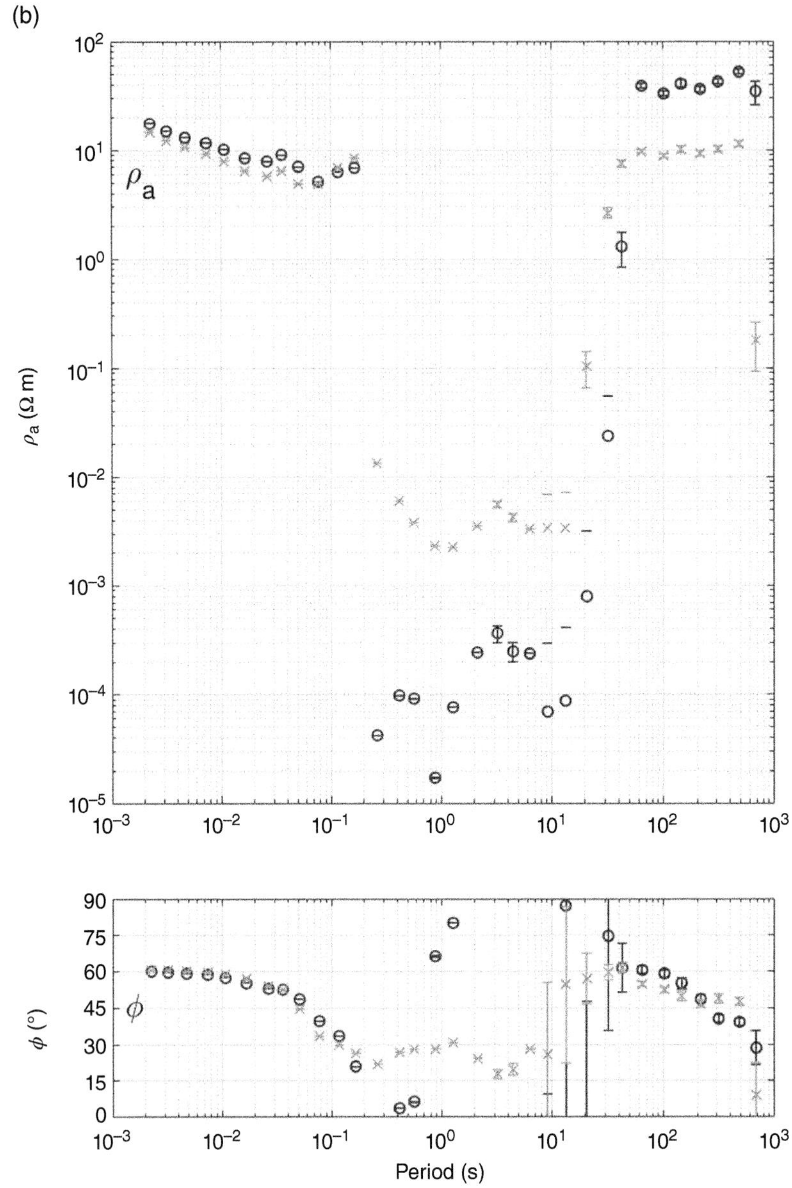

Figure 7.4 (Continued)

$$\mathbf{Z} = \langle \mathbf{E}\mathbf{H}_{\mathbf{R}}^{*} \rangle \langle \mathbf{H}\mathbf{H}_{\mathbf{R}}^{*} \rangle^{-1} \tag{7.20}$$

where $\mathbf{H}_{\mathbf{R}}$ is the remote reference magnetic field. Using remote referencing, incoherent noise averages to zero in the autopower term, eliminating the downward bias in impedance when the signal-to-noise ratio is poor. An example of remote referencing corrected noisy data is shown in Figure 7.4.

7.5.2. Inversion

The process used to generate models of the Earth from geophysical data is called inversion. Inversion is a major branch of geophysics whose full scope cannot be covered in this paper. This section will provide a short overview of MT inversion. Detailed examinations of 3D MT inversion are given by Egbert and Kelbert (2012) and Avdeev (2005). For a comprehensive review of inversion in general, see Aster et al. (2013). Inversion is an optimization procedure that seeks to minimize the misfit between the observed data and synthetic data calculated from a geophysical model, subject to constraints. The inversion routine then updates the model to reduce the residuals in subsequent iterations until a satisfactory misfit is reached. For 3D MT inversion, a popular inversion program is ModEM (Egbert & Kelbert, 2012;

Kelbert et al., 2014), which implements a nonlinear conjugate gradient algorithm (NLCG; Rodi & Mackie, 2001). The NLCG algorithm uses the gradient of a nonlinear function to find a local minimum. As implemented in ModEM, the function that is minimized (Equation (7.21)) is not only comprised of the data residuals, but it also includes a regularization term that minimizes the deviation from a reference model to avoid abrupt and physically unrealistic boundaries. The reference model is usually a very smooth and simple model, such as a uniform half-space. The inversion multiplies data residuals by an inverse data covariance matrix to downweight the influence of poor-quality data points on inversion results. Model smoothing is implemented by multiplying model parameters by a covariance matrix that controls coupling of model parameters so that model structures are spread out among adjacent cells and isolated cells of extreme contrast are avoided. The function Φ that is minimized is thus:

$$\Phi(\mathbf{m},\mathbf{d}) = \left(\mathbf{d} - f(\mathbf{m})\right)^{\mathrm{T}} \mathbf{C}_{\mathbf{d}}^{-1}\left(\mathbf{d} - f(\mathbf{m})\right) + \lambda(\mathbf{m} - \mathbf{m}_0)^{\mathrm{T}} \mathbf{C}_{\mathbf{m}}^{-1}(\mathbf{m} - \mathbf{m}_0) \tag{7.21}$$

where $\mathbf{m}$ is the model at the current iteration, $\mathbf{m}_0$ is the reference model, $\mathbf{d}$ is the observed data, f is a function that solves the MT forward problem to calculate synthetic data from the model, $\mathbf{C}_{\mathbf{m}}$ and $\mathbf{C}_{\mathbf{d}}$ are the model and data covariances, respectively, and λ is a tradeoff parameter that balances data misfit and model smoothness. During inversion, a larger value of λ is used in early iterations to promote model smoothness. As the inversion progresses and $\mathbf{m}$ becomes optimized to minimize $\mathbf{d} - f(\mathbf{m})$ gradually with each iteration, λ is reduced to a smaller value when subsequent iterations provide minimal improvement in data fitting, thereby shifting the priority of $\Phi(\mathbf{m}, \mathbf{d})$ toward minimizing $\mathbf{d} - f(\mathbf{m})$. Gradual reduction of λ helps guide $\Phi(\mathbf{m}, \mathbf{d})$ to a smaller minimum than the local minimum that would be reached using an initially small λ, while suppressing unrealistically large contrasts in model resistivity.

3D models are usually plotted either as cross sections, horizontal slices, isosurfaces, or a combination of these. Most academic researchers studying solid Earth problems use a convention of using warm colors to plot conductive features and cool colors for resistive features. However, in industry and hydrology it is often the opposite. Therefore, it is important to verify the convention used by looking at the plot's color scale. Color scales typically are logarithmic and usually range within −1 and 4, corresponding to resistivities of 0.1 Ωm to 10,000 Ωm. It is also helpful to adjust color scales to highlight certain features. An example of a resistivity model created by 3D inversion with ModEM is shown in Figure 7.5 (Bedrosian et al., 2018).

7.6. RESISTIVITY MODEL INTERPRETATION

MT inversion results in a resistivity model that represents the properties of subsurface materials, including rock, melt, and pore fluids (aqueous solutions or partial melt), but determining what exactly each model feature represents requires geologic interpretation. Many things affect material resistivity and this section will provide only a brief introduction to the topic. A comprehensive discussion is provided by Evans (2012). The resistivity of any single common material within the Earth's crust can be as low as 10^{-5} for metallic sulfide minerals (Pridmore & Shuey, 1976) to 10^{12} for a dry basalt (Evans, 2012). Earth materials that MT is sensitive to are never perfectly uniform and should be treated as multiphase materials where the bulk resistivity of most rocks ranges on the

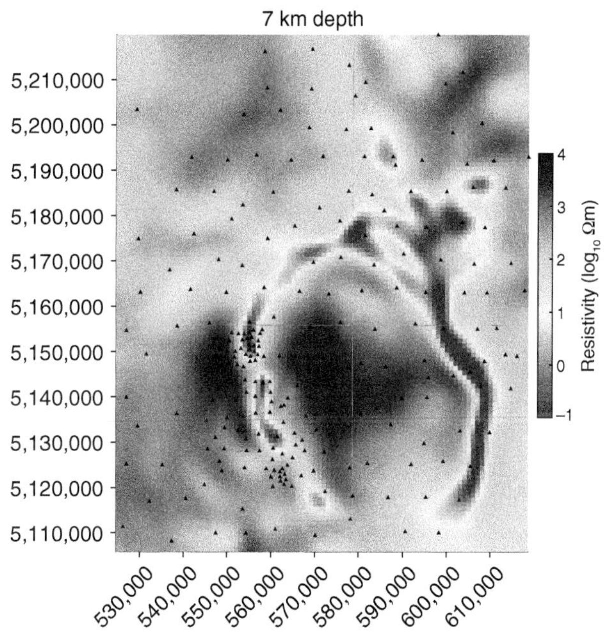

Figure 7.5 Example of a 3D resistivity model created from MT inversion with ModEM. The arcuate and irregularly-shaped features make it impossible to determine a single regional geoelectric strike direction, making this a good example of a region where 3D MT is highly advantageous. Bold shades represent extremely resistive or extremely conductive materials, with the black ring that dominates the lower half of the model being extremely conductive, encircling a region of highly resistive material. Faint shades are materials or moderate resistivity. Conductors in this model (Source: model used by Bedrosian et al. (2018). Reproduced with permission of John Wiley and Sons; see also IRIS DMC Website) are interpreted as metasediments in a volcanic region of the Washington Cascades (northwestern United States), while resistive features are plutonic rocks. Black markers represent station locations. A logarithmic color scale is used. Axes in this case are UTM distances, but it is also common to use latitude and longitude or distance from a certain location. (*See electronic version for color representation of this figure.*)

order of 10^{-1}–$10^{4}\,\Omega$m. Rocks are made up of crystals and interstitial fluid with varying levels of pore interconnectivity. Within rocks of the upper crust, the fluid component is generally far more conductive than the surrounding mineral grains, and the bulk resistivity depends primarily on the presence of fluids and how thoroughly they are interconnected to enable current flow. The relationship between resistivity of the two phases and their interconnectedness, or "cementation factor," was determined empirically by Archie (1942), in what is known as Archie's Law, given in Equation (7.22) (with complete fluid saturation assumed):

$$\sigma_{\text{bulk}} = \sigma_{\text{fluid}}\Phi^{m}, \tag{7.22}$$

where σ_{bulk} is the conductivity of the whole rock, σ_{fluid} is the conductivity of the pore fluid, Φ is the fluid fraction or porosity, and m is the cementation factor, which describes the tortuousity of fluid pathways. Archie's law requires the assumption that mineral grain conductivity is negligible compared to that of pore fluids. The value of m depends on the rock, and usually a value must be assumed. When $m = 1$, pore fluid interconnections are uninterrupted and current can flow unimpeded between pores. This is only achieved when crystals are suspended in fluid, such as a high degree fraction partial melt. Common values of m range from 1.5 to 1.8 in marine sands (Jackson et al., 1978) to 2.5–3 in igneous rocks (Evans, 1994). Unless geologic structures and tectonics suggest otherwise, it is common to interpret areas of low resistivity in the upper crust as highly permeable or having a high fracture density, but there is little that can be said about the rock type. Overall, crystalline rocks have very little pore space and low permeability and thus are highly resistive. Sedimentary rocks can be highly porous and thus have lower resistivities. Shales are also very conductive because of the extremely high surface areas of clay minerals. Charge can flow relatively easily along mineral surfaces, where crystal lattice defects create charge imbalances that are filled by cations that allow for the movement of charge (Evans, 2012). At mantle depths, conductivity can be high despite low fluid content because olivine behaves as a semiconductor, where high temperatures provide activation energy for solid state conduction to occur in olivine (Schock et al., 1989; Constable et al., 1992; Fullea et al., 2011).

Over the large distances that are of most interest in GICs, conductive regions tend to be in tectonically active areas where faulting and volcanism allow for increased fracture density, fluid transport, and clay alteration, in sedimentary basins where clays and water-saturated sediments accumulate, or in areas containing coal or shale beds, or certain ore formations such as banded irons. Resistive regions are tectonically stable parts of continents such as cratons, or areas where bedrock is exposed rather than buried beneath deep sediment. A further complication to interpreting MT data is that geologic materials (e.g., shales and metamorphic rocks) can have anisotropic conduction properties, which can make it difficult within the resolving power limitations of MT surveys to distinguish between the true dimensionality of the resistivity structure and its possible anisotropy. This could have some consequence for applying MT data to the problem of estimating GICs from geomagnetic disturbances. The impedance tensor contains all measured information on both the dimensionality and the possible anisotropy of the true Earth resistivity structure, so algorithms that project the ground magnetic fields directly through the impedance tensor to predict the ground electric fields used for GIC calculations intrinsically allow for anisotropic effects. Alternatively if anisotropic conditions exist, and ground resistivity models resulting from inversion of MT data that do not account for such anisotropy are used to calculate ground electric fields, it possible that bias might be introduced into the predicted ground electric fields.

7.7. CONCLUSIONS

Earth's time-varying magnetic field induces electric fields within the Earth through Faraday's Law and Ampere's Law. Magnetotelluric data is primarily used as a way to determine the electrical resistivity structure of the subsurface, and then making geologic interpretations from the resistivity model. MT uses the ratio of electric and magnetic fields to determine the impedance tensor, which is a transfer function through which the magnetic fields are projected to predict electric fields. Impedance tensors are inverted to produce resistivity models of the Earth. Of interest in GIC studies, the impedance and/or the resulting crust and upper mantle resistivity model is useful for predicting electric fields produced by GMDs that can damage power transmission and distribution networks. Accurate modeling of electric fields requires that 3D effects are taken into account, including distortion. Data acquisition instruments and survey design must be appropriate for the desired spatial and frequency coverage.

ACKNOWLEDGMENTS

The authors acknowledge the support of National Science Foundation grants EAR-1144353, EAR-1053632, and US Department of Energy and National Energy Technology Laboratory Award RES1100426, Task Release 004 for their support of E. B-M's PhD work. We also thank Kerry Key for a very helpful review of this manuscript.

REFERENCES

Archie, G. E. (1942). The electrical resistivity log as an aid in determining some reservoir characteristics. *Transactions of the American Institute of Mining, Metallurgical, and Petroleum Engineers*, *146*, 54–62.

Aster, R. C., Borchers, B., & Thurber, C. H. (Eds.) (2013). Iterative methods. In *Parameter Estimation and Inverse Problems* (2nd ed., pp. 141–168). Boston: Academic Press. doi:https://doi.org/10.1016/B978-0-12-385048-5.00006-9

Avdeev, D. B. (2005). Three-dimensional electromagnetic modelling and inversion from theory to application. *Surveys in Geophysics*, *26*(6), 767–799. doi:https://doi.org/10.1007/s10712-005-1836-x

Bahr, K. (1988). Interpretation of the magnetotelluric impedance tensor: regional induction and local telluric distortion. *Journal of Geophysics*, *62*(2), 119–127.

Beblo, M., & Liebig, V. (1990). Magnetotelluric measurements in Antarctica. *Physics of the Earth and Planetary Interiors*, *60*(1), 89–99. doi:https://doi.org/10.1016/0031-9201(90)90251-R

Bedrosian, P. A., & Love, J. J. (2015). Mapping geoelectric fields during magnetic storms: Synthetic analysis of empirical United States impedances. *Geophysical Research Letters*, *42*(23), 10160–10170. doi:https://doi.org/10.1002/2015GL066636

Bedrosian, P. A., Peacock, J. R., Bowles-Martinez, E., Schultz, A., & Hill, G. J. (2018). Crustal inheritance and a top-down control on arc magmatism at Mount St Helens. *Nature Geoscience*. doi:https://doi.org/10.1038/s41561-018-0217-2

Berdichevsky, M. N., Dmitriev, V. I., & Pozdnjakova, E. E. (1998). On two-dimensional interpretation of magnetotelluric soundings. *Geophysical Journal International*, *133*(3), 585–606. doi:https://doi.org/10.1046/j.1365-246X.1998.01333.x

Bernard, J., Vachette, C., & Valla, P. (1990). Deep groundwater survey with audio? Magnetotelluric soundings. In *SEG Technical Program Expanded Abstracts 1990* (Vols. *1–0*, pp. 528–531). Society of Exploration Geophysicists. doi:https://doi.org/10.1190/1.1890253

Bonner, L. R., & Schultz, A. (2017). Rapid prediction of electric fields associated with geomagnetically induced currents in the presence of three-dimensional ground structure: Projection of remote magnetic observatory data through magnetotelluric impedance tensors. *Space Weather*, *15*(1), 204–227. doi:https://doi.org/10.1002/2016SW001535

Booker, J. R. (2014). The magnetotelluric phase tensor: A critical review. *Surveys in Geophysics*, *35*(1), 7–40. doi:https://doi.org/10.1007/s10712-013-9234-2

Boteler, D. H. (2001). Space weather effects on power systems. In P. Song, H. J. Singer, & G. L. Siscoe (Eds.), *Space Weather* (pp. 347–352). Washington, DC: American Geophysical Union. doi:https://doi.org/10.1029/GM125p0347

Boteler, D. H. (2006). The super storms of August/September 1859 and their effects on the telegraph system. *Advances in Space Research*, *38*(2), 159–172. doi:https://doi.org/10.1016/j.asr.2006.01.013

Cagniard, L. (1953). Basic theory of the magneto-telluric method of geophysical prospecting. *Geophysics*, *18*(3), 605–635. doi:https://doi.org/10.1190/1.1437915

Caldwell, T. G., Bibby, H. M., & Brown, C. (2004). The magnetotelluric phase tensor. *Geophysical Journal International*, *158*(2), 457–469. doi:https://doi.org/10.1111/j.1365-246X.2004.02281.x

Carrington, R. C. (1859). Description of a singular appearance seen in the Sun on September 1, 1859. *Monthly Notices of the Royal Astronomical Society*, *20*(1), 13–15. doi:https://doi.org/10.1093/mnras/20.1.13

Chant, I. J., & Hastie, L. M. (1992). Time-frequency analysis of magnetotelluric data. *Geophysical Journal International*, *111*(2), 399–413. doi:https://doi.org/10.1111/j.1365-246X.1992.tb00586.x

Chave, A., & Jones, A. (2012). *The Magnetotelluric Method, Theory and Practice* (1st ed.). New York: Cambridge University Press. doi:https://doi.org/10.1017/CBO9781139020138

Chave, A. D., & Thomson, D. J. (1989). Some comments on magnetotelluric response function estimation. *Journal of Geophysical Research – Solid Earth*, *94*(B10), 14215–14225. doi:https://doi.org/10.1029/JB094iB10p14215

Constable, S., Shankland, T. J., & Duba, A. (1992). The electrical conductivity of an isotropic olivine mantle. *Journal of Geophysical Research – Solid Earth*, *97*(B3), 3397–3404. doi:https://doi.org/10.1029/91JB02453

Dong, S.-W., Li, T.-D., Lü, Q.-T., Gao, R., Yang, J.-S., Chen, X.-H., et al. (2013). Progress in deep lithospheric exploration of the continental China: A review of the SinoProbe. *Tectonophysics*, *606*, 1–13. doi:https://doi.org/10.1016/j.tecto.2013.05.038

Egbert, G. D., & Booker, J. R. (1986). Robust estimation of geomagnetic transfer functions. *Geophysical Journal of the Royal Astronomical Society*, *87*(1), 173–194. doi:https://doi.org/10.1111/j.1365-246X.1986.tb04552.x

Egbert, G. D., Eisel, M., Boyd, O. S., & Morrison, H. F. (2000). DC trains and Pc3s: Source effects in mid-latitude geomagnetic transfer functions. *Geophysical Research Letters*, *27*(1), 25–28. doi:https://doi.org/10.1029/1999GL008369

Egbert, G. D., & Kelbert, A. (2012). Computational recipes for electromagnetic inverse problems: Computational recipes for EM inverse problems. *Geophysical Journal International*, *189*(1), 251–267. doi:https://doi.org/10.1111/j.1365-246X.2011.05347.x

Egbert, G. D., & Livelybrooks, D. (1996). Single station magnetotelluric impedance estimation: Coherence weighting and the regression M-estimate. *Geophysics*, *61*(4), 964–970. doi:https://doi.org/10.1190/1.1444045

Evans, R. L. (1994). Constraints on the large-scale porosity and permeability structure of young oceanic crust from velocity and resistivity data. *Geophysical Journal International*, *119*(3), 869–879. doi:https://doi.org/10.1111/j.1365-246X.1994.tb04023.x

Evans, R. L. (2012). Earth's electromagnetic environment: 3A. Conductivity of earth materials. In A. D. Chave & A. G. Jones (Eds.), *The Magnetotelluric Method: Theory and Practice* (p. 552). New York: Cambridge University Press.

Fitterman, D. V. (1988). Overview of the structure and geothermal potential of Newberry Volcano, Oregon. *Journal of Geophysical Research – Solid Earth*, *93*(B9), 10059–10066. doi:https://doi.org/10.1029/JB093iB09p10059

Foster, J., Gjelde, E., Graham, W. R., Hermann, R. J., Kluepfel, H. M., Lawson, R. L., et al. (2004). *Report of the Commission to Assess the Threat to the United States from Electromagnetic Pulse (EMP) Attack. Volume 1: Executive Report*. National Research Council Washington DC, Committee on Electromagnetic Pulse Environment. Retrieved from http://www.dtic.mil/docs/citations/ADA484497

Foster, J., Gjelde, E., Graham, W. R., Hermann, R. J., Kluepfel, H. M., Lawson, R. L., et al. (2008). *Report of the Commission to Assess the Threat to the United States from Electromagnetic Pulse (EMP) Attack: Critical National Infrastructures*. McLean, VA: Electromagnetic Pulse Commission. Retrieved from http://www.dtic.mil/docs/citations/ADA484672

Fullea, J., Muller, M. R., & Jones, A. G. (2011). Electrical conductivity of continental lithospheric mantle from integrated geophysical and petrological modeling: Application to the Kaapvaal Craton and Rehoboth Terrane, southern Africa. *Journal of Geophysical Research*, *116*(B10). doi:https://doi.org/10.1029/2011JB008544

Gamble, T., Goubau, W., & Clarke, J. (1979). Magnetotellurics with a remote magnetic reference. *Geophysics*, *44*(1), 53–68. doi:https://doi.org/10.1190/1.1440923

Gregori, G. P., & Lanzerotti, L. J. (1980). Geomagnetic depth sounding by induction arrow representation: A review. *Reviews of Geophysics*, *18*(1), 203–209. doi:https://doi.org/10.1029/RG018i001p00203

Groom, R. W., & Bailey, R. C. (1989). Decomposition of magnetotelluric impedance tensors in the presence of local three-dimensional galvanic distortion. *Journal of Geophysical Research*, *94*(B2), 1913–1925.

Hibbs, A., Petrov, T., Pendleton, J., Milberger, S., Eiskamp, G., & Wilson, G. (2012). New electromagnetic sensors for magnetotelluric and induced polarization geophysical surveys. In *SEG Technical Program Expanded Abstracts 2012* (pp. 1–5). Society of Exploration Geophysicists. doi:https://doi.org/10.1190/segam2012-1478.1

Jackson, P., Smith, D., & Stanford, P. (1978). Resistivity-porosity-particle shape relationships for marine sands. *Geophysics*, *43*(6), 1250–1268. doi:https://doi.org/10.1190/1.1440891

Jiracek, G. R. (1990). Near-surface and topographic distortions in electromagnetic induction. *Surveys in Geophysics*, *11*(2–3), 163–203. doi:https://doi.org/10.1007/BF01901659

Jones, A. G., Chave, A. D., Egbert, G. D., Auld, D., & Bahr, K. (1989). A comparison of techniques for magnetotelluric response function estimation. *Journal of Geophysical Research – Solid Earth*, *94*(B10), 14201–14213. doi:https://doi.org/10.1029/JB094iB10p14201

Kappler, K. N., Morrison, H. F., & Egbert, G. D. (2010). Long-term monitoring of ULF electromagnetic fields at Parkfield, California. *Journal of Geophysical Research – Solid Earth*, *115*(B4), B04406. doi:https://doi.org/10.1029/2009JB006421

Kelbert, A., Meqbel, N., Egbert, G. D., & Tandon, K. (2014). ModEM: A modular system for inversion of electromagnetic geophysical data. *Computers & Geosciences*, *66*, 40–53. doi:https://doi.org/10.1016/j.cageo.2014.01.010

Ledo, J., Queralt, P., & Pous, J. (1998). Effects of galvanic distortion on magnetotelluric data over a three-dimensional regional structure. *Geophysical Journal International*, *132*(2), 295–301. doi:https://doi.org/10.1046/j.1365-246x.1998.00417.x

Love, J. J., Lucas, G. M., Kelbert, A., & Bedrosian, P. A. (2018). Geoelectric hazard maps for the mid-Atlantic United States: 100 year extreme values and the 1989 magnetic storm. *Geophysical Research Letters*, *45*(1), 5–14. doi:https://doi.org/10.1002/2017GL076042

Love, J. J., Pulkkinen, A., Bedrosian, P. A., Jonas, S., Kelbert, A., Rigler, E. J., et al. (2016). Geoelectric hazard maps for the continental United States. *Geophysical Research Letters*, *43*(18), 9415–9424. doi:https://doi.org/10.1002/2016GL070469

Lucas, G. M., Love, J. J., & Kelbert, A. (2018). Calculation of voltages in electric power transmission lines during historic geomagnetic storms: An investigation using realistic earth impedances. *Space Weather*, *16*(2), 185–195. doi:https://doi.org/10.1002/2017SW001779

Mackie, R. L., Smith, J. T., & Madden, T. R. (1994). Three-dimensional electromagnetic modeling using finite difference equations: The magnetotelluric example. *Radio Science*, *29*(4), 923–935. doi:https://doi.org/10.1029/94RS00326

McGary, R. S., Evans, R. L., Wannamaker, P. E., Elsenbeck, J., & Rondenay, S. (2014). Pathway from subducting slab to surface for melt and fluids beneath Mount Rainier. *Nature*, *511*(7509), 338–340. doi:https://doi.org/10.1038/nature13493

Meqbel, N., Egbert, G. D., Wannamaker, P. E., Kelbert, A., & Schultz, A. (2014). Deep electrical resistivity structure of the northwestern U.S. derived from 3-D inversion of USArray magnetotelluric data. *Earth and Planetary Science Letters*, *402*, 290–304. doi:https://doi.org/10.1016/j.epsl.2013.12.026

Mittet, R. (2015). Seismic wave propagation concepts applied to the interpretation of marine controlled-source electromagneticsSeismic wave propagation and marine CSEM. *Geophysics*, *80*(2), E63–E81. doi:https://doi.org/10.1190/geo2014-0215.1

Newman, G. A., Gasperikova, E., Hoversten, G. M., & Wannamaker, P. E. (2008). Three-dimensional magnetotelluric characterization of the Coso geothermal field. *Geothermics*, *37*(4), 369–399. doi:https://doi.org/10.1016/j.geothermics.2008.02.006

Nichols, E., Morrison, H., & Lee, S. (1994). Controlled source magnetotellurics for groundwater. In *SEG Technical Program Expanded Abstracts 1994* (pp. 553–554). Society of Exploration Geophysicists. doi:https://doi.org/10.1190/1.1932156

Parkinson, W. D. (1962). The influence of continents and oceans on geomagnetic variations. *Geophysical Journal of the Royal Astronomical Society*, *6*(4), 441–449. doi:https://doi.org/10.1111/j.1365-246X.1962.tb02992.x

Patro, P. K., Uyeshima, M., & Siripunvaraporn, W. (2013). Three-dimensional inversion of magnetotelluric phase tensor data. *Geophysical Journal International*, *192*(1), 58–66. doi:https://doi.org/10.1093/gji/ggs014

Peacock, J. R., Mangan, M. T., McPhee, D., & Wannamaker, P. E. (2016). Three-dimensional electrical resistivity model of the hydrothermal system in Long Valley Caldera, California, from magnetotellurics. *Geophysical Research Letters*, *43*(15), 2016GL069263. doi:https://doi.org/10.1002/2016GL069263

Peacock, J. R., & Selway, K. (2016). Magnetotelluric investigation of the Vestfold Hills and Rauer Group, East Antarctica. *Journal of Geophysical Research – Solid Earth*, *121*(4), 2258–2273. doi:https://doi.org/10.1002/2015JB012677

Peacock, J. R., Thiel, S., Reid, P., & Heinson, G. (2012). Magnetotelluric monitoring of a fluid injection: Example from an enhanced geothermal system. *Geophysical Research Letters*, *39*(18), L18403. doi:https://doi.org/10.1029/2012GL053080

Petiau, G. (2000). Second generation of lead-lead chloride electrodes for geophysical applications. *Pure and Applied Geophysics*, *157*(3), 357–382. doi:https://doi.org/10.1007/s000240050004

Pommier, A., & Le-Trong, E. (2011). "SIGMELTS": A web portal for electrical conductivity calculations in geosciences. *Computers & Geosciences*, *37*(9), 1450–1459. doi:https://doi.org/10.1016/j.cageo.2011.01.002

Pridmore, D. F., & Shuey, R. T. (1976). The electrical resistivity of galena, pyrite, and chalcopyrite. *American Mineralogist*, *61*(3–4), 248–259.

Pritchard, M. E., Silva, S. L. de, Michelfelder, G., Zandt, G., McNutt, S. R., Gottsmann, J., et al. (2018). Synthesis: PLUTONS: Investigating the relationship between pluton growth and volcanism in the Central Andes. *Geosphere*, *14*(3), 954–982. doi:https://doi.org/10.1130/GES01578.1

Pronenko, V., & Korepanov, V. (2009). Induction coil magnetometers – A review (Vol. *11*, p. 1381). *Presented at the EGU General Assembly Conference Abstracts*, Vienna, Austria. Retrieved from http://adsabs.harvard.edu/abs/2009EGUGA..11.1381P

Ripka, P. (2003). Advances in fluxgate sensors. *Sensors and Actuators A: Physical*, *106*(1), 8–14. doi:https://doi.org/10.1016/S0924-4247(03)00094-3

Robertson, K., Heinson, G., & Thiel, S. (2016). Lithospheric reworking at the Proterozoic–Phanerozoic transition of Australia imaged using AusLAMP Magnetotelluric data. *Earth and Planetary Science Letters*, *452*, 27–35. doi:https://doi.org/10.1016/j.epsl.2016.07.036

Rodi, W., & Mackie, R. L. (2001). Nonlinear conjugate gradients algorithm for 2-D magnetotelluric inversion. *Geophysics*, *66*(1), 174–187. doi:https://doi.org/10.1190/1.1444893

Sasaki, Y. (2004). Three-dimensional inversion of static-shifted magnetotelluric data. *Earth, Planets, and Space*, *56*(2), 239–248. doi:https://doi.org/10.1186/BF03353406

Schock, R. N., Duba, A. G., & Shankland, T. J. (1989). Electrical conduction in olivine. *Journal of Geophysical Research – Solid Earth*, *94*(B5), 5829–5839. doi:https://doi.org/10.1029/JB094iB05p05829

Schultz, A., Booker, J., & Larsen, J. (1987). Lake bottom magnetotellurics. *Journal of Geophysical Research – Solid Earth*, *92*(B10), 10639–10649. doi:https://doi.org/10.1029/JB092iB10p10639

Schultz, A., Egbert, G. D., Kelbert, A., Peery, T., Clote, V., Fry, B., et al. (2006–2018). USArray TA magnetotelluric transfer functions. doi:https://doi.org/10.17611/DP/EMTF/USARRAY/TA

Schultz, A., Kurtz, R. D., Chave, A. D., & Jones, A. G. (1993). Conductivity discontinuities in the upper mantle beneath a stable craton. *Geophysical Research Letters*, *20*(24), 2941–2944. doi:https://doi.org/10.1029/93GL02833

Shea, M. A., & Smart, D. F. (2006). Compendium of the eight articles on the "Carrington Event" attributed to or written by Elias Loomis in the American Journal of Science, 1859–1861. *Advances in Space Research*, *38*(2), 313–385. doi:https://doi.org/10.1016/j.asr.2006.07.005

Silverman, S. M. (2006). Comparison of the aurora of September 1/2, 1859 with other great auroras. *Advances in Space Research*, *38*(2), 136–144. doi:https://doi.org/10.1016/j.asr.2005.03.157

Simpson, F., & Bahr, K. (2005). *Practical Magnetotellurics*. New York: Cambridge University Press.

Siripunvaraporn, W., Egbert, G. D., Lenbury, Y., & Uyeshima, M. (2005). Three-dimensional magnetotelluric inversion: data-space method. *Physics of the Earth and Planetary Interiors*, *150*(1), 3–14. doi:https://doi.org/10.1016/j.pepi.2004.08.023

Smith, J. (1996). Conservative modeling of 3-D electromagnetic fields, Part I: Properties and error analysis. *Geophysics*, *61*(5), 1308–1318. doi:https://doi.org/10.1190/1.1444054

Stevens, K., & McNeice, G. (1998). On the detection of Ni?Cu Ore hosting structures in the Sudbury Igneous Complex using the Magnetotelluric Method. In *SEG Technical Program Expanded Abstracts 1998* (Vols. *1–0*, pp. 751–755). Society of Exploration Geophysicists. doi:https://doi.org/10.1190/1.1820581

Sutarno, D. (2005). Development of robust magnetotelluric impedance estimation: A review. *Indonesian Journal of Physics*, *16*(3), 11.

Thiel, S., Heinson, G., Reid, A., & Robertson, K. (2016). Insights into lithospheric architecture, fertilisation and fluid pathways from AusLAMP MT. *Australian Society of Exploration Geophysicists Extended Abstracts*, *2016*(1), 1–6. doi:https://doi.org/10.1071/aseg2016ab261

Tietze, K., Ritter, O., & Egbert, G. D. (2015). 3-D joint inversion of the magnetotelluric phase tensor and vertical magnetic transfer functions. *Geophysical Journal International*, *203*(2), 1128–1148. doi:https://doi.org/10.1093/gji/ggv347

Tikhonov, A. N. (1950). On determining electrical characteristics of the deep layers of the Earth's crust. *Doklady*, *73*(2), 295–297.

Tumanski, S. (2007). Induction coil sensors – A review. *Measurement Science and Technology*, *18*(3), R31. doi:https://doi.org/10.1088/0957-0233/18/3/R01

Tuncer, V., Unsworth, M. J., Siripunvaraporn, W., & Craven, J. A. (2006). Exploration for unconformity-type uranium deposits with audiomagnetotelluric data: A case study from the McArthur River mine, Saskatchewan, CanadaUranium exploration with AMT. *Geophysics*, *71*(6), B201–B209. doi:https://doi.org/10.1190/1.2348780

Tyburczy, J. A., & Fisler, D. K. (1995). Electrical properties of minerals and melts. In T. J. Ahrens (Ed.), *Mineral Physics & Crystallography: A Handbook of Physical Constants* (Vol. *2*, pp. 185–208). American Geophysical Union (AGU). doi:https://doi.org/10.1029/RF002p0185

Utada, H., & Munekane, H. (2000). On galvanic distortion of regional 3-D MT impedances on galvanic distortion of regional three-dimensional magnetotelluric impedances. *Geophysical Journal International*, *140*(2), 385–398. doi:https://doi.org/10.1046/j.1365-246x.2000.00014.x

Vozoff, K. (1991). The magnetotelluric method. In M. N. Nabighian (Ed.), *Electromagnetic Methods in Applied Geophysics* (pp. 641–712). Society of Exploration Geophysicists. doi:https://doi.org/10.1190/1.9781560802686.ch8

Wannamaker, P. E., Hohmann, G. W., & Ward, S. H. (1984) Magnetotelluric responses of three-dimensional bodies in layered earths. *Exploration Geophysics 15*, 190–191.

Wannamaker, P. E., Stodt, J. A., & Olsen, S. L. (1996). Dormant state of rifting below the Byrd Subglacial Basin, West Antarctica, implied by magnetotelluric (MT) profiling. *Geophysical Research Letters*, *23*(21), 2983–2986. doi:https://doi.org/10.1029/96GL02887

Wannamaker, P. E., Stodt, J. A., Pellerin, L., Olsen, S. L., & Hall, D. B. (2004). Structure and thermal regime beneath the

South Pole region, East Antarctica, from magnetotelluric measurements. *Geophysical Journal International*, *157*(1), 36–54. doi:https://doi.org/10.1111/j.1365-246X.2004.02156.x

Ward, S., & Hohmann, G. (1988). Electromagnetic theory for geophysical applications. In M. N. Nabighian (Ed.), *Electromagnetic Methods in Applied Geophysics* (Vol. *1*, pp. 130–311). Society of Exploration Geophysicists. doi:https://doi.org/10.1190/1.9781560802631.ch4

Weaver, J. T., & Agarwal, A. K. (1991). Is addition of induction vectors meaningful? *Physics of the Earth and Planetary Interiors*, *65*(3), 267–275. doi:https://doi.org/10.1016/0031-9201(91)90133-3

Yee, K. (1966). Numerical solution of initial boundary value problems involving Maxwell's equations in isotropic media. *IEEE Transactions on Antennas and Propagation*, *14*(3), 302–307. doi:https://doi.org/10.1109/TAP.1966.1138693

Yoshino, T. (2010). Laboratory electrical conductivity measurement of mantle minerals. *Surveys in Geophysics*, *31*(2), 163–206. doi:https://doi.org/10.1007/s10712-009-9084-0

Zhdanov, M. S. (2017). *Foundations of Geophysical Electromagnetic Theory and Methods* (2nd ed.). Elsevier. doi:https://doi.org/10.1016/C2014-0-02307-6

Zhdanov, M. S., Smith, R. B., Gribenko, A., Cuma, M., & Green, M. (2011). Three-dimensional inversion of large-scale EarthScope magnetotelluric data based on the integral equation method: Geoelectrical imaging of the Yellowstone conductive mantle plume. *Geophysical Research Letters*, *38*(8), L08307. doi:https://doi.org/10.1029/2011GL046953

8

The First 3D Conductivity Model of the Contiguous United States: Reflections on Geologic Structure and Application to Induction Hazards

Anna Kelbert[1,2], Paul A. Bedrosian[3,4], and Benjamin S. Murphy[2]

ABSTRACT

Estimation of ground level geoelectric fields has been identified by the National Space Weather Action Plan as a key component of assessment and mitigation of space weather impacts on critical infrastructure. Estimates of spatially and temporally variable electric fields are used to generate statistically based hazard maps and show promise toward monitoring and responding to geomagnetic disturbances in near real-time. One approach to geoelectric field estimation is to employ three-dimensional (3D) Earth conductivity models. These data-constrained conductivity models are the results of regional magnetotelluric inversions based primarily on NSF's Earthscope USArray impedances, which to date cover ~60% of the contiguous United States. Here, we present the first-ever composite conductivity model of the contiguous United States and describe its compilation from 3D regional conductivity models, a global mantle conductivity model, offshore bathymetry, and sediment thickness data. We discuss structures within the conductivity model and how they relate to the complex geologic tapestry of the continent. Finally, we discuss the utility of this synthesis model for estimation and mitigation of geomagnetically induced currents.

Key Points

- The first 3D conductivity model of the contiguous United States is developed and presented.
- The 3D electrical conductivity is discussed in terms of geology and tectonics.
- The utility of 3D electrical conductivity models to Induction hazards research is discussed.

[1] *Geomagnetism Program, United States Geological Survey, Denver, CO, USA*
[2] *College of Earth, Ocean, and Atmospheric Sciences, Oregon State University, Corvallis, OR, USA*
[3] *Geology, Geophysics, and Geochemistry Science Center, United States Geological Survey, Denver, CO, USA*
[4] *Crustal Geophysics, United States Geological Survey, Denver, CO, USA*

8.1. INTRODUCTION

Relative to the atmosphere, solid Earth is electrically conductive. Earth's electrical conductivity structure is highly variable; at the length scales of interest to the electric power industry, observed conductivity values span four to six orders of magnitude with spatial variations over kilometers to hundreds of kilometers (e.g., Yoshino, 2011; Evans, 2012). Within the context of this paper, we alternately refer to conductivity and resistivity, mathematical inverses of one another. To first order, the electrical conductivity of the Earth increases with depth (Evans, 2012), while laterally it has a spatially complicated distribution. Coherent structural features of enhanced or reduced electrical conductivity found deep in Earth's interior reflect its complex geological history and tectonic structure. This is particularly true in the

Geomagnetically Induced Currents from the Sun to the Power Grid, Geophysical Monograph 244, First Edition.
Edited by Jennifer L. Gannon, Andrei Swidinsky, and Zhonghua Xu.

lithosphere, the mechanically rigid uppermost layer of the Earth, which comprises the Earth's crust (~30–60 km thick, e.g., Chulick and Mooney (2002)) and uppermost mantle, typically ranging between 60 and 200 km in thickness.

In active tectonic regions, such as the western United States, lithospheric conductivity is most sensitive to partial melt and solute-bearing fluids (e.g., water, carbon dioxide), with bulk conductivity significantly enhanced by their presence at levels of a few percent or less. Such "geofluids" are commonly associated with subduction zones, regions of crustal extension, and hot spots, all typically accompanied by volcanism. In tectonically inactive regions, such as the central and eastern United States, crustal conductivity is strongly controlled by rock type, metamorphic grade, and the presence or absence of specific minerals, including metallic sulfides and graphite. Within the mantle part of the lithosphere, temperature and rock composition affect conductivity, although variations in water contained within nominally anhydrous minerals (at the ppm level) often dominate; these variations are a frozen-in relic of the last tectonic event to impact the region (Selway, 2014). Within the brittle upper crust, electrical conductivity is additionally sensitive to the quantity, salinity and temperature of pore fluids; the fraction and type of mineralogic clays; and the distributions of both phases within the subsurface. Mapping conductivity at near-surface scales (up to ~ 1 km) is therefore an important component of both hydrologic studies and energy resources investigations such as geothermal, oil and gas. Heterogeneity of the subsurface thus results in variations in electrical properties at a variety of spatial scales and over many orders of magnitude, from highly conducting to effectively insulating (resistive).

In geophysics, methods exist to probe the Earth's electrical conductivity structure to depths well beyond those accessible by drilling. These methods have varied depth and spatial resolution, but they all suffer from decreasing resolving power with increasing depth; hence, in examining geophysically derived resistivity models, Earth's upper crust often appears more heterogeneous than the lower crust and mantle lithosphere simply due to the fundamental physics behind these techniques, not necessarily due to intrinsic differences in the scale of heterogeneity at different depths. The underlying asthenosphere (as shallow as 60–70 km in tectonically active regions and 150–200 km otherwise, temperature >1300 °C) and the mantle transition zone (410–670 km, temperature ~ 1400–1600 °C) are also known to exhibit spatially varying conductivity, controlled primarily by water content and the accumulation of partial melt at material density interfaces (e.g., Toffelmier and Tyburczy, 2007; Kelbert et al., 2009; Munch et al., 2018). The mineral carrying capacity for hydration increases with depth into the asthenosphere (e.g., Hirschmann et al., 2005), a phenomenon that is partly responsible for the increase of electrical conductivity with depth in the upper mantle. Water storage capacity of the transition zone is much higher than that of the asthenosphere, but it greatly reduces in the mid-mantle below the 670 km interface (e.g., Ohtani et al., 2001).

The range of geophysical methods that are used to image Earth's conductivity is discussed in the following section; however, all are based upon geomagnetic induction, in which time- or frequency-dependent magnetic or electric fields induce electric currents in the subsurface. Both magnetic and electric fields at the surface of the Earth depend upon electrical conductivity in a non-linear way through Maxwell's equations and Ohm's law. Conductivity models are derived by solving the inverse problem of electromagnetic induction based upon measurements of these electric and/or magnetic fields. Inversion techniques use numerical optimization to calculate the best-fitting electrical conductivity distribution of the subsurface (e.g., see Kelbert et al. (2008b) and Kelbert et al. (2014) for example algorithms that achieve this aim). Depending on the particular method, the scale of investigation, the distribution of data, and the ultimate application, 1D, 2D, or 3D methods are employed. The resultant Earth conductivity models are typically the starting point for subsequent geophysical integration and interpretation.

We present herein a 3D conductivity model for the contiguous United States assembled from a collection of regional inverse models based primarily upon USArray magnetotelluric data (Schultz et al., 2006–2018; Schultz, 2010) of the U.S. National Science Foundation's EarthScope project (Williams et al., 2010). Such models are typically used for investigations into Earth structure, composition, and processes (e.g., Bedrosian, 2007). These models are also of increasing value for mineral resource evaluation, specifically at district and regional scales, in recognition of the critical role of deep lithospheric architecture on fluid delivery pathways (which are key to mineralization; see, e.g., Heinson et al. (2006) and Robertson et al. (2016)). Most recently, the importance of understanding electrical conductivity on regional or national scales is motivated by induction hazards research and the space weather community. Strong geomagnetic storms can disrupt the operation of electric power grids and damage critical infrastructure by inducing geoelectric fields in the ground that in turn enter grounded electric transmission lines. The amplitude and polarization of these induced geoelectric fields are strongly dependent upon subsurface conductivity structure (Bedrosian and Love, 2015). Hence, magnetotelluric data coverage of the United States is a prerequisite, whether using a data-driven approach (e.g., Bonner and Schultz,

2017; Lucas et al., 2018) or a model-driven approach, as described herein, to developing nationwide systems for geomagnetic hazard mitigation.

Here, we first outline the geophysical methods that allow us to constrain subsurface conductivity structure and compare the resolution and utility of these methods in Section 8.2. For subsequent discussion, we focus on the method of magnetotellurics, the most relevant technique at our depths and scales of interest (Section 8.3). In Section 8.4, we describe the MT transfer function (TF) database and data availability within the contiguous United States. We proceed with a literature review of electrical conductivity models at relevant spatial scales for the contiguous United States (Sections 8.5 and 8.6) and discuss the assembly of our merged 3D electrical conductivity model in Section 8.7. In Section 8.8, we examine the model in relation to the geologic and tectonic structure of North America and finally, explain its relevance to induction hazards research, in Section 8.9.

8.2. ELECTROMAGNETIC METHODS IN GEOPHYSICS

In geophysics, a wide range of methods (Table 8.1) are used to estimate subsurface electrical conductivity. We limit the discussion here to inductive techniques, but note that non-inductive techniques such as ground penetrating radar and electrical resistivity tomography are commonly used to constrain conductivity in near-surface environmental and geotechnical applications. Fundamental to all electromagnetic methods is the measurement of time-varying electric and/or magnetic fields at, above, or below the Earth's surface in response to currents induced in the subsurface. These induced currents may be due to natural sources, as in magnetotellurics, or generated via a controlled source (transmitter). For all inductive techniques, the depth-of-investigation is controlled by skin depth, a mathematical function of the subsurface conductivity and the frequency content of the inducing source. Both lower frequency and lower Earth conductivity result in increased depths-of-investigation. Additionally, the resolution of these techniques falls off with increasing depth due to the diffusive nature of the underlying physics.

Controlled-source techniques employ one or more transmitters (typically a grounded electric dipole or magnetic dipole source) and receivers to image the subsurface over small to modest areas (~10 to ~100,000 m^2). These systems commonly operate at frequencies of ~100 kHz to 10 Hz, and are capable of constraining conductivity at depths of a few meters to in excess of 500 m. Airborne variants of some systems are used primarily for groundwater and mineral resource investigations, where large survey areas can be quickly surveyed with high lateral resolution (on the order of 100 m). The magnetotelluric (MT) method uses large-scale natural sources to probe Earth structure on depth scales ranging from tens of meters to hundreds of kilometers. The depth resolution of MT depends not only on skin depth but also upon the aperture of the station array. In practice, MT is applied over survey areas ranging from tens of square kilometers (e.g., Rosenkjaer et al., 2015; Lindsay et al., 2017) to investigations in excess of 1,000,000 km^2 (e.g., AusLAMP, Robertson et al. (2016, 2017); SinoProbe, Dong et al. (2013, 2014)). The shallowest variant of MT, audio-frequency magnetotellurics, is limited to periods of 10 s or less, and typically used to

Table 8.1 Inductive Electromagnetic (EM) Methods for Constraining Subsurface Conductivity.

Method	Source	Production	Frequency/ period range	Spatial scale	Depth of investigation
Controlled-source EM	Magnetic or electric dipole transmitter	Hours/site	1 Hz to 100 kHz	Hundreds of meters to tens of kilometers	1 m to 1 km
Airborne EM	Transmitter 30–100 m in air	~100 line km/hour	100 Hz to 100 kHz	Tens to hundreds of kilometers	5–500 m
Radio-MT	Remote radio stations	Hours/survey	20–200 kHz	Several kilometers	1–100 m
Audio-frequency MT	Lightning	Hours/site	0.1 Hz to 100 kHz	Hundreds of meters to tens of kilometers	10 m to 1 km
Wideband MT	Lightning/magnetosphere–ionosphere interactions	Days/site	0.001–2000 s	Tens to hundreds of kilometers	200 m to 50 km
Long-period MT	Solar wind–magnetosphere interactions	Weeks/site	10–20,000 s	Hundreds to thousands of kilometers	5–400 km
Global EM	Ionospheric and magnetospheric currents	Years to decades	3–100 days	Continental to global	400–1600 km

image to depths of ~ 1 km. Wideband (1000 Hz to a few thousand seconds) MT is most commonly applied to investigate upper-crustal structure and is routinely used for geothermal and mineral-resource investigations. Long-period (10–20,000 s) MT is employed for deep investigations of regional geologic structure, including studies of the lithospheric mantle. At the largest of scales, global electromagnetics (periods of 3–100 days) utilize geomagetic observatory data and global ionospheric sources to constrain models of the deep asthenosphere on a global scale.

The wide variety of techniques listed in Table 8.1 all address the same conceptual problem: investigation of electrical conductivity of the Earth's subsurface. Indeed, the Earth's electrical conductivity varies over 4–6 orders of magnitude over all of the spatial scales outlined in Table 8.1. Most problems in electromagnetic geophysics may therefore be addressed by using one or a few of these techniques, and focusing on a specific area and spatial scale of electrical conductivity distribution. However, in compiling a continental scale electrical conductivity model for space weather research, one would benefit from collecting all available information obtained through these various techniques, combined with auxiliary geologic information, into a coherent electrical conductivity structure of the Earth's subsurface. Indeed, such was the goal of the earlier Fernberg (2012) study; however, the electrical conductivity information available at the time was mostly confined to indirect estimates based on surface geology, and the outcome was therefore by necessity limited to 1D profiles beneath broad physiographic provinces. Here, we are extending this approach to state-of-the-art 3D electrical conductivity models informed by measured data.

Taken to the extreme, this approach presents numerical challenges that are insurmountable in the framework of currently existing tools. Indeed, the wide variety of spatial scales and resolutions, which may range from meters in the near-surface, to thousands of kilometers in the Earth's mantle, requires that new tools are developed to allow for storage and numerical modeling at different resolutions (this is known as the multi-grid approach). Here, we are undertaking a simplified approach of only including the spatial scales of relevance to space weather research, and mapping all models to the same high-resolution grid covering the contiguous United States at 10 km spacing (Section 8.7).

8.3. MAGNETOTELLURICS AND GEOELECTRIC FIELD ESTIMATION

Magnetotellurics was developed as a geophysical technique, and is primarily used to determine the distribution of subsurface conductivity to depths of tens to hundreds of kilometers (Chave and Jones, 2012). It is by pure serendipity that the MT impedances, obtained from MT surveys and modeling, are exactly the type of data product necessary for space weather hazard mitigation. Indeed, MT impedance that relates geoelectric fields to geomagnetic fields at the Earth's surface provides a critical missing piece to estimate geomagnetically induced currents (e.g., Love et al., 2017) (see also Section 8.9, and other contributions in this monograph). For this reason, the need for nationwide MT impedance coverage is increasingly recognized, and such data now serve the electric power grid industry as well as the solid-Earth geophysical community. Various methods for estimation of local geoelectric fields using a geomagnetic field time series and a 3D MT impedance have been developed in the recent years (e.g., Kelbert et al., 2017; Weigel, 2017), and are being applied to analyze voltages in electric power grids (e.g., Bonner and Schultz, 2017; Lucas et al., 2018).

Magnetotelluric data measured at the Earth's surface consist of simultaneous measurements of natural, time-varying geomagnetic and geoelectric fields. Well-developed processing methods (most notably, Egbert and Booker, 1986; Chave et al., 1987; Smirnov and Egbert, 2012; Chave, 2014) allow for a reduction of electromagnetic field time series collected at a site into a set of so-called magnetotelluric transfer functions (MT TFs), most commonly impedances and vertical magnetic field transfer functions, at a discrete set of frequencies. An MT TF represents all of the information about Earth conductivity extracted from the raw measurements, and provides the key input to further inversion and interpretation. Specifically, the MT response tensor relates the horizontal components of the inducing geomagnetic field to the horizontal components of the induced geoelectric field (e.g., Weidelt and Chave, 2012),

$$\begin{pmatrix} E_x \\ E_y \end{pmatrix} = \begin{bmatrix} Z_{xx} & Z_{xy} \\ Z_{yx} & Z_{yy} \end{bmatrix} \begin{pmatrix} B_x \\ B_y \end{pmatrix}, \tag{8.1}$$

where

$$\mathbf{Z}(f,x,y) = \begin{bmatrix} Z_{xx} & Z_{xy} \\ Z_{yx} & Z_{yy} \end{bmatrix} (f,x,y) \tag{8.2}$$

defines the MT response tensor as a function of frequency f and Earth's surface location, (x, y). Note that MT response tensor tends to have the practical units of (mV/km)/nT, and is related to the MT impedance $\mu\mathbf{Z}$ (measured in ohms), where μ is the magnetic permeability, via a simple rescaling of units. Here, we note explicitly that this full impedance tensor reflects 3D Earth's structure, in the physical sense. Since the 1D MT impedance is represented by a frequency dependent scalar ($Z_{xx} = Z_{yy} = 0$, $Z_{xy} = -Z_{yx} = Z$), a collection of 1D electrical conductivity models, solved in the 1D sense (e.g., Ádám et al., 2012;

Fernberg, 2012; Blake et al., 2016), will always produce scalar impedances that cannot reflect the full 3D physics. On the other hand, a full 3D model corresponds to tensor MT impedances that cannot be, in general, reduced to scalars. Throughout this manuscript, we discuss electrical conductivity modeling that is performed by solving Maxwell's equations (and Ohm's law) in 3D space. These models and their responses at ground level are, therefore, irreducible to the 1D situation. Specifically, using a vertical profile from these 3D models to solve the equations in the 1D sense would not yield the correct local MT impedance.

Similarly, the vertical magnetic field TF, sometimes referred to as the tipper, relates the vertical and horizontal components of geomagnetic field at the Earth's surface,

$$B_z = \begin{bmatrix} T_{zx} & T_{zy} \end{bmatrix} \begin{pmatrix} B_x \\ B_y \end{pmatrix}. \tag{8.3}$$

While not of direct utility for geoelectric field estimation, tippers are often employed to further constrain magnetotelluric inversions, including some of the models we discuss in this manuscript (e.g., Table 8.3).

MT impedances can be either empirical, that is, statistically estimated from measured data, as in the Incorporated Research Institutions for Seismology Electromagnetic Transfer Functions (IRIS EMTF) database (Kelbert et al., 2011), or computed from an electrical conductivity model (which in turn may be obtained by inverting the empirical impedances). Traditionally, magnetotelluric methods have allowed for 1D or 2D interpretations in terms of the Earth conductivity, using the simplified versions of the MT impedance appropriate for these special cases. The past decade, however, has seen the advent of 3D magnetotellurics, which allows estimation of high-resolution 3D electrical conductivity variations in the Earth's deep interior. This dramatic transformation has been fueled primarily by the rapid growth in computational capabilities and the corresponding developments of academic 3D magnetotelluric inversion codes (e.g., Siripunvaraporn et al., 2005; Kelbert et al., 2014), as well as by the recent availability of high-quality 3D datasets (see Section 8.4).

Magnetotelluric inversion is a computationally demanding, non-linear, non-unique inverse problem. The latter implies that there exists an infinite variety of 3D models that fit the measured MT responses to within their estimated precision. Additionally, the spatial characteristics of 3D variability imaged by an inverse model can vary dramatically depending on the spatial coverage and frequency content of the inverted dataset (see Section 8.2). In the conductive Earth, the diffusive nature of electromagnetic energy at the long periods pertinent to magnetotellurics allows us to successfully average the effects of small-scale heterogeneities other than those found in the immediate vicinity of the site location. However, small-scale anomalies that are local to a site often affect the data across the frequency spectrum. This phenomenon is known as galvanic distortion (e.g., Groom and Bailey, 1989; Jones, 2012), and is a challenge in magnetotellurics as it can obscure the regional signal in the data. Some 3D MT inversion methods invert for the distortion effects explicitly (Avdeeva et al., 2015; Kordy et al., 2016), while others (e.g., Kelbert et al., 2014) resolve the problem implicitly by allowing small-scale near-surface structure to appear in the conductivity model. Either way, modern large-scale 3D MT inversion (see Section 8.5) has demonstrated the ability to overcome the potential problems of near-surface distortion and thereby produce geologically reasonable inverse solutions.

Galvanic distortion is one reason to use 3D electrical conductivity models to produce synthetic impedances in between the existing data sites. MT impedance data are commonly distorted by effects that are highly local. Interpolation of the empirical impedances to in-between site locations, or using a local MT impedance in a regional sense, runs the risk of erroneously extrapolating these very local effects to a significantly larger area, thus resulting in a poor estimate of the regional geoelectric field. In contrast, using the modeled impedances in-between sites is equivalent to a physics-based interpolation and could successfully filter these effects out. Even though the local near-surface Earth structure in between sites would still be unknown, we propose that it should not play a major role in geomagnetically induced current (GIC) estimation. This is a reasonable *a priori* assumption (see also Section 8.9), since GIC estimates are based upon voltages calculated along grounded power line segments via integration of the surface geoelectric field. However, a careful validation study against GIC measurements is warranted to ascertain this as truth.

The uncertainty in 3D MT models is very difficult to quantify. However, it is important to note that since the individual models are already constrained by data, the uncertainty in modeled impedances is somewhat better understood. For example, the variety of 3D MT models that we present in this manuscript (Section 8.5) all fit the data, in general, to within ~ 5% error bars in MT impedances at the existing site locations. The uncertainty in modeled impedances could be much higher in between the sites, perhaps translating to a 10–20% uncertainty in predicted MT impedances; precise estimates could be obtained by statistical analysis that would involve a comparison of modeled impedances to measured ("out-of-sample") data that were not taken advantage of in the respective MT inversions. Overall, quantification of model-related uncertainty is a subject of further study. However, we also know that electrical conductivity variations alone can cause approximately an order of magnitude variability in the electric fields at the Earth's surface (e.g., Bedrosian and Love, 2015). For models that are constrained by empirical data, model-related uncertainty is clearly a second-order effect relative to the use of a 3D conductivity model for geoelectric field estimation.

8.4. MAGNETOTELLURIC DATA IN THE CONTIGUOUS UNITED STATES

Magnetotellurics in the United States experienced a significant boost in 2006 when the National Science Foundation funded the MT component of the nationwide EarthScope geoscience data collection effort, USArray MT (Schultz et al., 2006–2018). The data set currently covers approximately two thirds of the contiguous United States at a quasi-regular 70 km spacing. An important feature of USArray MT project was the requirement that both the raw and the processed data were immediately available for public use (Kelbert et al., 2018a). To best accommodate this requirement, a comprehensive magnetotelluric database, IRIS EMTF, has been developed (Kelbert et al., 2011).

At the time of the writing of this article, exactly 1000 survey sites from USArray MT project have been archived in the IRIS EMTF database. Additionally, several complementary and/or historical MT surveys have been added to the database in the contiguous United States. The U.S. Geological Survey (USGS) has contributed 34 MT sites in Montana (Bedrosian et al., 2006) and 22 MT sites in Florida (Bedrosian et al., 2015). Community data contributions include 34 MT sites in Idaho from the Snake River Plain project (de Groot-Hedlin et al., 2003–2004), 57 MT sites in New Mexico from SAGE Summer Program (Pellerin et al., 2005–2035), and 21 MT sites in Washington from CAFE MT project (Wannamaker et al., 2010). Several additional community contributions are being finalized and archived. MT data are solicited from the community and are archived as they arrive. The IRIS EMTF data archive in its entirety, insofar as it is limited to the contiguous United States, is shown in Figure 8.1.

The original time series were collected in a variety of coordinate frames; usually, the horizontal geomagnetic field components are oriented to geomagnetic North and geomagnetic East, respectively, while the horizontal geoelectric field components may be collected at an arbitrary angle (not always at a right angle to each other), as field conditions permit. While the USArray MT time series are archived intact, the MT impedances and vertical field transfer functions archived at the IRIS EMTF database have all been rotated to geographic North and transformed into an orthogonal coordinate system. This allows for consistency when multiple sites are used jointly in a magnetotelluric inversion or for geoelectric field estimation. Note, however, that consistent use of the original USArray MT time series data (e.g., for validation of custom geoelectric field estimation algorithms) requires that the archived time series are also rotated to geographic North, or that the MT impedances found in the IRIS EMTF database are rotated back to the original local coordinate system.

8.5. 3D ELECTRICAL CONDUCTIVITY MODELS OF THE CONTIGUOUS UNITED STATES

3D electrical conductivity modeling has matured dramatically over the last decade, due to significant advancements in the speed and availability of non-commercial modeling and inversion codes (e.g., Siripunvaraporn et al., 2005; Kelbert et al., 2014). Coupled with these methodological advancements, the acquisition of spatially extensive MT data sets, appropriate for 3D inversion, has come into its own over this same time period. Nationwide data initiatives such as EarthScope in the United States, AusLamp in Australia, and SinoProbe in China are producing ever-expanding data sets ideally suited for 3D inversion modeling. At local scales, geothermal and mineral-resource investigations are acquiring dense 3D data sets over complex geologic environments.

A listing of regional 3D electrical conductivity models within the contiguous United States based primarily upon USArray MT (Schultz et al., 2006–2018) is provided in Table 8.2. To constrain our scope to scales of interest, we have included all published 3D conductivity models that utilize more than a few USArray MT sites, as well as four previously unpublished models, which we have included in this compilation (Table 8.3). We note that some of these previously published models are not available to the community, and their corresponding data fits cannot be assessed. Details of inversion configurations are data fits for the four previously unpublished models that are used in this manuscript are given in Table 8.3.

The models can be divided into three groupings, herein referred to as the western, central, and eastern models based upon the extent of USArray data they consider (Fig. 8.1). The 6 western models incorporate up to 325 stations covering part or all of 10 states in the northwestern United States. Studies based upon these models are focused on active tectonic processes within the western United States (subduction of the Juan de Fuca slab, Basin and Range extension, the Yellowstone hotspot) and their imprint upon the crust and mantle lithosphere. The three central models are based upon approximately 300 stations covering the upper and central mid-continent region and are focused on the ancient assembly of the North American continent and the 1.1 Ga Midcontinent Rift System. The five remaining eastern models span part or all of the eastern United States from Florida to Maine,

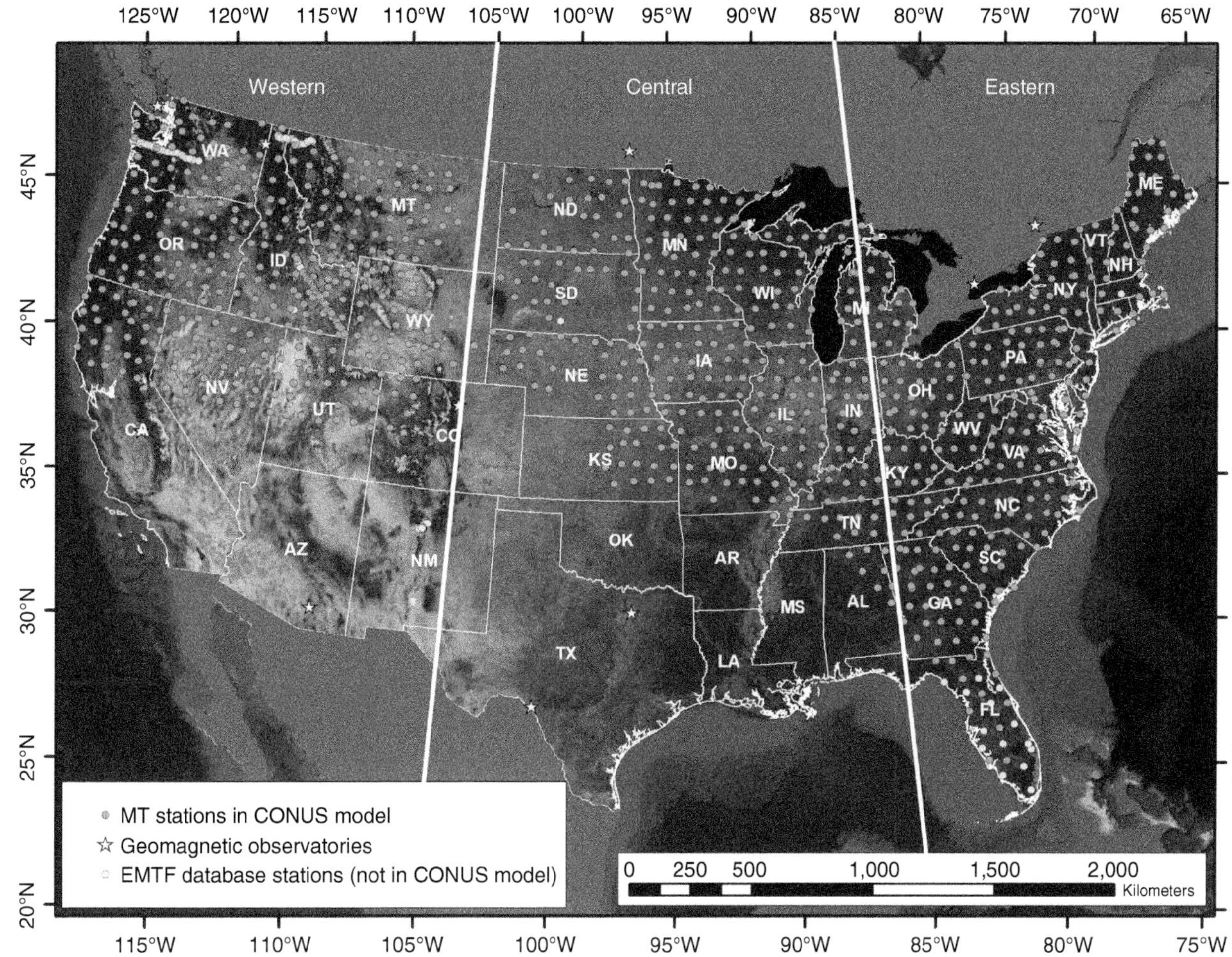

Figure 8.1 Locations of all magnetotelluric sites currently at IRIS EMTF database (Kelbert et al., 2011). Locations of magnetotelluric sites that were used to generate component inverse models merged to form the contiguous US (CONUS) model are shown as dots. Magnetotelluric sites not included in the component inverse models of CONUS are also shown. Additionally, locations of the historical geomagnetic observatories in the United States (USGS, 1901; Love and Finn, 2011) and Canada (Newitt and Coles, 2007) that were used to constrain the global electrical conductivity model of Sun et al. (2015), which forms the background for the CONUS model, are shown as stars. (*See electronic version for color representation of this figure.*)

Table 8.2 Regional 3D Electrical Conductivity Models in the Contiguous United States.

Model	Geophysical target	Latitude limits (°)	Longitude limits (°)	Nominal cell size (km^2)	Reference
PNW	Pacific Northwest/Cascadia	[42 49]	[−126 − 114]	13.3 × 10.2	Patro and Egbert (2008)
NWUS	Northwestern US	[35 52]	[−130 − 105]	12.5 × 12.5	Meqbel et al. (2014)
NWUS	Northwestern US	[36 49]	[−125 − 105]	15 × 15	Bedrosian and Feucht (2014)
NWUS	Northwestern US	[38 49]	[−125 − 105]	10 × 10	Zhdanov et al. (2012)
SRPY	Snake River Plane/Yellowstone	[41 47]	[−120 − 107]	10.2 × 10.2	Kelbert et al. (2012)
SRPY	Snake River Plane/Yellowstone	Unknown	Unknown	5 × 5	Zhdanov et al. (2011)
GP	The Great Plains	[39 51]	[−109 − 94]	12.5 × 12.5	B.S. Murphy, this chapter
MCR	The Midcontinent Rift	[34.5 51.4]	[−99.6 − 78.3]	20 × 20	Yang et al. (2015)
MCR	The Midcontinent Rift	[35 49]	[−99 − 82]	10 × 10	Bedrosian (2016)
OIINK	The Ozark Dome, MO, USA	[37 42]	[−96 − 82]	12.5 × 12.5	DeLucia et al. (2019)
SEUS	Southeastern US	[28 41]	[−91 − 74]	12 × 12	Murphy and Egbert (2017)
NEUS	Northeastern US	[37 47]	[−82 − 66]	12.5 × 12.5	B.S. Murphy, this chapter
APPL	The Appalachians	[27.5 48.5]	[−90 − 70]	15 × 15	B.S. Murphy, this chapter
ENAM	Eastern US	[23.0 54.6]	[−105 − 65]	30 × 30	B.S. Murphy, this chapter
ECUS	East-central US	[36 43]	[−100 − 75]	15 × 15	Gribenko and Zhdanov (2017)

Table 8.3 Details on Inversion Configurations and Model Performance for Regional 3D Electrical Conductivity Models in the Contiguous United States that were Previously Unpublished. All inversions were performed with ModEM software (Egbert and Kelbert, 2012a; Kelbert et al., 2014). As far as model covariance is concerned, "uniform" stands for the same level of smoothing applied in all three spatial directions at all model depths. For example, a "uniform 0.2" configuration with two smoothing passes spreads model structure over ~ 2.5 adjacent cells.

Model name	Nominal horizontal cell size	Vertical grid spacing	ModEM covariance	Prior/starting model	Coordinate system	Data setup	Data types and rotation	Error floors	nRMSE
ENAM	$30 \times 30\,km^2$	Logarithmically telescoping to 930 km; cell size ~ 250 m at surface, ~60 km at 300 km depth	Uniformly 0.2; 2 smoothing passes (frozen ocean)	Uniform 300 Ωm in upper 50 km with exponential decrease from 300 Ωm at 50 km to 30 Ωm at 200 km; includes 0.3 Ωm ocean	Lambert Conformal Conic Projection, NAD83 (origin at 37.5°N 85°W, standard parallels at 31.25°N and 43.75°N)	Sites: 324; Periods: 28 (11.6 s – 18725 s)	Full Impedance Tensor, Vertical Field Transfer Functions	Z: 5% of off-diagonal components, each row separately; T: 0.03	2.23
APPL	$15 \times 15\,km^2$	Logarithmically telescoping to 1000 km; cell size ~ 200 m at surface, ~50 km at 300 km depth	Uniformly 0.2; 2 smoothing passes (frozen ocean)	Uniform 1000 Ωm in upper 50 km with exponential decrease from ~ 1000 Ωm at 50 km to ~ 30 Ωm at 200 km; includes 0.3 Ωm ocean	Lambert Conformal Conic Projection, NAD83 (origin at 38°N 80°W, standard parallels at 34.25°N and 41.75°N)	Sites: 240; Periods: 30 (7.3 s – 18725 s)	Full Impedance Tensor (rotated from geographic to align exactly with grid)	Z: 5% of off-diagonal components, each row separately	1.44
NEUS	$12.5 \times 12.5\,km^2$	Logarithmically telescoping to ~ 1000 km; cell size ~ 200 m at surface, ~50 km at 300 km depth	Uniformly 0.2; 2 smoothing passes (frozen ocean)	Uniform 1000 Ωm in upper 50 km with exponential decrease from ~ 1000 Ωm at 50 km to ~ 30 Ωm at 200 km; includes 0.3 Ωm ocean	Lambert Conformal Conic Projection, NAD83 (origin at 42.5°N 73.25°W, standard parallels at 38.5°N and 46.5°N) Rotated 60°E of N	Sites: 115; Periods: 30 (7.3 s – 18725 s)	Full Impedance Tensor (rotated 60°E of N to align with grid)	Z: 5% of off-diagonal components, each row separately; T: 0.03	3.6
GP	$12.5 \times 12.5\,km^2$	Logarithmically telescoping to ~ 1000 km; cell size ~ 200 m at surface, ~50 km at 300 km depth	Uniformly 0.2; 2 smoothing passes	Uniform 1000 Ωm to 50 km with exponential decrease from ~ 1000 Ωm at 50 km through ~ 300 Ωm at 200 km to ~ 30 Ωm at 400 km	Lambert Conformal Conic Projection, NAD83 (origin at 44.75°N 101.5°W, standard parallels at 42.25°N and 47.25°N)	Sites: 85; Periods: 30 (7.3 s – 18725 s)	Full Impedance Tensor, Vertical Field Transfer Functions (rotated from geographic to align exactly with grid)	Z: 5% of off-diagonal components, each row separately; T: 0.03	1.13

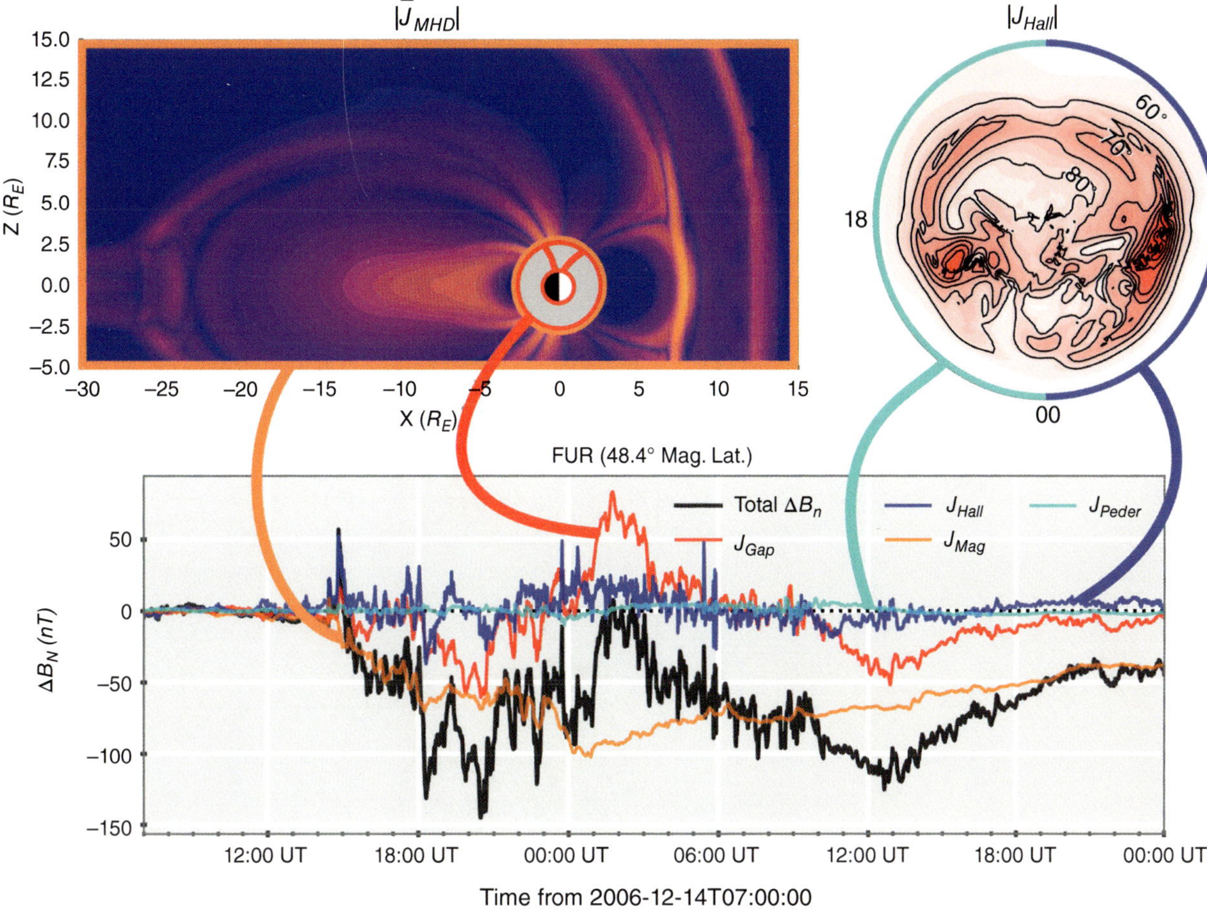

Figure 3.5 Contributions of different current density categories to the total virtual magnetometer ΔB for a mid-latitude station.

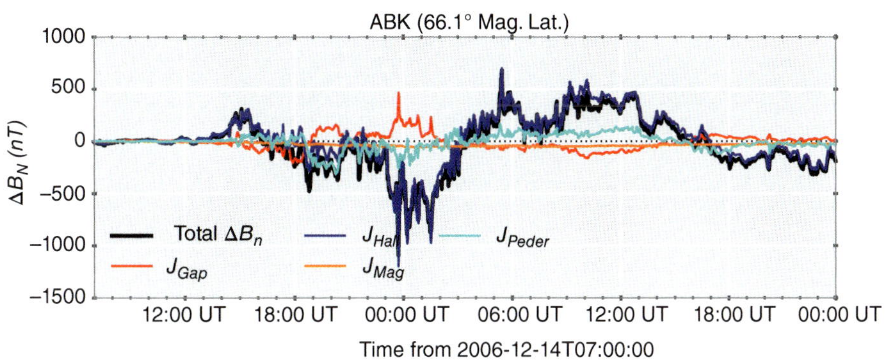

Figure 3.6 Similar to the bottom frame of Figure 3.5, but for a higher latitude station.

Geomagnetically Induced Currents from the Sun to the Power Grid, Geophysical Monograph 244. First Edition.
Edited by Jennifer Gannon, Andrei Swidinsky, and Zhonghua Xu.

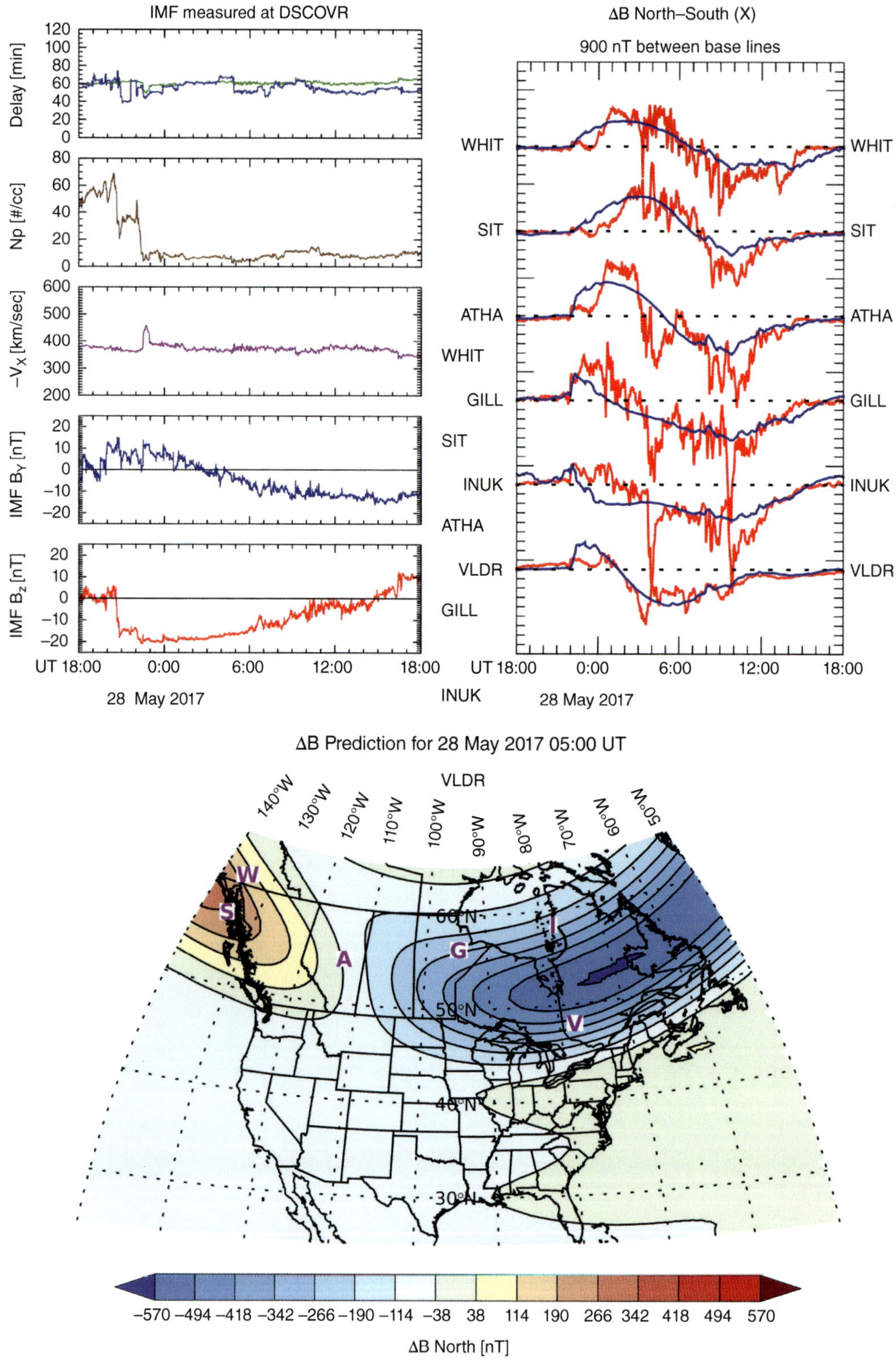

Figure 4.1 Empirical model prediction for 27–28 May 2017. The solar wind and IMF measured at the DSCOVR spacecraft are graphed in the upper left corner, predicted (blue) and measured (red) magnetic field perturbations in the North–South direction are shown in the upper right, and a map of these perturbations at 05:00 UT is shown at the bottom. The locations of the magnetometer stations are indicated with their first letter.

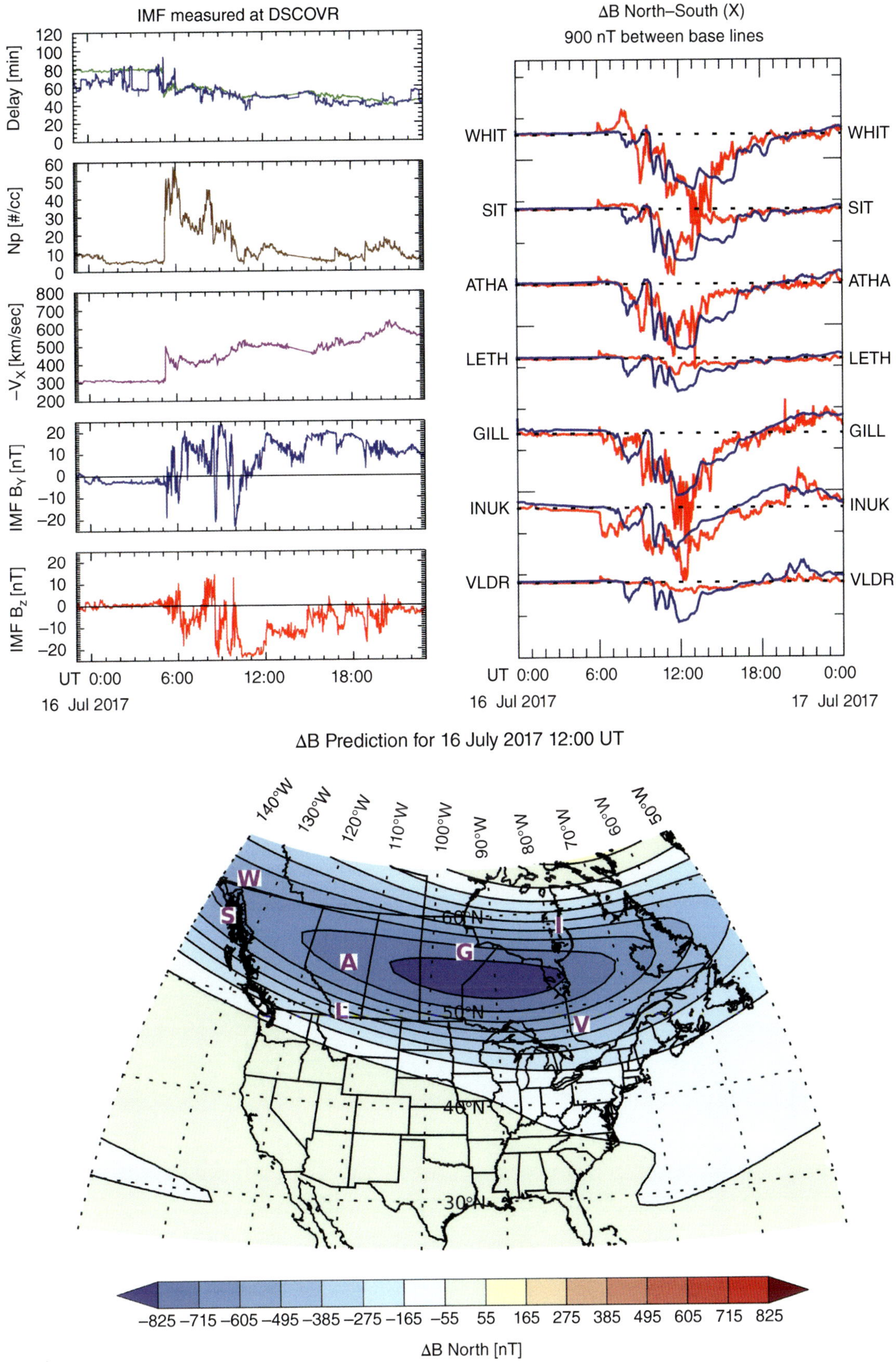

Figure 4.2 Empirical model prediction for 16 July 2017. The solar wind and IMF measured at the DSCOVR spacecraft are graphed in the upper left corner, predicted (blue) and measured (red) magnetic field perturbations in the North–South direction are shown in the upper right, and a map of these perturbations at 12:00 UT is shown at the bottom. The locations of the magnetometer stations are indicated with their first letter.

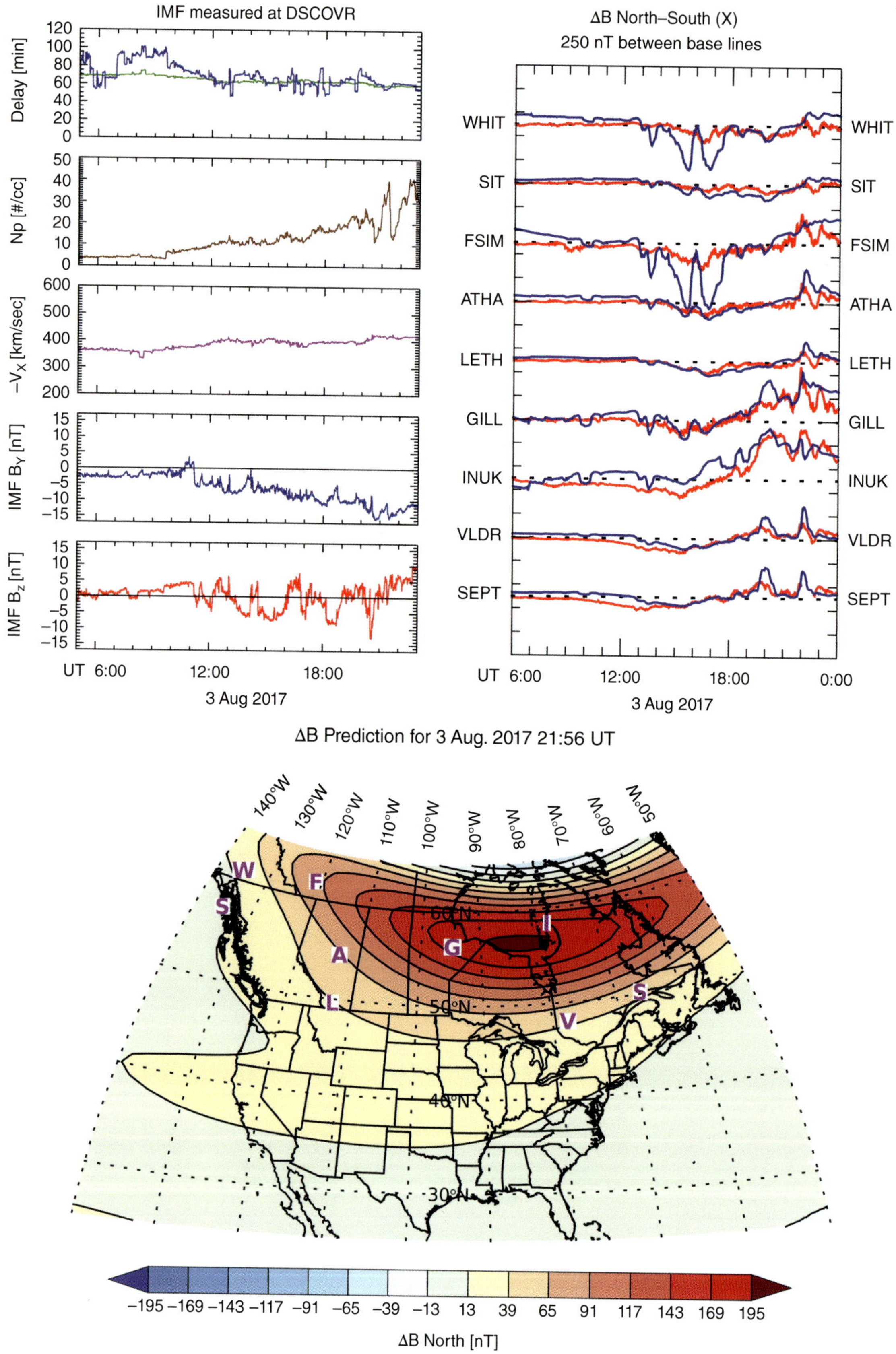

Figure 4.3 Empirical model prediction for 3 August 2017. The solar wind and IMF measured at the DSCOVR spacecraft are graphed in the upper left corner, predicted (blue) and measured (red) magnetic field perturbations in the North–South direction are shown in the upper right, and a map of these perturbations at 21:56 UT is shown at the bottom. The locations of the magnetometer stations are indicated with their first letter.

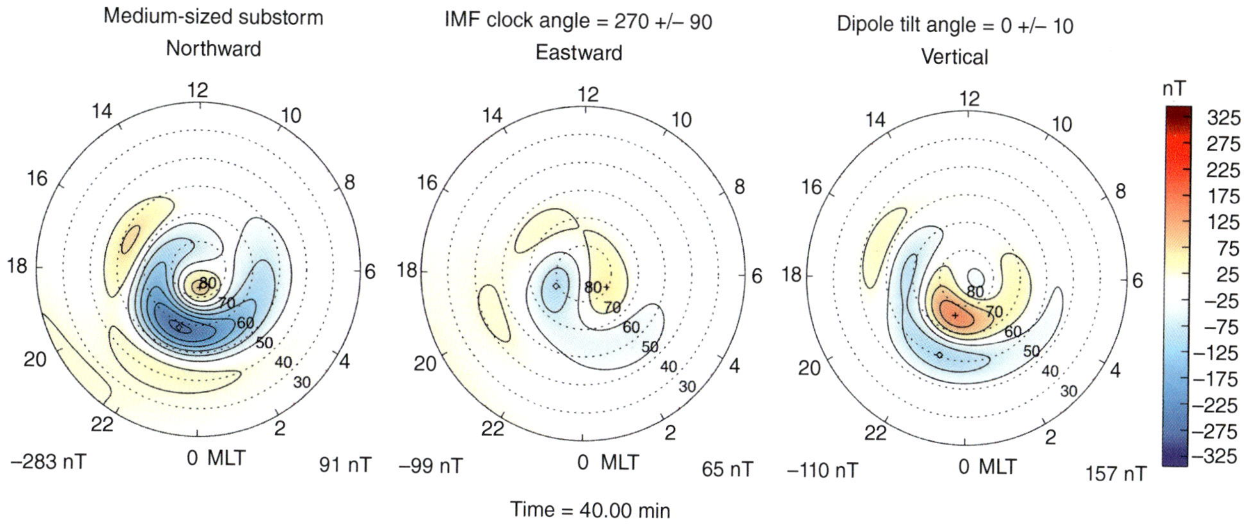

Figure 4.5 Maps of geomagnetic perturbations for a clock angle of 270° ± 90° ($-B_Y$ IMF) and dipole tilt angle of 0° ± 10° (near equinox). The measurements are from substorms having their greatest $|AL^*|$ index in the range of 300–500 nT and taken at 40 min after onset of the expansion phase. Each of the three vector components is shown. These plots show the results of least error fits using spherical cap harmonics, with the baseline (time = 0 min) removed.

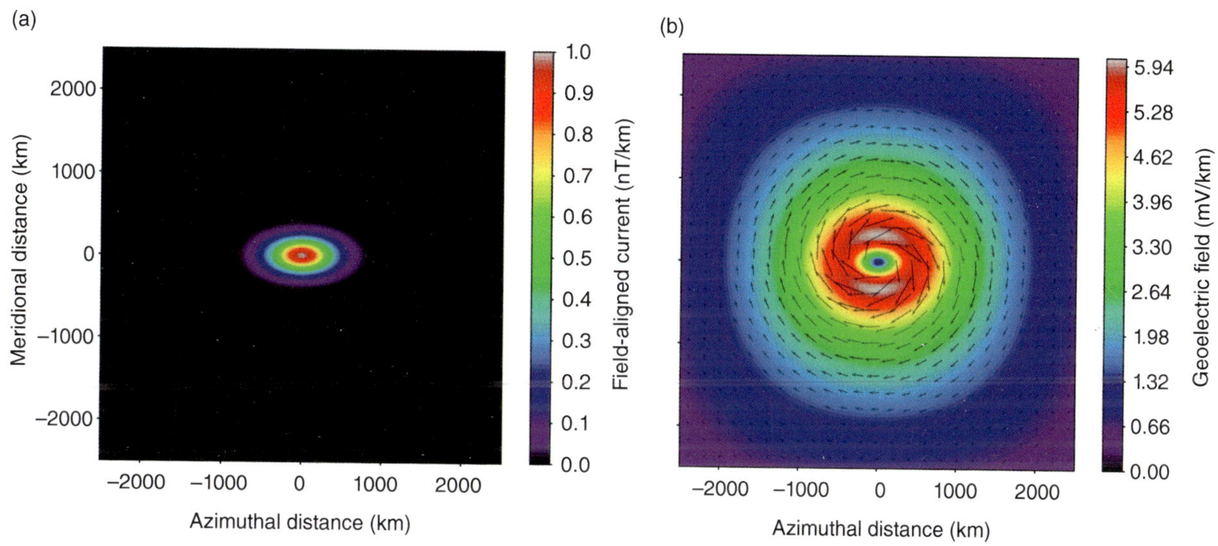

Figure 5.2 Two-dimensional profile of [a] the ionospherically-incident localized Gaussian FAC and [b] the resultant geoelectric fields. In [b], arrows show the direction and relative magnitude of the geoelectric fields.

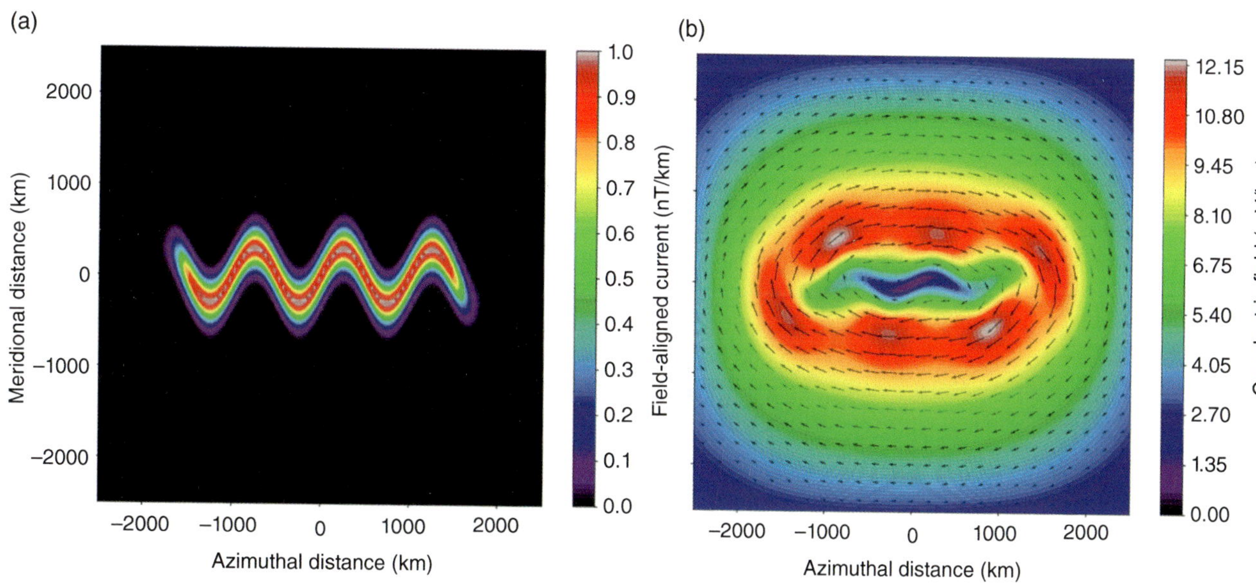

Figure 5.3 Two-dimensional profile of [a] the ionospherically-incident localized ribbon FAC and [B] the resultant geoelectric fields. In [b], arrows show the direction and relative magnitude of the geoelectric fields.

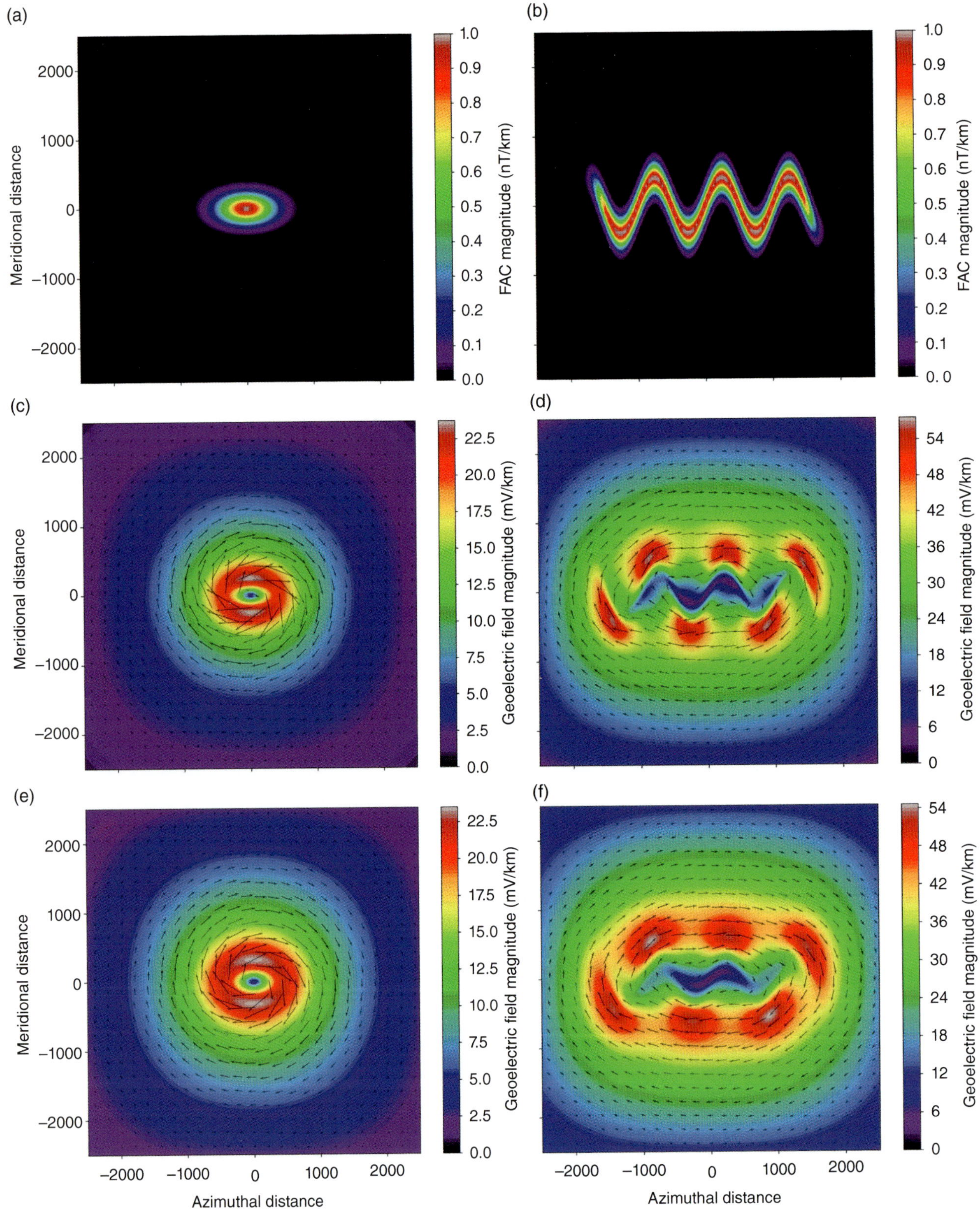

Figure 5.4 Spatial distribution of the ionospheric FACs and the resultant geoelectric fields. [a] Gaussian FAC structure; [b] Ribbon FAC structure; [c and d] Geoelectric fields from the Biot-Savart approach assuming an 18 mHz frequency; [e and f] Geoelectric fields from the wave method assuming a 22 mHz frequency. In [c–f], arrows indicate the local strength and direction of the geoelectric field.

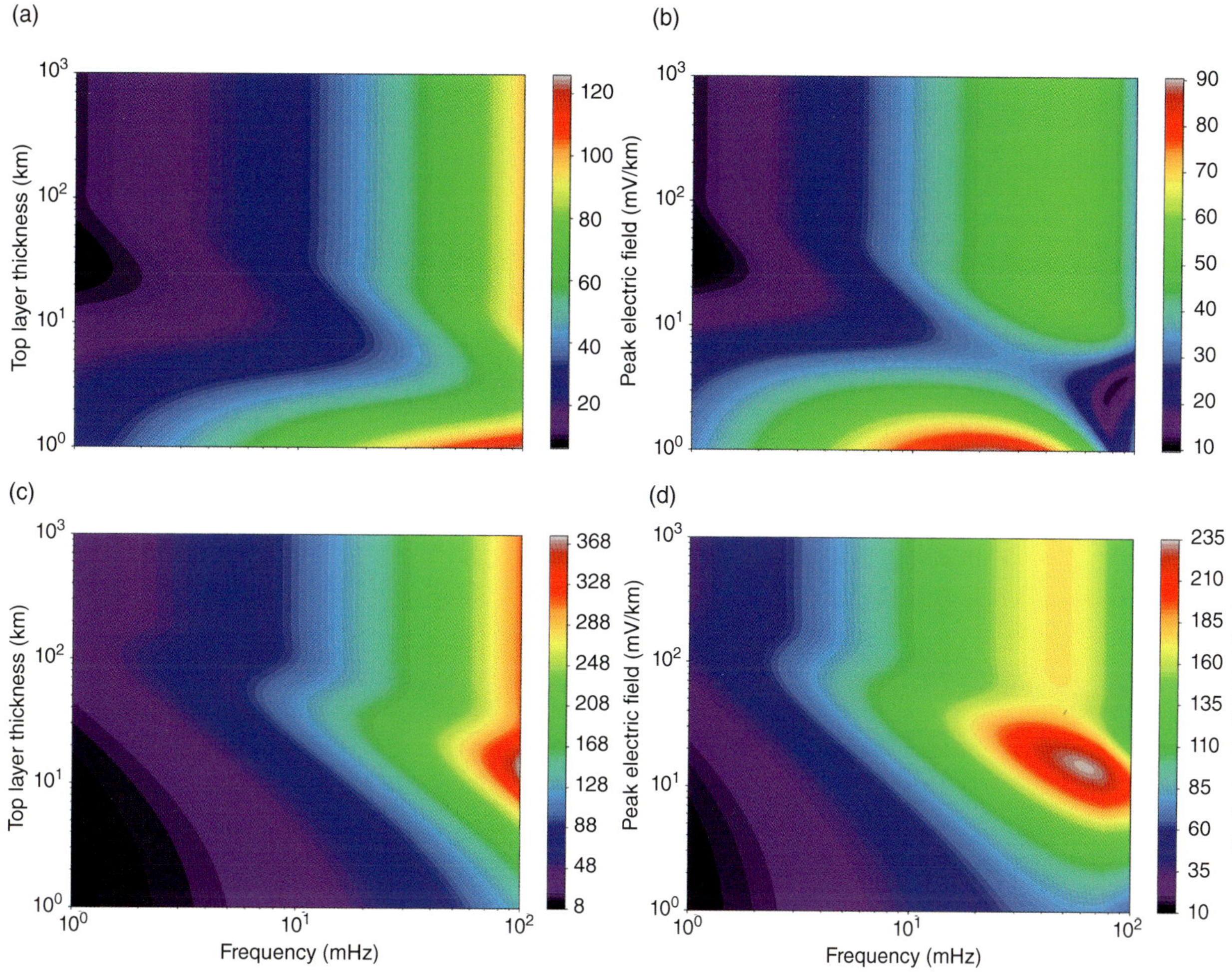

Figure 5.5 Comparison of peak geoelectric field magnitudes as a function of frequency and top-layer depth for (a) Biot–Savart approach with $\sigma_1 = 10^{-1}$ S/m and $\sigma_2 = 10^{-2}$ S/m; (b) Wave method with $\sigma_1 = 10^{-1}$ S/m and $\sigma_2 = 10^{-2}$ S/m; (c) Biot–Savart approach with $\sigma_1 = 10^{-2}$ S/m and $\sigma_2 = 10^{-1}$ S/m; (d) Wave method with $\sigma_1 = 10^{-2}$ S/m and $\sigma_2 = 10^{-1}$ S/m.

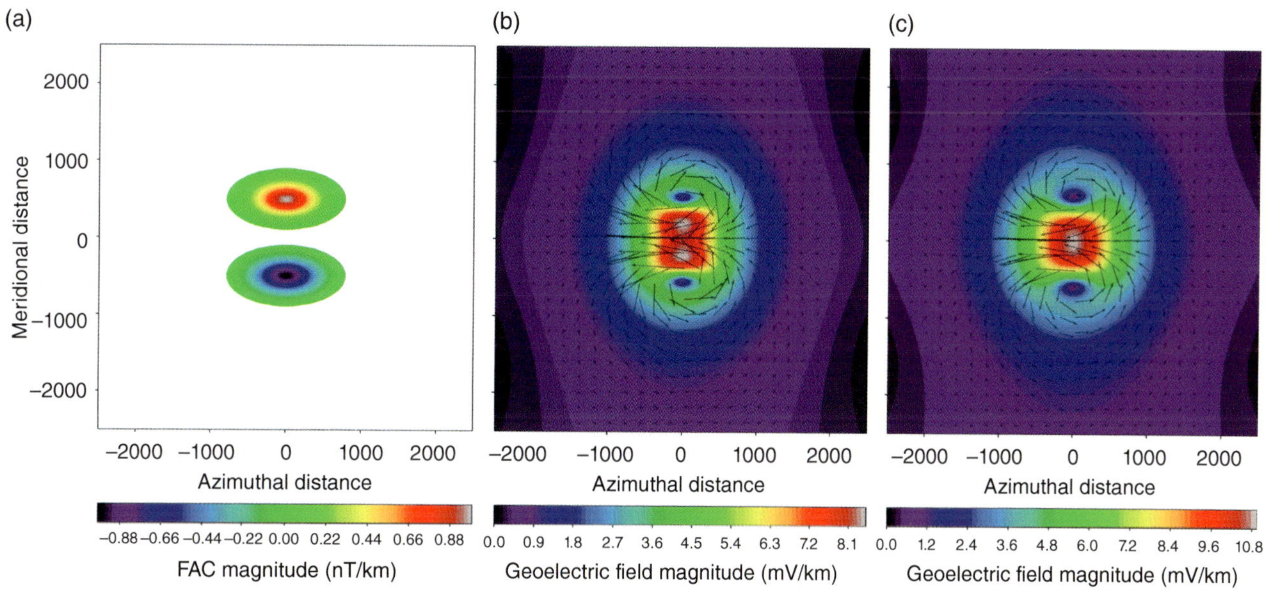

Figure 5.7 Spatial distribution of geoelectric fields for FACs that include a closure current. [a] Closed Gaussian FACs; [b] Geoelectric field calculated using the Biot-Savart approach; [c] Geoelectric field calculated using the wave method. In [a], the region of near-zero values beyond the FAC footprint is not plotted in order to highlight the geometry of the FACs; in [b and c], arrows indicate the direction and magnitude of the geoelectric field.

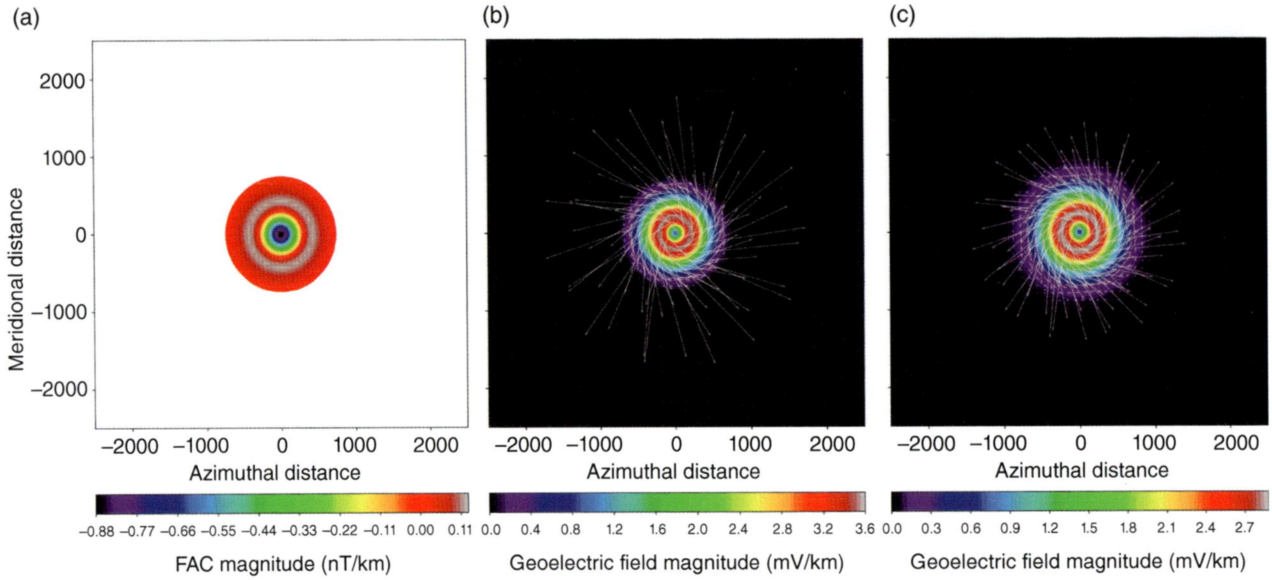

Figure 5.8 Geoelectric fields associated with a self-closing coaxial FAC. [a] Field aligned current; [b] Geoelectric field calculated using Biot-Savart approach; [c] Geoelectric field calculated using wave method. In [a], the region of near-zero values beyond the FAC footprint is not plotted in order to highlight the geometry of the FACs; in [b and c], arrows indicate the direction and magnitude of the geoelectric field.

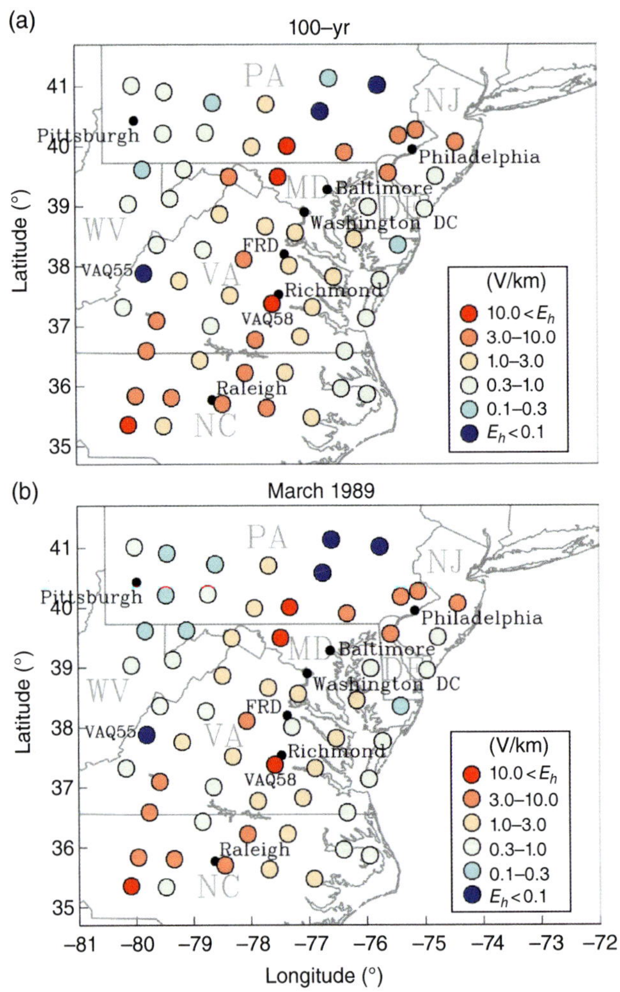

Figure 6.5 Maps showing (a) 100-year geoelectric exceedance amplitudes at the various EarthScope magnetotelluric survey sites across the Mid-Atlantic region of the United States, (b) maximum E_h^m that occurred during March 1989 magnetic storm.

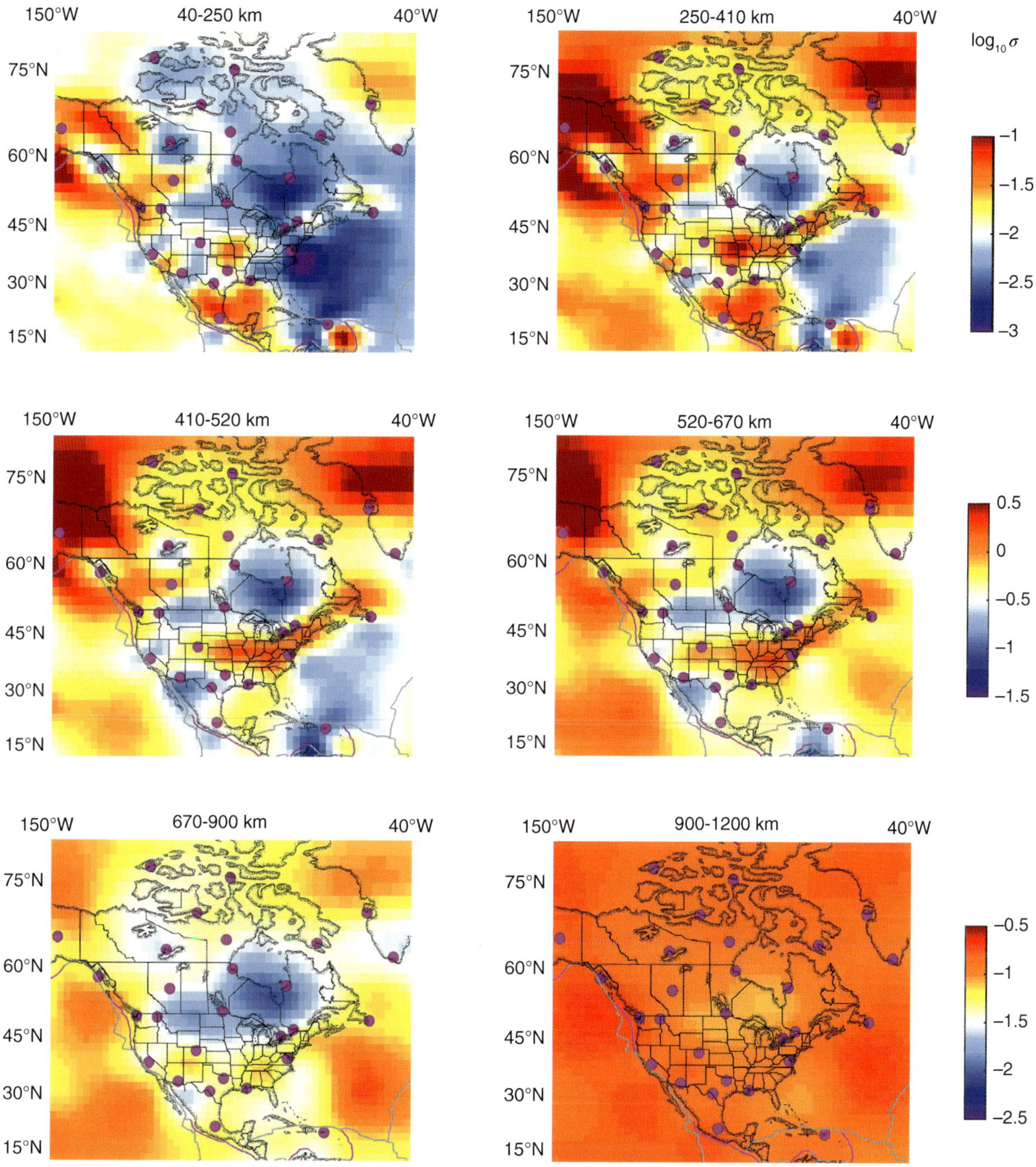

Figure 8.2 3 D electrical conductivity variations in Earth's mantle beneath North America based on the global model of Sun et al. [2015]. Shown is variability of two orders of magnitude in each of the layers of the Sun et al. [2015] published inverse model. Additionally, locations of the historical geomagnetic observatories from INTERMAGNET Geomagnetic Data [2013], primarily in the United States [USGS, 1901; Love and Finn, 2011] and Canada [Newitt and Coles, 2007], that were used to constrain this model (see also Figure 8.1) are shown in magenta.

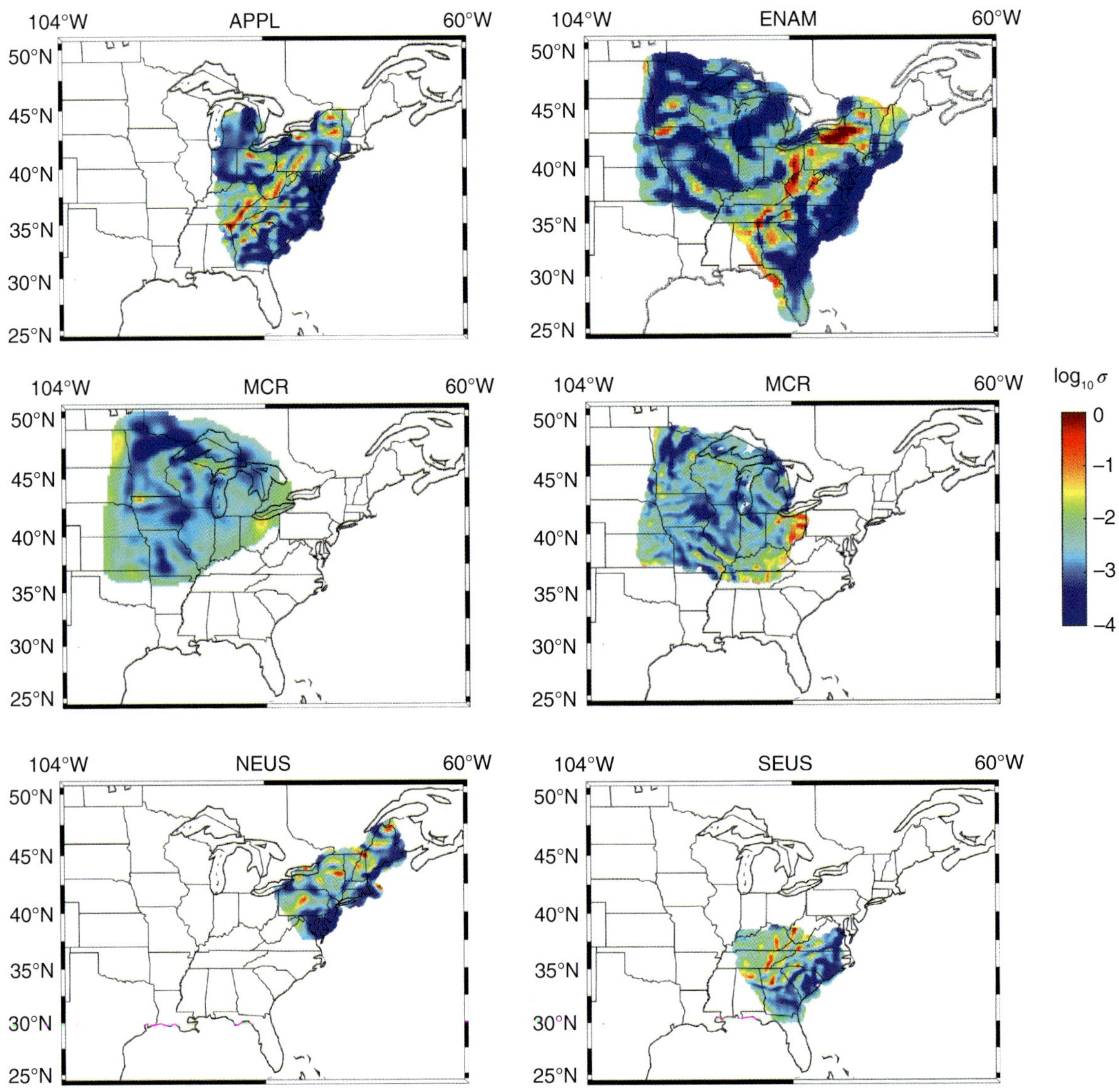

Figure 8.3 Relative locations of six regional models in the central and eastern US: MCR (Yang et al., 2015; left), MCR (Bedrosian, 2016; right), ENAM, APPL and NEUS (this work), and SEUS (Murphy and Egbert, 2017), shown at an upper mantle depth of 60 km. Note the differing spatial scales and resolutions.

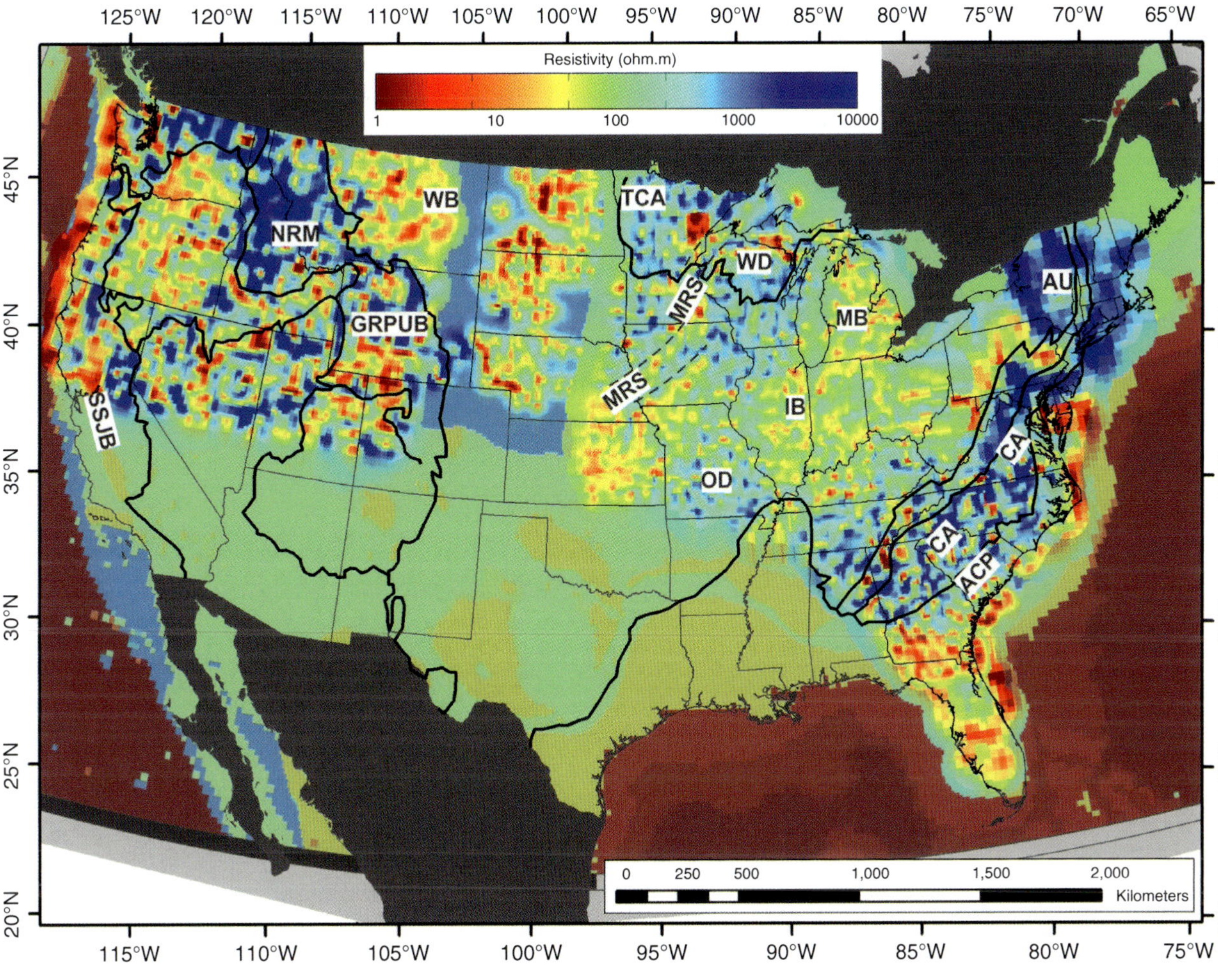

Figure 8.5 Electrical resistivity model at upper-crustal (3 km) depth. Labels and geologic provinces as in Figure 8.4 plus Northern Rocky Mountains (NRM), Sacramento-San Joachin Basin (SSJB), Trans-continental Arch (TCA), Wisconsin Dome (WD), Midcontinent Rift System (MRS), Ozark Dome (OD), Crystalline Appalachians (CA), Adirondack Uplift (AU), and Atalantic Coastal Plain (ACP).

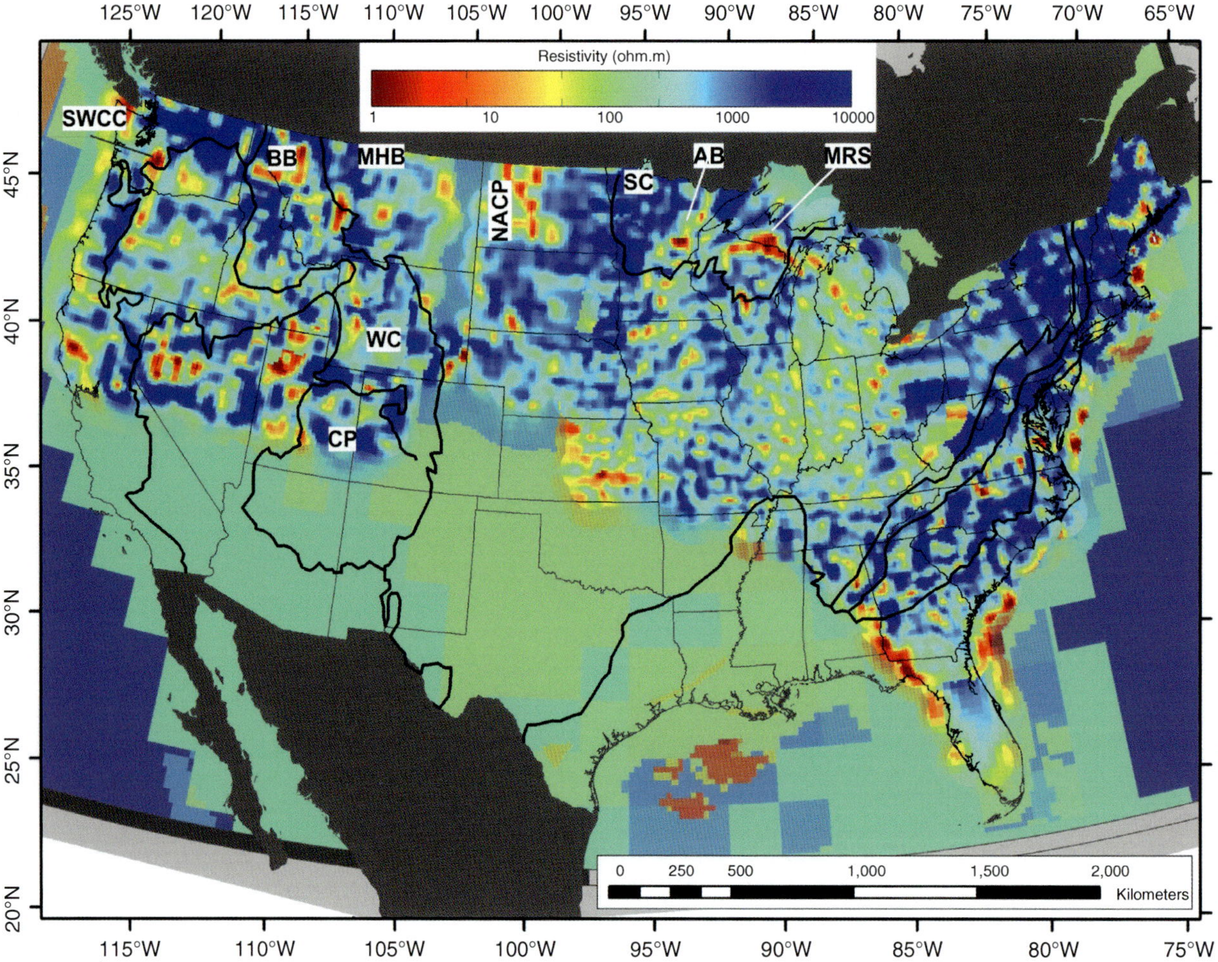

Figure 8.6 Electrical resistivity model at mid-crustal (15 km) depth. Geologic provinces as in Figure 4 plus Southern Washington Crustal Conductor (SWCC), Belt Basin (BB), Medicine Hat Block (MHB), Wyoming Craton (WC), Colorado Plateau (CP), North American Central Plains (NACP) anomaly, Superior Craton (SC), Animikie Basin (AB), and Marquette Range Supergroup (MRS).

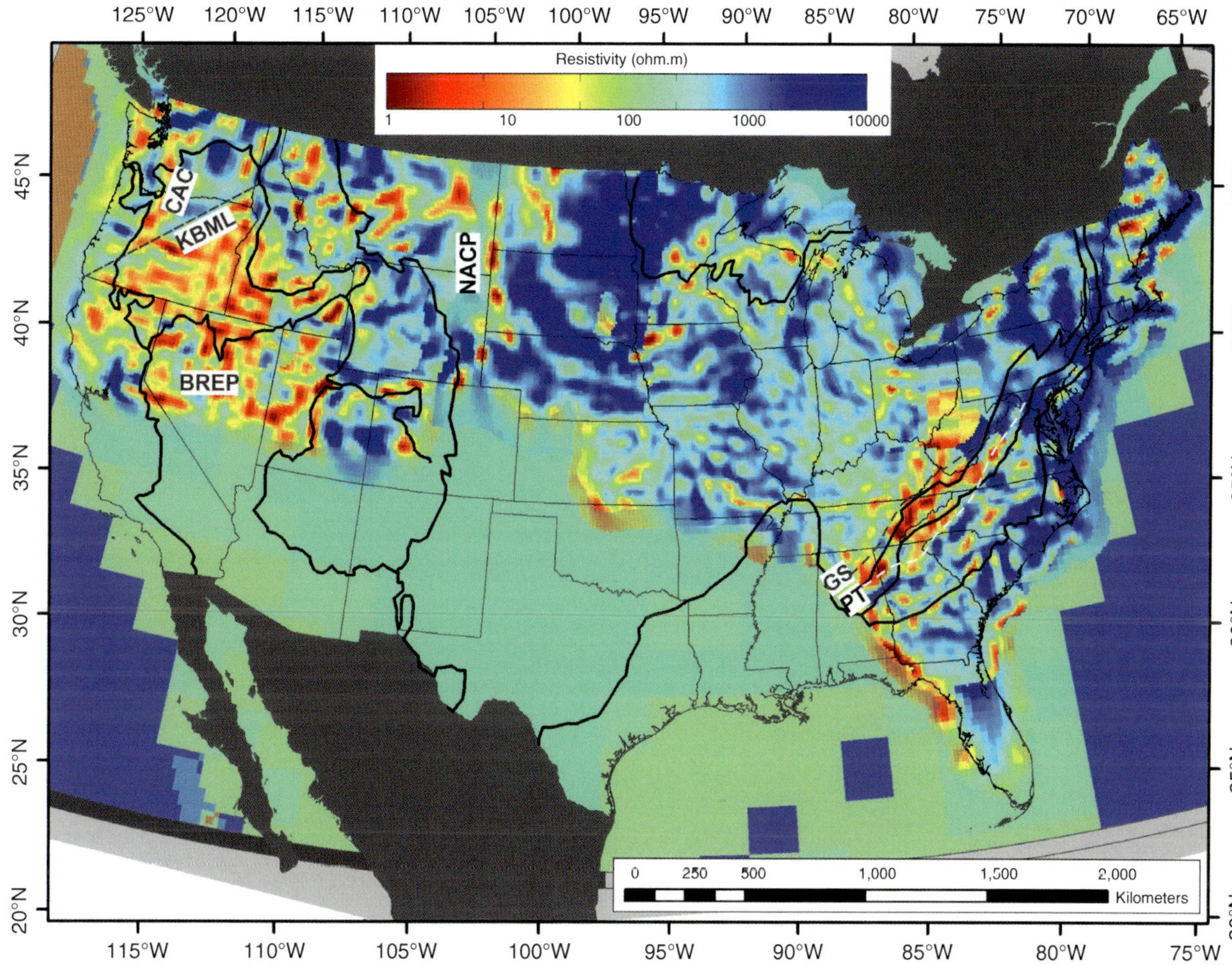

Figure 8.7 Electrical resistivity model at lower-crustal (30 km) depth. Geologic provinces as in Figure 8.4 plus Cascades arc conductor (CAC), Klamath-Blue-Mountains lineament (KBML), Basin and Range extensional province (BREP), Grenville suture (GS), and Piedmont terrane (PT).

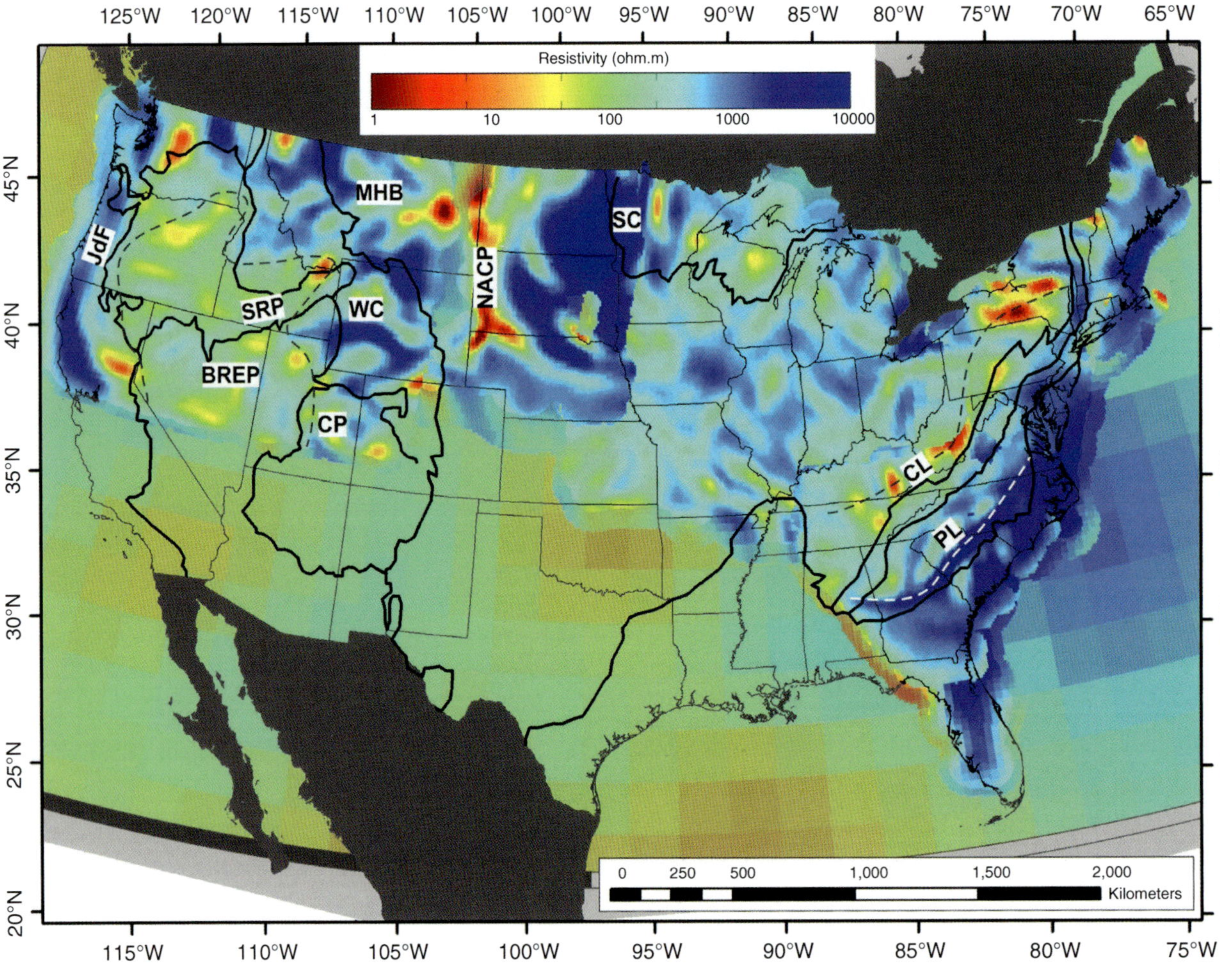

Figure 8.8 Electrical resistivity model predominantly within the mantle lithosphere (100 km) depth. Geologic provinces as in Figure 4 plus Juan de Fuca slab (JdF), Basin and Range extensional province (BREP), Snake River Plain (SRP), Medicine Hat Block (MHB), Wyoming Craton(WC), Colorado Plateau (CP), North American Central Plains (NACP) anomaly, Superior Craton (SC), Piedmont lithosphere (PL), and conductive lineament (CL).

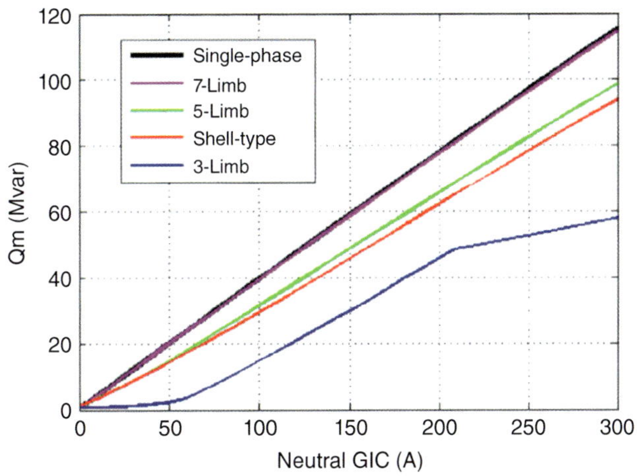

Figure 10.13 Relationship between transformer reactive power losses and GICs for different transformer types. Source: reproduced from Rezaei-Zare et al. (2016). Source: Reproduced with permission of IEEE.

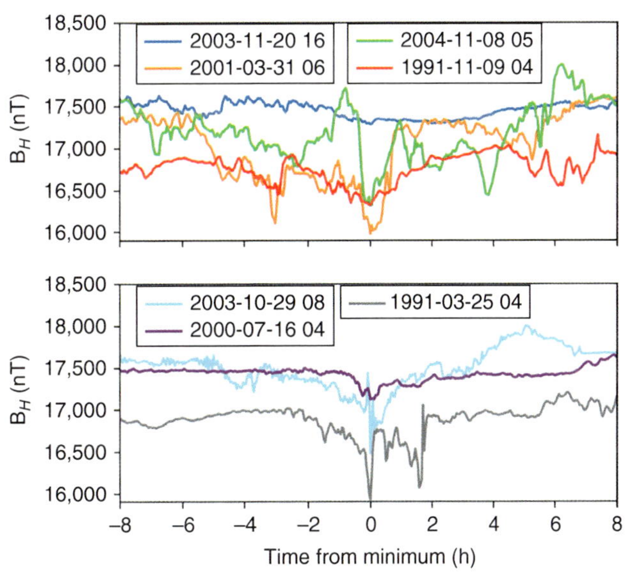

Figure 11.2 Horizontal component of the magnetic field measured in Ottawa, Canada for the modern storms listed in Table 11.1. The storms are ordered by minimum Dst index and the legend shows the time of minimum B_H. As shown, the profiles of each storm differ significantly, those believed to be most "geoeffective" exhibit a sudden storm commencement.

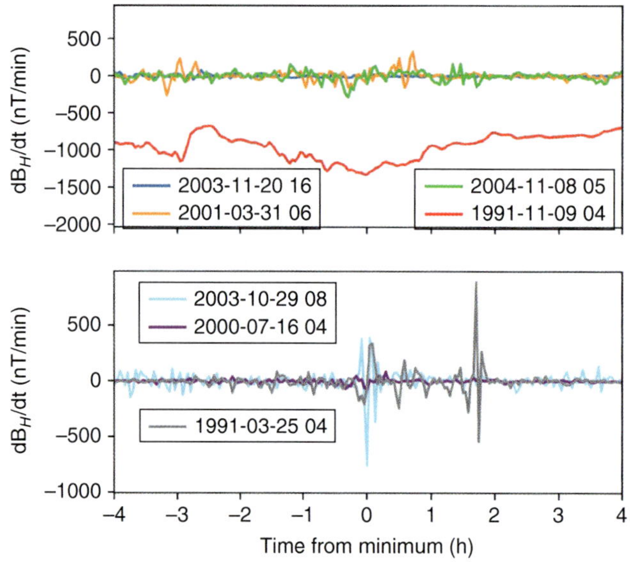

Figure 11.3 The time-derivative of the horizontal component of the magnetic field at the Ottawa, Canada station. The strongest GICs measured in the Northeastern US were from the March 1991 storm, which shows the largest changes in dB_H/dt. While the October 2003 storm has no known GIC effects, this storm also exhibited large differentials.

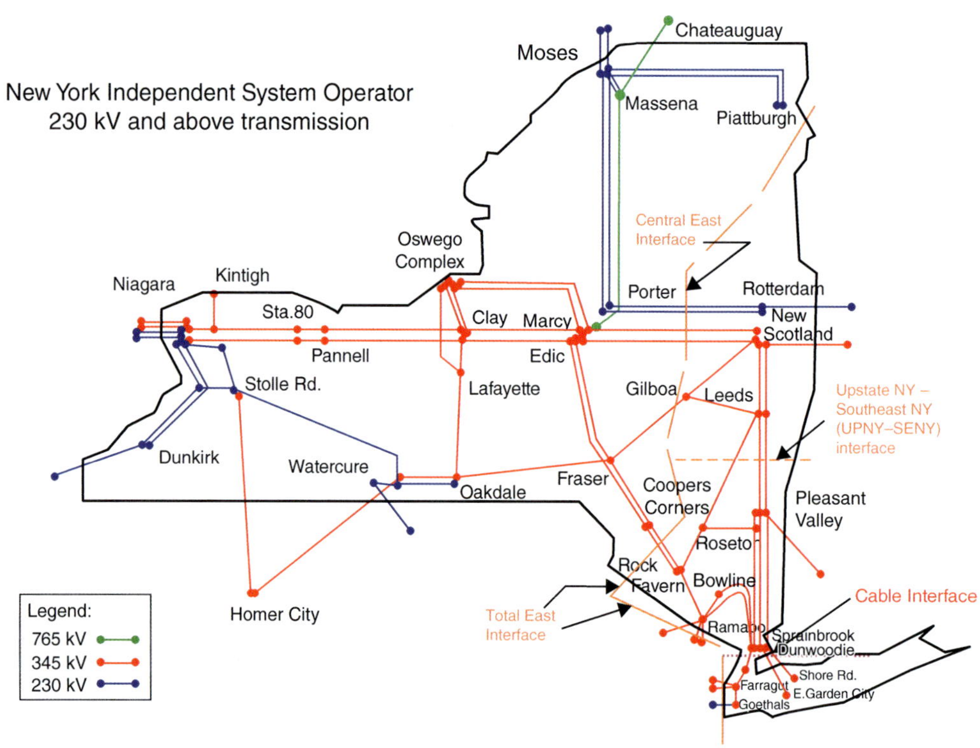

Figure 12.3 NYISO's high voltage transmission system.

and are concerned primarily with the Appalachian Mountains and the protracted set of continental collisions and rifting events that formed the region. Two of these models (ENAM and ECUS) span portions of both the central and eastern regions. Models spanning the western and central regions have not been attempted given the 800 km gap between them until recently. However, an unpublished inverse model obtained using the USArray MT data archived to date in the Great Plains is also included (GP). Ongoing data acquisition beneath the Great Plains is expected to close this gap by the end of 2018.

Of the models that are available and fit measured data reasonably well, when multiple overlapping models are present, we have chosen those that are of higher spatial resolution, or use a greater subset of data to constrain subsurface structure. For example, the MCR model from Bedrosian (2016) has a smaller cell size than that of Yang et al. (2015), resulting in somewhat higher resolution of upper-crustal structure. Similarly, the NWUS model of Meqbel et al. (2014) is included over that of Bedrosian and Feucht (2014), as the latter study did not incorporate vertical magnetic-field data in their inversion. A final consideration is extent of model structure contained within the lithospheric mantle. Dependent upon specifics of model regularization and applied data errors, different models contain varying degrees of structure.

We note that there exists a wealth of information contained within 1D and 2D inversion models throughout the contiguous United States, many of which are located within regions not currently covered by the USArray MT site coverage. Given uncertainties in how to laterally propagate the resistivity versus depth information into three dimensions, we do not incorporate these studies within the model assembly described in Section 8.7. The underlying MT transfer functions used to generate 1D and 2D models can, however, be incorporated into later 3D modeling efforts, which is part of the motivation behind the recent efforts to compile and archive community data sets (Kelbert et al., 2018a).

8.6. DEEP STRUCTURE: GLOBAL ELECTRICAL CONDUCTIVITY MODEL

It is a well-known fact that significant 3D variations in electrical conductivity are found deep in the Earth's mantle (e.g., Schultz, 1990). In parallel with the recent substantial advances in 3D magnetotelluric imaging, renewed interest in global electromagnetic induction methods has been spurred by the rise in computing capabilities and the corresponding progress in global seismic imaging, which demands additional information to allow a better constrained interpretation. Variations in mantle electrical conductivity have been probed using sets of 1D inversions (e.g., Semenov and Jozwiak, 2006; Khan et al., 2011; Munch et al., 2018), as well as fully 3D global and semi-global geomagnetic inversions (prior such efforts included Kelbert et al., 2009; Utada et al., 2009; Shimizu et al., 2010; Semenov and Kuvshinov, 2012). These studies have allowed imaging of large-scale lateral electrical conductivity variations deep in the Earth's mantle (~400–1600 km depths). However, multiple problems still remained, making interpretation ambiguous and causing discrepancies between some of these results.

The global electromagnetic induction method is based on inversion of very long-period (3–100 days) geomagnetic field data, obtained from decades of continuous geomagnetic field measurements at worldwide observatories such as USGS (1901). Continuous geomagnetic field measurements form an invaluable global magnetic observatory data set (Rasson, 2007; INTERMAGNET Geomagnetic Data, 2013), which is, however, relatively sparse, as well as clustered in several well-populated regions of the world (mostly Europe, Asia and North America). Magnetic satellite data (Olsen et al., 2006) are also available, and have the advantage of allowing a more uniform spatial data coverage, but come with their own spatial–temporal aliasing challenges that limit their resolution to long periods only. All of these data can provide information about the deep (upper to mid-mantle) electrical conductivity of the Earth. This technique is distinct from magnetotelluric inversion: while both methods solve Maxwell's equations in the quasi-static approximation, they are forced by different ionospheric source representations. To be used for estimation of deep Earth electrical conductivity, the dependence of the data on ionospheric and magnetospheric current distributions needs to be modeled with high accuracy, and the effects of their spatial variations either taken into account, or removed from the data before the inversion for electrical conductivity structure is performed. Methods for empirical correction of the data have been devised (Fujii and Schultz, 2002; Kelbert et al., 2009; Semenov and Kuvshinov, 2012), but they only recover small parts of the data sets and have not been shown to unambiguously work. The period range was also restricted by these considerations to over 5 days, and the inability to adequately represent the ocean/continent contrasts due to computational limitations required an ambiguous correction to the data. The same computational limitations imposed severe restrictions on the resolution of the inverse electrical conductivity models.

In Sun et al. (2015), we have successfully addressed many of these problems by schematically modeling the ionospheric and magnetospheric currents directly, inverting the global magnetic observatory data for these current distributions, and using these as the source in the global geomagnetic inversion for electrical conductivity.

This method does not fully resolve the inherent ambiguity in the data: that is, variations in the data set could be theoretically explained by either the source or conductivity. However, the source and conductivity variations have somewhat different physical signatures, which allowed us to impose reasonable *a priori* assumptions and regularization. For example, the source currents tend to follow latitudinal lines in the quasi-dipole coordinate system; they prevail (and have more structure) at auroral latitudes; and under the Earth's rotation, they wrap around it in "loops." The sources also vary with time (and are therefore frequency-dependent), while the same electrical conductivity structure has to explain all measured data. In Sun et al. (2015), we have carefully described all these considerations, and attempted multiple such inversions with varying regularization parameters. We found that while certain complicated source structures could fully explain all the variability in the data, such structures would not be physically viable. After performing a range of inversions for electrical conductivity distributions with varying, but physically consistent, source currents, we concluded that the resultant electrical conductivity structures were reasonably robust and interpretable. We therefore consider the global 3D electrical conductivity model published in Sun et al. (2015) a dramatically improved, higher resolution version of Kelbert et al. (2009) results. The Sun et al. (2015) mantle structure beneath North America is shown in Figure 8.2.

A great deal is known about the upper mantle structure beneath North America thanks to the high-resolution seismic and magnetotelluric imaging with USArray data (see also Sections 8.5 and 8.8). Unfortunately, the scale of our imaging does not allow a direct comparison with these high-resolution models; for instance, the whole of the northwestern United States imaged in detail all the way down to 410 km by Meqbel et al. (2014) magnetotelluric study shows up as a single shapeless conductor in the global model, located south of approximately 44°N. The magnitude of the upper mantle electrical conductivity in the area (0.03 S/m) is exactly the same as that inferred by the Meqbel et al. (2014). The resolving power of the Sun et al. (2015) model (Fig. 8.2) is the highest in and below the transition zone, at approximately 410–1200 km depths. While the lateral resolution is much lower than that of magnetotelluric models, it is notable that the mantle transition zone is imaged as relatively resistive beneath the active western United States, and relatively conductive in the central and eastern United States. This could be related to the dehydration processes occurring in the asthenospheric mantle of the moderately young and warm Cascadia subduction zone.

Here, we convert the model's spherical harmonics representation in geomagnetic coordinates to its native $2° \times 2°$ lateral grid resolution, rotate it to geographic coordinates, and use this model as the electrical conductivity background below 100 km depth, wherever deep electrical conductivity structure is not constrained by a large-scale magnetotelluric inversion. We note that all global electrical conductivity models have very limited resolution in the upper ~400 km. Methods to improve global model resolution at upper mantle depths would need to use realistic ionospheric current systems such as Alken et al. (2017) as forcing for global electromagnetic modeling.

8.7. MERGED 3D CONDUCTIVITY MODEL OF CONTIGUOUS UNITED STATES

To compile the wide variety of models described in Sections 8.5 and 8.6 into a common framework for contiguous United States, which we call CONUS, we created a 3D laterally uniform $0.1° \times 0.1°$ numerical grid. The grid includes the contiguous United States, and extends 650 cells in longitude between −130° and −65°, and 280 cells in latitude between 23° and 51°. Vertically, we started with a 0.1 km top layer, and increased the cell size via a recursive geometric relation with a common ratio of 1.065 until 100 km depth was reached, resulting in 67 vertical layers in the top ~ 100 km of the grid. Starting at 103.0564 km depth, we extended the grid farther down to ~ 410 km depth, the top of the mantle transition zone, now starting with a 6.2 km layer and using the same common ratio. This resulted in a total of 90 vertical layers. Finally, we extended the grid from 413.6655 km depth all the way down to 1277 km, using a starting layer of 28 km and the common ratio of 1.07, resulting in a total of 107 Earth layers. We completed the vertical setup by padding the model with four more layers using the common ratio of 1.5, which resulted in the bottom boundary of the model at a 2284.6 km depth. This is done for numerical reasons, so that any inaccuracy in the boundary conditions has minimal effect on the modeling results. Similarly, we also padded the grid laterally by six cells in the North–South directions and seven cells in the East–West directions, to account for the oceans which increase heterogeneity of the model, resulting in a $292 \times 664 \times 111$ numerical grid.

For background electrical conductivity in the top 100 km of the Earth's lithosphere, we used the Alekseev et al. (2015) global compilation, modified in the following ways. First, we reduced the background continental resistivity from 5000 Ωm (2×10^{-4} S/m) to 100 Ωm (10^{-2} S/m), which is more realistic for average lower crust and uppermost mantle in the contiguous United States. This includes any regions of young oceanic crust, which were left at 100 Ωm background value. We also incorporated the high-resolution sediment thickness map of Laske and

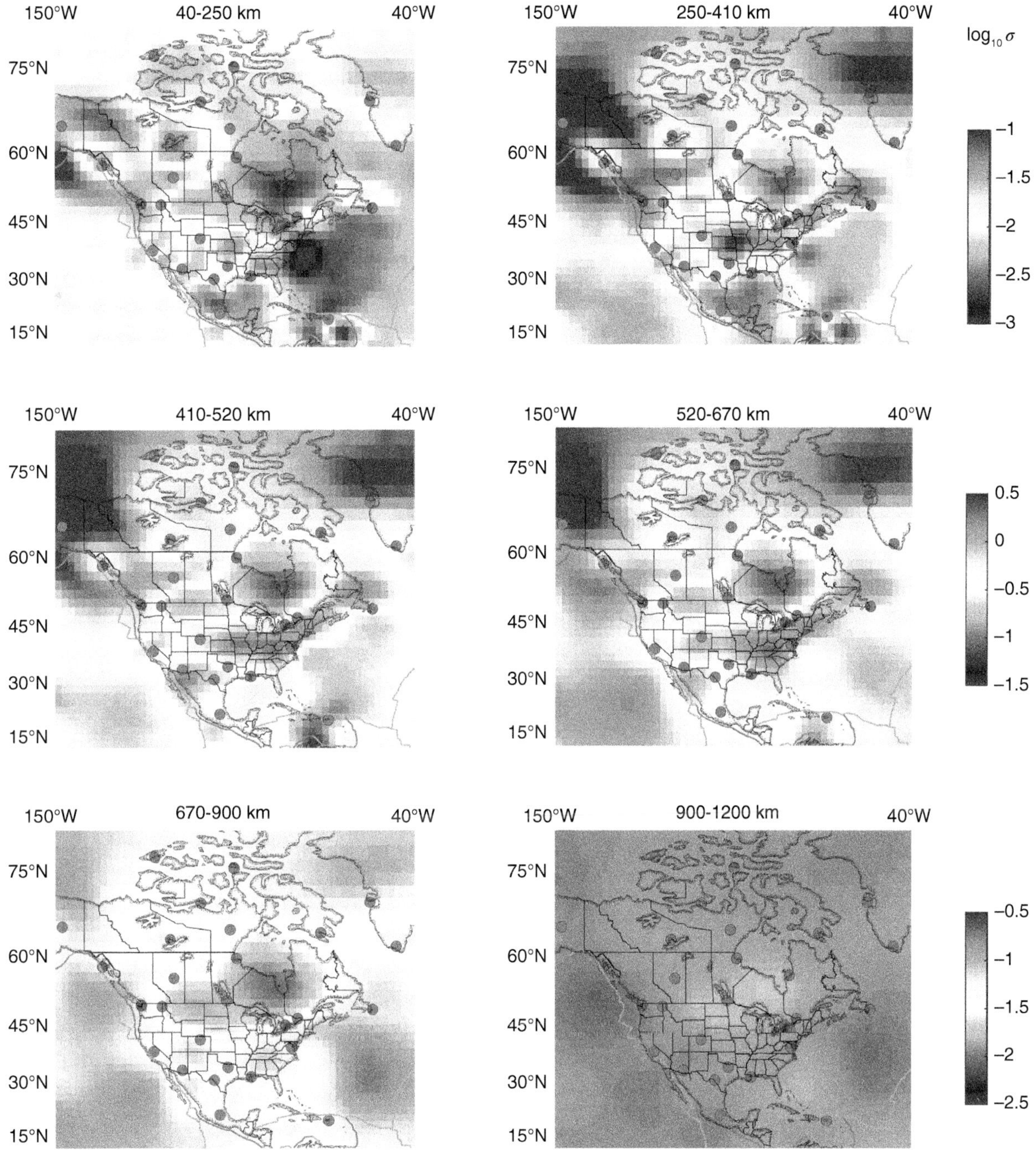

Figure 8.2 3D electrical conductivity variations in Earth's mantle beneath North America based on the global model of Sun et al. (2015). Shown is variability of two orders of magnitude in each of the layers of the Sun et al. (2015) published inverse model. Additionally, locations of the historical geomagnetic observatories from INTERMAGNET Geomagnetic Data (2013), primarily in the United States (USGS, 1901; Love and Finn, 2011) and Canada (Newitt and Coles, 2007), that were used to constrain this model (see also Fig. 8.1) are shown in magenta. *(See insert for color representation of the figure.)*

Masters (1997), setting the sediments to a default value of just over 33 Ωm (3×10^{-2} S/m). Then, this modified model was interpolated to our $0.1° \times 0.1°$ numerical grid with 67 vertical layers in the top 100 km. For background below 100 km depth, the global conductivity model (Sun et al., 2015; Section 8.6) was used.

We then proceeded to include MT inverse models, selected from Table 8.2 as described in Section 8.5.

Table 8.4 Regional 3D Electrical Conductivity Models used in the CONUS Compilation, their Vertical Limits in the Model, and their Original Cartesian Projections. Note that the original projection of the WUS model is uncertain; it was graciously georeferenced for us by the first author (N. Meqbel).

Model	Vertical limits	Cartesian coordinate system	Datum	Parallels	Origin latitude	Origin longitude	Reference
WUS	0–600 km	Equidistant Azimuthal	WGS84	N/A	43.905	−116.368	Meqbel et al. (2014)
SRPY	10–80 km	Equidistant Azimuthal	WGS84	N/A	42.016	−116.407	Kelbert et al. (2012)
GP	60–200 km	Lambert Conformal	NAD83	[42.25 47.25]	44.75	−101.5	B.S. Murphy, this chapter
MCR	0–500 km	UTM15 North	WGS84	N/A	42.6374	−90.6851	Bedrosian (2016)
SEUS	0–500 km	Lambert Conformal	NAD83	[30.25 38.25]	34.25	−82.5	Murphy and Egbert (2017)
NEUS	4–200 km	Lambert Conformal (rotated 60° N of E)	NAD83	[38.5 46.5]	42.5	−73.25	B.S. Murphy, this chapter
APPL	4–500 km	Lambert Conformal	NAD83	[34.25 41.75]	38.0	−80.0	B.S. Murphy, this chapter
ENAM	1–500 km	Lambert Conformal	NAD83	[31.25 43.75]	37.5	−85.0	B.S. Murphy, this chapter

These models were, first of all, georeferenced from the various Cartesian coordinate projections in which the magnetotelluric inversions have been performed. These selected inverse models and their projection details are further described in Table 8.4.

These inverse models were then interpolated to the $0.1° \times 0.1°$ computational grid using the nearest neighbor interpolation, laterally truncated to data locations, and incorporated into the 3D electrical conductivity background in the following order: ENAM, APPL, NEUS, MCR, SEUS in the central and eastern United States, followed by GP, then WUS and SRPY in the western United States. The truncations were defined based on the scale of the model; specifically, models of larger spatial scales and lower resolution (ENAM and MCR) were truncated to include structure to within 140 km from a site, equivalent to two times the nominal station spacing of USArray, and WUS was truncated to 105 km (1.5 time station spacing). Other 3D models focused on smaller regions (GP, YSRP, APPL, NEUS, and SEUS) were truncated to within 70 km from a site, which is equivalent to the nominal station spacing. In the present treatment, no smoothing was applied to model boundaries, to merge the models in places of overlap, other than truncation of all models at a fixed distance from their respective site locations. Thus, the fact that model boundaries generally do not appear as sharp discontinuities speaks to the consistency of these models.

In places where models overlap, preferential treatment is given those models that provide better resolution in the relevant area of the grid (e.g., in the near-surface) because of magnetotelluric inversion configurations that support higher sensitivity in that area (e.g., shorter periods included in the data set). For example, APPL is an outlier at the upper crustal depths, because shorter periods were excluded from the data set to accommodate lower spatial resolution. As is seen in Figure 8.3, MCR by Yang et al. (2015) seems to exhibit rather more subdued variability relative to its peers, including the MCR by Bedrosian (2016) at mid-mantle depths; variability at these depth might be real, but masked in the model by an unusually good data fit at the shorter periods. Between these two models, MCR by Bedrosian (2016) was selected for the CONUS model because of its higher resolution. YSRP was truncated to lower crust and upper mantle depths, to avoid a region of artificially subdued variability at around 5 km depth, an artefact of the inversion configuration that was used for that area. This trade off in how spatial heterogeneity is distributed over depth serves as a reminder that the 3D magnetotelluric inverse problem is non-unique, and that until methods are developed to explore the model space in a rigorous manner, certain choices in magnetotelluric inversion are subjective and are guided primarily by our overall understanding of geological context. While care was used to maintain consistency of the overall compilation, validation of the result against empirical magnetotelluric data, both in-sample (i.e., used for inversions in Table 8.4) and out-of-sample (i.e., data that were not utilized to obtain these models) is of primary importance, although beyond the scope of this contribution.

8.8. NORTH AMERICAN GEOLOGY REFLECTED IN THE MODEL

The North American continent has a complex geologic fabric, starting with ancient cratonic fragments, which over two billion years were accreted to and torn apart by numerous mountain building and rifting events (e.g., Dickinson, 2004; Whitmeyer and Karlstrom, 2007; Hatcher et al., 2010). Parts of North America have weathered multiple Wilson cycles – the formation and destruction of supercontinents such as Rodinia and Pangea. Some tectonic events, such as the opening of the Atlantic Ocean, retain a record in the lithosphere that is largely

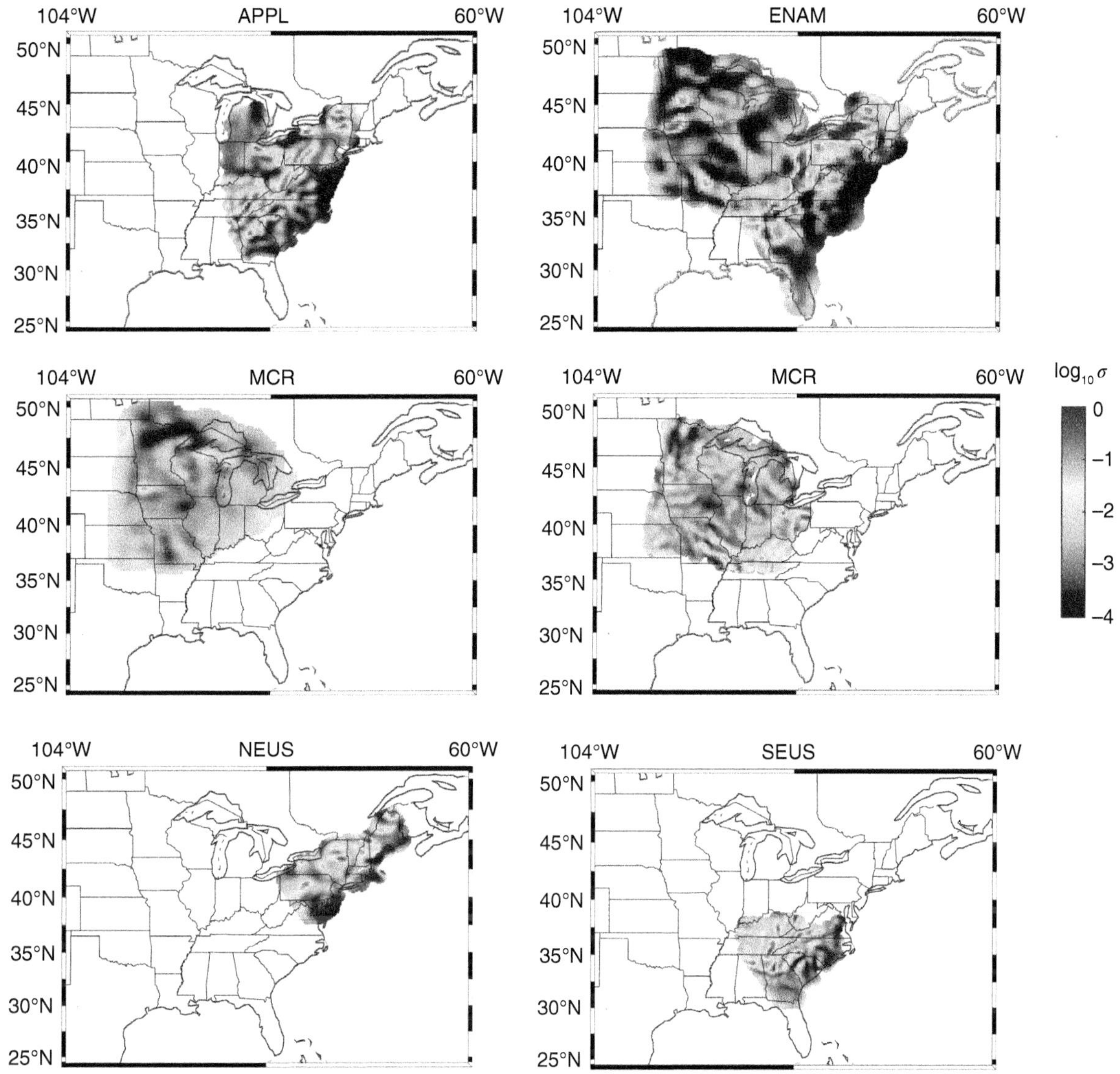

Figure 8.3 Relative locations of six regional models in the central and eastern United States: MCR (Yang et al., 2015; left), MCR (Bedrosian 2016; right), ENAM, APPL and NEUS (this work), and SEUS (Murphy and Egbert, 2017), shown at an upper mantle depth of 60 km. Note the differing spatial scales and resolutions. (*See insert for color representation of the figure.*)

untouched since it occurred. Others, including mountain ranges that absorbed the collisions between cratonic blocks, remain only as narrow, intensely deformed zones within the continental interior. Still others are reflected in deep sedimentary basins preserved along the periphery of ancient tectonic margins. The North American continent continues to evolve via subduction along its western margin (Alaska, the Northwestern United States, Central America), transform faulting in western Canada and California, extension both broad and narrow within the Basin and Range province and along the Rio Grande Rift, and a mantle plume beneath the Yellowstone hotspot that has left its mark in the lithosphere over the past 50 Ma.

We here examine key aspects of the 3D resistivity model in relation to geologic and tectonic structure, focusing on the nature of conductivity variations imaged at different spatial and depth scales within the model. We limit our discussion to the contiguous United States, and more specifically to those regions where the model has data support. What follows is not a complete discussion of the geologic model; a more comprehensive discussion of the geologic underpinnings of the model can be found in Bedrosian and Kelbert (2019).

We examine the resistivity through a series of depth slices at representative upper (3 km), middle (15 km), and lower (30 km) crustal depths as well as within the mantle

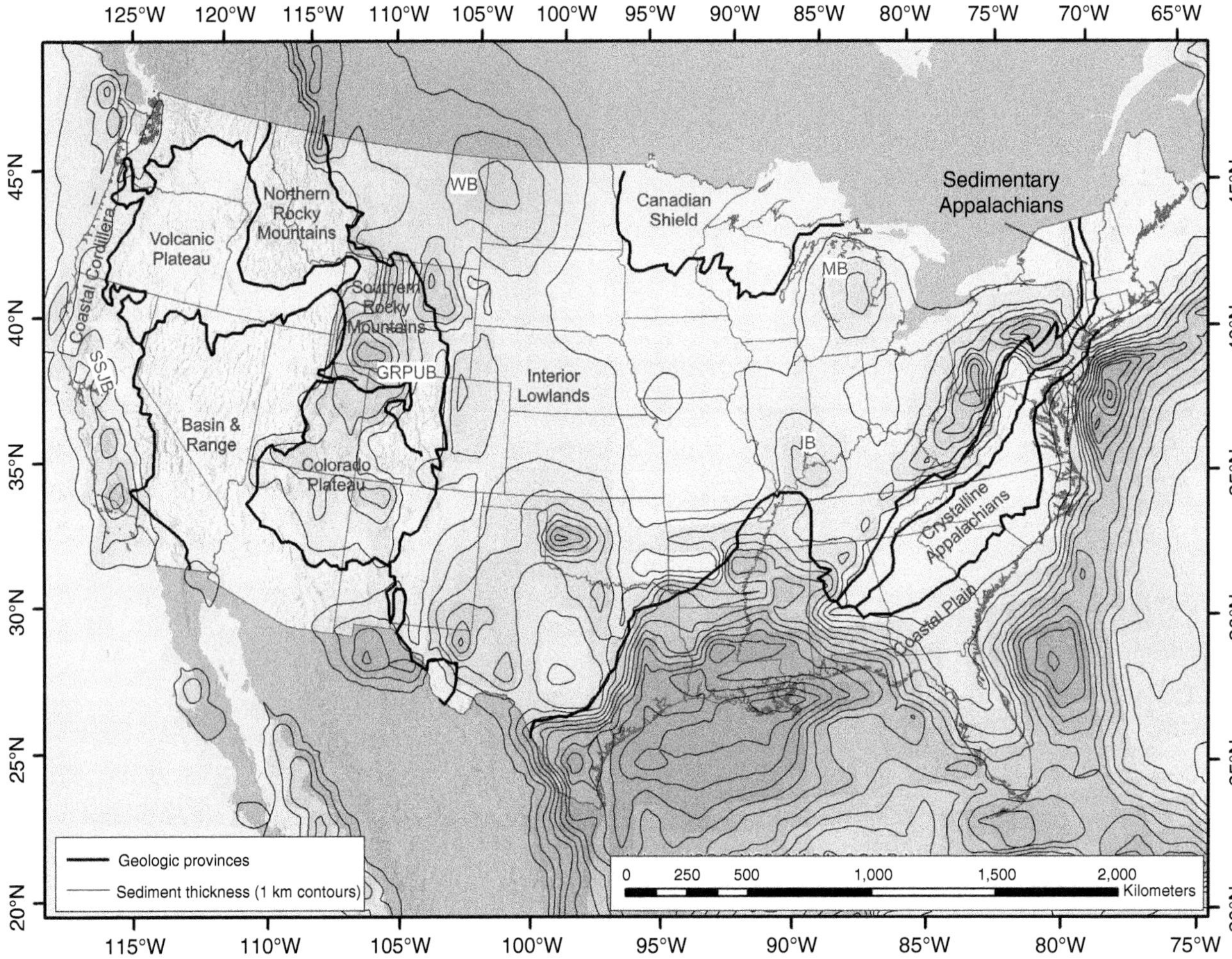

Figure 8.4 First-order geologic provinces within the CONUS adapted from Reed and Bush (2007). Background shading and contours represent estimated sediment thickness in kilometers, derived from Laske and Masters (1997). Labeled basins include the Williston Basin (WB), the Sacramento-San Joachin basins (SSJB), the Green River-Piceance-Uinta basins (GRPUB), the Michigan basin (MB), and the Illinois Basin (IB). (*See electronic version for color representation of this figure.*)

lithosphere (100 km). Much of the discussion centers upon nationwide geologic mapping compilations and studies of the evolution of the North American continent (Reed et al., 2005; Reed and Bush, 2007; Whitmeyer and Karlstrom, 2007; Horton et al., 2017). We also draw upon sediment thickness data from Laske and Masters (1997), a filtered version of which is presented in Figure 8.4. Our discussion focuses on areas where the resistivity model has crustal resolution, notably within the regions of EarthScope MT data coverage. To limit our interpretation to well-constrained regions, we have partially obscured those regions in the model at distances of >100 km from any MT station.

8.8.1. Upper-crustal Structure

Despite the 70-km station spacing of MT data used to generate the component models, the shortest periods inverted are sensitive to structure within the upper 5 km. At these relatively shallow depths, the model broadly reflects sedimentary basins and regions of uplift (Fig. 8.4). The former are relatively conductive due to shale, siltstone, sandstone, and conglomerate containing variable amounts of clay and interconnected pore fluids. The latter are resistive, reflecting shallow igneous and metamorphic basement rocks.

Sedimentary basins reflected in the mid-continent include thick accumulations of Phanerozoic sedimentary rock in the interior lowland, including the Michigan and Illinois basins (MB, IB) (Fig. 8.5). Older Precambrian sedimentary basins are also imaged in the mid-continent, including thick accumulations of sandstone within a series of rift basins extending more than 1000 km along the Midcontinent Rift System (MRS) which cuts across Minnesota, upper Michigan, Wisconsin, Iowa, Nebraska, and Kansas. In the

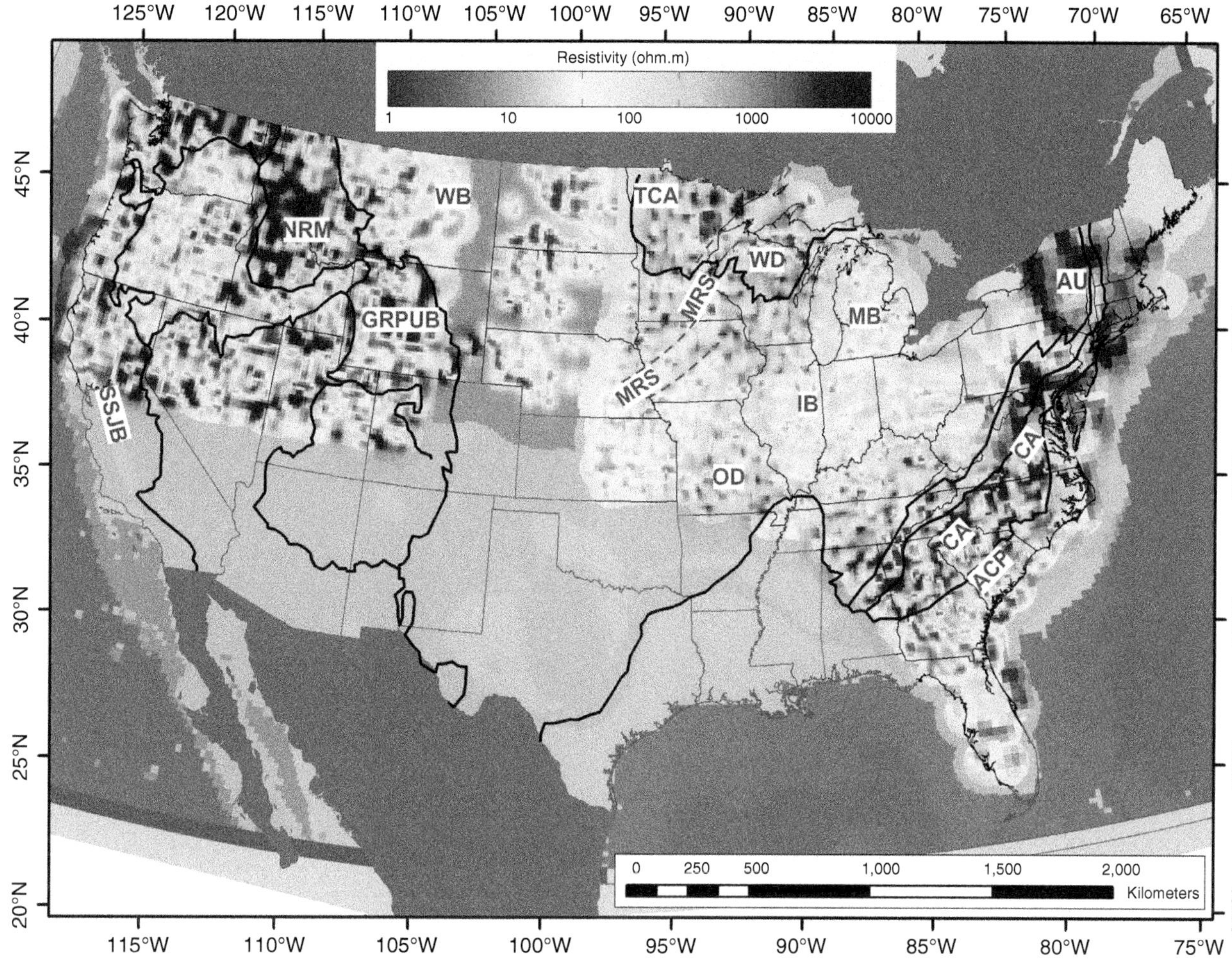

Figure 8.5 Electrical resistivity model at upper-crustal (3 km) depth. Labels and geologic provinces as in Figure 8.4 plus Northern Rocky Mountains (NRM), Sacramento-San Joachin Basin (SSJB), Trans-continental Arch (TCA), Wisconsin Dome (WD), Midcontinent Rift System (MRS), Ozark Dome (OD), Crystalline Appalachians (CA), Adirondack Uplift (AU), and Atalantic Coastal Plain (ACP). *(See insert for color representation of the figure.)*

southeastern United States, a wedge of high conductivity along the coast, widening to the south through the Carolinas and into Georgia, follows the Atlantic Coastal Plain (ACP), comprising a thick, young sedimentary sequence along the Atlantic passive margin (and extending offshore). In the western United States, thick sections of sedimentary rock derived from the rising Rocky Mountains occur within the Williston basin (WB) in Montana. A broad conductive region spanning parts of Utah, Wyoming, and Colorado is coincident with the Green River, Piceance, and Uinta basins (GRPUB), basins filled with material shed from regional uplifts and fine-grained deposits accumulated in ancient lakebeds. In California, a linear conductor marks the northern end of the Sacramento-San Joachin basin (SSJB), filled with a thick sequence of sediments shed from the adjacent Sierra Nevada Mountains.

In contrast to the conductive basins, shallow resistive areas indicate the relative absence of sediments and correspond to regions of crystalline basement uplift. Such areas include the Trans-continental Arch (TCA) in Minnesota, the Wisconsin and Ozark domes (WD, OD) in the mid-continent and the Adirondack uplift (AU) in upstate New York. In the east, a broad resistive region marks the crystalline Appalachians (CA), uplifted metamorphic and plutonic rocks. The Northern Rocky Mountains (NRM) in Idaho and western Montana are similarly resistive.

8.8.2. Middle-crustal Structure

Comparing the resistivity model at 3 and 15 km depth (Figs. 8.5 and 8.6), we observe a marked reduction in conductivity at greater depth. At mid-crustal levels, the conductive signature of the sedimentary basins vanishes, replaced by resistive igneous and metamorphic basement. Coherent resistive regions reflect fragments of Archean

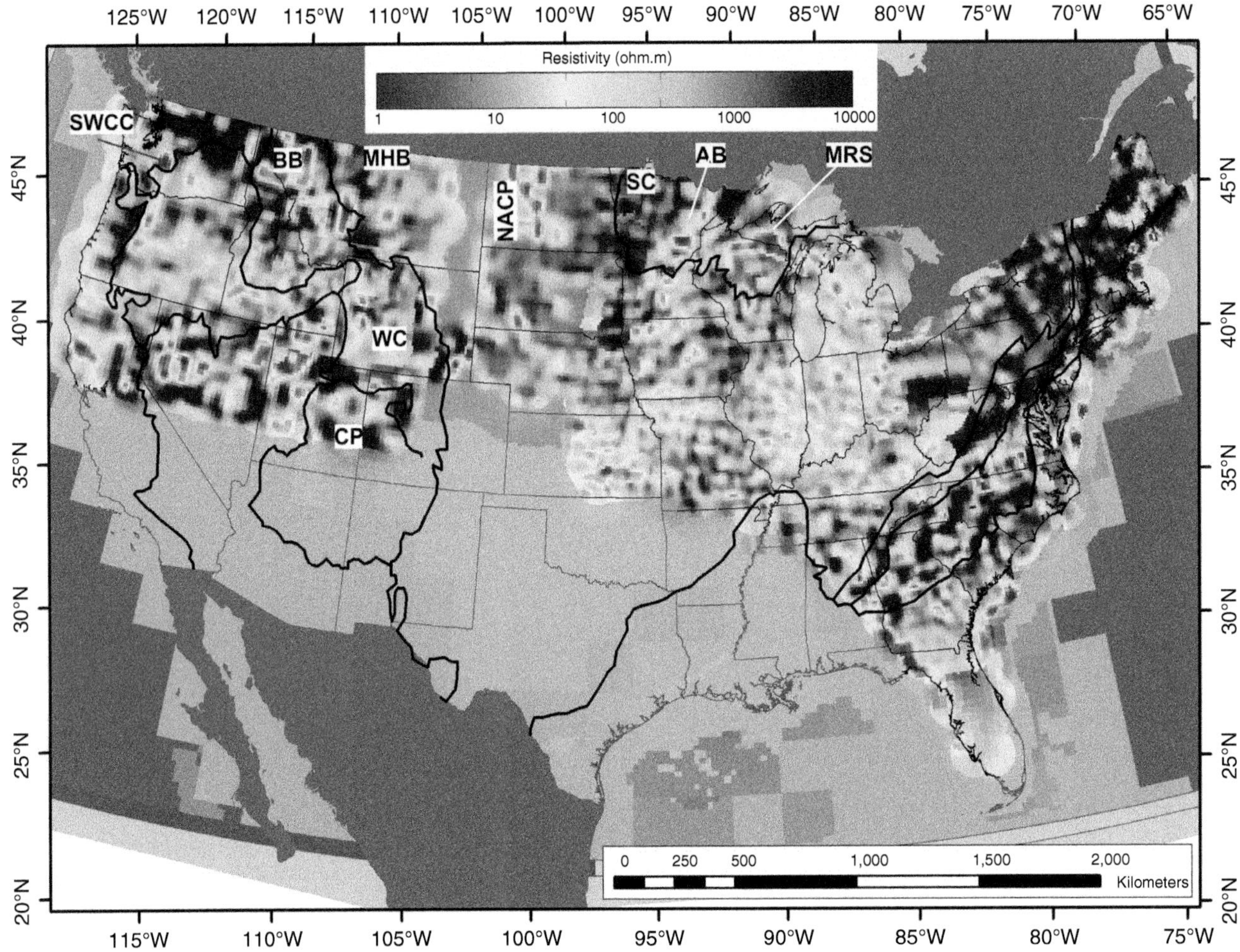

Figure 8.6 Electrical resistivity model at mid-crustal (15 km) depth. Geologic provinces as in Figure 8.4 plus Southern Washington Crustal Conductor (SWCC), Belt Basin (BB), Medicine Hat Block (MHB), Wyoming Craton (WC), Colorado Plateau (CP), North American Central Plains (NACP) anomaly, Superior Craton (SC), Animikie Basin (AB), and Marquette Range Supergroup (MRS). (*See insert for color representation of the figure.*)

cratons and Proterozoic crust. In the west, these include the Medicine Hat Block (MHB) in Montana, the Wyoming craton (WC) and the Colorado Plateau (CP) (Meqbel et al., 2014). In the mid-continent, resistive areas mark the Archean Superior Craton (SC) in Minnesota and upper Michigan (Bedrosian, 2016).

The conductivity anomalies imaged are attributed to conductive mineral phases formed during past tectonic events. Ancient suture zones may contain metallic sulfides and/or graphite, formed when deep marine sediments are deformed and metamorphosed during the collision of tectonic plates (Boerner et al., 1996). An example is the Southern Washington Crustal Conductor (Stanley et al., 1987), a circular conductive feature in southwest Washington (SWCC). Here, carbonaceous marine sedimentary rocks mark the suture between the resistive Siletz terrane and the former western margin of North America. Another example is the Belt Basin (BB), an ancient rift basin in northern Idaho and western Montana, where sulfide-bearing sedimentary rocks extend into the lower crust (Bedrosian and Box, 2016). Yet another example is found in North and South Dakota, where conductive zones are imaged through the entire crustal column along suture zones within the Trans-Hudson orogen (Whitmeyer and Karlstrom, 2007; Jones et al., 2005). Similarly, high conductivity within metasedimentary rocks of the Animikie Basin (AB) in Minnesota and the Marquette Range Supergroup (MRS) in the upper Michigan and northern Wisconsin are the result of metallic sulfides in the former, and graphite and sulfides in the latter (Bedrosian, 2016).

8.8.3. Lower-crustal Structure

At 30 km depth (Fig. 8.7), the resistivity model is representative of the lower crust and reflects a composite of both ancient and active tectonism. A first-order difference

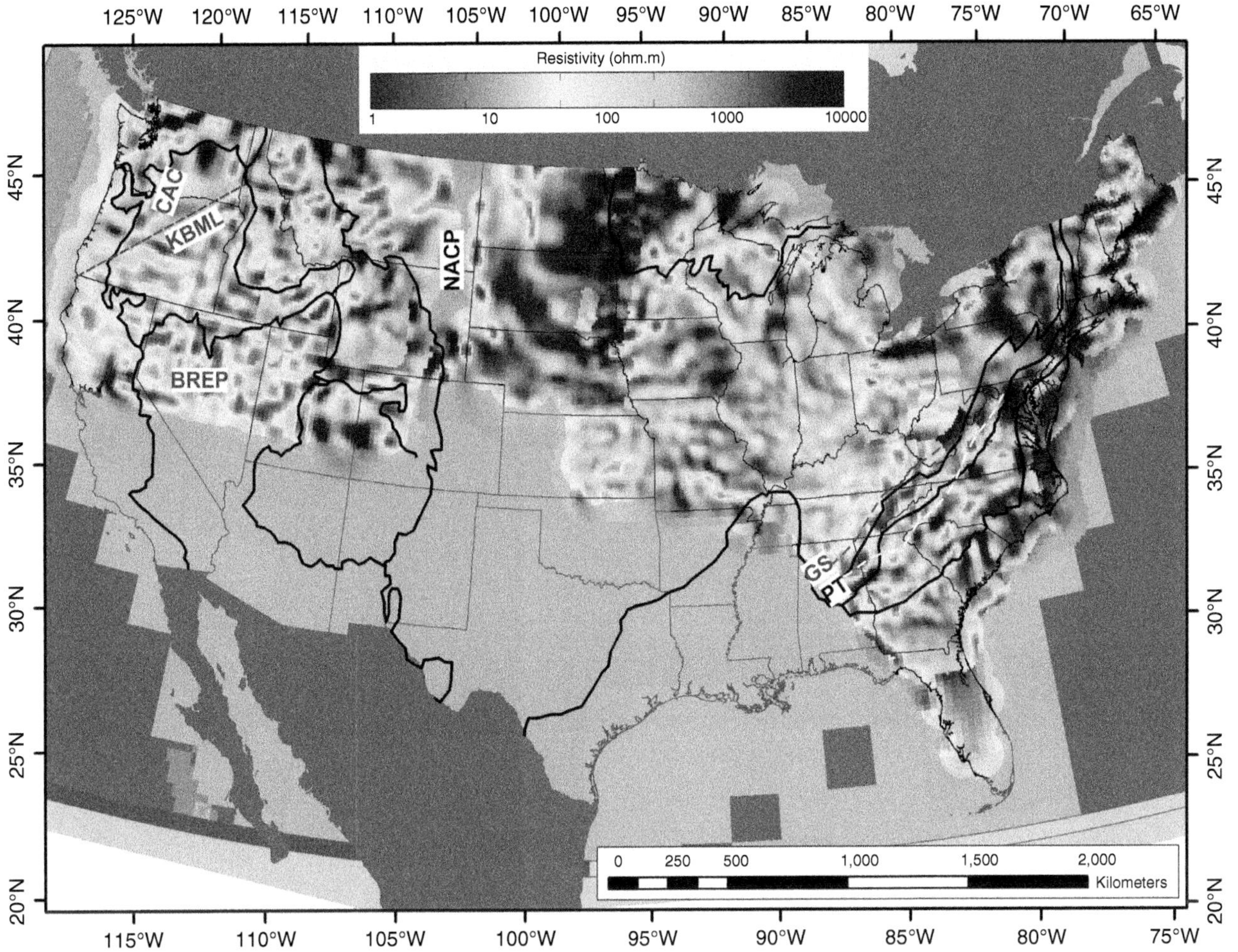

Figure 8.7 Electrical resistivity model at lower-crustal (30 km) depth. Geologic provinces as in Figure 8.4 plus Cascades arc conductor (CAC), Klamath-Blue-Mountains lineament (KBML), Basin and Range extensional province (BREP), Grenville suture (GS), and Piedmont terrane (PT). *(See insert for color representation of the figure.)*

is imaged between the western and eastern model domains, with higher conductivities on average in the tectonically active western United States. The bulk of the conductive region corresponds to the Basin and Range extensional province (BREP), where extension since 25 Ma has resulted in relatively thin, hot crust (Gilbert, 2012). The high conductivity in this region is interpreted to be due to a combination of mantle-sourced melts and saline fluids exsolved from them (Wannamaker et al., 2008). 3D resistivity models have significantly contributed to our understanding of extension in the western United States (Patro and Egbert, 2008; Bedrosian and Feucht, 2014; Meqbel et al., 2014). For example, models image a sharp conductivity contrast across the Klamath-Blue Mountains lineament (KBML; Riddihough et al., 1986) that divides Oregon in half. The lower-crustal extent of the BREP is significantly greater than that of surface extension, indicating that deformation is accommodated differently in the upper and lower crust.

At a smaller scale, a linear conductive feature parallels the Washington and Oregon coastline beneath the Cascades volcanic arc. The location of the Cascades arc conductor (CAC) marks where fluids from the subducting Juan de Fuca slab flux mantle melts that ascend within the overlying crustal column (Wannamaker et al., 2014). Outside of the Appalachian region, the lower crust of the central and eastern United States is resistive, indicating a lack of partial melt or aqueous fluids. Within the Appalachians, several linear conductive and resistive belts parallel the Appalachian Mountain chain. A linear resistor can be traced from Alabama through Virginia, following the crystalline Piedmont Terrane (PT) and extending deep into the mantle lithosphere (Wannamaker, 2005; Murphy and Egbert, 2017). Finally, a prominent linear conductor west of the PT marks the main Grenville suture (GS), a major crustal boundary from an ancient continental collision (Wannamaker, 2005; Fisher et al., 2010; Murphy and Egbert, 2017). Heightened electrical conductivity associated with this structure is likely due to the presence of graphite and sulfides throughout the crust that were trapped in this major zone of convergence.

8.8.4. Upper-mantle Structure

The 3D resistivity model shows significant heterogeneity within the mantle lithosphere, albeit at reduced spatial resolution. At 100 km depth (Fig. 8.8), the entire model lies within the rigid mantle lithosphere, with the exception of the BREP and the Snake River Plain (SRP), where we may be looking into asthenosphere that may be as shallow as 70 km (Kelbert et al., 2012; Meqbel et al., 2014). Conductivity is expected to be higher in the asthenosphere than in the lithospheric mantle, due to a greater degree of hydration in nominally-anhydrous mantle minerals and higher temperature.

Along the West Coast, the subducting Juan de Fuca plate (JdF) is imaged as a coast-parallel resistor. At this depth, the eastern edge of the resistor marks the approximate location of the Cascades volcanic arc and lines up with the Cascades arc conductor in Figure 8.7. The Medicine Hat Block (MHB), Wyoming Craton (WC), and Colorado Plateau (CP) are also still visible at this depth as major resistive domains (Meqbel et al., 2014). These features form lithospheric resistors due to their lower mantle temperatures and relatively dry composition. The central and eastern United States show minimal variation in mantle structure, but several key features stand out. The generally resistive nature of the mantle lithosphere beneath the central United States reflects the low temperatures and relatively dry composition associated with this ancient geologic domain (Yang et al., 2015; Bedrosian, 2016). In the north-central United States (western North Dakota and South Dakota), the well-studied North American Central Plains (NACP)

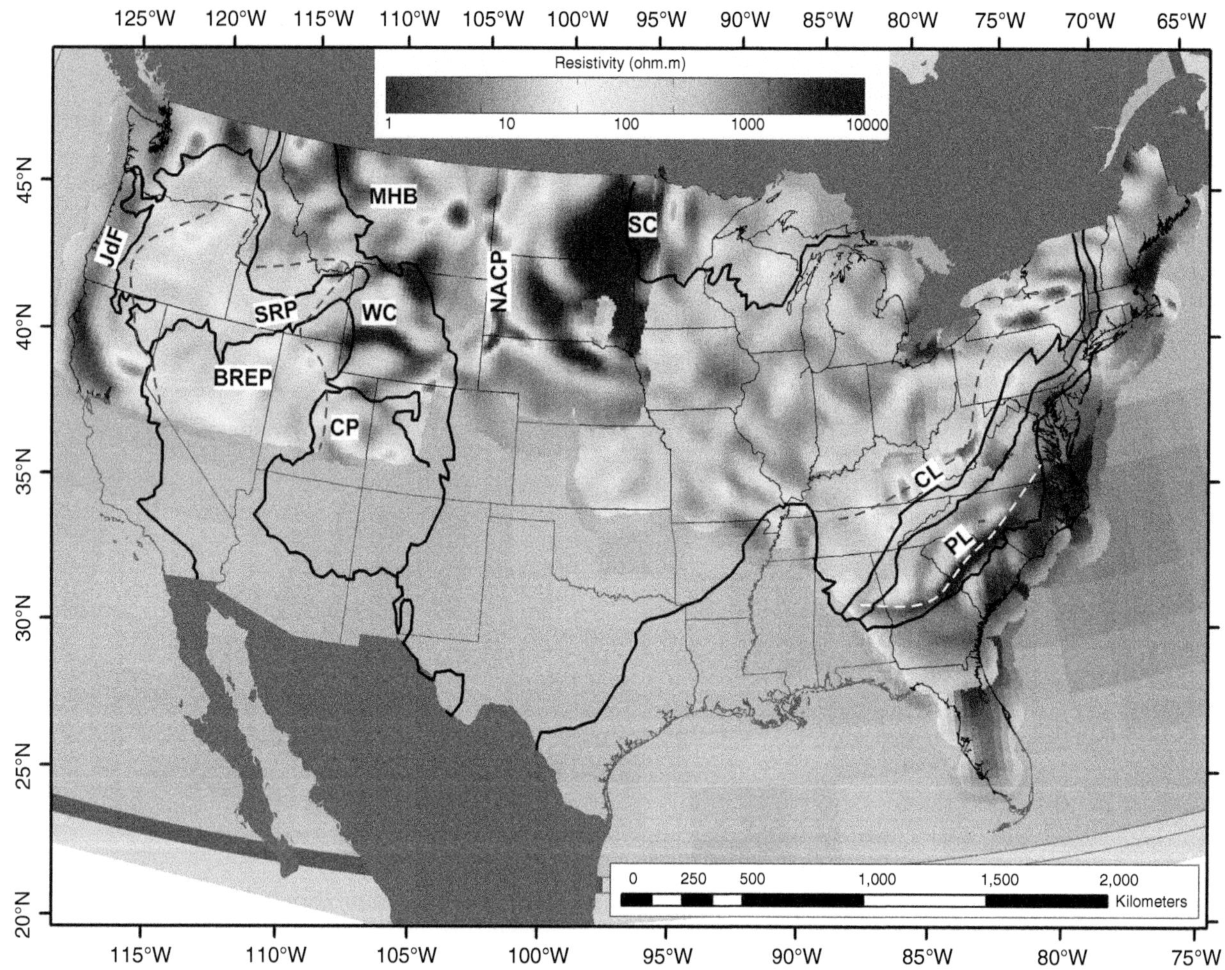

Figure 8.8 Electrical resistivity model predominantly within the mantle lithosphere (100 km) depth. Geologic provinces as in Figure 8.4 plus Juan de Fuca slab (JdF), Basin and Range extensional province (BREP), Snake River Plain (SRP), Medicine Hat Block (MHB), Wyoming Craton (WC), Colorado Plateau (CP), North American Central Plains (NACP) anomaly, Superior Craton (SC), Piedmont lithosphere (PL), and conductive lineament (CL). *(See insert for color representation of the figure.)*

geomagnetic anomaly appears as a major conductive feature in the mantle lithosphere. This anomaly was first detected with magnetovariational surveys (Reitzel et al., 1970; Alabi et al., 1975; Camfield and Gough, 1977) and has been interpreted to be associated with the suture zone of the Trans-Hudson orogeny (Jones et al., 2005). Along the eastern continental margin, mantle images reveal an anomalous coast-parallel resistor (PL), which Murphy and Egbert (2017) suggest marks a step in lithospheric thickness that has influenced tectonic processes in this area over the last 200 million years. A conductive lineament (CL) cuts through Kentucky, West Virginia, Pennsylvania, and New York. Gribenko and Zhdanov (2017) speculate that a portion of this deep conductor may be related to an ancient hotspot track, although other ancient tectonic processes may alternatively be responsible for this feature.

8.9. UTILITY FOR INDUCTION HAZARDS RESEARCH AND OPERATIONS

GICs flow in all long technological conductors that are grounded, including pipelines, telecommunication networks, railways and, most notably, electric power grids (e.g., Lanzerotti, 1983; Barnes et al., 1991; Trichtchenko, 2016). In discretely earthed technological systems such as an electric power grid, GICs are most commonly measured and modeled as the quasi-DC earthing currents flowing in or out of the system at substation locations. As described in the rich literature related to the topic (most notably, Lehtinen and Pirjola, 1985; Viljanen and Pirjola, 1994; Overbye et al., 2012; Boteler and Pirjola, 2017), GICs at a substation location depend linearly on line voltages, which are obtained as

$$V_L = \int_L \mathbf{E} \cdot \mathbf{dl}. \tag{8.4}$$

for each power line L. A gridded map of ground level geoelectric fields is therefore required for accurate estimation of GICs at every substation. For international applications as well as in the United States, it has been typical to use a 1D approximation for geoelectric fields (e.g., Pulkkinen et al., 2007), which has its limits of applicability. More recently, more comprehensive approaches to GIC modeling have emerged, for example, in the United Kingdom (Beggan et al., 2013) and New Zealand (Divett et al., 2017), which use a thin-sheet approach to model spatially variable gridded geoelectric fields. These field estimates are then used to compute line voltages, and subsequently, GICs in their respective power grids. To date, no nation has a fully 3D electrical conductivity model that could be used to incorporate the Earth's 3D structure into the GIC computations in a rigorous manner. We refer the reader to a recent review of Kelbert (2018) for detailed discussion.

For an area with good MT data coverage and/or a complete 3D electrical conductivity model, accurate gridded geoelectric field maps may be obtained. The flowchart presented in Figure 8.9 illustrates two alternative workflows for estimation of gridded geoelectric fields.

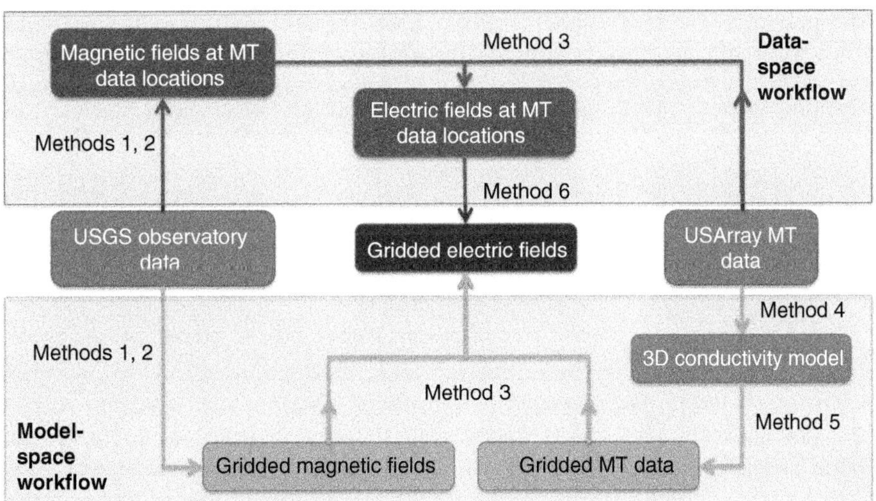

Figure 8.9 Two alternative workflows for estimation of gridded (geo-)electric fields are illustrated in this flowchart. Starting with the empirical geomagnetic and magnetotelluric data, one could either use interpolation methods to estimate real-time gridded geoelectric fields ("data-space workflow"), or a conductivity model could be used for a physics-based variant of the interpolation ("model-space workflow"). Method 1: Interpolation using multi-station transfer functions; not based on physics. Method 2: Spherical elementary currents (SECs) or other physics-based interpolation. Method 3: Multiplication in frequency domain or convolution in time domain. Method 4: MT inversion. Method 5: Numerical modeling based on Maxwell's equations. Method 6: Interpolation, for example, using Delaunay triangulation; not based on physics.

Indeed, estimation of real-time, ground level geoelectric fields starts with two empirical data types: a time series of geomagnetic field measurements (such as the U.S. geomagnetic observatory data; USGS (1901)) and magnetotelluric transfer functions (such as the USArray MT data; Schultz et al. (2006–2018)). Using the geomagnetic field time series at a limited set of locations, some version of interpolation may be used to estimate geomagnetic fields at the locations of the MT survey sites. These estimates may be multiplied (in the frequency domain) or convolved (in the time domain; e.g., Kelbert et al., 2017) with the MT transfer functions to obtain geoelectric field time series at the locations of MT sites. Then, another interpolation technique may be applied to estimate geoelectric fields on a grid. This method does not require an electrical conductivity model, but may suffer from possible drawbacks due to the fact that the geoelectric field interpolation it requires is not based on the underlying physics. Furthermore, interpolation of the geoelectric field, which varies spatially more rapidly than the geomagnetic field, cannot account for lateral variations in subsurface conductivity structure that occur between MT sites. Specifically, in this approach, local distortion effects are likely to have a regional influence on geoelectric fields – a side effect that could skew estimated GICs. Geomagnetic fields are not affected by this complication, and are generally much smoother in space than geoelectric fields, justifying the use of interpolation methods.

Alternatively, using the geomagnetic field time series measured at geomagnetic observatories, one could interpolate directly to a grid, to obtain gridded magnetic fields at ground level. Independently, one would use the MT transfer functions to obtain a 3D electrical conductivity model via the method of MT inversion (Section 8.3), and use the numerical MT forward-modeling technique to obtain MT transfer functions directly on the regular grid. This would amount to a rigorous, physics-based interpolation method. The gridded magnetic fields and gridded MT transfer functions would then be multiplied (in frequency domain) or convolved (in the time domain) to estimate ground-level geoelectric fields directly on the regular grid.

These model-derived impedances would accurately capture regional geoelectric structure, but they may not accurately capture the response from near-surface and/or highly localized structures between the original MT site locations that were used to derive this 3D conductivity model. For example, between primary field observation points, this method cannot accurately predict the so-called "static shift" that occurs due to highly localized near-surface structures at spatial scales much smaller than the scale of interest. However, it is still unclear whether these highly localized structures are important in estimating GIC hazards, or whether it is better to strip off the near-surface "distortions" in favor of effects due to more regional structure. Indeed, at the regional scale of a power grid, ignoring the highly localized structure and focusing on regional structures is likely more appropriate. This physics-based interpolation technique therefore may have a significant advantage over the direct field interpolation method, which suffers from smearing and erroneous integration of highly localized structural heterogeneities in space. Validation against measured GICs is required to test this hypothesis.

Both pathways provide real-time gridded geoelectric fields to the power grid industry, an operational goal identified by the interagency Space Weather Operations, Research, and Mitigation (SWORM) Task Force in support of the power grid industry in the United States. Ground-level geoelectric fields could be used as the input to power grid flow models, both operationally, for situational awareness in the control room, as well as for historical scenario analysis that allows a power grid to investigate potential points of failure in their grid in advance of a space weather event. The use of 3D MT transfer functions as part of the workflow has been shown to provide an improvement in ground-level geoelectric fields relative to the use of 1D conductivity models (e.g., Bonner and Schultz, 2017; Kelbert et al., 2017; Weigel, 2017; Lucas et al., 2018). We suggest that interpolation based on a numerical solution of Maxwell's equations using a 3D conductivity model such as CONUS (Section 8.7) could provide a further improvement against the utilization of empirical MT transfer functions interpolated to the locations of interest. Cross-validation of these two alternative techniques against measured GIC data is necessary to better understand smoothing effects provided by the use of a 3D conductivity model for physics-based interpolation.

8.10. CONCLUSIONS

We have presented the first ever comprehensive 3D electrical conductivity model of the contiguous United States and explored its geological underpinnings. The model is a high-resolution, $0.1° \times 0.1°$ spherical coordinate compilation of regional 3D conductivity models obtained previously by authors and colleagues in Cartesian coordinates at differing resolutions, projections and spatial limits, and now pieced together into a consistent framework. As such, though novel, this conductivity model is far from definite.

First of all, the data set that jointly constrains this conductivity model is as yet incomplete. For reasons of numerical efficiency, most but not all existing MT data in the United States were utilized (Fig. 8.1). Moreover, additional MT data are being obtained in the framework of the USArray MT (Schultz et al., 2006–2018) data collection project to fill in the mid-continental gap, and additional historical and community data are still coming in to the IRIS EMTF archive. A joint inversion of all these data in a single high-resolution grid would further improve the consistency of the model and dramatically improve the resolution of deep lithospheric and asthenospheric structure of the contiguous United States. Such an inversion has not yet been attempted, and not only for the lack of data.

Indeed, numerical MT techniques are yet to mature to the state which would allow a single high-resolution inversion of all long-period MT data in the United States. A spherical coordinate version of MT forward modeling, coupled with an efficient domain-based parallelization, would need to be developed and validated to make this next stage of MT inversion a reality. Additional work is needed to estimate the magnitude of errors introduced upon modeling regional and continental scale data sets using Cartesian coordinate approximations. In the meantime, as new data are coming in and regional inversion results are obtained, the 3D electrical conductivity compilation can be updated. We envision yearly USGS data releases of the 3D electrical conductivity model of contiguous United States.

Having said that, this initial release of the 3D electrical conductivity model is well constrained by empirical data in areas where data were utilized, and is a dramatic improvement over the largely heuristic Fernberg (2012) 1D electrical conductivity compilation. We envision that this model can be directly used for geoelectric field estimation, albeit in a different way than a compilation of 1D models would be utilized. The 3D model can be used by an MT professional in conjunction with a magnetotelluric forward modeling code such as ModEM (Egbert and Kelbert, 2012b; Kelbert et al., 2014) to compute modeled MT impedance data at any set of points at the Earth's surface. At the USGS Geomagnetism Program, we intend to utilize this model and its future releases to provide high-resolution gridded, modeled MT impedances in all areas of the United States. By using the USGS gridded impedances, pre-computed over the footprint of their grid, a power grid utility eliminates the need to compute the Earth impedance recursively using a regional 1D conductivity model (e.g., from the Fernberg (2012) compilation). Then, any analysis of local and regional space weather hazards may utilize existing modeled impedances available to within a $\sim$ 10 km distance from any location in the contiguous United States. Of course, these modeled impedances will only be as good as the data that ultimately constrain them. This effort, therefore, does not eliminate the need for complete EarthScope-style MT survey coverage in the United States.

Furthermore, in the framework of an ongoing collaboration (Love et al., 2018) between the USGS and the National Atmospheric and Oceanic Administration Space Weather Prediction Center (NOAA SWPC), the gridded MT impedances will be used to provide near real-time electric field maps of the United States to the public. The 3D Geoelectric Field Maps data product is currently in development at NOAA SPWC, and we hope that it could prove a useful starting point for space weather situation awareness in the power grid industry of the United States.

ACKNOWLEDGMENTS

We thank Phil Wannamaker and an anonymous reviewer, as well as Carol A. Finn, Andrei Swidinsky, Jeffrey J. Love, Jill McCarthy, and Janet L. Slate for their valuable comments on a draft manuscript. We are grateful to our colleagues Naser Meqbel and Bo Yang for graciously providing their inverse models in a digital form. This work was supported by the U.S. Geological Survey Geomagnetism Program. USArray MT TA project was led by PI Adam Schultz and Gary Egbert. They would like to thank the Oregon State University MT team and their contractors, lab and field personnel over the years for assistance with data collection, quality control, processing and archiving. They also thank numerous districts of the U.S. Forest Service, Bureau of Land Management, the U.S. National Parks, the collected State land offices, and the many private landowners who permitted access to acquire the MT TA data. USArray TA was funded through NSF grants EAR-0323311, IRIS Subaward 478 and 489 under NSF Cooperative Agreement EAR-0350030 and EAR-0323309, IRIS Subaward 75-MT under NSF Cooperative Agreement EAR-0733069 under CFDA No. 47.050, and IRIS Subaward 05-OSU-SAGE under NSF Cooperative Agreement EAR-1261681.

REFERENCES

Ádám, A., E. Prácser, and V. Wesztergom (2012), Estimation of the electric resistivity distribution (EURHOM) in the European lithosphere in the frame of the EURISGIC WP2 project, *Acta Geodaetica et Geophysica Hungarica*, *47*(4), 377–387, doi:10.1556/AGeod.47.2012.4.1.

Alabi, A., P. Camfield, and D. Gough (1975), The north american central plains conductivity anomaly, *Geophysical Journal International*, *43*(3), 815–833.

Alekseev, D., A. Kuvshinov, and N. Palshin (2015), Compilation of 3D global conductivity model of the Earth for space weather applications, *Earth, Planets and Space*, *67*(1), 108.

Alken, P., A. Maute, A. Richmond, H. Vanhamäki, and G. Egbert (2017), An application of principal component analysis to the interpretation of ionospheric current systems, *Journal of Geophysical Research: Space Physics*, *122*, 5687–5708, doi:10.1002/2017JA024051.

Avdeeva, A., M. Moorkamp, D. Avdeev, M. Jegen, and M. Miensopust (2015), Three-dimensional inversion of magnetotelluric impedance tensor data and full distortion matrix, *Geophysical Journal International*, *202*(1), 464–481.

Barnes, P. R., D. T. Rizy, B. W. McConnell, F. M. Tesche, and E. R. J. Taylor (1991), Electric Utility Experience with Geomnagnetic Disturbances, *Tech. rep.*, Oak Ridge National Laboratory.

Bedrosian, P., and S. Box (2016), Highly conductive horizons in the Mesoproterozoic Belt-Purcell Basin: Sulfidic early basin strata as key markers of Cordilleran shortening and Eocene extension, *Geological Society of America Special Papers*, *522*, 305–339.

Bedrosian, P., and A. Kelbert (2019), Reflections on the geological structure of North America based on magnetotelluric models, *Earth and Planetary Science Letters*. Manuscript in preparation.

Bedrosian, P., S. Box, and L. Pellerin (2006), Magnetotelluric transect between the Selkirk Crest, Idaho and the Whitefish Range, Montana to probe deep crustal structure, doi:10.17611/DP/EMTF/USGS/CGGSC/BB06, retrieved from the IRIS database on March 20, 2018.

Bedrosian, P., A. Kelbert, B. Burton, J. Morris, and C. Blum (2015), Long-period magnetotelluric transfer functions from the Florida Peninsula, doi:10.17611/DP/EMTF/USGS/GEOMAG/FL15, retrieved from the IRIS database on March 20, 2018.

Bedrosian, P. A. (2007), MT+, integrating magnetotellurics to determine earth structure, physical state, and processes, *Surveys in Geophysics*, *28*(2–3), 121–167.

Bedrosian, P. A. (2016), Making it and breaking it in the midwest: Continental assembly and rifting from modeling of earthscope magnetotelluric data, *Precambrian Research*, *278*, 337–361.

Bedrosian, P. A., and D. W. Feucht (2014), Structure and tectonics of the Northwestern United States from EarthScope USArray magnetotelluric data, *Earth and Planetary Science Letters*, *402*, 275–289, doi:10.1016/j.epsl.2013. 07.035.

Bedrosian, P. A., and J. J. Love (2015), Mapping geoelectric fields during magnetic storms: Synthetic analysis of empirical United States impedances, *Geophysical Research Letters*, *42*(23), 10160–10170, doi:10.1002/2015GL066636.

Beggan, C. D., D. Beamish, A. Richards, G. S. Kelly, and A. W. P. Thomson (2013), Prediction of extreme geomagnetically induced currents in the UK high-voltage network, *Space Weather*, *11*, 407–419, doi:10.1002/swe.20065.

Blake, S. P., P. T. Gallagher, J. McCauley, A. G. Jones, C. Hogg, J. Campanyà, C. D. Beggan, A. W. P. Thomson, G. S. Kelly, and D. Bell (2016), Geomagnetically induced currents in the Irish power network during geomagnetic storms, *Space Weather*, *14*, 1136–1154, doi:10.1002/2016SW001534. Received.

Boerner, D., R. Kurtz, and J. Craven (1996), Electrical conductivity and paleo-proterozoic foredeeps, *Journal of Geophysical Research: Solid Earth*, *101*(B6), 13775–13791.

Bonner, L. R., and A. Schultz (2017), Rapid prediction of electric fields associated with geomagnetically induced currents in the presence of three-dimensional ground structure: Projection of remote magnetic observatory data through magnetotelluric impedance tensors, *Space Weather*, doi:10.1002/2016SW001535, 2016SW001535.

Boteler, D. H., and R. J. Pirjola (2017), Modeling geomagnetically induced currents, *Space Weather*, *15*, 258–276, doi:10.1002/2016SW001499.

Camfield, P., and D. Gough (1977), A possible Proterozoic plate boundary in North America, *Canadian Journal of Earth Sciences*, *14*(6), 1229–1238.

Chave, A. D. (2014), Magnetotelluric data, stable distributions and impropriety: An existential combination, *Geophysical Journal International*, *198*, 622–636, doi:10.1093/gji/ggu121.

Chave, A. D., and A. G. Jones (Eds.) (2012), *The Magnetotelluric Method*, pp. 1–552, Cambridge University Press, Cambridge, UK.

Chave, A. D., D. J. Thomson, and M. E. Ander (1987), On the robust estimation of power spectra, coherences, and transfer functions, *Journal of Geophysical Research*, *92*(B1), 633–648.

Chulick, G. S., and W. D. Mooney (2002), Seismic structure of the crust and uppermost mantle of North America and adjacent oceanic basins: A synthesis, *Bulletin of the Seismological Society of America*, *92*(6), 2478–2492.

de Groot-Hedlin, C., S. Constable, and K. Weitemeyer (2003–2004), Transfer functions for deep magnetotelluric sounding along the yellowstone-snake river hotspot track, doi:10.17611/DP/EMTF/YSRP/2004, retrieved from the IRIS database on March 20, 2018.

DeLucia, M. S., B. S. Murphy, S. Marshak, and G. D. Egbert (2019), The Missouri High-Conductivity Belt, revealed by magnetotelluric imaging: Evidence of a trans-lithospheric shear zone beneath the Ozark Plateau, Midcontinent USA? *Tectonophysics*, *753*, 111–123, doi:10.1016/j.tecto.2019.01.011.

Dickinson, W. R. (2004), Evolution of the North American cordillera, *Annual Review of Earth and Planetary Sciences*, *32*, 13–45.

Divett, T., M. Ingham, C. D. Beggan, G. S. Richardson, C. J. Rodger, A. W. P. Thomson, and M. Dalzell (2017), Modeling geoelectric fields and geomagnetically induced currents around New Zealand to explore GIC in the South Island s electrical transmission network, *Space Weather*, *15*, 1396–1412.

Dong, H., W. Wei, G. Ye, S. Jin, A. G. Jones, J. Jing, L. Zhang, C. Xie, F. Zhang, and H. Wang (2014), Three-dimensional electrical structure of the crust and upper mantle in ordos block and adjacent area: Evidence of regional lithospheric modification, *Geochemistry, Geophysics, Geosystems*, *15*(6), 2414–2425.

Dong, S.-W., T.-D. Li, Q.-T. Lü, R. Gao, J.-S. Yang, X.-H. Chen, W.-B. Wei, Q. Zhou, et al. (2013), Progress in deep lithospheric exploration of the continental china: A review of the sinoprobe, *Tectonophysics*, *606*, 1–13.

Egbert, G. D., and J. R. Booker (1986), Robust estimation of geomagnetic transfer functions, *Geophysical Journal of the Royal Astronomical Society*, *87*(1), 173–194.

Egbert, G. D., and A. Kelbert (2012a), Computational recipes for electromagnetic inverse problems, *Geophysical Journal International*, *189*, 251–267, doi:10.1111/j.1365-246X.2011. 05347.x.

Egbert, G. D., and A. Kelbert (2012b), Computational recipes for electromagnetic inverse problems, *Geophysical Journal International*, *189*, 251–167.

Evans, R. (2012), Earths electromagnetic environment. 3a. Conductivity of earth materials, in *The Magnetotelluric Method: Theory and Practice*, edited by A. D. Chave and A. G. Jones, pp. 50–93, Cambridge University Press.

Fernberg, P. (2012), *One-Dimensional Earth Resistivity Models for Selected Areas of Continental United States & Alaska*, pp. 1–190, EPRI Technical Update 1026430, Palo Alto, CA.

Fisher, C. M., S. L. Loewy, C. F. Miller, P. Berquist, W. R. Van Schmus, R. D. Hatcher Jr, J. L. Wooden, and P. D. Fullagar (2010), Whole-rock Pb and Sm-Nd isotopic constraints on the growth of southeastern Laurentia during Grenvillian orogenesis, *Bulletin*, *122*(9–10), 1646–1659.

Fujii, I., and A. Schultz (2002), The {3D} Electromagnetic response of the {E}arth to ring current and auroral oval excitation, *Geophysical Journal International*, *151*(3), 689–709.

Gilbert, H. (2012), Crustal structure and signatures of recent tectonism as influenced by ancient terranes in the western united states, *Geosphere*, *8*(1), 141–157.

Gribenko, A., and M. Zhdanov (2017), 3-D inversion of the MT EarthScope data, collected over the East Central United States, *Geophysical Research Letters*, *44*, 11800–11807, doi:10.1002/2017GL075000.

Groom, R. W., and R. C. Bailey (1989), Decomposition of magnetotelluric impedance tensors in the presence of local three-dimensional galvanic distortion, *Journal of Geophysical Research*, *94*(B2), 1913–1925.

Hatcher, R. D., R. Tollo, M. Bartholomew, J. Hibbard, and P. Karabinos (2010), The Appalachian orogen: A brief summary, *From Rodinia to Pangea: The Lithotectonic Record of the Appalachian Region: Geological Society of America Memoir*, *206*, 1–19.

Heinson, G. S., N. G. Direen, and R. M. Gill (2006), Magnetotelluric evidence for a deep-crustal mineralizing system beneath the olympic dam iron oxide copper-gold deposit, southern australia, *Geology*, *34*(7), 573–576.

Hirschmann, M. M., C. Aubaud, and A. C. Withers (2005), Storage capacity of h2o in nominally anhydrous minerals in the upper mantle, *Earth and Planetary Science Letters*, *236*(1–2), 167–181.

Horton, J., C. San Juan, and D. Stoeser (2017), The State Geologic Map Compilation (SGMC) geodatabase of the conterminous United States (ver. 1.1, August 2017), in *U.S. Geological Survey Data Series 1052*, p. 46, U.S. Geological Survey, doi:10.3133/ds1052.

INTERMAGNET Geomagnetic Data (2013), International Real-time Magnetic Observatory Network.

Jones, A. G. (2012), Distortion of magnetotelluric data: Its identification and removal, in *The Magnetotelluric Method*, edited by A. D. Chave and A. G. Jones, pp. 219–302, Cambridge University Press, Cambridge, UK.

Jones, A. G., J. Ledo, and I. J. Ferguson (2005), Electromagnetic images of the Trans-Hudson orogen: The North American Central Plains anomaly revealed, *Canadian Journal of Earth Sciences*, *42*(4), 457–478.

Kelbert, A. (2018), The role of global/regional earth conductivity models in natural geomagnetic hazard mitigation, *Surveys in Geophysics*. Manuscript in preparation.

Kelbert, A., S. Erofeeva, C. Trabant, R. Karstens, and M. Van Fossen (2018a), Taking magnetotelluric data out of the drawer, *Eos*, *99*, doi:10.1029/2018EO112859

Kelbert, A., G. D. Egbert, and A. Schultz (2008b), Non-linear conjugate gradient inversion for global EM induction: Resolution studies, *Geophysical Journal International*, *173*(2), 365–381, doi:10.1111/j.1365-246X.2008.03717. x.

Kelbert, A., A. Schultz, and G. D. Egbert (2009), Global electromagnetic induction constraints on transition-zone water content variations, *Nature*, *460*(7258), 1003–1006.

Kelbert, A., G. D. Egbert, and A. Schultz (2011), IRIS DMC data services products: EMTF, *The Magnetotelluric Transfer Functions*, doi:10.17611/DP/EMTF.1.

Kelbert, A., G. D. Egbert, and C. DeGroot-Hedlin (2012), Crust and upper mantle electrical conductivity beneath the Yellowstone Hotspot Track, *Geology*, *40*(5), 447–450, doi:10.1130/G32655.1.

Kelbert, A., N. Meqbel, G. D. Egbert, and K. Tandon (2014), ModEM: A modular system for inversion of electromagnetic geophysical data, *Computers & Geosciences*, *66*, 40–53, doi:10.1016/j.cageo.2014.01.010.

Kelbert, A., C. C. Balch, A. Pulkkinen, G. D. Egbert, J. J. Love, E. J. Rigler, and I. Fujii (2017), Methodology for time-domain estimation of storm time geoelectric fields using the 3-d magnetotelluric response tensors, *Space Weather*, *15*(7), 874–894.

Khan, A., A. Kuvshinov, and A. Semenov (2011), On the heterogeneous electrical conductivity structure of the Earth??s mantle with implications for transition zone water content, *Journal of Geophysical Research*, *116*(B1), 1–19, doi:10.1029/2010JB007458.

Kordy, M., P. Wannamaker, V. Maris, E. Cherkaev, and G. Hill (2016), 3-Dimensional magnetotelluric inversion including topography using deformed hexahedral edge finite elements and direct solvers parallelized on symmetric multiprocessor computers–Part II: Direct data-space inverse solution, *Geophysical Journal International*, *204*(1), 94–110.

Lanzerotti, L. J. (1983), Geomagnetic induction effects in ground-based systems, *Space Science Reviews*, *34*, 347–356.

Laske, G., and G. Masters (1997), A global digital map of sediment thickness, *Eos, Transactions, American Geophysical Union*, *78*(F483), 19–97.

Lehtinen, M., and R. Pirjola (1985), Currents produced in earthed conductor networks by geomagnetically-induced electric fields, *Annales Geophysicae*, *3*(4), 479–484.

Lindsay, M., J. Spratt, S. Occhipinti, A. Aitken, M. Dentith, J. Hollis, and I. Tyler (2017), Identifying mineral prospectivity using 3d magnetotelluric, potential field and geological data in the east Kimberley, *Australia, Geological Society, London, Special Publications*, *453*, SP453–8.

Love, J. J., and C. A. Finn (2011), The USGS Geomagnetism Program and its role in space weather monitoring, *Space Weather*, *9*, S07001, doi:10.1029/2011SW000684.

Love, J. J., P. A. Bedrosian, and A. Schultz (2017), Down to Earth with an electric hazard from space, *Space Weather*, *15*(5), 658–662.

Love, J. J., E. J. Rigler, A. Kelbert, C. A. Finn, P. A. Bedrosian, and C. C. Balch (2018), On the feasibility of real-time mapping of the geoelectric field across North America, *Tech. rep.*, U.S. Geological Survey Open File Report 2018-1043.

Lucas, G. M., J. J. Love, and A. Kelbert (2018), Calculation of voltages in electric power transmission lines during historic geomagnetic storms: An investigation using realistic earth impedances, *Space Weather*.

Meqbel, N. M., G. D. Egbert, P. E. Wannamaker, A. Kelbert, and A. Schultz (2014), Deep electrical resistivity structure of the northwestern U.S. derived from 3-D inversion of USArray magnetotelluric data, *Earth and Planetary Science Letters*, *402*, 290–304, doi:10.1016/j.epsl.2013.12.026.

Munch, F., A. Grayver, A. Kuvshinov, and A. Khan (2018), Stochastic inversion of geomagnetic observatory data including rigorous treatment of the ocean induction effect with implications for transition zone water content and thermal structure, *Journal of Geophysical Research: Solid Earth*, *123*(1), 31–51.

Murphy, B. S., and G. D. Egbert (2017), Electrical conductivity structure of southeastern North America: Implications for lithospheric architecture and Appalachian topographic rejuvenation, *Earth and Planetary Science Letters*, *462*, 66–75.

Newitt, L. R., and R. Coles (2007), Observatories in Canada, in *Encyclopedia of Geomagnetism and Paleomagnetism*, edited by D. Gubbins and E. Herrero-Bervera, pp. 726–727, Springer.

Ohtani, E., M. Toma, K. Litasov, T. Kubo, and A. Suzuki (2001), Stability of dense hydrous magnesium silicate phases and water storage capacity in the transition zone and lower mantle, *Physics of the Earth and Planetary Interiors*, *124*(1–2), 105–117.

Olsen, N., H. Luhr, T. J. Sabaka, M. Mandea, M. Rother, L. Toffner-Clausen, and S. Choi (2006), CHAOS – A model of the Earth's magnetic field derived from CHAMP, Ørsted, and SAC-C magnetic satellite data, *Geophysical Journal International*, *166*(1), 67–75, doi:10.1111/j.1365-246X.2006.02959.x.

Overbye, T. J., T. R. Hutchins, K. Shetye, J. Weber, and S. Dahman (2012), Integration of geomagnetic disturbance modeling into the power flow: A methodology for large-scale system studies, in *Proc. 2012 North Am. Power Symp.*, September, Champaign, IL, USA.

Patro, P. K., and G. D. Egbert (2008), Regional conductivity structure of Cascadia: Preliminary results from 3D inversion of USArray transportable array magnetotelluric data, *Geophysical Research Letters*, *35*, L20311, doi:10.1029/2008GL035326.

Pellerin, L., J. Peacock, D. Jones, P. Bedrosian, and SAGE students and faculty (2005–2035), Summer of Applied Geophysical Experience (SAGE): Magnetotelluric Transfer Functions in Santa Fe, New Mexico, USA, doi:10.17611/DP/EMTF/SAGE.MT, retrieved from the IRIS database on March 20, 2018.

Pulkkinen, A., R. Pirjola, and A. Viljanen (2007), Determination of ground conductivity and system parameters for optimal modeling of geomagnetically induced current flow in technological systems, *Earth, Planets and Space*, *59*, 999–1006.

Rasson, J. L. (2007), INTERMAGNET, in *Encyclopedia of Geomagnetism and Paleomagnetism*, edited by D. Gubbins and E. Herrero-Bervera, pp. 715–717, Springer, Dordrecht, The Netherlands.

Reed, J. C., and C. A. Bush (2007), *About the Geologic Map in the National Atlas of the United States of America*, Citeseer.

Reed, J. C., J. O. Wheeler, and B. E. Tucholke (2005), *Geologic Map of North America: Guide*, vol. *1*, Geological Society of America.

Reitzel, J., D. Gough, H. Porath, and C. Anderson III (1970), Geomagnetic deep sounding and upper mantle structure in the western united states, *Geophysical Journal International*, *19*(3), 213–235.

Riddihough, R., C. Finn, and R. Couch (1986), Klamath-Blue Mountain lineament, Oregon, *Geology*, *14*(6), 528–531.

Robertson, K., G. Heinson, and S. Thiel (2016), Lithospheric reworking at the proterozoic–phanerozoic transition of australia imaged using auslamp magnetotelluric data, *Earth and Planetary Science Letters*, *452*, 27–35.

Robertson, K., G. Heinson, D. Taylor, and S. Thiel (2017), The lithospheric transition between the Delamerian and Lachlan orogens in western Victoria: New insights from 3D magnetotelluric imaging, *Australian Journal of Earth Sciences*, *64*(3), 385–399.

Rosenkjaer, G. K., E. Gasperikova, G. A. Newman, K. Arnason, and N. J. Lindsey (2015), Comparison of 3D MT inversions for geothermal exploration: Case studies for Krafla and Hengill geothermal systems in Iceland, *Geothermics*, *57*, 258–274.

Schultz, A. (1990), On the vertical gradient and associated heterogeneity in mantle electrical-conductivity, *Physics of the Earth and Planetary Interiors*, *64*(1), 68–86.

Schultz, A. (2010), EMscope: A continental scale magnetotelluric observatory and data discovery resource, *Data Science Journal*, *8*, IGY6–IGY20.

Schultz, A., G. D. Egbert, A. Kelbert, T. Peery, V. Clote, B. Fry, staff of the National Geoelectromagnetic Facility, and their contractors (2006–2018), USArray TA Magnetotelluric Transfer Functions, doi:10.17611/DP/EMTF/USARRAY/TA, retrieved from the IRIS database on March 20, 2018.

Selway, K. (2014), On the causes of electrical conductivity anomalies in tectonically stable lithosphere, *Surveys in Geophysics*, *35*(1), 219–257.

Semenov, A., and A. Kuvshinov (2012), Global 3-D imaging of mantle conductivity based on inversion of observatory C - responses-II. Data analysis and results, *Geophysical Journal International*, doi:10.1111/j.1365-246X. 2012.05665.x.

Semenov, V. Y., and W. Jozwiak (2006), Lateral variations of the mid-mantle conductance beneath Europe, *Tectonophysics*, *416*, 279–288, doi:10.1016/j. tecto.2005.11.017.

Shimizu, H., H. Utada, K. Baba, T. Koyama, M. Obayashi, and Y. Fukao (2010), Three-dimensional imaging of electrical conductivity in the mantle transition zone beneath the North Pacific Ocean by a semi-global induction study, *Physics of the Earth and Planetary Interiors*, *183*(1–2), 252–269, doi:10.1016/j.pepi.2010.01.010.

Siripunvaraporn, W., G. D. Egbert, Y. Lenbury, and M. Uyeshima (2005), Three-dimensional magnetotelluric inversion: Data-space method, *Physics of the Earth and Planetary Interiors*, *150*(1–3), 3–14.

Smirnov, M. Y., and G. D. Egbert (2012), Robust principal component analysis of electromagnetic arrays with missing data, *Geophysical Journal International*, *190*, 1423–1438, doi:10.1111/j.1365-246X.2012.05569.x.

Stanley, W. D., C. Finn, and J. L. Plesha (1987), Tectonics and conductivity structures in the southern washington cascades, *Journal of Geophysical Research: Solid Earth*, *92*(B10), 10,179–10,193.

Sun, J., A. Kelbert, and G. D. Egbert (2015), Ionospheric current source modeling and global geomagnetic induction using ground geomagnetic observatory data, *Journal of Geophysical Research: Solid Earth*, *120*(10), 6771–6796.

Toffelmier, D. A., and J. A. Tyburczy (2007), Electromagnetic detection of a 410-km-deep melt layer in the southwestern United States, *Nature*, *447*(7147), 991.

Trichtchenko, L. (2016), Modelling natural electromagnetic interference in man-made conductors for space weather applications, *Annales Geophysicae*, *34*, 427–436, doi:10.5194/angeo-34-427-2016.

USGS (1901), USGS Geomagnetism Program Definitive Data. INTERMAGNET. Other/Seismic Network, doi:10.7914/SN/NT.

Utada, H., T. Koyama, M. Obayashi, and Y. Fukao (2009), A joint interpretation of electromagnetic and seismic tomography models suggests the mantle transition zone below Europe is dry, *Earth and Planetary Science Letters*, *281*(3–4), 249–257, doi:10.1016/j.epsl.2009.02.027.

Viljanen, A., and R. Pirjola (1994), Geomagnetically induced currents in the finnish high-voltage power system, *Surveys in Geophysics*, *15*, 383–408.

Wannamaker, P., R. Evans, R. McGary, V. Maris, and Q. G. Inc. (2010), Long Period Magnetotelluric Transfer Functions from the CAFE-MT experiment in central Washington, doi:10.17611/DP/EMTF/CAFE-MT.LP, retrieved from the IRIS database on March 20, 2018.

Wannamaker, P. E. (2005), Anisotropy versus heterogeneity in continental solid earth electromagnetic studies: Fundamental response characteristics and implications for physicochemical state, *Surveys in Geophysics*, *26*(6), 733–765.

Wannamaker, P. E., D. P. Hasterok, J. M. Johnston, J. A. Stodt, D. B. Hall, T. L. Sodergren, L. Pellerin, V. Maris, W. M. Doerner, K. A. Groenewold, et al. (2008), Lithospheric dismemberment and magmatic processes of the Great Basin–Colorado Plateau transition, Utah, implied from magnetotellurics, *Geochemistry, Geophysics, Geosystems*, *9*(5), 38, doi:10.1029/2007gc001886.

Wannamaker, P. E., R. L. Evans, P. A. Bedrosian, M. J. Unsworth, V. Maris, and R. S. McGary (2014), Segmentation of plate coupling, fate of subduction fluids, and modes of arc magmatism in cascadia, inferred from magnetotelluric resistivity, *Geochemistry, Geophysics, Geosystems*, *15*(11), 4230–4253.

Weidelt, P., and A. D. Chave (2012), The magnetotelluric response function, in *The Magnetotelluric Method*, edited by A. D. Chave and A. G. Jones, pp. 122–164, Cambridge University Press, Cambridge, UK.

Weigel, R. S. (2017), A comparison of methods for estimating the geoelectric field, *Space Weather*, *15*(2), 430–440, doi:10.1002/2016SW001504, 2016SW001504.

Whitmeyer, S. J., and K. E. Karlstrom (2007), Tectonic model for the Proterozoic growth of North America, *Geosphere*, *3*(4), 220–259.

Williams, M. L., K. M. Fischer, J. T. Freymueller, B. Tikoff, A. M. Tréhu, and others (2010), *Unlocking the Secrets of the North American Continent: An EarthScope Science Plan for 2010-2020*, pp. 1–78, EarthScope.

Yang, B., G. D. Egbert, A. Kelbert, and N. M. Meqbel (2015), Three-dimensional electrical resistivity of the north-central USA from EarthScope long period magnetotelluric data, *Earth and Planetary Science Letters*, *422*, 87–93.

Yoshino, T. (2011), Electrical properties of rocks, in *Encyclopedia of Solid Earth Geophysics*, edited by H. K. Gupta, pp. 270–276, Springer-Verlag, Dordrecht, The Netherlands.

Zhdanov, M. S., A. Gribenko, M. Čuma, and M. Green (2012), Geoelectrical structure of the lithosphere and asthenosphere beneath the northwestern United States, *Journal of Geology & Geophysics*, *1*(2), 106, doi:10.4172/2329-6755.1000106.

Zhdanov, M. S., R. B. Smith, A. Gribenko, M. Cuma, and M. Green (2011), Three-dimensional inversion of large-scale earthscope magnetotelluric data based on the integral equation method: Geoelectrical imaging of the yellowstone conductive mantle plume, *Geophysical Research Letters*, *38*(8), L08307, doi:10.1029/2011GL046953.

9

A Data-Driven Approach to Estimating Geoelectric Fields: Comparison, Validation, and Discussion of Geomagnetic Hazard Assessment Within Common Physiographic Zones

Stephen W. Cuttler

ABSTRACT

Geomagnetic field time series are convolved with magnetotelluric surface impedances to estimate the geoelectric variations during a magnetic storm. Magnetotelluric survey sites are selected from two physiographic zones in Minnesota and Virginia, United States. A magnetic time series from 22 to 26 June 2015 is considered that encompasses a magnetic storm of moderate size recorded at the Brandon, Manitoba, and Fredericksburg, Virginia. This study produces estimated time series of the geoelectric field throughout the magnetic storm, demonstrating that two sites 125 km apart can differ by up to two orders of magnitude (4484 mV/km at one site and 41 mV/km at another site 125 km away). These estimations are compared a one-dimensional conductivity models representative of its physiographic zone. Finally, a range of geoelectric estimations are presented to examine variety of geoelectric estimations present throughout two adjoining physiographic zones. This study demonstrates that there may be significant local variability in the geo-electric field magnitude and phase that regional 1D models do not capture. This shows that an evaluation of the realistic conductivity distribution through a data-driven approach is necessary to produce a reliable estimation of how the geoelectric field evolves over the course of a magnetic storm.

Key Points

- Estimations of the geoelectric field are produced from measurements of the geomagnetic field during a magnetic storm convolved with impedances.
- Geoelectric field estimations differ greatly across small distances and within common physiographic zones.
- Geoelectric field estimations produced from one-dimensional conductivity models do not represent the magnitude and phase of the geoelectric response at multiple locations within a common physiographic zone.

Department of Geophysics, Colorado School of Mines, Golden, CO, USA

9.1. INTRODUCTION

Geomagnetic storms can produce large potential differences across power transmission systems through geoelectric currents induced in the Earth's electrically conducting interior (e.g., Molinski, 2002; Pirjola, 2002). These storms represent a hazard to power transmission infrastructure through their potential to damage transformers and other critical components of power transmissions systems. Infrastructure damage due to magnetic storms has occurred on numerous occasions, such as the March 1989 storm that caused blackouts in the Hydro-Quebec power grid of Canada (e.g., Allen et al., 1989; Béland and Small, 2005) as well as power failures from the August 1972 storm (e.g., Anderson et al., 1974). Interest in averting the deleterious effects of magnetic storms has motivated

Geomagnetically Induced Currents from the Sun to the Power Grid, Geophysical Monograph 244, First Edition.
Edited by Jennifer L. Gannon, Andrei Swidinsky, and Zhonghua Xu.

regulatory agencies to issue instructions to power-grid companies to assess the resilience of their operations and infrastructure in preparation for future magnetic storms (FERC, 2013) (Order No. 77) (e.g., Veeramany et al., 2016), academic work on magnetic storm assessment (e.g., Schrijver et al., 2015), and risk assessments in the private sector (e.g., Riswadkar and Dobbins, 2010; AonBenfield, 2013; Lloyds, 2013).

In order to prepare for a magnetic storm, those who maintain power systems need an estimation of the induced geoelectric field at the locations of their power transmission infrastructure. Local assessments of geoelectric field estimation have been performed at various places around the world (e.g., Boteler, 2001; Viljanen et al., 2013; Torta et al., 2014). In addition to direct monitoring of the geoelectric field, one can estimate the geoelectric field though characterizing the behavior of the Earth's ground-level geomagnetic field across broad physiographic zones (e.g., Love et al., 2016b) and characterizing an impedance (e.g., Chave, 2012) at a given location, which is dependent on the three-dimensional conductivity distribution of the Earth's interior (e.g., Bedrosian and Feucht, 2014; Meqbel et al., 2014; Alekseev et al., 2015). Subsurface conductivity can also be characterized with active sources, as described in works such as (Spies and Frischknecht, 1991) on land or (Swidinsky et al., 2015) in the marine environment; such conductivity models can be used to supplement impedances typically constructed using passive magnetotelluric methods.

Geoelectric mapping studies incorporate impedance tensors derived from magnetotelluric measurements of the geomagnetic and geoelectric fields (e.g., Egbert, 2007) such as those collected from the USArray program (e.g., Schultz, 2010; Williams et al., 2010). These include (Bedrosian and Love, 2015), who used three-dimensional magnetotelluric EarthScope impedances to estimate the geoelectric response of a simulated time-varying geomagnetic field across northern North America. Love et al. (2016a) used these impedances to produce a hazard map of half of the United States by convolving these impedances against latitude-dependent maps of extreme geomagnetic activity. Torta et al. (2017) improved the estimate of the phase and magnitude of measured geomagnetically induced currents (GICs) in power transmission systems by modeling local two- and three-dimensional conductivity structures with their impedances.

Two studies in particular consider the use of three-dimensional EarthScope impedances for the use of estimating the geoelectric field for use in predicting GIC hazard assessment. Bonner and Schultz (2017) demonstrates the significant differences in geoelectric field estimations between one-dimensional conductivity models and three-dimensional EarthScope impedances. Lucas et al. (2018) also uses EarthScope impedances to produces estimations of the direct-current (DC) voltage offset experienced across power transmission lines, as well as examining how variances in the estimated geoelectric field along a line influences this DC voltage offset.

Recognizing that many ongoing efforts to estimate geoelectric hazards utilize conductivity models representing broad physiographic zones through one-dimensional conductivity models, there is a need to improve estimations of the geoelectric hazards as discussed in the U.S. Space Weather Action Plan (NSTC, 2015). Following the work of Bonner and Schultz (2017) and Lucas et al. (2018), I present a case study where I use similar methods to investigate how an estimation of the geoelectric field can differ between two closely spaced sites within the same physiographic zone in order to highlight the importance of using accurate Earth impedance tensors for geoelectric field estimation.

9.2. INDUCTION

I am concerned with the relationship between a ground-level geomagnetic field $\mathbf{B}(t, x, y)$ generated as a result of magnetic storm activity and the induced geoelectric field variations $\mathbf{E}(t, x, y)$ at a given location (x, y) on the Earth as a function of time t. I consider the inducing geomagnetic field $\mathbf{B}(t, x, y)$ to be a locally planar field, with time-varying horizontal geomagnetic fields (B_x, B_y) and geoelectric fields (E_x, E_y) (e.g., Cagniard, 1953). In addition, I am interested in the frequency band that produces the greatest hazard risk for power systems: 10^{-1} to 10^{-4} Hz (periods of 10 to 10,000 s) (Barnes et al., 1991; NERC, 2014a, 2014b). This range allows us to invoke the quasi-static approximation, thus electromagnetic induction in the Earth becomes a diffusive process that ignores displacement currents.

The Fourier transform relates the geomagnetic and geoelectric fields in time to the frequency domain according to

$$\mathcal{F}\{\mathbf{B}(t)\} = \tilde{\mathbf{B}}(f) \quad \text{and} \quad \mathcal{F}\{\mathbf{E}(t)\} = \tilde{\mathbf{E}}(f), \tag{9.1}$$

where f is the frequency of sinusoidal variation. In magnetotellurics, one uses $\tilde{\mathbf{B}}(f, x, y)$ and $\tilde{\mathbf{E}}(f, x, y)$ to determine a complex impedance tensor $\overline{\overline{\mathbf{Z}}}(f, x, y \,|\, \sigma(\mathbf{r}))$ having units of ohms (Ω), where $\sigma(\mathbf{r})$ is the subsurface conductivity distribution, and $\mathbf{r}$ is a position vector (Weidelt and Chave, 2012, chapter 4.1.2). The impedance tensor can be written as

$$\overline{\overline{\mathbf{Z}}}(f, x, y \,|\, \sigma(\mathbf{r})) = \begin{bmatrix} Z_{xx} & Z_{xy} \\ Z_{yx} & Z_{yy} \end{bmatrix} (f, x, y \,|\, \sigma(\mathbf{r})), \tag{9.2}$$

and is the result of induction within the three-dimensional conductivity distribution of the subsurface corresponding to the geology of the volume of Earth at a given geographic location (x, y). In this three-dimensional case, each element of the tensor (Z_{xx}, Z_{xy}, Z_{yx}, and Z_{yy}) is typically non-zero and independent of other components, and the subscripts (x, y) correspond to the northing and easting of the geographic location, respectively. The impedance tensor is related to the geomagnetic field $\tilde{\mathbf{B}}(f, x, y)$ and the geoelectric field $\tilde{\mathbf{E}}(f, x, y)$ via

$$\tilde{\mathbf{E}}(f, x, y) = \frac{1}{\mu} \bar{\bar{\mathbf{Z}}}(f, x, y \,|\, \sigma(\mathbf{r})) \cdot \tilde{\mathbf{B}}(f, x, y), \quad (9.3)$$

where μ is magnetic permeability. Equation (9.3) can be used to estimate the geoelectric field in the frequency domain at a given survey site.

9.3. DATA

The USArray EarthScope program supported by the National Science Foundation (NSF) consists of a continent-wide survey of seismic and magnetotelluric sites conducted throughout the United States beginning in 2006 (Kelbert et al., 2011). Geomagnetic vector field time series sampled at 1 Hz were collected via fluxgate magnetometers and geoelectric field time series sampled at 1 Hz were collected via pairs of lead–lead chloride non-polarizing electrodes arranged in a standard magnetotelluric array (Schultz, 2010). These time series were processed to produce impedances at 30 discrete periods from $T = 7.3143$ s to $T = 1.8725 \cdot 10^4$ s (Egbert, 2007). These impedances are the result of the three-dimensional conductivity distribution of the subsurface in the three-dimensional volume below survey site as well as in the region around the survey site, and therefore are representative of its corresponding geologic structure.

For the main part of this study, I selected four survey sites within two physiographic zones presented in Fernberg (2012) for comparison: two sites each within the physiographic zones SU-1 in Minnesota, Wisconsin, and Michigan, United States, and PT-1 in Virginia, United States (Fig. 9.1). I selected two impedances located 125 km apart within in the SU-1 physiographic zone in Minnesota, United States: RED36 (Fig. 9.2) (Schultz et al., 2006–2018a) and MNB36 (Fig. 9.3) (Schultz et al., 2006–2018b), with MNB36 almost directly north of RED36. The geologic setting in this area is controlled by a mid-continent rift system, which produced a series of intracontinental sedimentary basins juxtaposed against older cratonic rock. These juxtapositions are further exaggerated by faulting that places conductive, sedimentary rock characteristic of basin fill against more resistive igneous and metamorphic cratonic rock (Bally and Palmer, 1989). MNB36 is located within the Archean Superior province to the north while RED36 is located within the Animikie Basin to the south. In addition, both sites show the influence of the footprint from a conductive anomaly observed to well up below Wisconsin (Bowles-Martinez and Schultz, 2015). This conductive anomaly can be attributed to interconnected sulfides deposited as a result the fossil trace of the Keewenaw hotspot, which is likely the result of a failed rifting in the metoproterozoic era (Schultz et al., 2014; Bowles-Martinez and Schultz, 2015). This 1-D conductivity model representing the SU-1 physiographic zone has shown little resemblance to existing 3-D models of the region (Schultz et al., 2014; Bowles-Martinez and Schultz, 2015), highlighting the need within the region for additional examination relating to geoelectric hazards. To address this, regional model updates to these models are underway (NERC, 2017).

Figure 9.2 shows the impedance for RED36 which, as we shall see, corresponds to smaller estimations of $E(x, y, t)$. Figures 9.2a through 9.2d correspond to the Z_{xx}, Z_{xy}, Z_{yx}, and Z_{yy} components of the impedance tensor, respectively. Figure 9.3 represents the impedance for MNB36 similarly, which I found corresponds to a larger geoelectric response. In Figure 9.2, we see that the diagonal components of the impedance response (Z_{xx} and Z_{yy}) are very small but of differing magnitude, indicating that the conductivity distribution of the subsurface is likely two-dimensional. The off-diagonal components (Z_{xy} and Z_{yx}), though unique from one another, only reach 2Ω in each component at shorter periods. Comparing Figure 9.2 to Figure 9.3, MNB36 demonstrates a much greater scaling of the geomagnetic to the geoelectric field compared with RED36. The plots for MNB36 further show a much larger difference in Z_{xx} and Z_{yy}, indicating that the subsurface conductivity distribution is more representative of a three-dimensional conductivity model.

Discrete time series of the near-surface geomagnetic field (Love and Chulliat, 2013) were obtained through the Brandon (BRD) magnetic observatory located in Brandon, Manitoba, operated by National Resources Canada (NRCan) (Fig. 9.4). This observatory was considered due to its proximity to the locations of the selected EarthScope survey sites in Minnesota (Fig. 9.1). The geomagnetic time series gathered at this observatory is precise to within 0.1 nT and have undergone digital and analog filtering.

In addition, the magnetic field time series have undergone tapering at the end of the time series using the cosine taper

$$c(t) = \begin{cases} 0.5\left(1 - \cos\frac{\pi}{a}t\right) & 0 \le t \le a \\ 1 & a \le t \le (1-a) \\ 0.5\left(1 - \cos\frac{\pi}{a}(1-t)\right) & (1-a) \le t \le 1 \end{cases} \quad (9.4)$$

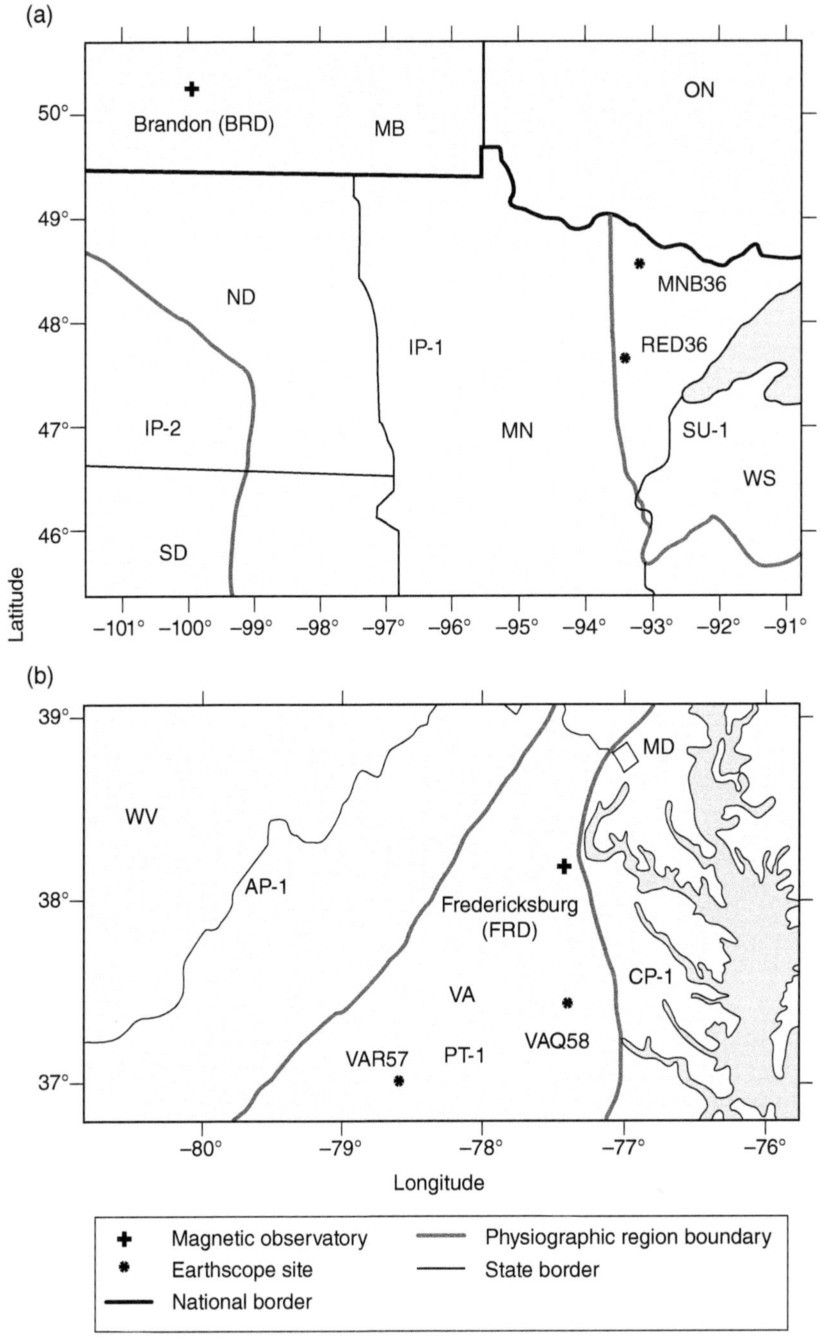

Figure 9.1 Maps showing the locations of the magnetic observatories (available through the INTERMAGNET program managed by the USGS and NRCan) and the EarthScope USArray sites referenced to physiographic zones are marked by the thicker lines (derived from Fernberg (2012). (a) The NRCanBrandon (BRD) magnetic observatory in Manitoba and the EarthScope sites MNB36 and RED36 in northern Minnesota. Both sites fall within the SU-1 physiographic zone. (b) The USGS Fredericksburg (FRD) magnetic observatory in Virginia and the EarthScope sites VAQ58 and VAR57 in southern and eastern Virginia which fall within the PT-1 zone. (*See electronic version for color representation of this figure.*)

where t is time, a is a cutoff time to begin the taper on each end of the time series (chosen to be 3 h in this study), and both a and t are normalized to 1. This serves to reduce the time series to 0 at the end of each window to reduce leakage into other frequencies when a Fourier transform is performed. Prewhitening has also been applied to the geomagnetic field time series to reduce the covariance matrix of the time series to an identity matrix,

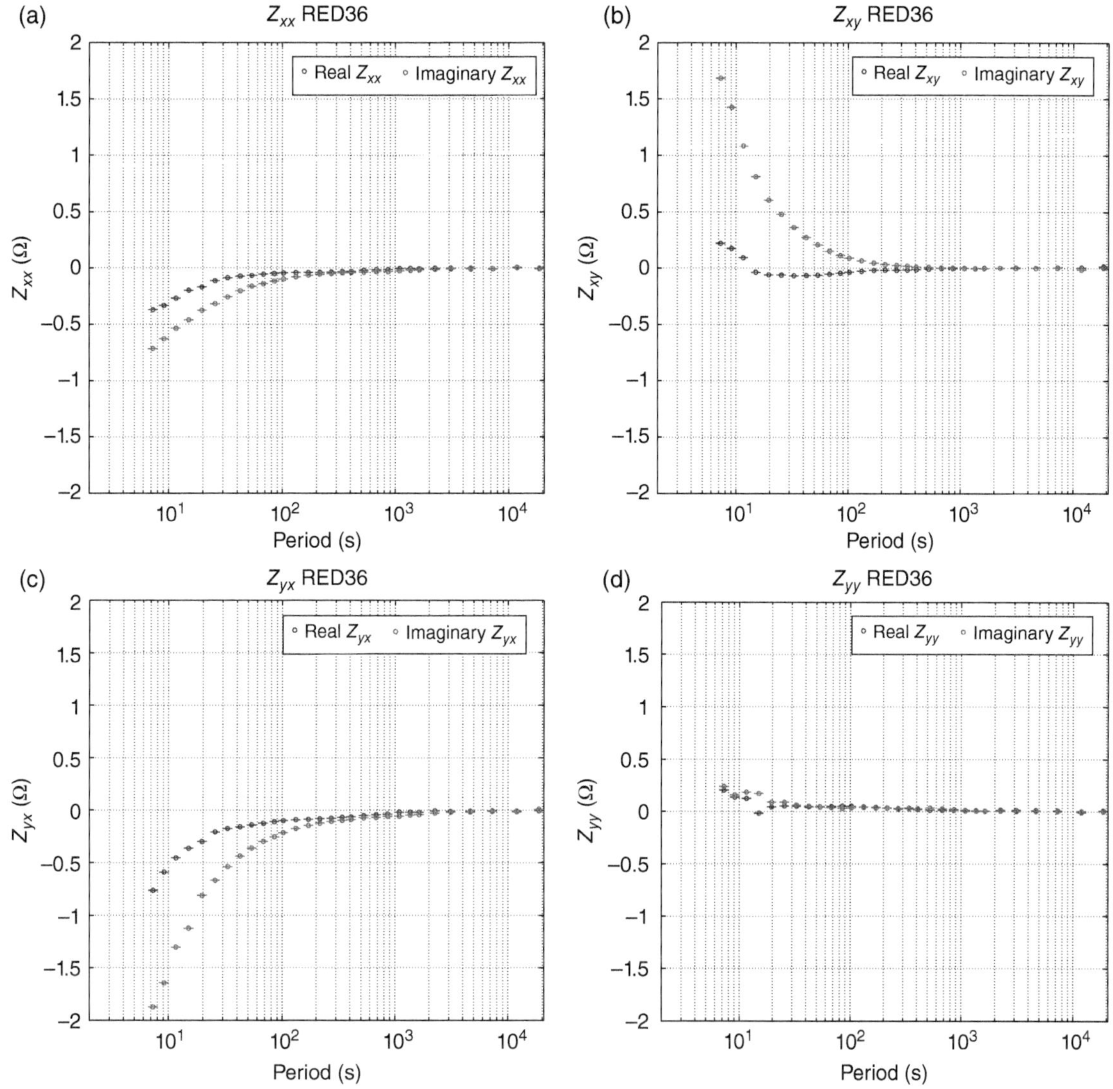

Figure 9.2 Real (left panels, upper lines; right panels, lower lines) and imaginary (left panels, lower lines; right panels, upper lines) impedance plots from the EarthScope survey site RED36, which were used to produce estimations of the geoelectric field with the Brandon (BRD) observatory geomagnetic time series. (*See electronic version for color representation of this figure.*)

removing all first-order structure from the time series in order to minimize data leakage.

The magnetic storm I selected for this study occurred on 22 June 2015 and was recorded at a sampling frequency of 1 Hz over three days beginning at 00:00 on 22 June 2015, encompassing the entire evolution of the storm. The storm's onset begins at approximately 18:00 on 22 June 2015 and ceases at approximately 18:00 on 24 June 2015. The storm reaches its largest magnitude of 1207 nT in $B_x(t)$ and 725 nT in $B_y(t)$ during the second day of the storm, with the largest geomagnetic variation occurring between 18:00 on 22 June 2015 and 12:00 on 23 June 2015 into the time series. After 18:00 on 24 June 2015, geomagnetic activity concludes and the geomagnetic field returns to a baseline value.

I also considered impedances from two survey sites located 200 km apart in Virginia, United States: VAR57 (Schultz et al., 2006–2018c) and VAQ58 (Schultz et al., 2006–2018d), with VAR57 100 km west-southwest of VAQ58 (Fig. 9.1). Both VAR57 and VAQ58 are in the Piedmont province of the southern Appalachian Mountains, which is mainly composed of accreted sedimentary material thrust upon the North American plate. Three-dimensional magnetotelluric inversions of this area indicate a large resistive anomaly beneath the Piedmont province and its adjacent coastal plain (Murphy and Egbert, 2017). Though debate continues as to the nature of this resistive anomaly, the magnetotelluric inversions compared against seismic studies suggest that this anomaly could be produced by a void initially caused

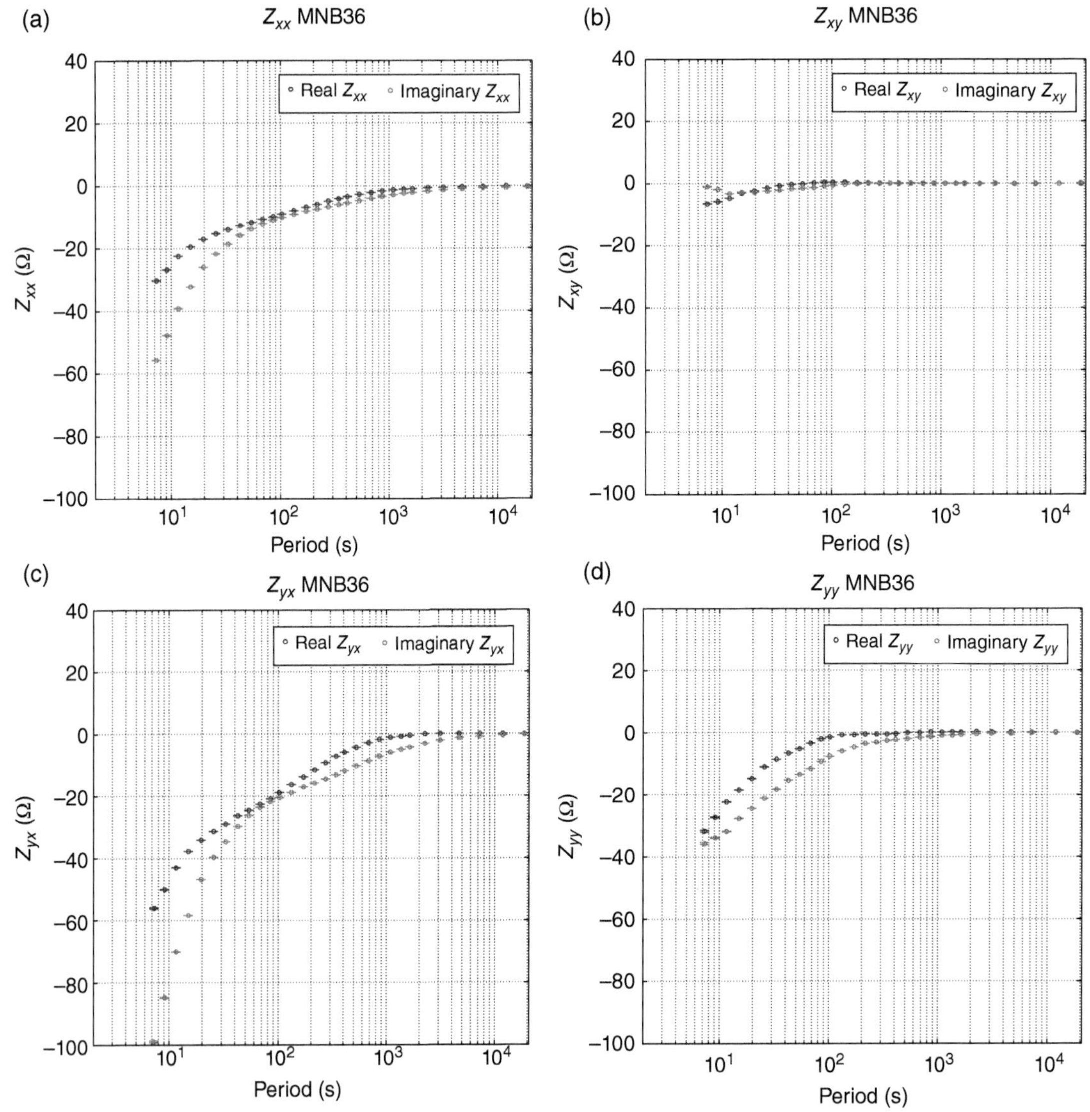

Figure 9.3 Real (left panels, upper lines; right panels, lower lines) and imaginary (left panels, lower lines; right panels, upper lines) impedance plots from the EarthScope survey site MNB36, which were used to produce estimations of the geoelectric field with the Brandon (BRD) observatory geomagnetic time series. (*See electronic version for color representation of this figure.*)

by a major orogen-parallel, post-Alleghenian delamination event, which was then filled with trapped partial melt (Whalen et al., 2015). These stations fall within the PT-1 physiographic zone for the purposes of geoelectric field estimation. Figure 9.5 shows the impedance for VAQ58, while Figure 9.6, shows the impedance for VAR57. Figure 9.5a through 9.5d correspond to the Z_{xx}, Z_{xy}, Z_{yx}, and Z_{yy} components of the impedance tensor, respectively, as do the corresponding plots for VAR57 in 9.6. Similar to RED36, I found that VAQ58 corresponds to larger estimations of $E(x, y, t)$, while VAR57 corresponds to a smaller geoelectric response. The magnitudes of each component of VAR57 are very small, while VAQ58 has much larger responses in all four components. The behavior of each component of the impedance at both sites indicates that the subsurface conductivity distribution is likely to be three-dimensional. Although we cannot say from this study whether the magnitude of variability shown between these impedances is typical, this region highlights the need to consider local enhancements in geoelectric field hazard analyses.

Similarly, I obtained discrete time series of the near-surface geomagnetic field through the Fredericksburg (FRD) magnetic observatory located in Fredericksburg, Virginia, operated by the USGS (Figure 9.7). Data from this observatory follow the same specifications as time series obtained from Brandon (BRD) magnetic observatory, and the time-series considered here covers the same

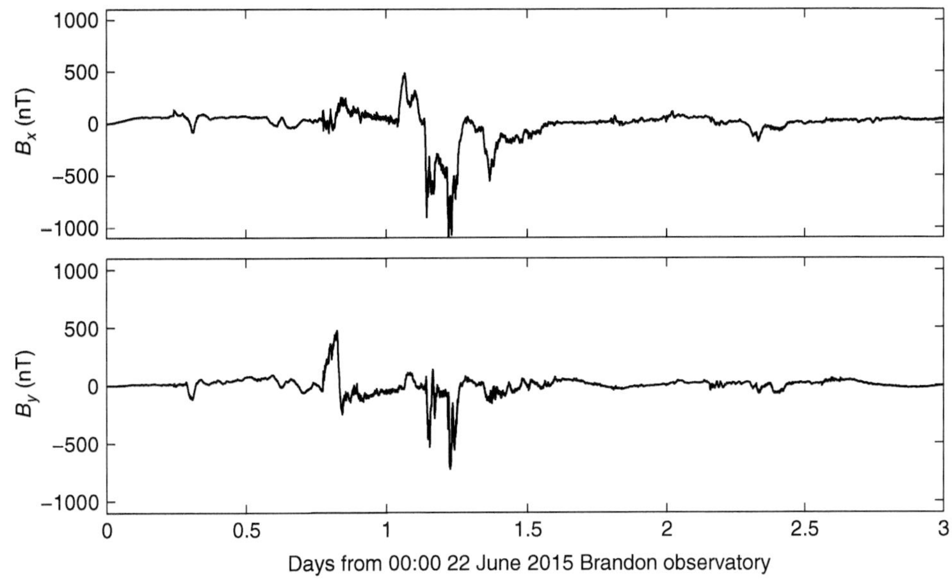

Figure 9.4 Full time series for geomagnetic fields observed at the Brandon (BRD) magnetic observatory for the 22 June 2015 storm for 3 days beginning at 00:00 22 June 2015.

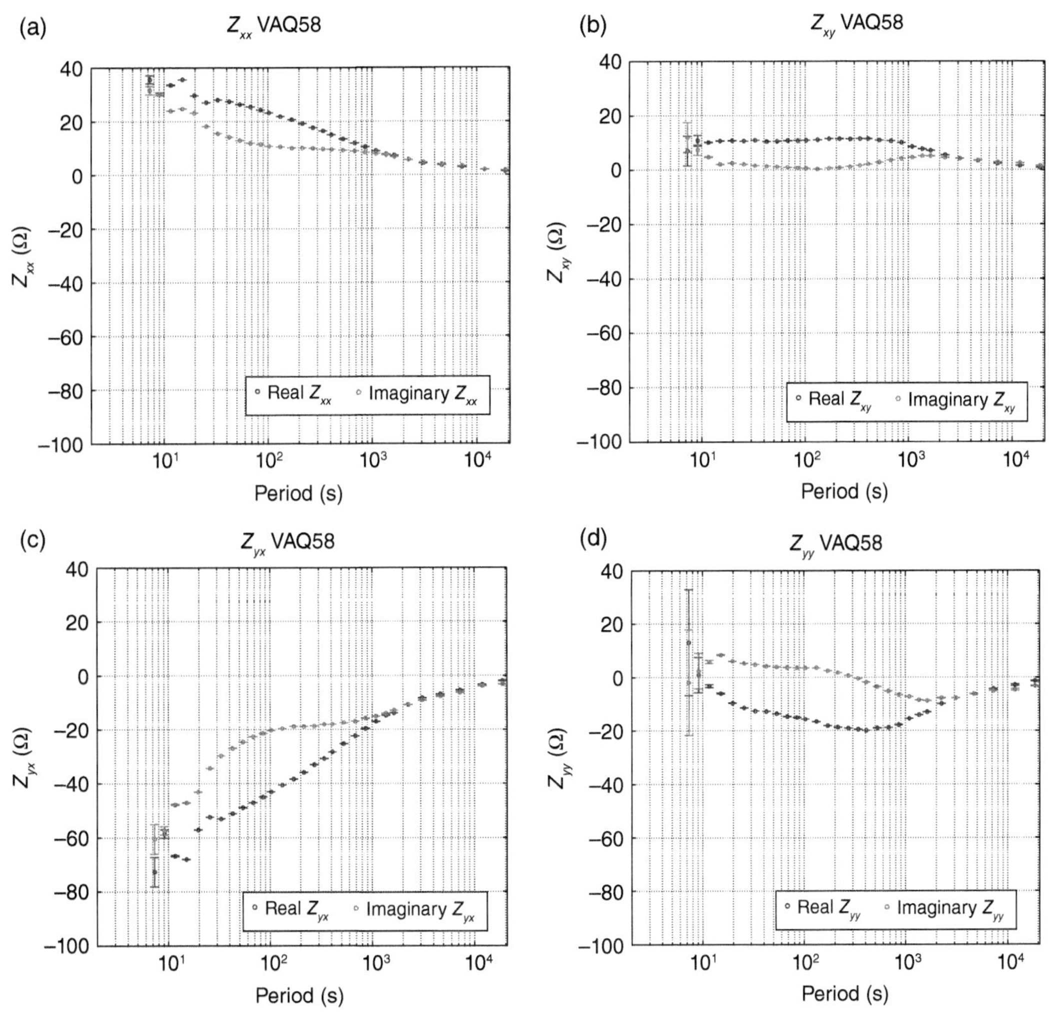

Figure 9.5 Discrete real (top panels, upper lines; bottom panels, lower lines) and imaginary (top panels, lower lines; bottom panels, upper lines) impedance plots from the EarthScope survey site VAQ58, which is associated with the Fredericksburg (FRD) observatory magnetic time series. Data quality is considered reliable between 10 and 10,000 s. (*See electronic version for color representation of this figure.*)

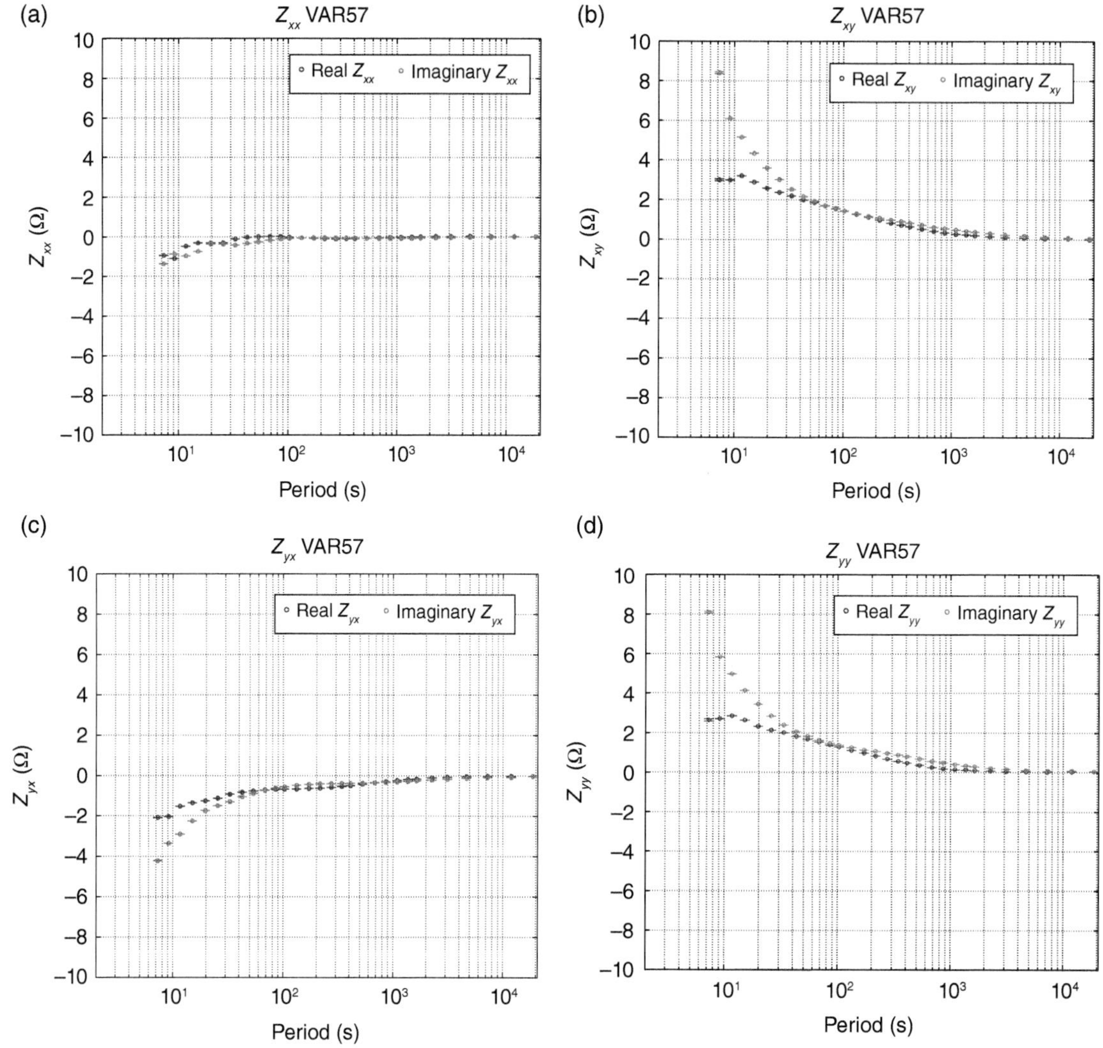

Figure 9.6 Discrete real (left panels, upper lines; right panels, lower lines) and imaginary (left panels, lower lines; right panels, upper lines) impedance plots from the EarthScope survey site VAR57, which is associated with the Fredericksburg (FRD) observatory magnetic time series. Data quality is considered reliable between 10 and 10,000 s. (*See electronic version for color representation of this figure.*)

time interval as that considered for the BRD geomagnetic time series for comparison with a common storm. Though I show the same 22 June 2015 magnetic storm over the same time interval, the response of $B_x(t)$ and $B_y(t)$ measured at FRD is smaller in magnitude, reaching 213 nT in $B_x(t)$ and 150 nT in $B_y(t)$ during the second day of the storm. As with BRD, the largest geomagnetic variations occur between 18:00 on 22 June 2015 and 12:00 on 23 June 2015. The magnitude of the geomagnetic field at the Virginia sites is reduced compared to the Minnesota sites, likely due to the difference in latitude between the two magnetic observatories: BRD is located near to the auroral oval at 49.8° latitude (50° magnetic latitude), while FRD is located much further below the auroral oval at 38.2° latitude (40° magnetic latitude).

9.4. 1-D CONDUCTIVITY MODELS AND PHYSIOGRAPHIC ZONES

Fernberg (2012) characterizes several physiographic zones within the continental United States by a one-dimensional conductivity model of the subsurface for geomagnetic hazard assessment. RED36 and MNB36 in Minnesota fall within the physiographic zone described by the 1-D conductivity model SU-1 or the Superior Uplands that characterize Minnesota, Wisconsin, and Upper Michigan, which is characterized by 10 km of low conductivity (~7000 Ωm) upper crust and increasing conductivities approaching the upper mantle (Fig. 9.8). VAR57 and VAQ58 fall within the physiographic zone described by the conductivity model PT-1 in eastern

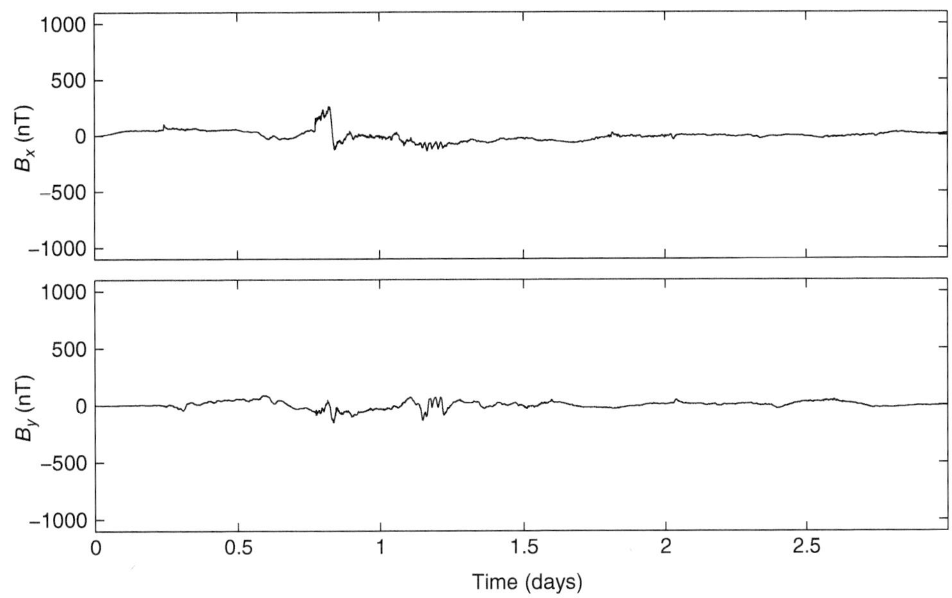

Figure 9.7 Full time series for geomagnetic fields at the Fredericksburg (FRD) magnetic observatory for the 22 June 2015 storm for 3 days beginning at 00:00 22 June 2015.

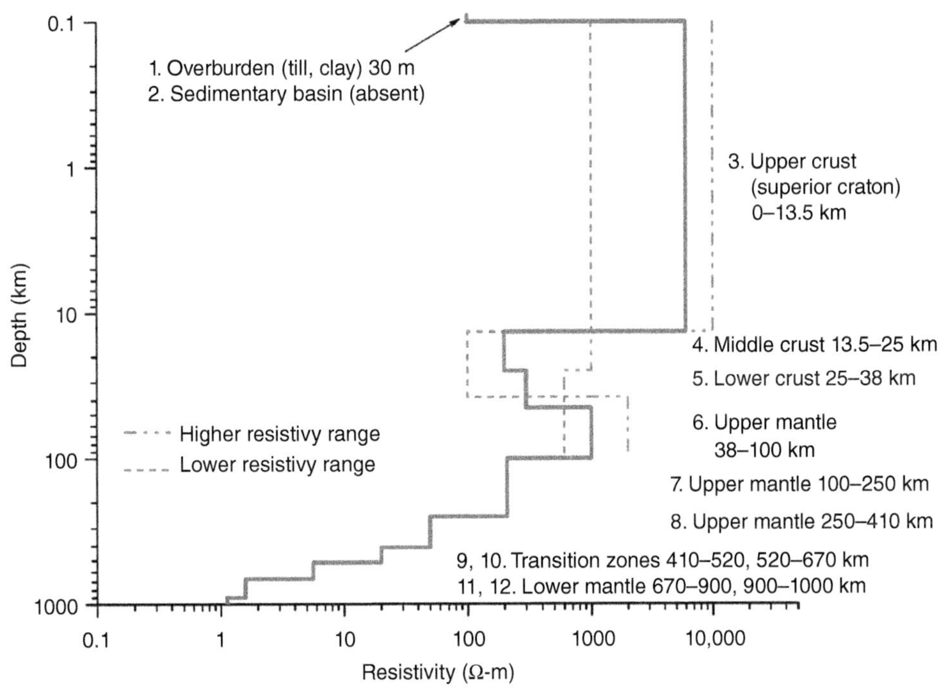

Figure 9.8 1D Resistivity model for SU-1, indicating a one-dimensional conductivity profile representative of the SU-1 physiographic zone in northeastern Minnesota as well as parts of upper Wisconsin and Michigan. Originally presented in Fernberg, (2012).

Virginia. PT-1 is characterized by comparatively higher conductivity (1000 Ωm) sedimentary basin for the first 10 km depth, and then decreasing in conductivity near to the upper mantle (8000 Ωm at 100 km depth) before increasing in conductivity at depths of 1000 km (Fig. 9.9).

These one-dimensional conductivity models encompass large physiographic zones of the United States which demonstrate considerable geologic complexity. Bedrosian and Love (2015) have demonstrated that such broad applications of these models to estimating the geoelectric field

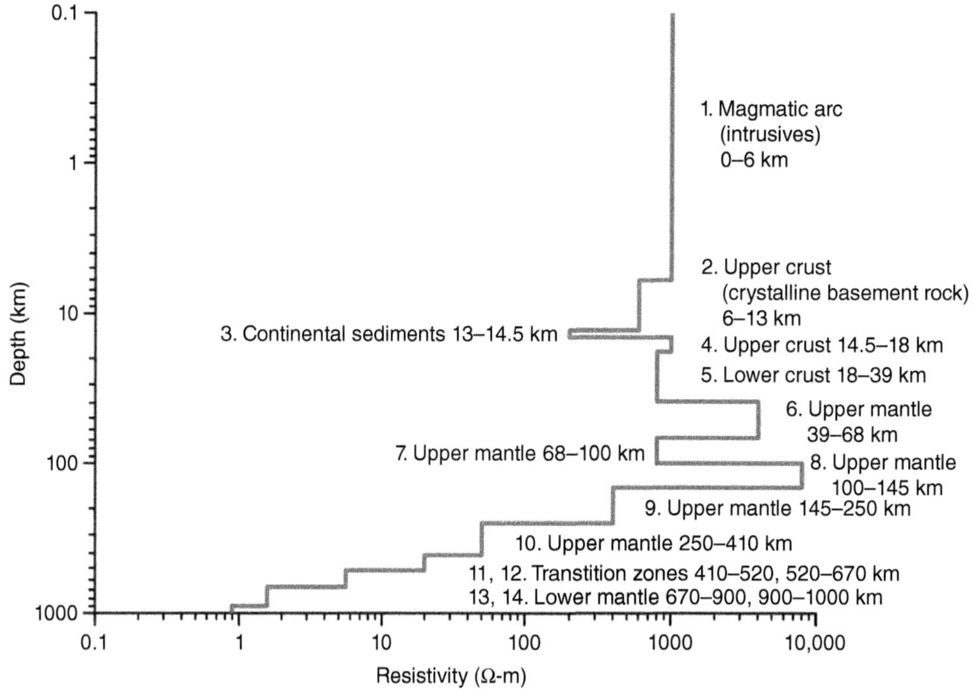

Figure 9.9 1D Resistivity model for PT-1, indicating a one-dimensional conductivity profile representative of the PT-1 physiographic zone in central Virginia. Originally presented in Fernberg, (2012).

to geologically complex zones do not remain consistent when evaluated against EarthScope impedances. These models do not consider the three-dimensional effects of geoelectric induction when estimating the geoelectric field.

9.5. METHODS

In this study, I begin with a geomagnetic time series $\mathbf{B}(x, y, t)$ and a series of impedance transfer functions $\overline{\overline{\mathbf{Z}}}(\omega, x, y \mid \sigma(\mathbf{r}))$. Because I use a measured geomagnetic field in the time domain yet derive magnetotelluric impedances in the frequency domain, I apply a discrete Fourier transform (e.g., Press et al., 1992) to the geomagnetic time series in order to compare directly the two according to Equation (9.3). In the time domain, I can represent the relationship between the geoelectric field, the geomagnetic field, and its corresponding transfer function through the expression

$$\mathbf{E}(t, x, y) = \frac{1}{\mu} \overline{\overline{\mathbf{Z}}}(t, x, y \mid \sigma(\mathbf{r})) \star \mathbf{B}(t, x, y), \qquad (9.5)$$

which is the time-domain analogue of Equation (9.3). There are several approaches for relating the geomagnetic field in the time domain $\mathbf{B}(x, y, t)$ and the impedance in the frequency domain $\overline{\overline{\mathbf{Z}}}(\omega, x, y \mid \sigma(\mathbf{r}))$ to produce an estimation of the geoelectric field in the time domain. These approaches include applying an inverse fast Fourier transform to represent the impedances in the time domain. Alternatively, one can apply a forward Fourier transform to the measured $\mathbf{B}(x, y, z, t)$ time series for a direct application of Equation (9.3), as originally presented in Bonner and Schultz (2017) and also applied by Lucas et al. (2018)) to calculate the direct-current (DC) voltages across power transmission lines during magnetic storm events. In this study, I apply the latter approach, followed by an inverse Fourier transform to estimate $\mathbf{E}(t, x, y)$.

The impedances $\overline{\overline{\mathbf{Z}}}(f, x, y \mid \sigma(\mathbf{r}))$ are defined according to Equation (9.2) and are interpolated using a cubic spline over the equivalent frequency band represented by $\mathbf{b}(f, x, y)$. When performing the Fourier Transform described in Equation (9.1) using the geomagnetic time series from the Brandon and Fredericksburg observatories, $1.72793 \cdot 10^5$ unique frequencies are generated from $2.89 \cdot 10^{-6}$ Hz to 0.5 Hz, which exceeds the frequency band of the EarthScope impedances ($5.34 \cdot 10^{-5}$ Hz to 0.1367 Hz). Because this frequency band exceeds the range of sampled frequencies represented in the impedance transfer functions, I must also extrapolate the impedances such that $\mathbf{B}(x, y, z, t)$ and $\overline{\overline{\mathbf{Z}}}(f, x, y \mid \sigma(\mathbf{r}))$ encompass the same frequency band and frequency discretization. In order to minimize the influence of extrapolated $\overline{\overline{\mathbf{Z}}}(f, x, y \mid \sigma(\mathbf{r}))$ beyond the frequency band where impedance information is present, a low-pass Butterworth filter is applied,

$$|H(f)|^2 = \frac{1}{1+\left(\frac{f}{f_c}\right)^{2N}}, \quad (9.6)$$

where N represents the order of the filter and f_c is the cutoff frequency, so that frequencies $f > f_c$ are attenuated, where $N = 3$. I chose cutoff frequencies to attenuate all frequencies larger than 10^{-1} Hz (an equivalent period T of 10 s) and all frequencies smaller than 10^{-4} Hz (an equivalent period T of 10,000 s) because electrical power grids are primarily susceptible to varying geoelectric fields between periods of 10 and 10,000 s (Barnes et al., 1991; NERC, 2014a, 2014b) and because the impedance transfer functions are limited to this frequency band. This Butterworth filter is then applied to the extrapolated impedance transfer functions according to

$$\overline{\overline{\mathbf{Z}}}_{filtered}(f,x,y\,|\,\sigma(\mathbf{r})) = \overline{\overline{\mathbf{Z}}}(f,x,y\,|\,\sigma(\mathbf{r}))\cdot H(f), \quad (9.7)$$

where $\overline{\overline{\mathbf{Z}}}_{filtered}$ is the filtered impedances. Thus, I can produce an estimation of the geoelectric field $\mathbf{E}(t, x, y)$ by combining the above methods to produce the integrated equation

$$\mathbf{E}(t,x,y) = \mathcal{F}^{-1}\left[\frac{1}{\mu}\overline{\overline{\mathbf{Z}}}_{filtered}(f,x,y\,|\,\sigma(\mathbf{r}))\cdot\widetilde{\mathbf{B}}(f,x,y)\right], \quad (9.8)$$

which produces a time-series of the geoelectric field as it evolves over the duration of a geomagnetic storm (Bonner and Schultz, 2017).

To compare the estimation of the time-series of the geoelectric field to the one-dimensional resistivity models presented in Fernberg (2012), I use the following algorithm to compute a synthetic one-dimensional impedance that can be used to produce its own estimation of the geoelectric field given the same magnetic field time series $B(t, x, y)$. I consider the resistivity distribution as a model of n layers (where n is the total number of layers present within the resistivity model) each with a uniform resistivity ρ and thickness d. In a one-dimensional earth, these layers are considered to be perfectly horizontal and are perpendicular to the incoming plane wave of $E(t, x, y)$ and $B(t, x, y)$.

The synthetic one-dimensional impedance calculated at each frequency f through a recursive relationship (Simpson and Bahr, 2005) is represented by

$$Z_i(f) = \frac{\alpha_i}{f_i}\left[\frac{\tanh(\alpha_i d_i) + \frac{\alpha_i}{f} Z_{i+1}(f)}{1 - \frac{\alpha_i}{f_i} Z_{i+1}(f)\tanh(\alpha_i d_i)}\right], \quad (9.9)$$

where Z_{i+1} is the calculated impedance of the layer immediately below, f_i is a discrete frequency, d_i is the thickness of the current layer, and α_i is represented by

$$\alpha_i = i2\pi\mu\sigma_i, \quad (9.10)$$

where σ is the conductivity of a given layer. The bottommost layer is considered to be an infinite halfspace sphere. Therefore, I represent the bottommost layer as

$$Z_n(f) = \frac{\alpha_n}{f_n}, \quad (9.11)$$

where n is the bottommost layer. Once $Z(f)$ has been calculated for the appropriate band of f, the result replaces $Z(f, x, y)$ in Equation (9.3) in order produce an estimation of the geoelectric field as a function of time.

This method is then verified against time series collected through the USArray EarthScope program, specifically survey site MNB36. To perform this assessment, I examined the measured geomagnetic and geoelectric time series obtained from the MNB36 survey site and selected a 1 day time interval that demonstrated standard geomagnetic activity. I then produced estimations of the geoelectric field using this magnetic time series and Equation (9.8). Figure 9.10 shows the result of this assessment, where the measured magnetic field is presented above the estimations of the geoelectric field. The residual difference at each time step is represented in green (light gray in print). This assessment showed good agreement between the measured and estimated geoelectric time series throughout the time interval. This yielded an RMS percent error for E_x of 17.51 mV/km and for E_y of 13.88 mV/km and maximum values of measured E_x to be 166.6 mV/km and estimated E_x to be 114.8 mV/km, while for measured E_y to be 240.8 mV/km and estimated E_y to be 247.8 mV/km for the MNB36 site.

9.6. RESULTS

Figure 9.11b and c show the geoelectric response in time associated with the same geomagnetic time series for the 22 June 2015 storm for $E_x(t)$ and $E_y(t)$, respectively. Each panel demonstrates the important effect on the induced geoelectric field of solid-Earth impedance. Despite being only 125 km apart, the geoelectric response for MNB36 is several orders of magnitude higher than the geoelectric response of RED36, particularly in E_y. In E_x, the maximum magnitude reached for MNB36 is 2275 mV/km, while RED36 only reaches a maximum magnitude of 29 mV/km. In E_y, this difference in magnitude is much larger, with the geoelectric response of MNB36 reaching a maximum magnitude of 4484 mV/km while RED36 only reaches a maximum magnitude of 41 mV/km.

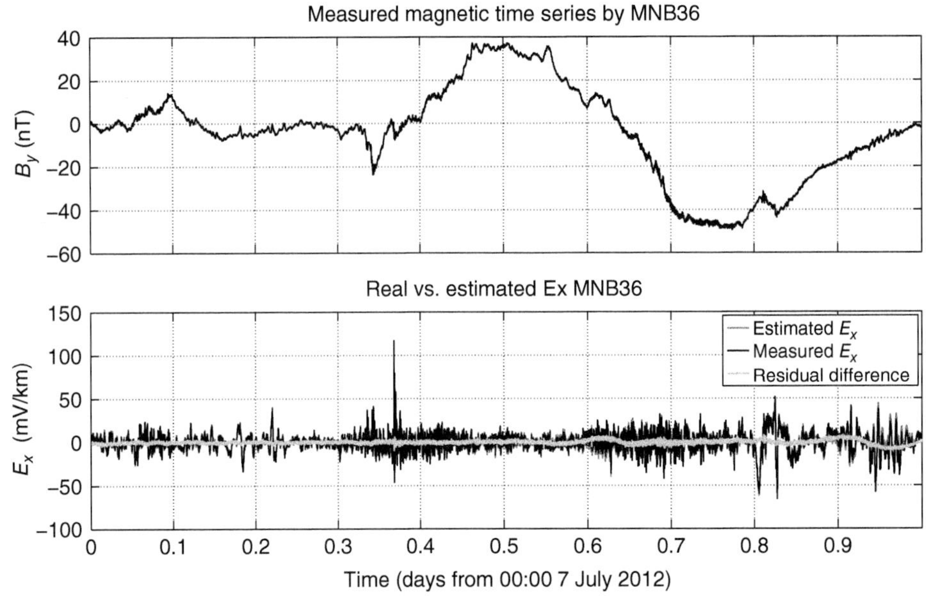

Figure 9.10 Validation geoelectric field estimations showing a comparison between the measured geoelectric field over 2 days at MNB36 and the estimation of the geoelectric field for the same time interval. (Top) The measured magnetic field from the MNB36 site over the selected time interval. (Bottom) Time series of the measured geoelectric field taken from MNB36, overlaid on estimations of the geoelectric field (red), and the residual difference between the two (thick overlaid line). (*See electronic version for color representation of this figure.*)

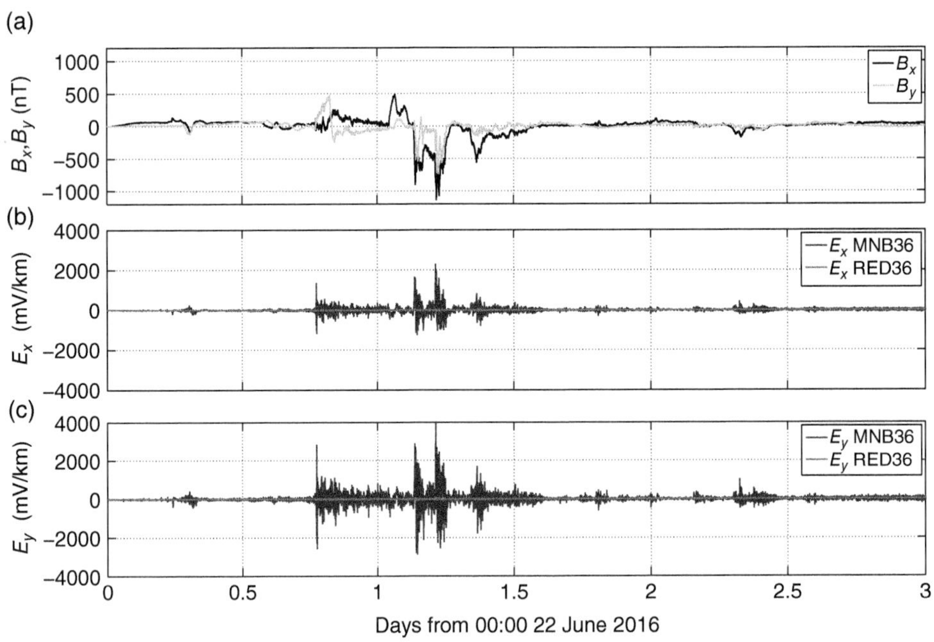

Figure 9.11 Time series for geomagnetic fields at the Brandon (BRD) magnetic observatory for the 22 June 2015 storm for 3 days beginning at 00:00 22 June 2015 and corresponding estimated geoelectric fields. (a) 1-second resolution geomagnetic field $B_x(t)$ (bottom curve) and By(t) (top curve) in nT defined relative to a constant baseline, (b) estimations of the geoelectric field Ex for EarthScope impedance MNB36 (larger variations in each plot) and RED36 (smaller variations in each plot) in mV/km, (c) equivalent time series to (b) for estimated E_y. (*See electronic version for color representation of this figure.*)

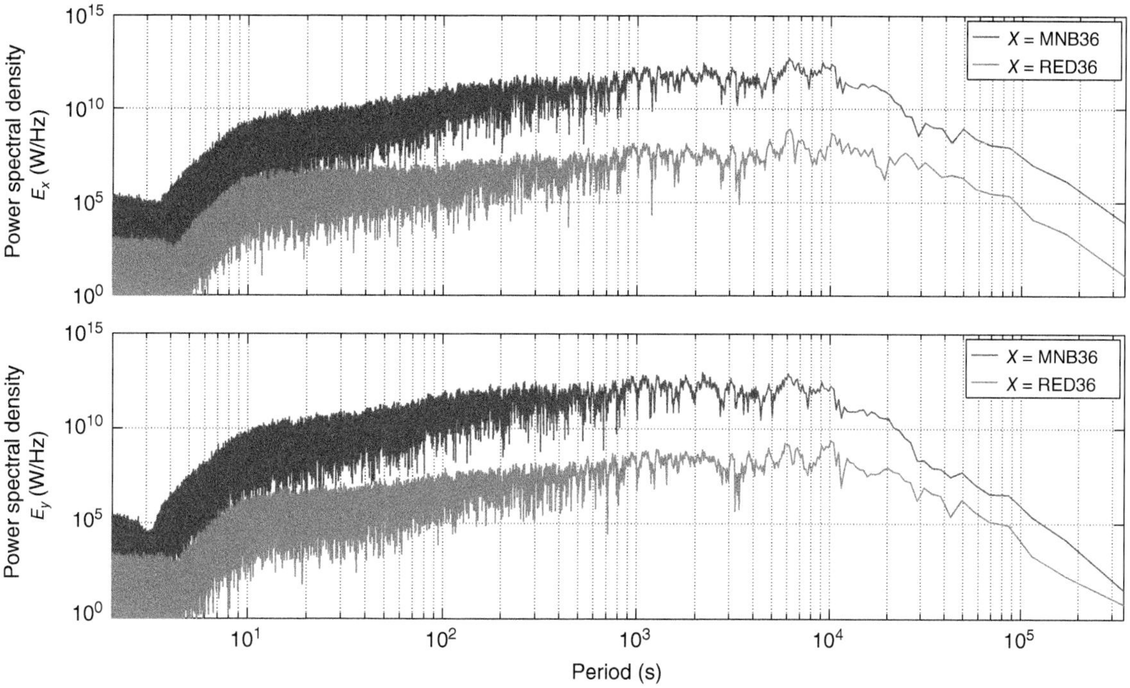

Figure 9.12 Power spectra derived from estimated geoelectric fields during the 22 June 2015 storm recorded at the Brandon (BRD) magnetic observatory for EarthScope impedances MNB36 (top curves) and RED36 (bottom curves). (*See electronic version for color representation of this figure.*)

As with the geomagnetic field, the largest geoelectric variation occurs between the onset of the storm at approximately 0.75 days into the time series and concludes shortly after 1.5 days, with the maximum magnitudes occurring within this period. Some smaller induced geoelectric fields occur in the days following, but these responses are an order of magnitude smaller than during the initial storm period, with E_y of MNB36 reaching a maximum magnitude of 400 mV/km during these periods. I note that E_y is on average twice the magnitude of E_x for both survey sites. This is related to the response of E_x being coupled to the inducing geomagnetic field B_y; a large B_y produces a correspondingly large response in E_x.

Figure 9.12 shows a comparison of the power spectra of the estimated geoelectric fields associated with RED36 and MNB36 for the x-and y-components of the geoelectric field. Because we are considering impedances with information primarily between periods of 10 and 10,000 s, the majority of the power spectral energy of the estimated geoelectric field shown in this figure is contained within this band. Responses from frequencies outside of this band are attenuated by the low-pass Butterworth filter shown in Equation (9.6), and thus are not considered significant for geomagnetic hazard assessment (Barnes et al., 1991; NERC, 2014a, 2014b). For the power spectra of the y-component of the geoelectric field, this power spectral density differs by five orders of magnitude (10^{12} W/Hz for MNB36 and 10^{7} W/Hz for RED36), particularly for longer periods (10^3 to 10^4 s). This difference is less pronounced for the x-component of the geoelectric field, with the difference only being two orders of magnitude for periods between 10^1 and 10^3 s (10^8 W/Hz for MNB36 and 10^6 W/Hz for RED36) and three orders of magnitude for periods between 10^3 and 10^4 s. We can also see that, for MNB36, the power spectral density of the x-component of the geoelectric field is smaller than the y-component by four to five orders of magnitude throughout the periods of interest, while the difference between the x-component and the y-component of the geoelectric field for RED36 is much smaller, differing on average by only one order of magnitude throughout this frequency band.

In Figure 9.13, I show similar estimations of the geoelectric fields at two survey sites in Virginia, United States, which are within the physiographic zone PT-1 according to Fernberg (2012). Specifically, Figure 9.13b and c show the geoelectric response in time associated with the 22 June 2015 storm for $E_x(t)$ and $E_y(t)$ respectively. Each plot shows the effect of two impedances: EarthScope USArray VAQ58 in blue (black in print) and VAR57 in red (gray in print). These two survey sites are farther apart than the Minnesota survey sites at 200 km. As with the Minnesota example, the geoelectric response for VAQ58 is several orders of magnitude higher than the geoelectric response of VAR57, particularly in E_y. In E_x, the maximum magnitude reached for VAQ58 is 1402 mV/km, while VAR57

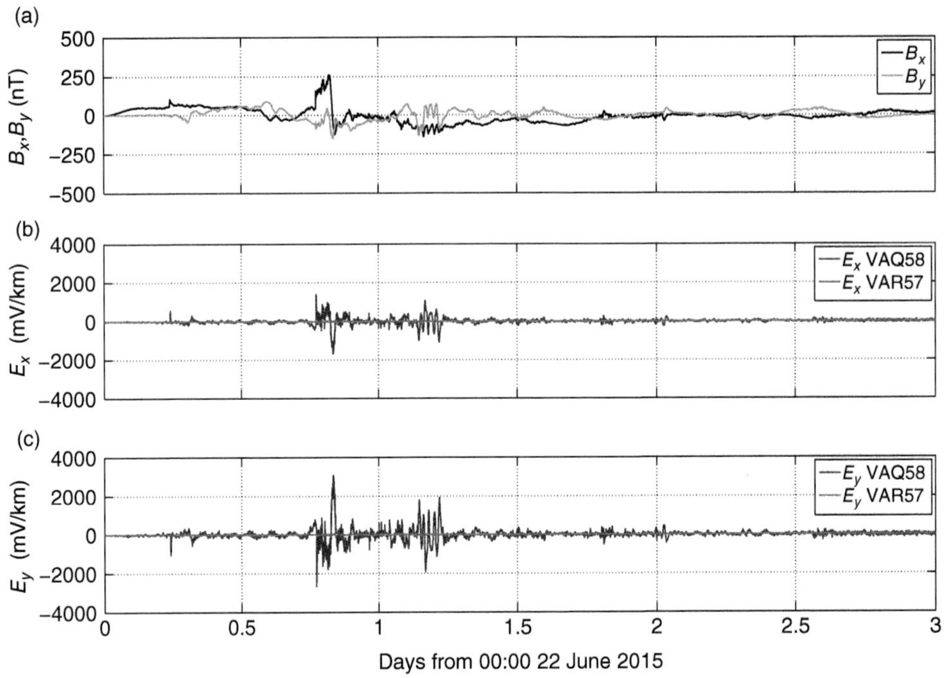

Figure 9.13 Time series for geomagnetic fields at the Fredericksburg (FRD) magnetic observatory for the 22 June 2015 storm for 3 days beginning at 00:00 22 June 2015 and corresponding estimated geoelectric fields. (a) 1-second resolution geomagnetic field $B_x(t)$ (darker curve) and $B_y(t)$ (lighter curve) in nT defined relative to a constant baseline, (b) estimations of the geoelectric field E_x for EarthScope impedance VAQ58 (larger variations in both plots) and VAR57 (smaller variations in both plots) in mV/km, (c) equivalent time series to (b) for the estimated E_y. (*See electronic version for color representation of this figure.*)

only reaches a maximum magnitude of 93 mV/km. In E_y, this difference in magnitude is much more drastic, with the geoelectric response of VAQ58 reaching a maximum magnitude of 3862 mV/km while VAR57 only reaches 148 mV/km.

Figure 9.14 shows a comparison of the power spectra of the estimated geoelectric fields associated with VAR57 and VAQ58 for the x-and y-components of the geoelectric field. As with the power spectral densities associated with the Minnesota survey sites (MNB36 and RED36), we see significant differences in the power spectral density for the geoelectric responses of VAQ58 and VAR57. The power spectra of both the x and y-component of the geoelectric field differs by five orders of magnitude between the two survey sites, particularly for longer periods (10^2 to 10^4 s). Unlike the Minnesota survey sites, however, the power spectral density between both sites differs little in magnitude between the x and y-components of the geoelectric field, differing on average one order of magnitude throughout the periods of interest (10^1 to 10^4 s). As with the power spectral densities considered in Figure 9.12, the power spectral density observed at periods less than 10^1 s and greater than 10^4 s is not considered to be relevant in this study.

I then compared my estimations of the geoelectric field to those made from the one-dimensional conductivity models representing the physiographic region SU-1. I generated a synthetic impedance through the process illustrated by Equations (9.9)–(9.11), then generated an estimation of the geoelectric field in the same manner as my previous estimations. Figure 9.15 shows the result of this comparison for the SU-1 physiographic zone, which contains MNB36 and RED36 in Minnesota, United States. The SU-1 estimation reaches a maximum magnitude of the geoelectric field of 273 mV/km, which overestimates the response from RED36, but greatly underestimates the response from MNB36 by an order of magnitude. This shows us that an approximation of the subsurface conductivity distribution to a one-dimensional conductivity model for such a large zone is insufficient to make reliable estimations of the geoelectric field for induction hazard assessment, as is consistent with the findings shown in Bonner and Schultz (2017) and Lucas et al. (2018).

9.7. EXAMINING AN ARRAY OF IMPEDANCES IN MINNESOTA

Figure 9.16 shows the locations of five survey sites in Minnesota, United States which fall within the SU-1 and IP-1 physiographic zones. As in my previous assessment, I will associate the impedances from these survey sites

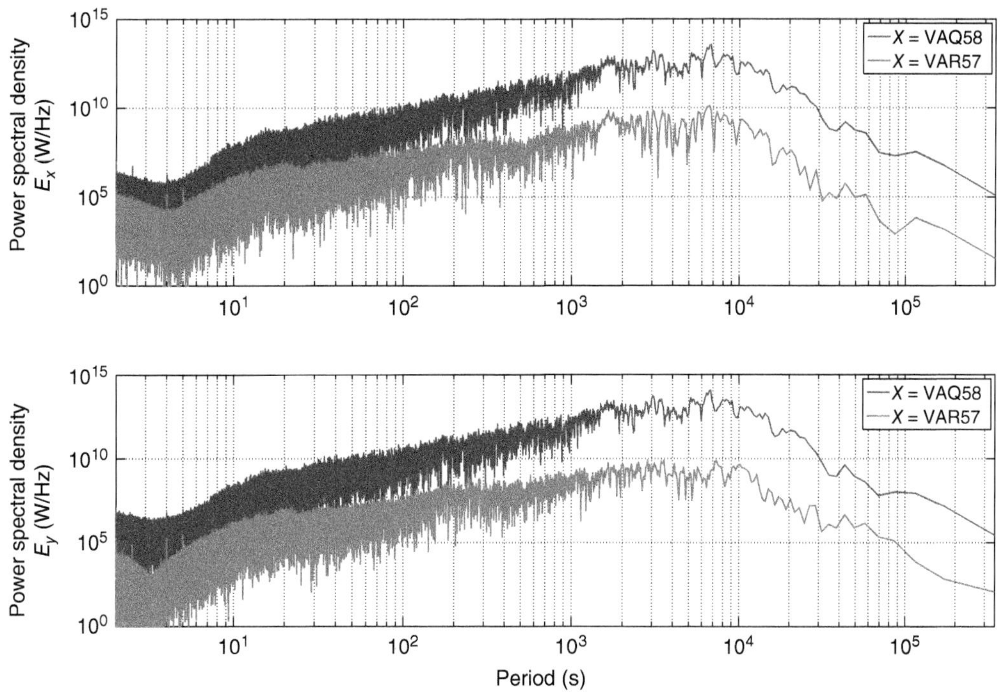

Figure 9.14 Power spectra derived from estimated geoelectric fields during the 22 June 2015 storm recorded at the Fredericksburg (FRD) magnetic observatory for EarthScope impedances VAQ58 (top curve) and VAR57 (bottom). (*See electronic version for color representation of this figure.*)

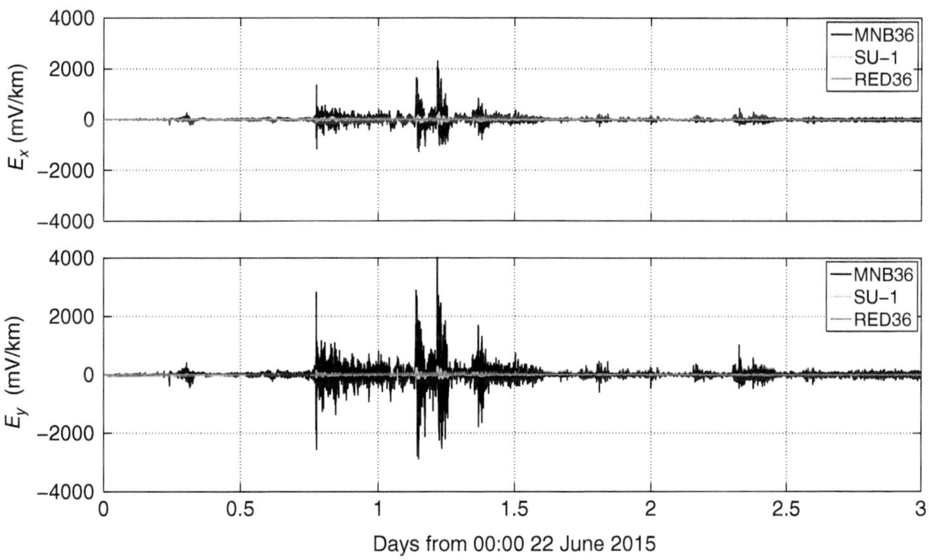

Figure 9.15 Estimations of the electric field E_x (top) and E_y (bottom) for EarthScope impedance MNB36 (larger variations) RED36 (smallest variations), from the SU-1 resistivity model (moderate variations) in mV/km for a period of 3 days calculated from the geomagnetic field time series observed at the Brandon (BRD) magnetic field observatory (for more information, see Fig. 9.4). (*See electronic version for color representation of this figure.*)

with the BRD magnetic observatory in Manitoba, Canada. This map shows the locations of the survey sites MNB36 and RED36 as shown in Figure 9.1 considered previously to determine estimations of the geoelectric field in Figure 9.11. This map also shows three additional survey sites within the state of Minnesota, United States: MNG35 (Schultz et al., 2006–2018e), approximately 100 km west-northwest of Minneapolis, Minnesota; MND33 (Schultz et al., 2006–2018f), approximately 80 km northwest of Fargo, South Dakota; and MND35

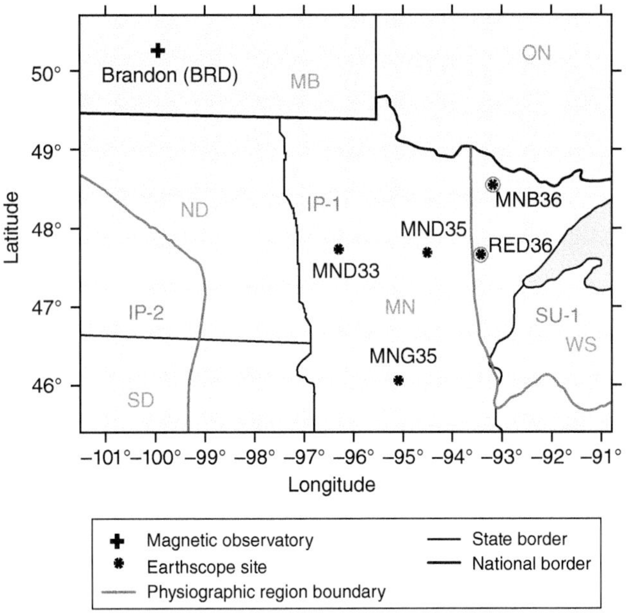

Figure 9.16 Map showing the location of NRCan Brandon (BRD) magnetic observatory in Manitoba and a selection of five EarthScope USArray survey sites throughout the state of Minnesota representative of the full spectrum of the maximum magnitude of the estimated geoelectric fields throughout the state of Minnesota and hazards zones IP-1 and SU-1. (*See electronic version for color representation of this figure.*)

(Schultz et al., 2006–2018g), approximately 30 km west of Grand Rapids, Minnesota. All three impedance stations fall within the zone IP-1 (Fernberg 2012) which is represented by its own one-dimensional conductivity model. These survey sites were chosen from the impedance stations present in the state of Minnesota in order to represent a full set of responses in the estimated geoelectric response between the two extremes presented by MNB36 and RED36 as shown in Figure 9.11.

Figure 9.17 shows the estimated y-component of the geoelectric response for each of the five impedance stations for the 22 June 2015 geomagnetic storm: MNB36 and RED36 as shown in Figure 9.11, and MND33, MNG35, and MND35. The x-component of the geomagnetic field is shown above the estimations of the y-component of the geoelectric field. I consider this component on its own because Figure 9.11 demonstrates that the y-component of the estimation of the geoelectric field had the most significant response between the components of the geoelectric field in this physiographic zone. We can see that, among these five survey sites, the estimation of the geoelectric field shows regular increments in magnitudes of the geoelectric field, demonstrating that the full range of geoelectric responses (from the maximum response of MNB36 to the minimum response of RED36) are present within the two hazard

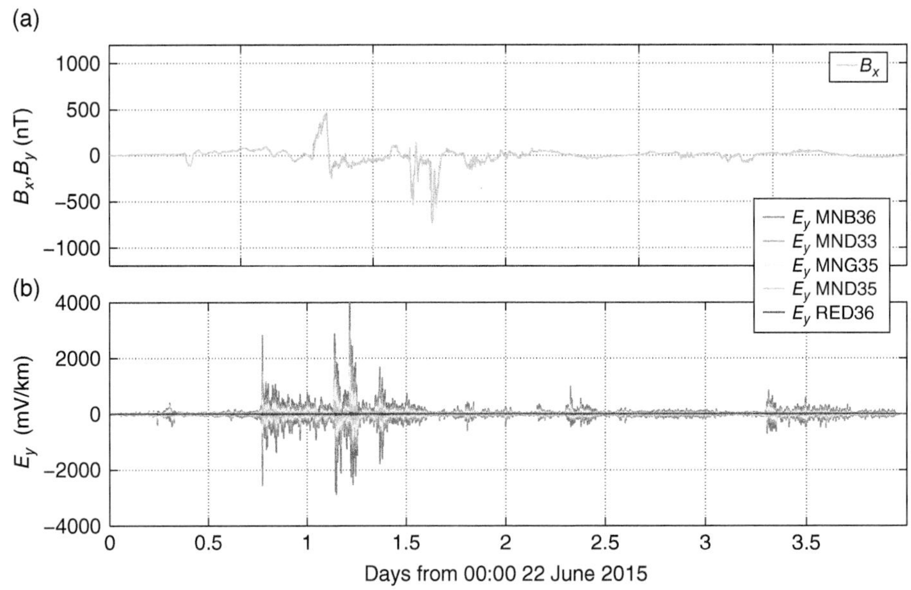

Figure 9.17 (a) Full time series for the x-component (northing) of the geomagnetic fields at the Brandon (BRD) magnetic observatory for the 22 June 2015 storm for 3 days beginning at 00:00 22 June 2015. (b) Estimated geoelectric time series for the y-component of the geoelectric field for five impedances taken from survey sites throughout the state of Minnesota, overlaid in the following order of magnitude, from largest to smallest: MNB36 (largest variations), MND33, MNG35, MND35, and RED36 (smallest variations). (*See electronic version for color representation of this figure.*)

zones that characterize the state of Minnesota. For example, the maximum magnitude of the geoelectric field occurs 1.21 days into the time series, with MNB36 reaching 4484 mV/km for the maximum extreme and RED36 reaching 41 mV/km for the minimum extreme. To compare, MND33 reaches 2962 mV/km, MNG35 reaches 1681 mV/km, and MND35 reaches 734 mV/km at the same instance. Had the geoelectric time series been estimated from the 1-dimensional models SU-1 (for MND36 and RED36) and IP-1 (for MND33, MNG35, and MND35), the same geoelectric response would have been estimated for each location within their respective zones, thus the wide differences in the magnitude of the geoelectric response between each location would not have been accounted for when assessing the geomagnetic hazards within these physiographic zone.

9.8. IMPLICATIONS FOR GEOHAZARD ASSESSMENT

My results show that geoelectric fields within the same physiographic zone can be several orders of magnitude different even during the same magnetic storm. This suggests that, in general, there may be significant local differences between the geo-electric fields produced by 1D models of large physiographic zones and those estimated from full 3D impedance tensors during magnetic storms. In both zones considered here (Minnesota and Virginia), I estimated the maximum geoelectric field at two survey sites to be two orders of magnitude different despite being located in the same physiographic zones and proximal to one another. While Bedrosian and Love (2015) used a simple sinusoidal source, I have shown a similar result derived from observatory geomagnetic fields. Therefore, by considering the measured impedance at each site rather than applying a one-dimensional conductivity model to a wide zone, I can improve estimates of the behavior and amplitude of the geoelectric field and the corresponding hazard to electric power grids.

The assessment demonstrates that physiographic zones with complex, three-dimensional distributions in the subsurface are associated with differing responses in the estimated geoelectric time series between different locations, but other variables are present that may also have significant influence. Some of these variables include the latitude of the site with respect to the auroral oval and the development of the magnetic storm itself with time. Specifically, future study will need to examine the geoelectric response associated with sharp lateral conductivity gradients, such as those at the ocean–land interface (Torta et al., 2017). It has also been considered that, when considering three-dimensional conductivity models, the geomagnetic and geoelectric fields need not be orthogonal to one another, as such parameters as the magnitude of the geoelectric field and the direction of the geoelectric field vector relative to the power transmission line has a significant effect on the estimated GIC. This includes causing phase differences and delays between peaks in the geomagnetic field and GICs which are not observable when using one-dimensional conductivity models. Such factors will require further study to determine how each specifically influences estimates of the behavior of the geoelectric field at a given location.

ACKNOWLEDGMENTS

I gratefully acknowledge the support provided by the National Science Foundation (NSF) under Award Number 15-20864. I thank A. Swidinsky for reading the manuscript and his useful input and conversations. I thank J.J. Love, C. A. Finn, G. M. Lucas A. Kelbert for useful conversations. I also thank two anonymous reviewers for thier helpful comments, critiques, and suggestions that have served to greatly improve this chapter. The results presented in this paper rely on data collected at magnetic observatories. I thank the national institutes that support them and INTERMAGNET for promoting high standards of five magnetic observatory practice (www.intermagnet.org). USArray MT TA project was led by PI Adam Schultz and Gary Egbert. They would like to thank the Oregon State University MT team and their contractors, lab and field personnel over the years for assistance with data collection, quality control, processing, and archiving. I also thank numerous districts of the U.S. Forest Service, Bureau of Land Management, the U.S. National Parks, the collected State land offices, and the many private landowners who permitted access to acquire the MT TA data. USArray TA was funded through NSF grants EAR-0323311, IRIS Subaward 478 and 489 under NSF Cooperative Agreement EAR-0350030 and EAR-0323309, IRIS Subaward 75-MT under NSF Cooperative Agreement EAR-0733069 under CFDA No. 47.050, and IRIS Subaward 05-OSU-SAGE under NSF Cooperative Agreement EAR-1261681.

REFERENCES

Alekseev, D., A. Kuvshinov, and N. Palshin (2015), Compilation of 3-d global conductivity model of the Earth for space weather applications, *Earth, Planets, and Space*, *67* 1–11, doi:10.1186/s40623-015-0272-5.

Allen, J., L. Frank, H. Sauer, and P. Reiff (1989), Effects of the March 1989 solar activity, *Eos, Transactions, American Geophysical Union*, *70*(46)(1479), 1486–1488.

Anderson, C. W., L. J. Lanzerotti, and G. MacLennan (1974), Outage of the L4 system and the geomagnetic disturbances of 4 August 1972, *Bell System Technical Journal*, *53*(9), 1817–1837.

AonBenfield (2013), *Geomagnetic Storms*, AonBenfield, Sydney, Australia, pp. 1–12.

Bally, A. W., and A. R. Palmer (1989), The geology of North America: An overview, *Geological Society of America*, 1–629, doi:10.1130/DNAG-GNA-A.

Barnes, P. R., D. T. Rizy, B. W. McConnell, F. M. Tesche, and E. R. Taylor Jr. (1991), Outage of the L4 system and the geomagnetic disturbances of 4 August 1972, *Electric Utility Industry Experience With Geomagnetic Disturbances, vol. ORNL6665*, pp. 1–78, Oak Ridge National Laboratory.

Bedrosian, P. A., and D. W. Feucht (2014), Structure and tectonics of the northwestern United States from EarthScope USArray magnetotelluric data, *Earth and Planetary Science Letters*, *402*, 275–289, doi:10.1016/j.epsl.2013.07.035.

Bedrosian, P. A., and J. J. Love (2015), Mapping geoelectric fields during magnetic storms: Synthetic analysis of empirical United States impedances, *Geophysical Research Letters*, *42*(23), pp. 10160–10170, doi:10. 1002/2015GL066636, 2015GL066636.

Béland, J., and K. Small (2005), Space weather effects on power transmission systems: The cases of Hydro-Québec and Transpower New Zealand Ltd, in, edited by I. A. Daglis, *Effects of Space Weather on Technology Infrastructure*, pp. 287–299, Springer, Dordrecht, Netherlands.

Bonner, L. R., and A. Schultz (2017), Rapid prediction of electric fields associated with geomagnetically induced currents in the presence of three-dimensional ground structure: Projection of remote magnetic observatory data through magnetotelluric impedance tensors, *Space Weather*, *15*(1), 204–227, doi:10.1002/2016SW001535.

Boteler, D. H. (2001), Assessment of geomagnetic hazard to power systems in Canada, *Natural Hazards*, *23*, 101–120.

Bowles-Martinez, E., and A. Schultz (2015), Midcontinent rift and remnants of initiating mantle plume imaged with magnetotellurics, *Abstract T11D-2922 presented at 2015 Fall Meeting AGU, San Francisco, California* (14–18 December).

Cagniard, L. (1953), Basic theory of the magneto-telluric method of geophysical prospecting, *Geophysics*, *18*(3), pp. 605–635.

Chave, A. D. (2012), Estimation of the magnetotelluric response function, in, edited by A. D. Chave and A. G. Jones, *The Magnetotelluric Method*, pp. 165–218, Cambridge University Press, Cambridge, UK.

Egbert, G. D. (2007), Robust electromagnetic transfer functions estimates, in, edited by D. Gubbins and E. Herrero-Bervera, *Encyclopedia of Geomagnetism and Paleomagnetism*, pp. 866–870, Springer, Dordrecht, Netherlands.

FERC (2013), Reliability standards for geomagnetic disturbances, *Federal Energy Regulatory Commission Federal Regulatory Rules Regulations*, *78*(100), 30747 – 30762.

Fernberg, P. (2012), One-dimensional earth resistivity models for selected areas of continental United States and Alaska, *EPRI Technical Update 1026430*, pp. 1–190.

Kelbert, A., S. Erofeeva, C. Trabant, R. Karstens, M. V. Fossen, G. Egbert, and A. Schultz (2011), Iris dmc data services products: Emtf, the magnetotelluric transfer functions, *U.S. Geological Survey*, doi:10.17611/DP/EMTF.1.

Lloyds (2013), Emerging risk report: Solar storm risk to the North American electric grid, *Lloyds of London*, pp. 1–22.

Love, J., A. Pulkkinen, P. A. Bedrosian, S. Jonas, A. Kelbert, E. J. Rigler, C. A. Finn, C. C. Balch, R. Rutledge, R. M. Waggle, A. T. Sabata, J. U. Kozyra, and C. E. Black (2016a), Geoelectric hazard maps for the continental United States, *Geophysical Research Letters*, *43*, 1–10, doi:10.1002/2016 GL070469.

Love, J. J., and A. Chulliat (2013), An international network of magnetic observatories, *Eos, Transactions, American Geophysical Union*, *94*(42), 373–384, doi:10.1002/2013EO420001.

Love, J. J., P. Coisson, and A. Pulkkinen (2016b), Global statistical maps of extreme-event magnetic observatory 1 min first differences in horizontal intensity, *Geophysical Research Letters*, *43*, 4126–4135, doi:10.1002/2016GL068664.

Lucas, G. M., J. Love, and A. Kelbert (2018), Calculation of voltages in electric power transmission lines during historic geomagnetic storms: An investigation using realistic earth impedances, *Space Weather*, *16*(2), 185–195, doi:10.1002/2017SW001779.

Meqbel, N. M., G. D. Egbert, P. E. Wannamaker, A. Kelbert, and A. Schultz (2014), Deep electrical resistivity structure of the northwestern U.S. derived from 3-D inversion of USArray magnetotelluric data, *Earth and Planetary Science Letters*, *402*, 290–304, doi:10.1016/j.epsl.2013.12.026.

Molinski, T. S. (2002), Why utilities respect geomagnetically induced currents, *Journal of Atmospheric and Solar-Terrestrial Physics*, *64*, 1765–1778.

Murphy, B. S., and G. D. Egbert (2017), Electrical conductivity structure of southeastern north america: Implications for lithospheric architecture and appalachian topographic rejuvenation, *Earth and Planetary Science Letters*, *462*, 6675, doi:10.1016/j.epsl.2017.01.009.

NERC (2014a), *Benchmark Geomagnetic Disturbance Event Description*, North American Electric Reliability Corporation (NERC), Atlanta, GA, pp. 1–26.

NERC (2014b), Transformer thermal impact assessment: Project 2013-03 (geomagnetic disturbance mitigation), North American Electric Reliability Corporation (NERC), Atlanta, GA, pp. 1–16.

NERC (2017), *Geomagnetic Disturbance Research Work Plan of the North American Electric Reliability Corporation*, North American Electric Reliability Corporation (NERC), Atlanta, GA, pp. 1–28.

NSTC (2015), *National Space Weather Action Plan*, National Science and Technology Council (U.S.) United States. Executive Office of the President, pp. 1–12.

Pirjola, R. (2002), Review on the calculation of surface electric and magnetic fields and of geomagnetically induced currents in ground-based technological systems, *Surveys in Geophysics*, *23*, 71–90.

Press, W., S. Teukolsky, W. Vetterling, and B. Flannery (1992), *Numerical Recipes in Fortran 77 The Art of Scientific Computing*, Second Edition, Cambridge University Press, New York City, NY, USA, pp. 490–525.

Riswadkar, A. V., and B. Dobbins (2010), Solar storms: Protecting your operations against the suns dark side, *Zurich Services Corporation*, *1–12*, doi:10.1029/2004SW000123.

Schrijver, C., K. Kauristie, A. Aylward, C. Denardini, S. Gibson, A. Glover, N. Gopalswamy, M. Grande, M. Hapgood,

D. Heynderick, N. Jokowski, V. Kalegaev, G. Lapenta, J. Linker, S. Liu, C. Mandrini, I. Mann, T. Nagatsuma, D. Nandy, T. Obara, T. O'Brien, T. Onsager, H. Opgenoorth, M. Terkildsen, C. Valladares, and N. Vilmer (2015), Understanding space weather to shield society: A global road map for 2015 to 2025 commissioned by COSPAR and ILWS, *Advances in Space Research*, *55*(12), 2745–2807, doi:10.1016/j.asr.2015.03.023.

Schultz, A. (2010), A continental scale magnetotelluric observatory and data discovery resource, *Data Science Journal*, *8*, IGY6IGY20.

Schultz, A., G. Egbert, A. Kelbert, T. Peery, V. Clote, B. Fry, S. Erofeeva, staff of the National Geoelectromagnetic Facility, and their contractors (2006–2018a), Usarray ta magnetotelluric transfer functions: Red36, doi:10.17611/DP/14954909.

Schultz, A., G. Egbert, A. Kelbert, T. Peery, V. Clote, B. Fry, S. Erofeeva, staff of the National Geoelectromagnetic Facility, and their contractors (2006–2018b), Usarray ta magnetotelluric transfer functions: Mnb36, doi:10.17611/DP/14948263.

Schultz, A., G. Egbert, A. Kelbert, T. Peery, V. Clote, B. Fry, S. Erofeeva, staff of the National Geoelectromagnetic Facility, and their contractors (2006–2018c), Usarray ta magnetotelluric transfer functions: Var57, doi:10.17611/DP/15015681.

Schultz, A., G. Egbert, A. Kelbert, T. Peery, V. Clote, B. Fry, S. Erofeeva, staff of the National Geoelectromagnetic Facility, and their contractors (2006–2018d), Usarray ta magnetotelluric transfer functions: Var58, doi:10.17611/DP/15014571.

Schultz, A., G. Egbert, A. Kelbert, T. Peery, V. Clote, B. Fry, S. Erofeeva, staff of the National Geoelectromagnetic Facility, and their contractors (2006–2018e), Usarray ta magnetotelluric transfer functions: Mng35, doi:10.17611/DP/14921249.

Schultz, A., G. Egbert, A. Kelbert, T. Peery, V. Clote, B. Fry, S. Erofeeva, staff of the National Geoelectromagnetic Facility, and their contractors (2006–2018f), Usarray ta magnetotelluric transfer functions: Mnd33, doi:10.17611/DP/14950039.

Schultz, A., G. Egbert, A. Kelbert, T. Peery, V. Clote, B. Fry, S. Erofeeva, staff of the National Geoelectromagnetic Facility, and their contractors (2006–2018g), Usarray ta magnetotelluric transfer functions: Mnd35, doi:10.17611/DP/14919251.

Schultz, A., P. Bedrosian, K. Key, D. Livelybrooks, G. D. Egbert, E. Bowles-Martinez, and P. E. Wannamaker (2014), Magnetotelluric investigations of convergent margins and of incipient rifting: Preliminary results from the earthscope mt transportable array and mt flexarray deployments in cascadia and in the north american mid-continent region, *Abstract GP31A-3676 presented at 2014 Fall Meeting, AGU, San Francisco, California* (15–19 December).

Simpson, F., and K. Bahr (2005), *Practical Magnetotellurics*, Cambridge University Press, Cambridge, UK, pp. 1–254.

Spies, B., and F. Frischknecht (1991), Electromagnetic Sounding. In *Electromagnetic Methods in Applied Geophysics*, eISBN: 978-1-56080-268-6, print ISBN: 978-1-56080-022-4, doi.org/10.1190/1.9781560802686.ch5, pp. 285–425.

Swidinsky, A., S. Hölz, and M. Jegen (2015), Rapid resistivity imaging for marine controlled-source electromagnetic surveys with two transmitter polarizations: An application to the North Alex mud volcano, West Nile Delta, *Geophysics*, *80*(2), E97–E110, doi.org/10.1190/geo2014-0015.1.

Torta, J. M., S. Marsal, and M. Quintana (2014), Assessing the hazard from geomagnetically induced currents to the entire high-voltage power network in Spain, *Earth, Planets, and Space*, *66*, 87, doi:10.1186/1880-5981-66-87.

Torta, J. M., A. Marcuello, J. Campany, S. Marsal, P. Queralt, and J. Ledo (2017), Improving the modeling of geomagnetically induced currents in Spain, *Space Weather*, *15*, 691–703, doi:10.1002/2017SW001628.

Veeramany, A., S. D. Unwin, G. A. Coles, J. E. Dagle, D. W. Millard, J. Yao, C. S. Glantz, and S. N. G. Gourisetti (2016), Framework for modeling high-impact, low-frequency power grid events to support risk-informed decisions, *International Journal of Disaster Risk Reduction*, *18*, 125–137, doi:10.1016/j.ijdrr.2016.06.008.

Viljanen, A., R. Pirjola, E. Prcser, S. Ahmadzai, and V. Singh (2013), Geomagnetically induced currents in Europe: Characteristics based on a local power grid model, *Space Weather*, *11*, 575–584, doi:10.1002/swe.20098.

Weidelt, P., and A. D. Chave (2012), The magnetotelluric response function, in *The Magnetotelluric Method*, pp. 122–164, Cambridge University Press, Cambridge, UK.

Whalen, L., E. Gazel, C. Vidito, J. Puffer, M. Bizimis, W. Henika, and M. Caddick (2015), Supercontinental inheritance and its influence on supercontinental breakup: The central atlantic magmatic province and the breakup of pangea, *Geochemistry, Geophysics, Geosystems*, *16*(10), 3532–3554, doi:10.1002/2015GC005885.

Williams, M. L., K. Fischer, J. Freymueller, B. Tikoff, A. Trehu, et al. (2010), Unlocking the secrets of the North American continent: An EarthScope science plan for 2010 to 2020, *EarthScope*, 1–78, doi:10.1002/swe.20073.

Part III
Power System Impacts

10

An Overview of Modeling Geomagnetic Disturbances in Power Systems

Komal S. Shetye and Thomas J. Overbye

ABSTRACT

Research on geomagnetic disturbances (GMDs) and geomagnetically induced currents (GICs) in the power grid has increased in prominence recently, although their detrimental effects on power systems have been known since several decades. It is a highly inter-disciplinary problem, which involves monitoring solar weather and the earth's magnetic field, modeling deep earth resistivity, and finally evaluating the impacts on the power grid in order to mitigate them. Hence, it often requires knowledge of domain(s) that may be outside one's area of expertise. This chapter describes the basics of modeling GMDs in power systems, not assuming that the reader is a power engineer. It contains background on common power system components and terminology. It also shows a basic GMD simulation using a uniform electric field, walking the reader through examples which they can access and solve alongside. It talks about the DC GICs, the AC power grid, their interaction, and how that leads to two solution processes; one uses just the DC model with resistances, while the voltage stability study uses the full AC model of the grid. Wide-area visualizations of the results are shown. The chapter concludes with a brief discussion on the more realistic time-varying electric fields, and remarks on model validation.

Key Points

- The basic concepts regarding the modeling of geomagnetic disturbances in power systems are discussed.
- Transformers and substations are the most important elements from a data collection perspective; they also act as gateways for geomagnetically induced currents into the grid.
- There are two main aspects to power system geomagnetic disturbance analysis: the DC component which just calculates the transformer DC currents, and the AC portion which calculates reactive power losses and bus AC voltages for voltage stability assessment.

Department of Electrical and Computer Engineering, Texas A&M University, College Station, TX, USA

10.1. INTRODUCTION

The effects of geomagnetic disturbances (GMDs) on ground-based electrical systems have been known for over 170 years (Boteler et al., 1998a), with the first impacts observed on telegraph systems in the 1840s. The power grid came into existence afterwards, in the 1880s. GMD-induced blackouts and other impacts were experienced later in the twentieth century. The systems that GMDs affect thus include those consisting of electrical conductors, such as power transmission lines, telegraph lines, and even pipelines and railway tracks whose main purpose may not (always) be to transmit signals.

While research on GIC impacts on the power grid has been happening for several decades (Albertson et al., 1974; Aspnes et al., 1981; Balma, 1992; Bozoki et al., 1996; Kappenman, 1996), the issue gained prominence more recently in the early 2010s following a joint report by the North American Electric Reliability Corporation (NERC) and the Department of

Geomagnetically Induced Currents from the Sun to the Power Grid, Geophysical Monograph 244, First Edition.
Edited by Jennifer L. Gannon, Andrei Swidinsky, and Zhonghua Xu.

Energy (DOE) on the "High Impact, Low Frequency (HILF) Event Risk to the North American Bulk Power System" (NERC and DOE, 2010). Electromagnetic events manifesting naturally such as geomagnetic disturbances or man-made attacks such as electromagnetic pulses (EMPs), were included as one of the three types of HILF events that pose a major threat to power systems.

This was followed by a special NERC assessment focusing on just the impacts of GMDs (NERC, 2012). This report determined that voltage collapse was more likely to be the cause of a blackout rather than transformers damage due to overheating. In fact, voltage collapse caused by tripping of reactive power support devices known as static var compensators (SVCs) was the cause of the March 1989 in Hydro-Quebec blackout (NERC, 1990). Following the 2012 report, the Federal Energy Regulatory Commission (FERC) issued orders to NERC to develop reliability standards for planning and operating the grid under a GMD (FERC, 2013, 2016). NERC is a regulatory body tasked with ensuring the reliability of the power grid through the enforcement of standards. These standards can be related to planning (i.e., offline studies), operating, modeling, among other topics. According to the planning standards developed (TPL-007-1 and TPL-007-2), utilities are required to collect data and maintain system models required to perform GMD assessments, for which benchmark events in terms of electric fields have been defined (NERC, 2014, 2017).

Accordingly, in this chapter we introduce the basic modeling of the GMD phenomenon into the power system analysis context. This is a fundamental step needed for further analyses such as (i) model validation (e.g., ground conductivity, models of grid components such as transformers, etc.), (ii) developing mitigation measures, and (iii) modeling enhancements/additions to name a few. That being said, one has to keep in mind the multi-disciplinary nature of the GMD problem. Engineers and scientists from different fields have to rely on each other for expertise and state-of-the-art models, data, and tools to predict, simulate, and assess impacts. Space scientists, geophysicists and power engineers must work together, and the some of the key challenges lie at the interfaces between them, that is, starting from the sun to the earth's magnetosphere and surface, to the deep earth and finally to the grid. It is becoming increasingly important to obtain cross-domain knowledge, in order to better understand and model these interactions. This could involve starting from the basics of a topic, in some cases.

The GMD-enhanced power system modeling described in this chapter is geared toward a general scientific audience, who may not necessarily be from a power engineering background, and, because of this, it may elaborate on some basic power system concepts for completeness. These include an overview of the power grid, components important for GIC modeling, and understanding the power system GIC model and calculations by walking through simulation steps and solutions of test systems. The examples are modeled and solved using PowerWorld Simulator, a commercial software package which comes with a GIC Calculation Module. This off-the shelf tool was chosen because (i) of the availability to the authors, and (ii) the similar availability to readers of the demo version, which can be used to reproduce the examples described in this chapter.[1] The basic solution concepts of this chapter could also be followed using other tools as available to the reader. The point of this chapter is not to highlight the capabilities of any one tool or package, but to explain the results through studies and visualizations.

It is organized into the following main sections. Section 10.2 presents an overview of the power grid and its major components, while Section 10.3 goes deeper into discussing transformer characteristics such as winding configurations, types such as autotransformers and generator step-up transformers, and core types, all of which play a key role in modeling GICs. In Section 10.4, we begin with the modeling of the electric field, and how GIC flow through the system is calculated. The rest of this section goes through simulation examples in a GIC simulation tool to explain the basic inputs, outputs, options, and settings involved in a GIC study. An important power system concept of "per unit" is also introduced in this section. In Section 10.5, GIC-induced reactive power losses occurring in transformers are modeled into the GIC power flow problem using the popularly-known "K factors" to assess the voltage stability of the system. This is also supported with an example, with visualization of voltages using contours. Finally, the chapter concludes with Section 10.6 introducing the modeling of time series electric fields with validation results, followed by a summary in Section 10.7.

10.2. MAJOR POWER SYSTEM COMPONENTS

10.2.1. Overview of the Power Grid

The delivery of electricity to consumers takes places through a very complex machine, which we know as the power grid. Figure 10.1 shows the basic structure of the grid (DOE, 2015). It is a 50 or 60 Hz alternating current (AC), three-phase power system, hence three "wires" or conductors are shown at each stage in the grid. Through the rest of the chapter, we assume that the three phases are balanced, that is, all three phases have the same voltage magnitude, and there is a phase shift of 120° between the voltage

[1] The test system sizes are small enough to be solvable in the demo version of the software.

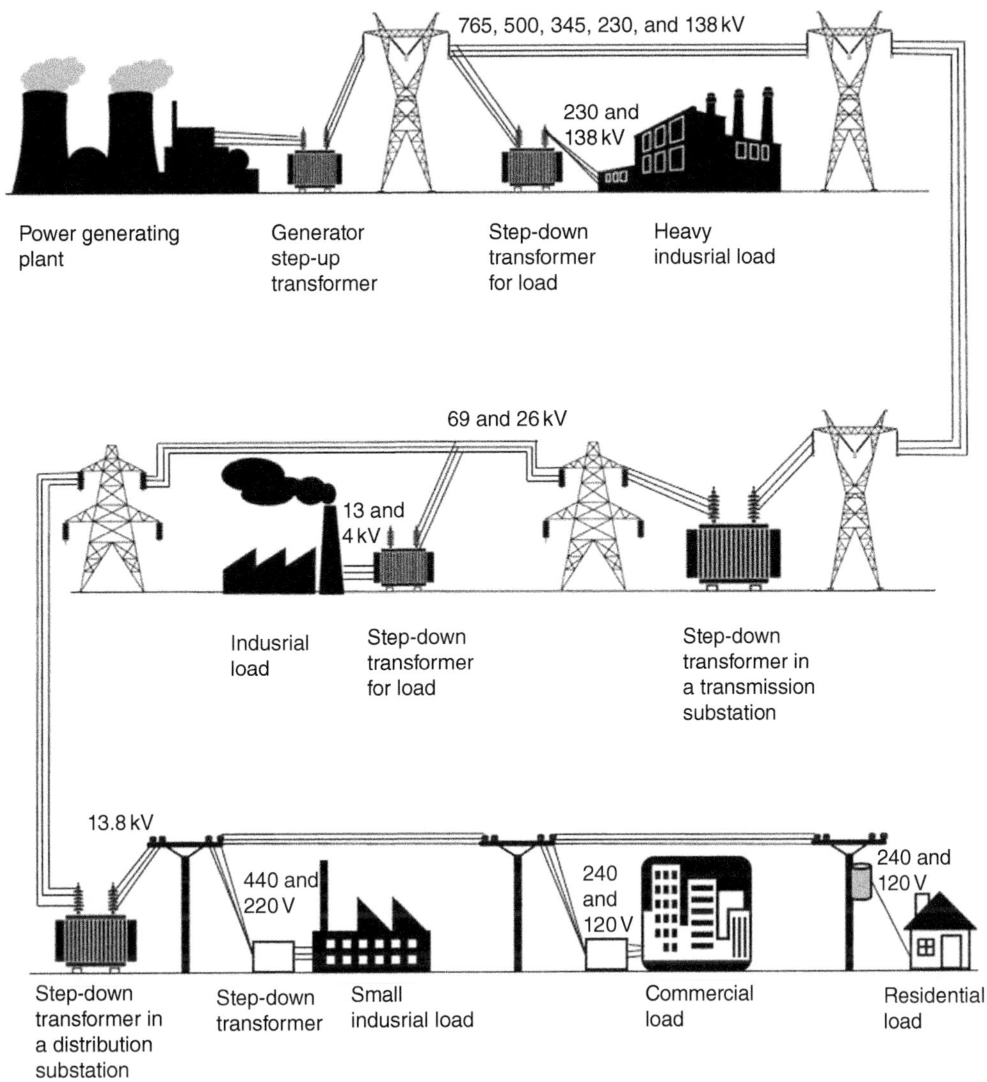

Figure 10.1 Basic structure of a power grid reproduced from DOE (2015).

angles of the consecutive phases. These assumptions are used in the "positive sequence analysis" of a power system. This allows us to model large systems which can consist of large amount of data in a concise manner, in addition to overcoming computational challenges. Commercial GIC software such as PowerWorld Simulator, GE PSLF, and Siemens PTI PSS/E are examples of positive sequence software packages. Tools such as Electromagnetic Transient Program (EMTP) can model unbalanced networks, with the detailed three-phase representation. These are also useful for GIC studies, especially for harmonic analysis.

Electricity is generated at lower voltages, in the United States, for instance, these voltage levels are 6.9 kilo-Volts (kV), 13.8 kV, 34.5 kV, etc. Often, generators are located in remote areas, away from load centers, which could be a coal or gas plant, nuclear units, hydroelectric station, or even renewable generation. To transmit power over long distances through transmission lines, the voltage is stepped up at the generating end using generator step up (GSU) transformers. For a certain amount of active power, increasing the voltage V reduces the current I, which reduces the active power losses (i.e., I^2R) on the lines, where R denotes line resistance. Figure 10.1 also shows one of the typical values that the voltage is stepped up to for transmission. 765 kV lines are found in the Eastern Interconnect,[2] whereas both the Eastern and

[2] The North American Power Grid consists of three large power grids, each of which works in synchronism on its own like one large machine. They are connected to each other in places by DC lines. They are the Eastern Interconnect which is the largest, the Western Interconnect, and Texas.

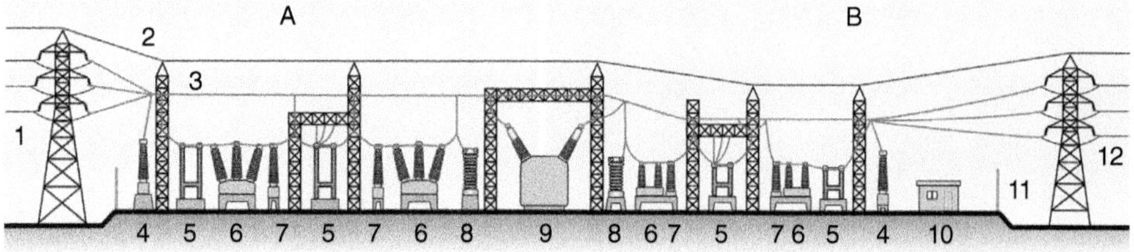

Figure 10.2 Transmission substation illustration. (A) Primary Side. 1. Incoming (primary) power lines. 2. Shield wire. 3. Overhead lines. 4. Instrument Transformer. 5. Disconnect Switch. 6. Circuit Breaker. 7. Current transformer. 8. Lightning arrester. 9. Main (power) transformer. (B) Secondary Side. 10. Control building. 11. Security fence. 12. Outgoing (secondary) power lines. Image Source: CC-https://kids.kiddle.co/Image:Electrical_substation_model_(side-view).PNG.Licensed under CCBy3.0

Western Interconnects contain 500 and 345kV lines in their high voltage transmission networks.

Closer to the loads, or at some junctions before that, the voltage is stepped down to a subtransmission level in subtransmission substations. Substations thus consist of one or more transformers, incoming and outgoing power lines, buses, as well as switching and protection equipment; see Figure 10.2. Even closer to the customers are distribution substations which further step down the voltage. The last step consists of distribution transformers, which can be either be mounted on poles, or pad-mounted, that is, placed in steel cases mounted on concrete pads for underground or ground level distribution lines. These are the ones that usually power homes.

At every stage of the transmission, depending on factors such as voltage level and rated current, different conductors may be used, thus yielding different properties such as resistance and reactance to the lines. Also line lengths make an impact on these parameters. These factors are even more crucial for GIC analysis, as we will observe in the subsequent sections.

10.2.2. Substations and Transformers

Having introduced the main components of the power grid, we now discuss the details of two that are important from the GIC analysis perspective. They are important because they act as the "gateways" for GICs to enter the grid, that is, GICs flow from the earth into the grid (or vice versa) through "grounded" connections of substations and transformer neutrals.

As mentioned earlier and denoted in Figure 10.2, substations contain protection elements such as surge arrestors, and circuit breakers. A key part of the overall protection of the substation is the grounding of the substation into the earth. It protects personnel and substation equipment from being exposed to dangerously high voltages. The grounding system consists of an underground mesh or grid with interconnected bare conductors and ground rods. The structures of all substation equipment are bonded to them. Substation grounding resistance is an important field used in GIC calculations. It is the effective grounding resistance of the substation, consisting of its grounding mat and the ground paths emanating out from the substation such as due to the grounding of shield wires (NERC, 2012; Zheng et al., 2014).

Substations also contain transformers, and each transformer is also grounded through its neutral. The grounding depends on the winding connections of the transformer. Transformers form an important part in the overall GIC modeling process; their construction and configuration affects the flow and magnitude of GICs in terms of DC resistances, and their response to GICs such as heating, saturation, and harmonics plays a major role in evaluating grid impacts. Hence we discuss them in more detail in the next section.

10.3. TRANSFORMER: A CLOSER LOOK

10.3.1. Windings and GIC Flows

Figure 10.3 shows the different types of typical transformers and connections (Torta et al., 2014). The common element in all the subfigures is that every transformer neutral, marked "N," is grounded through a resistance R_g, which is the effective grounding resistance. This includes the substation grounding resistance, as well as any resistance of the transformer neutral itself (e.g., if the transformer is resistance grounded). In this case, the two "resistors" can be considered to be in series. It is useful to note here that a transformer may not always have a neutral on both sides. This happens when one of the sides is "delta" connected, whereas "wye" connected (also known as "star" because of the shape) windings have a neutral which can be grounded if desired. In the latter case, they are referred to as grounded-wye or GWye in short.

As mentioned earlier, a GSU transformer steps up voltage at the generator terminals to transmit power over long distances. This is typically a Delta-Wye transformer, with Delta on the low voltage side, shown in Figure 10.3a.

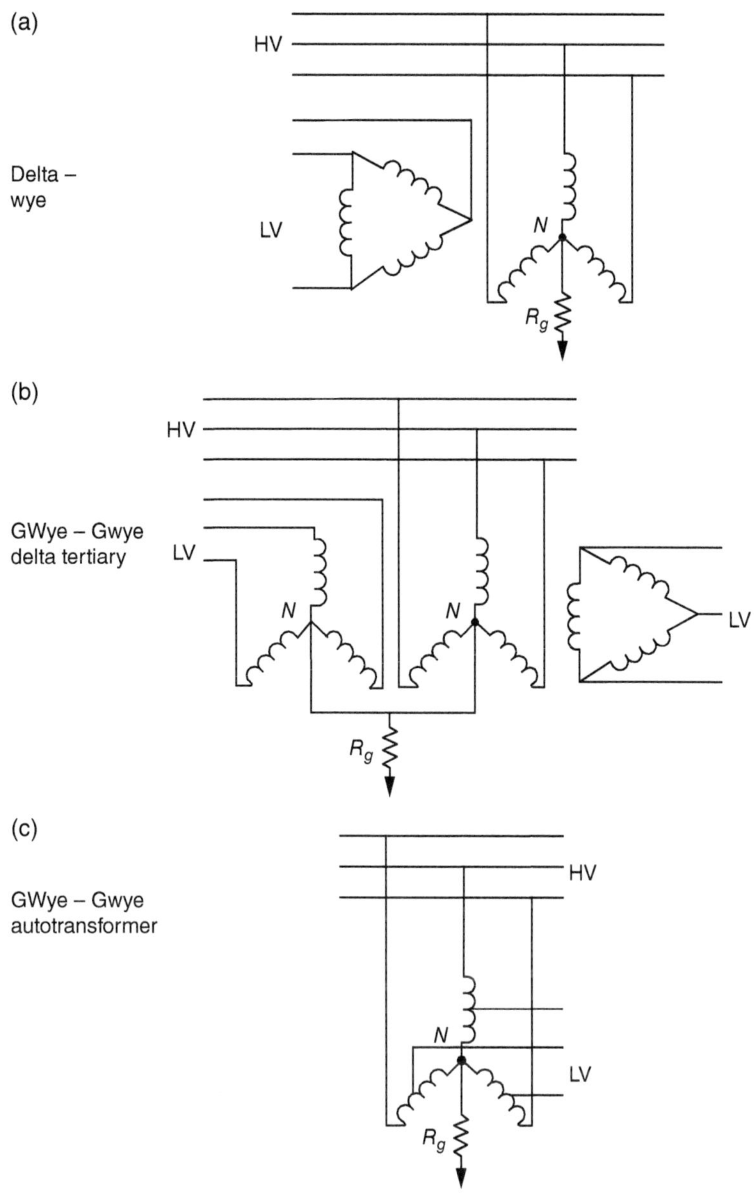

Figure 10.3 Transformer winding connections. Source: image from Torta et al. (2014). Torta, https://link.springer.com/article/10.1186/1880-5981-66-87#copyrightInformation. Licensed under CCBY 2.0.

Because of the structure of the delta connection, there is no neutral point and hence no scope for a connection to ground. Hence GIC does not flow in the delta winding of a transformer, and so on the low side of GSUs. This is a useful point to keep in mind for GIC modeling.

The GWye – GWye transformer configuration is very common in transmission systems. For safety reasons, typically the transformer neutral is always grounded. Both the high and low sides are wye connected. The high and low voltage sides of a transformer are also referred to as the primary and secondary. Sometimes these transformers also have a third, tertiary winding, whose nominal voltage can be an order of magnitude lower than the primary. The tertiary winding is typically delta connected, therefore it does not affect GIC distribution. The tertiary winding helps in (i) preventing harmonic currents from flowing into the transmission system, (ii) limiting fault current, and (iii) stabilizing the neutral point when the load is unbalanced.

Another common type of transformer, which needs special consideration in GIC modeling is the autotransformer. Unlike electrically isolated transformers where the primary and secondary windings are wound separately around the iron core (see Fig. 10.4a), autotransformers consist of a series and a common winding, which are parts of a single winding. This reduces material usage and hence also the size, overall costs as well as power losses. It has other advantages such as voltage regulation. On the other hand, the lack of isolation between the high and low voltage side can cause hazards. Figure 10.4 compares the winding structures, current flows, and DC circuits between isolation transformers and autotransformers. Single phase transformers are shown here to explain the concept; three phase transformers have similar characteristics and distinctions across the two types.

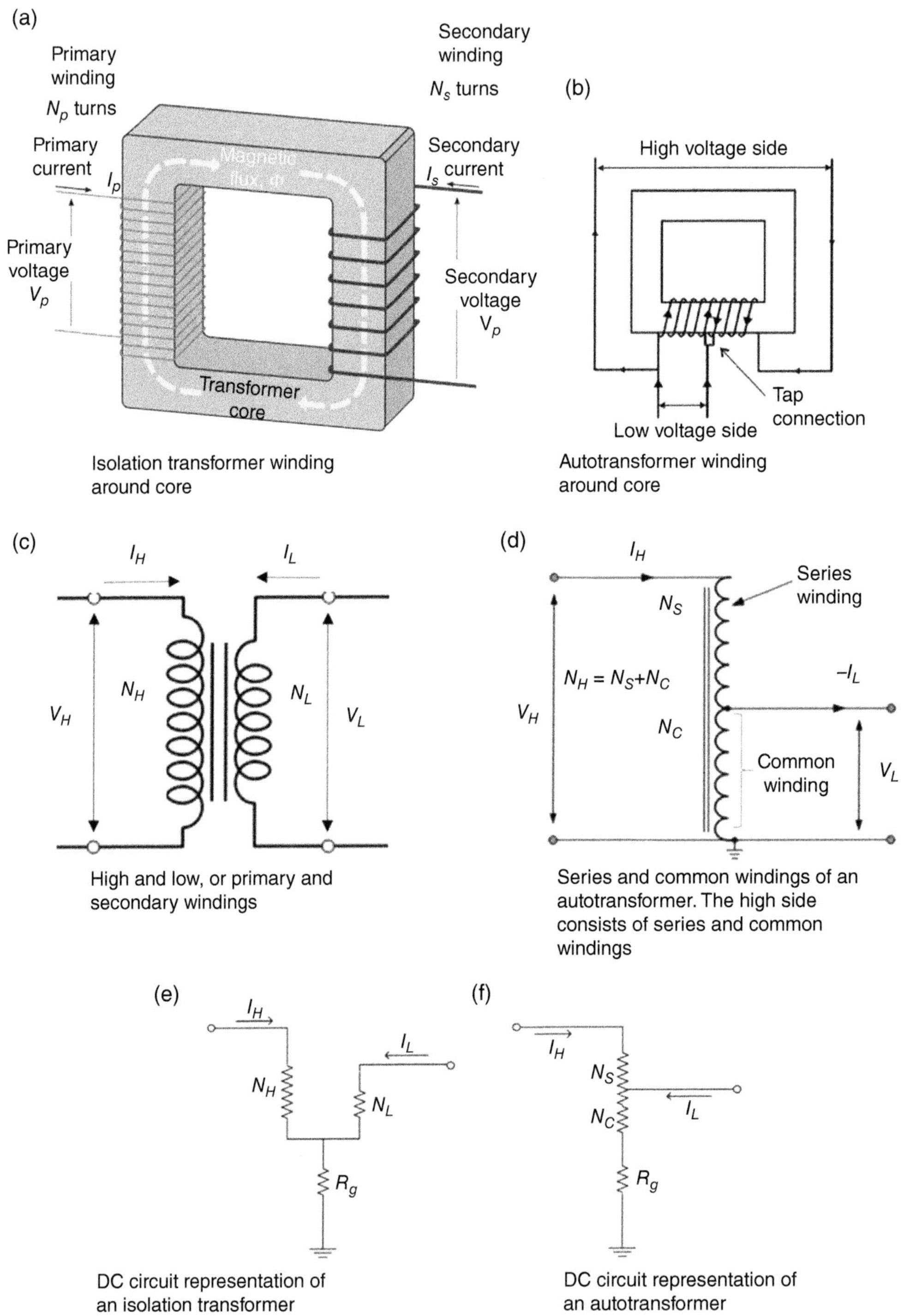

Figure 10.4 Differences between isolation transformer (left column) and autotransformers. Source: image sources: https://commons.wikimedia.org/wiki/File:Transformer3d_col3.svg. Licensed under CCBY SA 3.0 (Zheng et al., 2014). Reproduced with permission of Electrical Engineering Portal and IEEE.

Referring to Figure 10.4c and d, N_H = number of turns of the primary or high side winding, N_L = number of turns of the secondary or low side winding, and N_S (N_C) = number of turns of the series (common) winding of the autotransformer. The voltages and currents in the windings are as shown. The basic function of a transformer is to "transform" the voltage from one magnitude level to the other, which is governed by this fundamental relationship:

$$\frac{V_H}{V_L} = \frac{I_L}{I_H} = \frac{N_H}{N_L} = \frac{N_s + N_C}{N_C} = a_t \tag{10.1}$$

where a_t is known as the turns ratio.

The reason these different transformer winding types and winding connections were described is because they affect GIC flow through the transformers, and also determine the value of what we call the "Effective GIC". This depends on the GIC flowing in both the windings of a transformer (Albertson et al., 1981). The effective GIC (I_{GIC}), and not the current through the transformer neutral (I_N), is what determines the reactive power losses of the transformer. Most GIC monitors, however, are set up to measure the transformer neutral current.

For a delta-wye transformer, the effective GIC is simply the GIC through the wye winding, since delta is not grounded and hence has no GIC flowing. For a grounded wye – grounded wye transformer (e.g., Figure 10.4b without the delta tertiary), the effective GIC needs to be calculated from the unequal GIC flow in the primary and secondary, or series and common windings if it is an autotransformer. This is done by equating the MMF's produced by them to that produced by the effective GIC in the high side winding only (Zheng et al., 2014). Referring to Figure 10.4e and f we get,

$$I_H N_H + I_L N_L = I_{GIC} N_H. \tag{10.2}$$

Since $N_H/N_L = a_t$ which is the turns ratio, we get,

$$I_{GIC} = \left| \frac{a_t I_H + I_L}{a_t} \right|. \tag{10.3}$$

If we use the same definition of turns ratio as in (10.1), I_{GIC} for an autotransformer can be found using (10.3) as well.

10.3.2. DC and AC Resistance

So far we have discussed substation grounding and transformer winding resistances. The reason they are important is because for the most part, direct current (DC) conditions are assumed in GIC calculations. This is because of the frequencies associated with them (<0.1 Hz) making them "quasi-dc" compared to the 60 (or 50) Hz grid.

The difference between the AC and DC resistance values in power systems arises from skin and proximity effects (Dixon, 2003). While the details of these phenomena involving self and mutual induction are interesting and suggested for further reading, the gist is that with the increase in frequency, the resistance of a conductor tends to increase. Resistance also depends on the ambient temperature; for example, copper has a resistivity of $1.68 \times 10^{-8}\,\Omega$.m (Ohm-meter) at 20°C, with a temperature gradient of 0.004 per °C.

Hence while collecting transformer data, it is important to check whether it is indeed a DC resistance, and also the temperature at which it was measured, since different transformer test reports may have these values at different temperatures, while for consistency in results it is useful to use a standard value in a study. For example, NERC suggests winding resistances be normalized to 75°C (NERC, 2013).

Where actual winding resistance values are not available, these can be estimated using AC data available in actual power grid models. Examples of such methods include table 2 from NERC (2013), and Equations (9) and (10) from Overbye et al. (2012). For example, any power flow case has a transformer represented by an impedance, which has a series reactance (X) and resistance (R) value. Now for an actual case representing a real system, such as one of the North American interconnects, the transformer series resistance values present in the case originate from the entity that owns and/or is responsible for these transformers, who would have most likely referred to transformer test reports to extract these values.

It may be difficult for non-utility entities such as researchers to get access to these actual power system models due to confidentiality concerns. In such cases, one may find publicly available test systems to be helpful. However, they should be able to reflect the complexity, scale, and characteristics of actual systems in order for the research and development to be applicable to the real world. An example of such systems, which are synthetic and hence available publicly can be found at Xu (2016). They include GMD analysis parameters, and are realistic cases since they are derived from statistical analysis of actual power grid models (Birchfield et al., 2017a, 2017b).

For transmission lines, data such as in Ramos (2018) has shown that the DC and AC resistance values compare very well. The substation grounding resistance values are typically harder to find. A transformer test report is available during its purchase, or during maintenance. For substations, one may have to go around physically and measure the resistance values using the Fall of Potential Method and procedures given in standards such as IEEE (1991, 2012), which may involve significant time and effort. However with the NERC transmission planning standard TPL-007-1 in effect since July 2017 (NERC, 2017), maintaining system models and data is required and hence more entities may be allocating resources to gathering and recording such data to ensure accurate studies. Even though substation grounding resistance values have previously been estimated using heuristics (Overbye et al., 2012), subsequent research (Bui et al., 2013) has shown that having actual values is important since it can significantly affect GIC results.

10.4. GIC CALCULATIONS: DC ANALYSIS

10.4.1. Simple GIC Circuit

Following the discussion so far, Figure 10.5 shows a simple power system, and its DC representation for GIC analysis. The image on the top is a one-line diagram of a two-bus system. The three phases are represented by a single line for simplicity, hence the term "one-line". There is a generator at each bus, with a transmission line between them. There are two delta wye-grounded transformers at each generator, indicating that they are GSUs. Although not explicitly depicted here, each transformer is also in a substation. The neutral of the wye of each transformer is connected to its substation ground, with resistance values of R_{gnd1} and R_{gnd2} respectively. The delta is not shown since it has no connection to ground. Hence its resistance values also do not come into the picture.

We also have to consider the DC resistance of the transmission line. The three phases are usually referred to as "a", "b", and "c". Referring to Figure 10.5, usually $Ra = Rb = Rc$. Let us call this R_l. The total resistance is hence $R_l/3$ since they are in parallel. In general, R_l can be found as,

$$\frac{1}{R_l} = \frac{1}{R_a} + \frac{1}{R_b} + \frac{1}{R_c}. \tag{10.4}$$

The same applies for the winding resistances, that is, the total winding resistance at each of the transformers is $R_{w1}/3$ and $R_{w2}/3$.

Next, we go into the details of the electric field modeling in the grid, and the subsequent GIC flows, with the help of a simple walk-through example.

10.4.2. The GIC Model

There have been several models proposed to simulate the effect of GMD induced electric fields in the power grid. Earlier the Earth Surface Potential (ESP) method was widely used (Albertson and Van Bealen, 1970; Prabhakara et al., 1992). However, it was later found that this approach is not sufficient to model realistic, that is, non-uniform electric fields. This is because such electric fields are vectors whereas the ESP approach considers them as potential gradients (Boteler and Pirjola, 1998). To model these, the GMD-induced electric fields need to be included as DC voltage sources in series with the transmission lines. The DC voltages are found by integrating

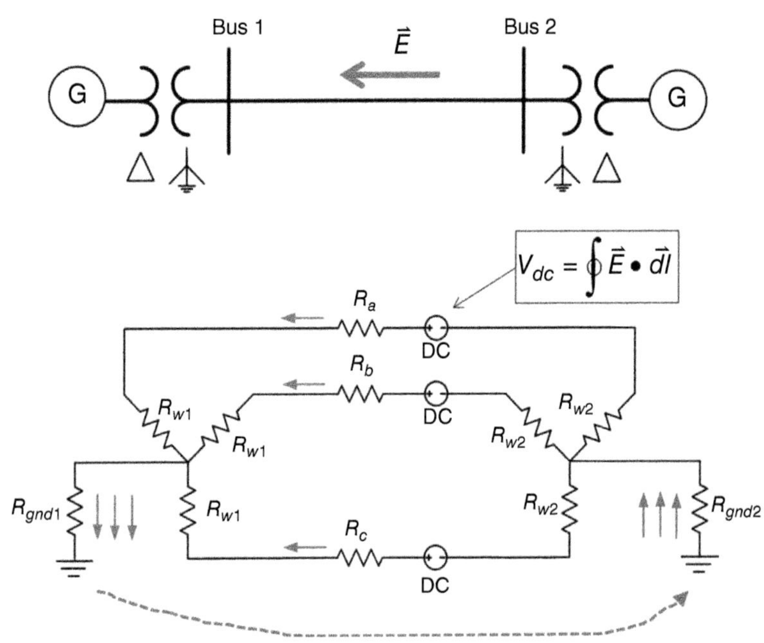

Figure 10.5 Simple power system and its three phase, DC representation for GIC analysis.

the electric field over the path and length of the transmission lines as,

$$U = \int_R \overline{E} \cdot \overline{dl} \tag{10.5}$$

where R is the geographic route of the line, $\overline{E}$ is the electric field along the route, and $\overline{dl}$ is the incremental line segment (Horton et al., 2012; Overbye et al., 2013). The “bars” on top imply that they are vectors. The integration of the electric field is path dependent.

If the applied electric field is uniform over the length of the transmission line, U can also be found using,

$$U = E_N L_N + E_E L_E \tag{10.6}$$

where E_N (E_E) is the northward (eastward) electric field, and L_N (L_E) is the northward (eastward) length of the transmission line. For a uniform electric field, the route of the line is not important, just the end points are sufficient. Equation (10.6) can also be used in the case of non-uniform electric fields, by dividing the line into small enough segments and assuming a uniform field over each segment, and summing the results.

The DC voltages induced in the lines are then converted into their Norton equivalent current sources I_{inj} (injection) at the buses. Following this conversion, multiple approaches are available in literature that model the GIC flow; notable examples include the Nodal Admittance Matrix method and the Lehtinen–Pirjola technique (Lehtinen and Pirjola, 1985). In this chapter, we discuss the former in more detail. The two methods are mathematically equivalent, and have some differences in the number of steps and the complexity of the matrix inversion involved (Boteler and Pirjola, 2014).

Coming back to the current sources, consider a vector of these across all the buses as I_{INJ}. This conversion helps in utilizing sparse matrix methods for quick GIC calculations, especially for large systems. The power flow problem has long been solved using sparsity techniques, specifically in building what we call as the nodal admittance matrix (Y_{BUS}). In a system with N buses, it forms an $N \times N$ matrix, so the index of each matrix element is a pair of nodes (nodes means buses in this context). The value of each element is the sum of all the admittance values at the node (for diagonal elements) or between the two nodes, with the off diagonal elements having a negative sign. The fact that for large systems, many of the buses are not connected to each other, contributes to the sparsity. Y_{BUS} elements are influenced by the conductance (g) and susceptance (b) values of components such as lines, capacitors, transformers, reactors, etc. Readers are encouraged to refer to Glover et al. (2017) for more details on the power flow problem, and building the Y_{BUS}.

Transmission lines may have series capacitors installed for reactive power support. Since capacitors block DC current, lines with series capacitors are as good as open for GIC calculation. However, they will certainly come into consideration in the AC power flow part of the GMD analysis done for voltage stability assessment.

Analogous to the Y_{BUS}, in the GIC flow problem we have the G matrix, which is also sparse. In a system with N buses and S substations, G forms an $M \times M$ matrix where $M = N + S$. The elements of this matrix consist of only conductance values. As mentioned earlier, grounded transformers and substations provide a path for GICs to flow into or out of the ground. Including substation neutral buses in the G matrix takes into account the conductance of the substation ground, and that between the transformer windings to their connected substation neutrals. G also depends on the topology of the system, and changes when lines go in and out of service, etc.

With this G matrix, and the Norton current vector I_{INJ} representing the effect of the electric field, the vector of nodal DC voltages, V, is first calculated as,

$$V = G^{-1} I_{INJ}. \tag{10.7}$$

Thus, V is the solution of the GIC problem, with (10.7) being the Nodal Admittance Matrix formulation of this problem. The actual GIC flow through the system I_{GIC} can then be found using, in simple words, Ohm’s law. More specifically, transformer effective GICs, I_{GIC} can be found by defining an M dimensional vector C_t, as

$$I_{GIC,t} = \left| C_t V \right| \tag{10.8}$$

where t = transformer number (Overbye et al., 2013). C_t is a very sparse vector and consists of conductance values, which relate the solved voltages at the buses and substation neutrals, to the Norton currents.

Equations (10.5)–(10.7) are quite useful while dealing with larger systems. For a simple example, such as the one in Figure 10.5, one can solve it easily without needing the G matrix.

10.4.3. Simulation Examples

To better understand the GIC model discussed so far and the simulation results expected, we now look at some examples of solving GICs in an off-the-shelf package. As mentioned earlier, while PowerWorld Simulator was available to the authors, readers may choose to apply the principles of the following simple studies in other tools. To go through the next two cases, one needs the free, demo 42-bus version of PowerWorld Simulator, which is available at https://www.powerworld.com/gloveroverbyesarma. The example cases discussed in this chapter can be downloaded from https://my.syncplicity.com/share/flgbtqs65fzorvj/AGU%20Chapter%20Cases.

10.4.3.1. Four Bus Example

Readers are encouraged to open the "GIC_FourBus" case in PowerWorld Simulator, and follow the steps to input an electric field. This example was also discussed in Shetye and Overbye (2015). The case consists of two generators with GSUs, and a 765 kV transmission line. Buses 1 and 3 belong to Substation A, whereas Buses 2 and 4 are in Substation B.

To perform GIC analysis in PowerWorld Simulator, click on the "Add-Ons" menu item, and select "GIC" to display the GIC Analysis Form. To model a uniform electric field, the two inputs required are the electric field magnitude and direction, as shown in Figure 10.6. The direction is specified in degrees where 0° means north, 90° implies east, and so on. For an example study, specify the maximum field as "1 V/km", and the storm direction as "90°". Then click on "Calculate GIC Values."

For this eastward, 1 V/km electric field, E_E = 1 V/km. The transmission line is 170.8 km long, and is oriented west–east (Bus 1 is the "From Bus" and Bus 2 is the "To Bus"), which implies that the eastward length L_E = 170.8 km. Hence, using Equation (10.6), $U = 1 * 170.8 = 170.8$ V. This is displayed in the box in Figure 10.7, which shows the one-line diagram for this case. Note that the DC representation of this test system is similar to the circuit shown in Figure 10.5. Another point worth mentioning is that the line length is calculated based on the geo-coordinates of its terminal buses. A straight line path is then assumed between the substations these buses lie in. In this example this means that substations A and B are 170.8 km apart. This assumption can be used when the exact, detailed line path is not available and is in fact commonly used. Hence, substation latitude and longitude are another data set being collected for GIC studies.

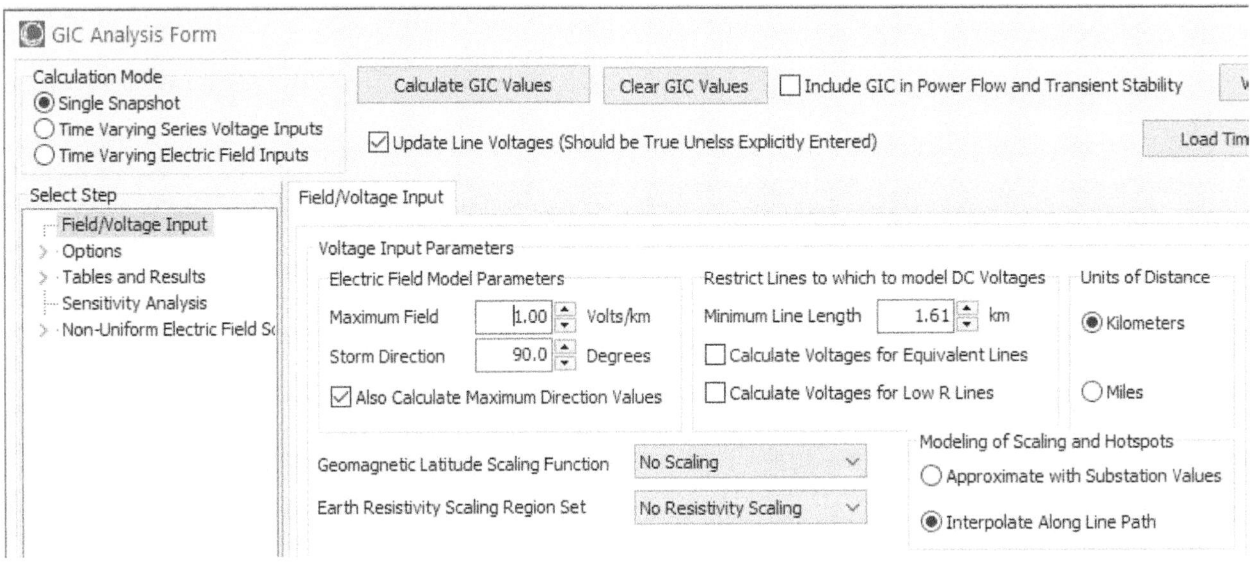

Figure 10.6 GIC analysis form, with the electric field input dialog.

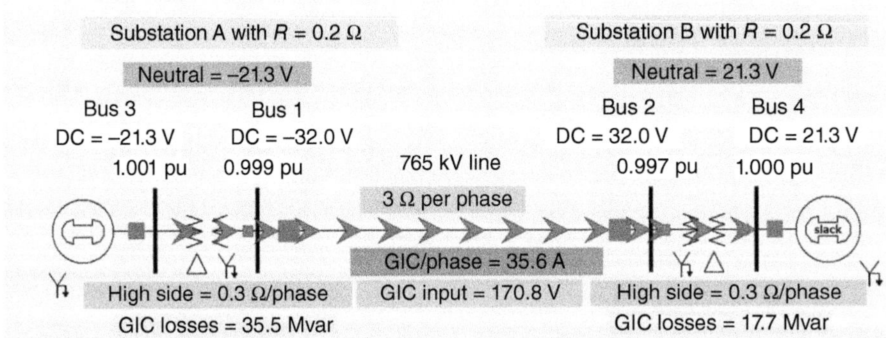

Figure 10.7 Four-bus test case with GICs visualized for a 1 V/km Eastward electric field.

The actual GICs in this four-bus system are easy to calculate for a 1 V/km Eastward electric field, which induced 170.8 V in the transmission line. The DC resistance values of the line, transformer windings, and substation grounding have been shown in the one-line diagram. The three phases for the transmission line and transformers are in parallel, so the total three phase resistance of the line is (3/3) Ω = 1 Ω, and (0.3/3) Ω = 0.1 Ω for each of the transformers. Only the high side winding resistance is considered here, since the low side is delta connected. These resistances are connected in series with the substation A and B grounding resistances, which are 0.2 Ω each, resulting in,

$$I_{\mathrm{GIC,per\,phase}} = \frac{170.8}{(1+0.1+0.1+0.2+0.2)} * \frac{1}{3} = 35.58\,\mathrm{A}$$

with the GIC flow direction being from the "From Bus" (Bus 1), to the "To Bus" (Bus 2). An important point in interpreting the results is to differentiate between the per phase GICs in transmission lines and transformers, and the total three phase GICs. Since the three phases are in parallel, the total current is just three times the per phase current. The convention commonly used for GIC analysis is to use the per phase current for transformers and transmission lines, and the total three phase current for the substation neutral.

10.4.3.2. Forty Bus Example

Moving on to a slightly larger system consisting of 40 buses, Figure 10.8 shows GICs for 1 V/km eastward electric field. The transmission lines are at nominal value of 345 kV and 161 kV. Only GICs have been calculated for this scenario, and not the GIC-induced transformer reactive power losses. Hence the "System Losses" metric on the one-line shows 0 Mvar (Mega-var, i.e., Volt-Ampere Reactive is a unit of reactive power, whereas Watt is a measure of real power). In power systems, we usually speak in terms of MW and Mvar.

Open "Transformers" tab under "Tables and Results" to see the input data as well as the GIC results of the applied electric field. Click "Calculate GIC Values" if the transformer GICs appear to be zero. Figure 10.9 shows some of the estimated data for the GIC calculation. There is more data associated with the transformers, which will be discussed in Section 10.5. There is an option to manually enter coil (i.e., winding) resistance. If this is not available, it is estimated from the series AC resistance in the power flow case (Overbye et al., 2012). There are also options to specify winding configurations such as wye and delta. Last, one can specify a transformer as being an autotransformer, or if this is not known, the software estimates it using heuristics such as ratio of high to low side voltage. These

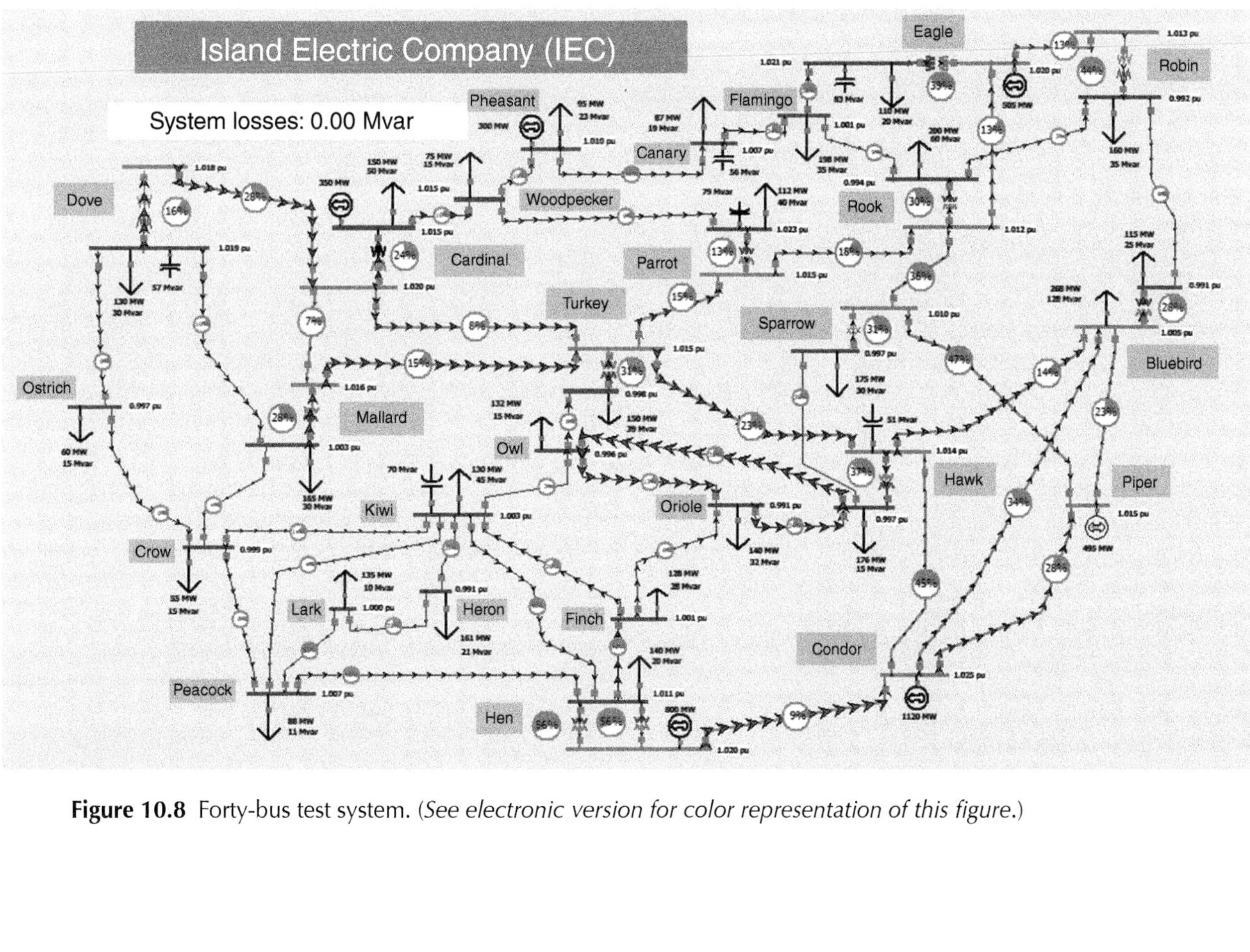

Figure 10.8 Forty-bus test system. (*See electronic version for color representation of this figure.*)

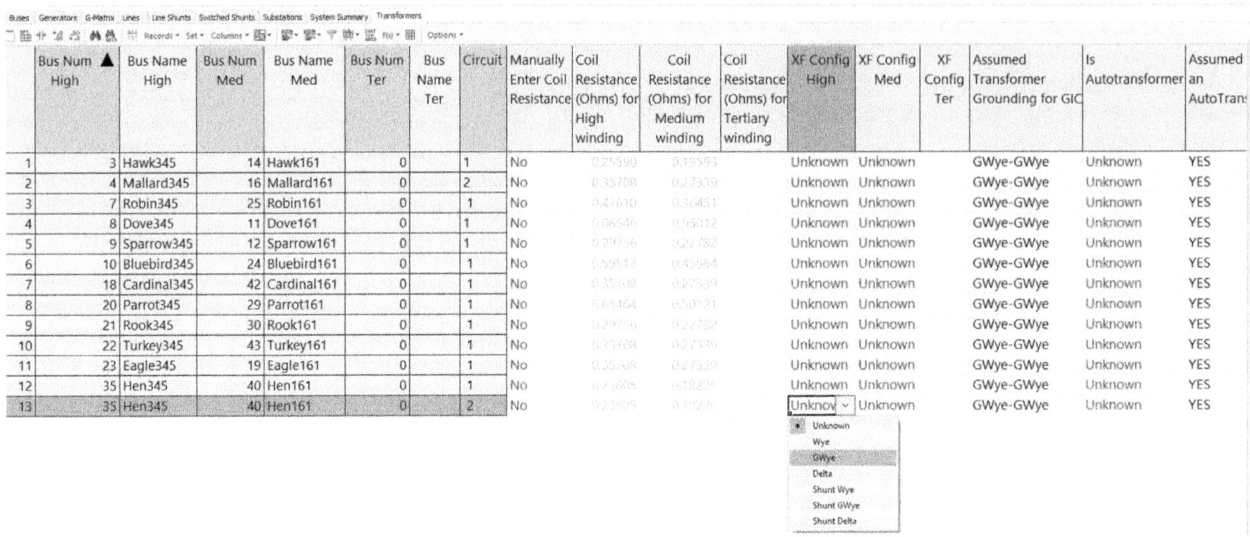

	Bus Num High	Bus Name High	Bus Num Med	Bus Name Med	Bus Num Ter	Bus Name Ter	Circuit	Manually Enter Coil Resistance	Coil Resistance (Ohms) for High winding	Coil Resistance (Ohms) for Medium winding	Coil Resistance (Ohms) for Tertiary winding	XF Config High	XF Config Med	XF Config Ter	Assumed Transformer Grounding for GIC	Is Autotransformer	Assumed an AutoTran:
1	3	Hawk345	14	Hawk161	0		1	No	0.25590	0.19593		Unknown	Unknown		GWye-GWye	Unknown	YES
2	4	Mallard345	16	Mallard161	0		2	No	0.35708	0.27339		Unknown	Unknown		GWye-GWye	Unknown	YES
3	7	Robin345	25	Robin161	0		1	No	0.47610	0.36451		Unknown	Unknown		GWye-GWye	Unknown	YES
4	8	Dove345	11	Dove161	0		1	No	0.06546	0.05012		Unknown	Unknown		GWye-GWye	Unknown	YES
5	9	Sparrow345	12	Sparrow161	0		1	No	0.29756	0.22782		Unknown	Unknown		GWye-GWye	Unknown	YES
6	10	Bluebird345	24	Bluebird161	0		1	No	0.59513	0.45564		Unknown	Unknown		GWye-GWye	Unknown	YES
7	18	Cardinal345	42	Cardinal161	0		1	No	0.35708	0.27339		Unknown	Unknown		GWye-GWye	Unknown	YES
8	20	Parrot345	29	Parrot161	0		1	No	0.65464	0.50121		Unknown	Unknown		GWye-GWye	Unknown	YES
9	21	Rook345	30	Rook161	0		1	No	0.29756	0.22782		Unknown	Unknown		GWye-GWye	Unknown	YES
10	22	Turkey345	43	Turkey161	0		1	No	0.35708	0.27339		Unknown	Unknown		GWye-GWye	Unknown	YES
11	23	Eagle345	19	Eagle161	0		1	No	0.35708	0.27339		Unknown	Unknown		GWye-GWye	Unknown	YES
12	35	Hen345	40	Hen161	0		1	No	0.23805	0.18226		Unknown	Unknown		GWye-GWye	Unknown	YES
13	35	Hen345	40	Hen161	0		2	No	0.23805	0.18226		Unknown	Unknown		GWye-GWye	Unknown	YES

Figure 10.9 Transformer input and estimated data for DC analysis.

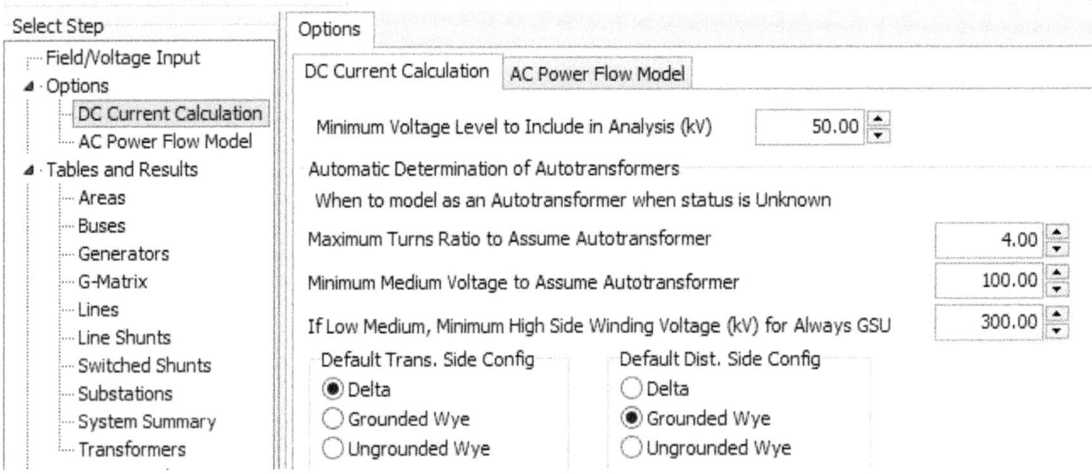

Figure 10.10 DC current calculation options for transformers.

can also be adjusted under "Options → DC Current Calculation" as shown in Figure 10.10.

Returning to the Transformers tab, we now look at the results of the 1 V/km eastward field shown in Figure 10.11. The transformer effective GIC has been expressed in two units, Amperes and Per Unit (pu). Power systems make use of the concept of per unit extensively, primarily because of the several voltage levels involved, and different devices having their own nominal values and ratings, etc. It is easier to refer everything to a common "base" to perceive relative magnitudes, and detect potential problems if any.

10.4.3.3. Per Unit System

A good example of per unit is expressing (AC) voltages. Power systems are operated to keep bus voltages around 1 pu, with 0.95–1.05 pu being an acceptable range in most cases. Anything significantly beyond these values indicates undervoltage or overvoltage problems in the system, which are often tied to insufficient or excessive reactive power supply, respectively. For a transformer, the nominal voltage of the high side and low side winding (e.g., a 500/345 kV transformer) become their respective base values for voltage magnitude. An MVA base is also used for power. The power base could either be the device's rated power such as a generator or transformer, or what we term as a "system base." A value of 100 MVA is commonly used for system base. This helps to normalize quantities on a single reference.

The pu effective GIC shown here is calculated using the transformer apparent power rating (measured in Volt-Amperes or VA), and the high side voltage as the base, to

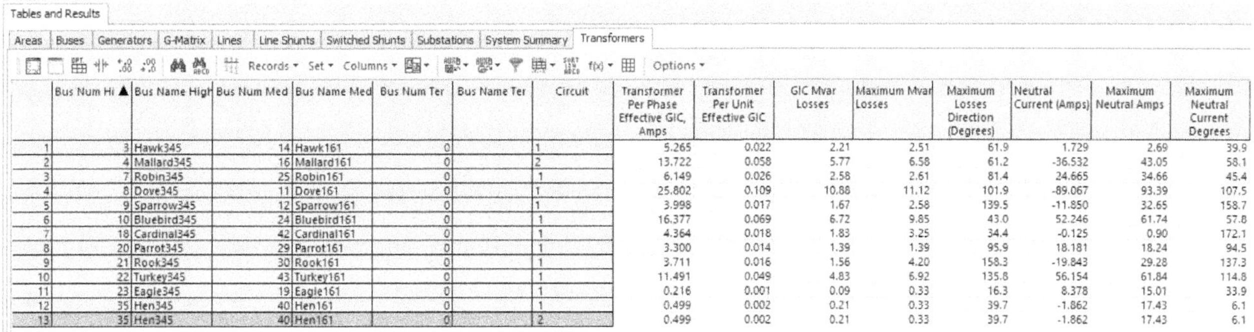

	Bus Num Hi ▲	Bus Name High	Bus Num Med	Bus Name Med	Bus Num Ter	Bus Name Ter	Circuit	Transformer Per Phase Effective GIC, Amps	Transformer Per Unit Effective GIC	GIC Mvar Losses	Maximum Mvar Losses	Maximum Losses Direction (Degrees)	Neutral Current (Amps)	Maximum Neutral Amps	Maximum Neutral Current Degrees
1	3	Hawk345	14	Hawk161	0		1	5.265	0.022	2.21	2.51	61.9	1.729	2.69	39.9
2	4	Mallard345	16	Mallard161	0		2	13.722	0.058	5.77	6.58	61.2	-36.532	43.05	58.1
3	7	Robin345	25	Robin161	0		1	6.149	0.026	2.58	2.61	81.4	24.665	34.66	45.4
4	8	Dove345	11	Dove161	0		1	25.802	0.109	10.88	11.12	101.9	-89.067	93.39	107.5
5	9	Sparrow345	12	Sparrow161	0		1	3.998	0.017	1.67	2.58	139.5	-11.850	32.65	158.7
6	10	Bluebird345	24	Bluebird161	0		1	16.377	0.069	6.72	9.85	43.0	52.246	61.74	57.8
7	18	Cardinal345	42	Cardinal161	0		1	4.364	0.018	1.83	3.25	34.4	-0.125	0.90	172.1
8	20	Parrot345	29	Parrot161	0		1	3.300	0.014	1.39	1.39	95.9	18.181	18.24	94.5
9	21	Rook345	30	Rook161	0		1	3.711	0.016	1.56	4.20	158.3	-19.843	29.28	137.3
10	22	Turkey345	43	Turkey161	0		1	11.491	0.049	4.83	6.92	135.8	56.154	61.84	114.8
11	23	Eagle345	19	Eagle161	0		1	0.216	0.001	0.09	0.33	16.3	8.378	15.01	33.9
12	35	Hen345	40	Hen161	0		1	0.499	0.002	0.21	0.33	39.7	-1.862	17.43	6.1
13	35	Hen345	40	Hen161	0		2	0.499	0.002	0.21	0.33	39.7	-1.862	17.43	6.1

Figure 10.11 GIC analysis results.

yield the current base. Apparent power takes into account both real and reactive power. In a three phase system, apparent power is usually denoted by S, and is calculated as,

$$S = \sqrt{3} \times V_L \times I_L = 3 \times V_{ph} \times I_{ph} \tag{10.9}$$

where the subscript "L" denotes line voltage or current, and "ph" represents phase quantities. Phase voltage is measured between the terminals of any one of three phases (a, b, or c) and the neutral, whereas line voltage is measured between two phases. Hence, for a delta configuration, the line voltage equals the phase voltage, whereas in a wye connection the line voltage equals $\sqrt{3}$ times the phase voltage.

With this quick background on three-phase power, consider the last row in Figure 10.11 which is the transformer at Substation Hen, Circuit 2. We can calculate its current base as,

$$I_{Base} = \frac{(VA)_{Base}}{V_{Base} \times \sqrt{3}} = \frac{100 \times 10^6}{345 \times 10^3 \times \sqrt{3}} = 167.348\,\text{A}$$

Hence, $I_{GIC,\,pu}$ at Transformer Hen 345–161, 2 = $\frac{I_{GIC,A}}{I_{Base}}$ $= \frac{0.499}{167.348} = 0.002982\,\text{pu}$.

10.4.3.4. Additional Simulation Options

We look at some more options and settings, which are also important, involved in a GIC study. Go to the "Areas" tab, which looks something like Figure 10.12. This case has only one area, but the features listed here such as "Ignore GIC DC Volts" and "Ignore GIC Losses" are quite useful while working with large scale, interconnect level cases.

Actual power system models or cases of the "planning" type are created on an interconnect level. These contain power flow data, as well that for performing economic studies, dynamic simulations, etc., looking out several years. With the GMD transmission planning standards in effect, we now have GIC parameters being added to these large, interconnect-level planning models.

In grid operations, a large interconnected power system is split up into several Balancing Areas which are connected to each other by transmission lines called Tie Lines. These areas are also present in the planning models. Each area is responsible for balancing its total generation with the load and losses within it along with the power transfers along its tie lines. Usually there are power transactions scheduled between areas. To give an idea of the scale involved, a planning model of the North American Eastern Interconnect consists of approximately 70,000 buses and 130 Areas.

Often an entity will be performing a GIC study for a much smaller portion of an interconnect case, such as a utility which may be one or a few more Areas, or on a larger scale a transmission system operator (TSO) comprising of dozens of Areas. These are what the "Areas" refer to in Figure 10.12. By setting the option to "NO" for certain areas for "Ignore GIC DC Volts" we can include the effect the electric field in that region. This could cover the GIC study footprint plus some Areas around it as desired, depending on how large (spatially) a GMD storm needs to be applied for the study. For the other Areas marked YES, the induced voltage will be ignored, that is, effectively no electric field will be assumed. In short, this option helps to assume the GMD-induced electric field over a certain footprint.

Similarly, setting "Ignore GIC Losses" to NO includes the GIC-induced reactive power losses of all the transformers in those Areas. This can be done for the Areas in, for instance, the utility footprint under study, and their tie-line connected neighboring Areas if desired. This will account for the reactive power losses and flows from the neighbors, which in reality may have an impact on the system under consideration. Since both GICs and reactive power are more local in nature and do not travel very far in the system, including just a small buffer region around the main study footprint, both for GIC DC Volts

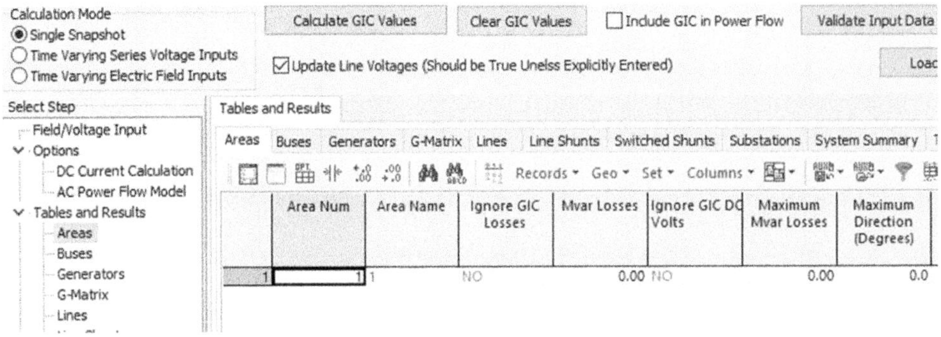

Figure 10.12 Options for selecting "Areas" to include their (i) GMD-induced DC voltages, and (ii) Transformer reactive power losses.

and reactive losses is sufficient, to consider the impact of GICs and losses flowing in from outside or flowing out to other areas.

10.5. REACTIVE POWER LOSSES AND VOLTAGE STABILITY: AC ANALYSIS

In the previous section, we looked at the DC aspects of a power system GMD study, that is, just the calculation of GICs. So far, only the DC representation of the power system (e.g., Fig. 10.5) was used. In this section, we look at how the transformer reactive power losses are calculated from their GICs, and how they impact the voltages in the system. For the latter half, we make use of the full AC model of the grid, with reactances, capacitors, and the Y_{BUS}, and perform power flow solution(s) including the GIC-induced transformer reactive power losses to determine the bus AC voltages.

GICs, being quasi-dc, can cause half-cycle saturation in AC transformers, giving rise (i) to harmonic currents (i.e., multiples of the fundamental grid frequency of 60 (or 50) Hz), and (ii) increased reactive power absorption. This phenomenon is quite involved and warrants its own chapter for discussion. It is explained well in Walling and Khan (1991) and NERC (2012) with several illustrations. It also needs EMTP and harmonic analysis tools with the detailed three-phase representation for simulation as mentioned earlier, and hence cannot be directly applied to very large system analyses. Some studies have coupled these for about medium sized systems using interfaces such as MATLAB (Adhikari et al., 2016).

The harmonics can cause power system protection devices such as relays to trip inadvertently, or under a "false alarm," which may take devices such as capacitors or Static Var Compensators (SVC) out of service. These two devices especially are important during a GMD since they provide reactive power support, at a time when transformers are demanding more of it. A key cause of the March 1989 blackout in Hydro-Quebec was the tripping or shutting down of seven such SVCs within quick succession, caused by harmonics generated by saturated transformers (NERC, 1990).

For the most general and simple case, the losses are linearly dependent on the transformer GICs (Dong et al., 2001), and its terminal (i.e., bus) voltage (Walling and Khan, 1991). If we express the losses and the bus voltage in pu (high side is used), this is given by

$$Q_{loss,\mathrm{pu}} = K V_{\mathrm{pu}} I_{GIC,\mathrm{pu}} \qquad (10.10)$$

where K is a scalar known colloquially as the transformer "K factor." This a relatively new parameter that is gaining importance for transformers, so much so that manufacturers are planning to provide their values in transformer test reports, mostly for GMD studies.

The K factor depends on several parameters such as (i) the number of phases of the transformers, (ii) the core type/construction, (iii) number of transformer "legs," and if these are not known, assumptions can be made based on the voltage level. This is due to the difference in the flux distribution in the core of these different types, which leads to different levels of saturation across the types for the same value of GIC, leading to different quantities of losses.

The linear relationship holds true for most of the types of transformers; see Figure 10.13. However, it also shows the non-linear or rather piece-wise linear relationship for three-phase three-limb (or three-leg) transformers. This non-linearity exists due to the core and tank structure, and how the flux return path is external to the core, and hence faces higher reluctance. Hence a three-leg transformer also needs more GIC to go into half cycle saturation.

This is the reason why data related to the number of phases and core types needs to be collected for transformer GMD studies. Based on these figures, it is evident that single phase transformers are the most susceptible to high reactive power loading under a GMD, the least being the three-phase three-leg type. Hence, sometimes when the core type is not known, one many assume the worst case scenario, that is, the K factor of a single phase transformer, being cognizant that the voltage stability results may be too pessimistic.

From the simulation perspective, these values can be viewed under the "Transformers" tab of the GIC Module. Using the 40-bus case again, Figure 10.14 shows the input data for calculating losses. There are core types to choose from to specify, while "GIC Model" denotes what relationship is used for mapping GICs to the losses. If left as Default, it takes on the values shown in Figure 10.15, which can be found under Options → AC Power Flow Model. These values have been derived from literature. If

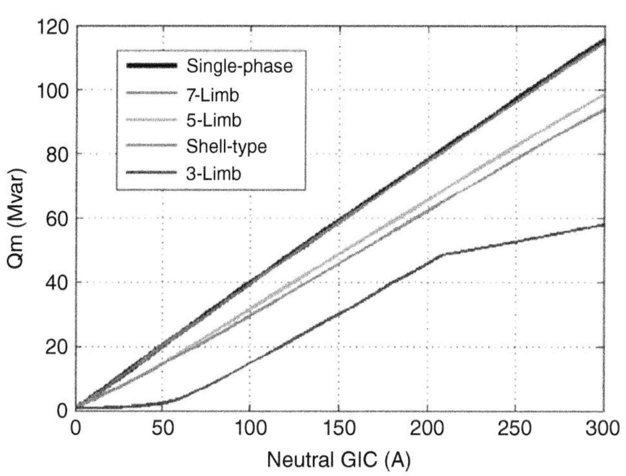

Figure 10.13 Relationship between transformer reactive power losses and GICs for different transformer types. Source: reproduced from Rezaei-Zare et al. (2016). Source: Reproduced with permission of IEEE. (*See insert for color representation of the figure.*)

the GIC Model is toggled to "Piecewise Linear" the user can enter the parameters of this linear function, shown in the highlighted line in Figure 10.14b. Note that one should try to acquire the actual K values for the transformers in the study system for the best results, otherwise these suggested defaults can be used.

Leaving the parameters of this case to default, let us run a basic voltage stability study. Clear the GIC values calculated so far by clicking on the said button shown in Figure 10.12 if the "Include GIC in Power Flow" button was unchecked. Otherwise just close the case without saving changes, and open it again. Now solve the power flow solution once; the easiest way to do this is to navigate to the main menu Ribbon of Simulator, click on Tools, and then on "Single Solution – Full Newton." This is the base case power flow solution with no GMD.

A power flow solution, in simple words, takes the load and generation values (real and reactive power), and voltage set points specified at generators and other buses, and calculates bus voltage magnitudes and angles, and power flows along the lines. The voltages can be viewed in a table (Case Information → Model Explorer → Buses), or visualized with the help of a contour as shown in Figure 10.16a. A contour can be quickly generated by right clicking on the one-line, and selecting the Contouring option. A dialog will pop up. Note that this is a Bus, Voltage, Per Unit Magnitude Contour, so those parameters can be selected from the Contouring dialog if they are not already populated by default.

Now returning to the GIC Add-On, increase the electric field to 5 V/km at 90°. This time, checkmark the option "Include GIC in Power Flow" and click on Calculate GIC Values. Then, solve the power flow solution again using the same "Single – Solution – Full Newton" button. After updating the contour (again by right clicking and following previous steps, and clicking "Recalculate

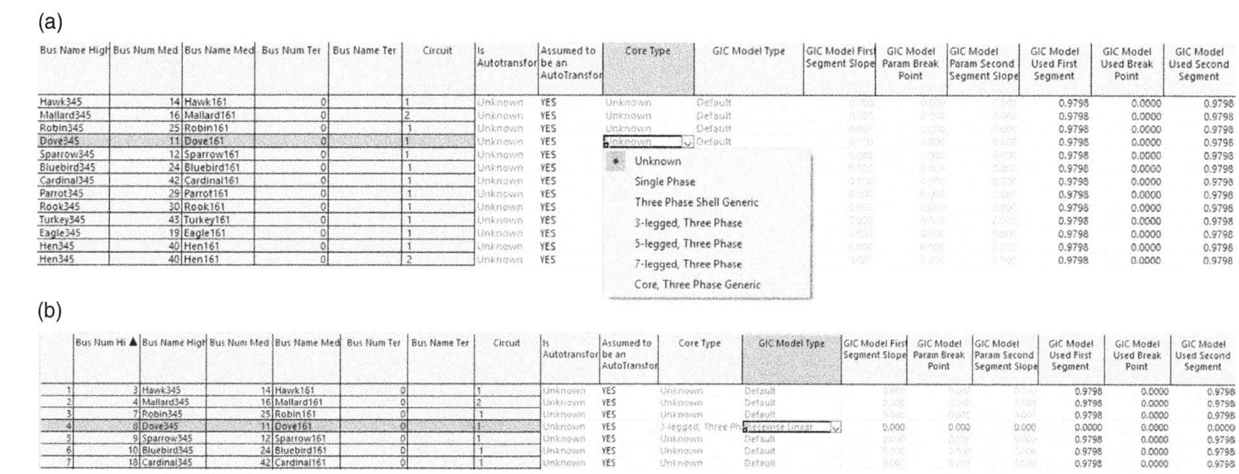

(a)

Bus Name High	Bus Num Med	Bus Name Med	Bus Num Ter	Bus Name Ter	Circuit	Is Autotransfor	Assumed to be an AutoTransfor	Core Type	GIC Model Type	GIC Model First Segment Slope	GIC Model Param Break Point	GIC Model Param Second Segment Slope	GIC Model Used First Segment	GIC Model Used Break Point	GIC Model Used Second Segment
Hawk345	14	Hawk161	0		1	Unknown	YES	Unknown	Default	0.000	0.000	0.000	0.9798	0.0000	0.9798
Mallard345	16	Mallard161	0		2	Unknown	YES	Unknown	Default	0.000	0.000	0.000	0.9798	0.0000	0.9798
Robin345	25	Robin161	0		1	Unknown	YES	Unknown	Default	0.000	0.000	0.000	0.9798	0.0000	0.9798
Dove345	11	Dove161	0		1	Unknown	YES	Unknown	Default	0.000	0.000	0.000	0.9798	0.0000	0.9798
Sparrow345	12	Sparrow161	0		1	Unknown	YES			0.000	0.000	0.000	0.9798	0.0000	0.9798
Bluebird345	24	Bluebird161	0		1	Unknown	YES			0.000	0.000	0.000	0.9798	0.0000	0.9798
Cardinal345	42	Cardinal161	0		1	Unknown	YES			0.000	0.000	0.000	0.9798	0.0000	0.9798
Parrot345	29	Parrot161	0		1	Unknown	YES			0.000	0.000	0.000	0.9798	0.0000	0.9798
Rook345	30	Rook161	0		1	Unknown	YES			0.000	0.000	0.000	0.9798	0.0000	0.9798
Turkey345	43	Turkey161	0		1	Unknown	YES			0.000	0.000	0.000	0.9798	0.0000	0.9798
Eagle345	19	Eagle161	0		1	Unknown	YES			0.000	0.000	0.000	0.9798	0.0000	0.9798
Hen345	40	Hen161	0		1	Unknown	YES			0.000	0.000	0.000	0.9798	0.0000	0.9798
Hen345	40	Hen161	0		2	Unknown	YES			0.000	0.000	0.000	0.9798	0.0000	0.9798

(b)

	Bus Num Hi ▲	Bus Name High	Bus Num Med	Bus Name Med	Bus Num Ter	Bus Name Ter	Circuit	Is Autotransfor	Assumed to be an AutoTransfor	Core Type	GIC Model Type	GIC Model First Segment Slope	GIC Model Param Break Point	GIC Model Param Second Segment Slope	GIC Model Used First Segment	GIC Model Used Break Point	GIC Model Used Second Segment
1	3	Hawk345	14	Hawk161	0		1	Unknown	YES	Unknown	Default	0.000	0.000	0.000	0.9798	0.0000	0.9798
2	4	Mallard345	16	Mallard161	0		2	Unknown	YES	Unknown	Default	0.000	0.000	0.000	0.9798	0.0000	0.9798
3	7	Robin345	25	Robin161	0		1	Unknown	YES	Unknown	Default	0.000	0.000	0.000	0.9798	0.0000	0.9798
4	8	Dove345	11	Dove161	0		1	Unknown	YES	3-legged, Three Ph	Piecewise Linear	0.000	0.000	0.000	0.0000	0.0000	0.0000
5	9	Sparrow345	12	Sparrow161	0		1	Unknown	YES	Unknown	Default	0.000	0.000	0.000	0.9798	0.0000	0.9798
6	10	Bluebird345	24	Bluebird161	0		1	Unknown	YES	Unknown	Default	0.000	0.000	0.000	0.9798	0.0000	0.9798
7	18	Cardinal345	42	Cardinal161	0		1	Unknown	YES	Unknown	Default	0.000	0.000	0.000	0.9798	0.0000	0.9798

Figure 10.14 (a) Transformer core types in PowerWorld Simulator. (b) Entering K values and core type.

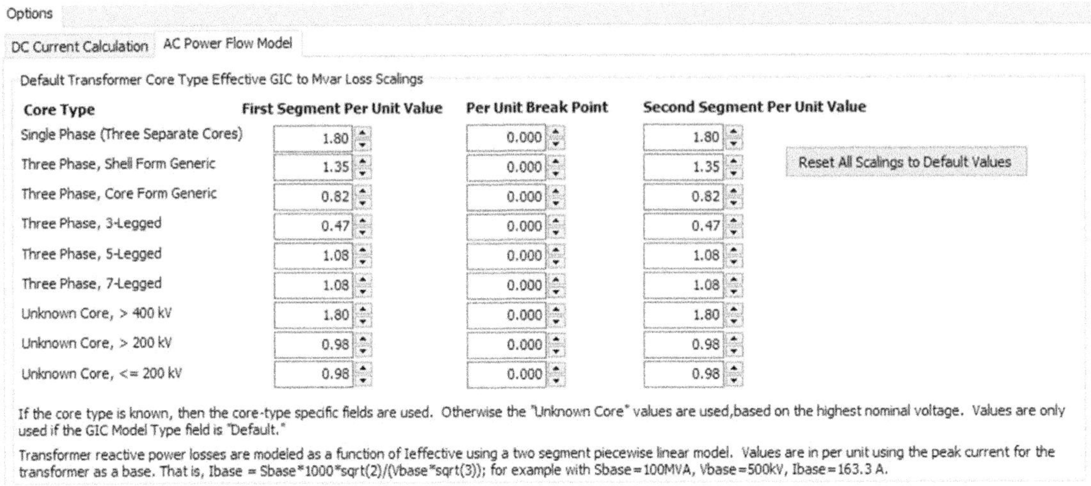

Figure 10.15 Default *K* factor settings (can be modified).

Contour"), we see in Figure 10.16b the GIC flows on the lines, and that the system voltages have now decreased. The one-line also shows the total reactive losses because of the GMD.

Of course, this is a very simple study example of a very small system, and the lowest bus voltage in this case is only 0.976 pu, which is not necessarily a cause for concern. Ultimately it is up to operating entity to determine safe voltage limits. Such studies can also help plan for reactive power support in the event of GMDs of a certain strength, especially when capacitors may be tripping. While we cannot model harmonic tripping of devices here, those studies could be performed in EMTP type tools and the results translated here to perform large system voltage stability assessments.

10.6. DYNAMIC ELECTRIC FIELDS

We have discussed static electric fields so far to calculate a solution "snapshot" of the system (see Fig. 10.6, top left corner); however a GMD electric field is usually a time series (i.e., $E(t)$) expressed over several hours, or at least minutes. While $E(t)$ has been used or suggested for use in individual transformer hotspot heating analysis, a large system GIC study can also be done as a time step simulation to calculate the changing GICs and reactive power losses over the duration of a storm. This could help us determine properties such as maximum rate of change of voltage during intense periods of the storm, or implement such analyses in a real-time environment to simulate grid operations and mitigation actions in the presence of a GMD. Figure 10.6 also shows the options to load time varying electric fields for simulation.

While we cannot delve into the details of these options due to space limitations, the underlying principles are more or less similar to the static field, that is Equations (10.5)–(10.7). Time domain or time series power system simulations are commonly performed, hence there exists a capability to include electric field modeling in there. To calculate GICs we still only consider the DC representation of the system. A time-varying electric field can be included in two ways, (i) including a piece-wise linear function from the tool itself (Time varying series voltage inputs), or (ii) including an external file which has the $E(t)$ data (Time varying electric field input). With the latter we can define a spatio-temporally varying electric field as well.

It is well known by now that the uniform field assumption especially over a large footprint is not realistic due to non-uniformities caused by local earth conductivity structures (Schultz, 2010; Love et al., 2018). Some of these non-uniformities have been addressed in benchmark GMD scenarios such as the one in NERC (2017). In that a 1D-layered earth conductivity model is assumed, which yields much less spatial variation than the electric field calculated using the 3D electromagnetic transfer functions (EMTF's). These detailed, spatio-temporally varying electric fields, which constitute a large volume of data are being applied to validate historical storm events using actual, large-system models (Gannon et al., 2017). For example, Figure 10.17 shows the result of a time series comparison of an actual measured GIC (magnitude) to the simulated value calculated in PowerWorld Simulator (Shetye et al., 2018). The spatial resolution for the electric field was $0.5° \times 0.5°$ latitude and longitude, with a sampling rate of 10 s. The results demonstrate the existing quality of the magnetic field data, ground and power grid models, as well as the calculation methods involved in utilizing them.

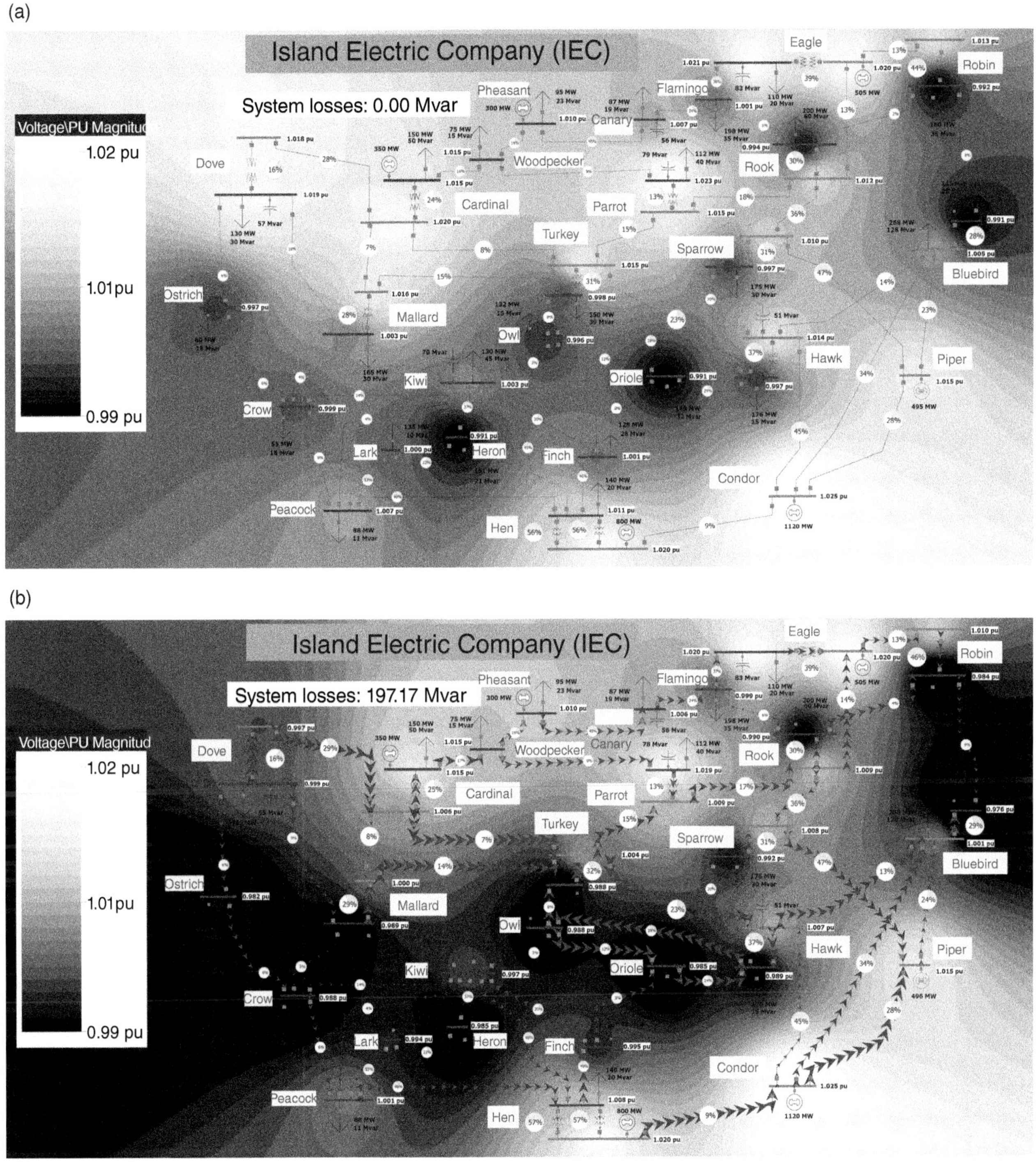

Figure 10.16 (a) Bus voltages of the 40-bus case without a GMD. (b) Bus voltages of the 40-bus case with a 5V/km eastward electric field. (*See electronic version for color representation of this figure.*)

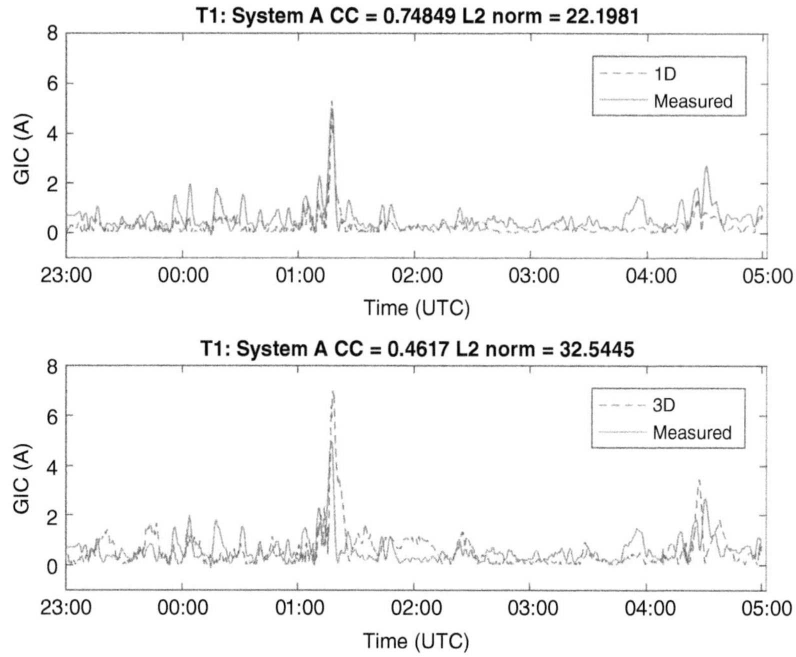

Figure 10.17 Comparison between measured and simulated GIC (using 1D conductivity model) at an actual transformer, for a GMD event that occurred in May 2016.

10.7. CONCLUDING REMARKS

There is certainly more that can be covered on GMD modeling in power systems; this chapter just scratches the surface of the power system basics in theory and simulations, and is intended as an introduction to the topic for non-expert professionals in fields involved in the interdisciplinary problem of GIC-modeling. There are many more advanced simulation capabilities available in existing tools, as well as being developed and implemented in research. Validating models and simulation results with GIC measurements for more trustworthy system assessments, has been driving a lot of these advancements. For instance, models are now available to study electric field "hotspots," that is local intensifications, whereas on the transformer side there are quite detailed and complex models for magnetic (i.e., saturation), harmonic, and thermal modeling in the presence of GICs.

The goal of this chapter was to introduce the reader to the fundamental pieces of a power system GMD model, while walking through simple examples to get familiar with the tools and common simulation steps, etc. The idea is that with the help of these basics, the reader may be able to build on them by better understanding power system related terminology, devices, parameters, data, models, simulations, and measurements, which may help foster and improve collaborative efforts in this multidisciplinary research area.

ACKNOWLEDGMENTS, SAMPLES, AND DATA

There are no conflicts of interest to declare.

The supporting data such as the two power system case examples can be obtained from a web link (https://my.syncplicity.com/share/flgbtqs65fzorvj/AGU%20Chapter%20Cases).

APPENDIX

Autotransformer: A type of transformer where there is no physical separation between the "high" voltage side and "low" side. There is a series and common winding. The common winding is shared by the "effective" high and low sides.

Delta: The windings of a three-phase transformer are connected in the shape of capital Greek letter "Delta." Looks like a triangle.

Effective GIC: The GIC flowing in a transformer through both the windings, and not the total flowing into the neutral. This is especially different in case of an autotransformer.

EMTP: Electromagnetic transient program.

GSU: Generator step-up (transformer). It increases the voltage at a generator to make it suitable for transmission (higher voltage, less I^2R powet losses).

K-factor: A scalar that relates transformer GICs to its reactive power losses.

Per Unit: A unit-less system of expressing power system quantities such as voltage, current, and power, normalized to a certain "base" value.

Wye: The windings of a three-phase transformer are connected in the shape of the letter "Y." Also known as "Star."

Ybus: Nodal admittance matrix. Consists of admittance values of various power system elements at or between certain nodes. Commonly used in power flow solutions, etc.

REFERENCES

Adhikari, S., D. Mueller, R. Walling, and A. J. O'Laughlin (2016), A comprehensive study of geomagnetic disturbance (GMD) system impact, *IEEE PES General Meeting,* July 2016.

Albertson, V. D., J. G. Kappenman, N. Mohan, and G. A. Skarbakka (1981), Load-flow studies in the presence of geomagnetically induced currents, *IEEE Transactions on Power Apparatus and Systems*, *100*(2), 594–607, doi:https://doi.org/10.1109/TPAS.1981.316916.

Albertson, V. D., J. M. Thorson, and S. A. Miske (1974), The effects of geomagnetic storms on electric power systems, *IEEE Trans. Power App. & Sys.*, *PAS-93*, 1031–1044.

Albertson, V. D., and J. A. Van Baelen (1970), Electric and magnetic fields at the Earth's surface due to auroral currents, *IEEE Transactions on Power Apparatus and Systems*, *89*(4), 578–584, doi:0.1109/TPAS.1970.292604.

Aspnes, J. D., R. P. Merritt, and S. Akasofu (1981), Effects of geomagnetically induced current on electric power systems, *The Northern Engineer*, *13*(3), 34–38.

Balma, P. M. (1992), Geomagnetic effects on a bank of single phase generator step-up transformers, *Proceedings of Geomagnetically Induced Currents Conference,* California, Nov 8–10, 1989, EPRI TR-lOO450.

Birchfield, A. B., K. M. Gegner, T. Xu, K. S. Shetye, and T. J. Overbye (2017b), Statistical considerations in the creation of realistic synthetic power grids for geomagnetic disturbance studies, *IEEE Transactions on Power Systems*, *32*(2), 1502–1510, doi:https://doi.org/10.1109/TPWRS.2016.2586460

Birchfield, A. B., T. Xu, K. M. Gegner, K. S. Shetye, and T. J. Overbye (2017a), Grid structutal characteristics as validation criteria for synthetic networks, *IEEE Transactions on Power Systems*, *32*(4), 3258–3265, doi:https://doi.org/10.1109/TPWRS.2016.2616385

Boteler, D. H., and R. J. Pirjola (1998), Modeling geomagnetically induced currents produced by realistic and uniform electric fields, *IEEE Transactions on Power Delivery*, *13*(4), 1303–1308, doi:https://doi.org/10.1109/61.714500.

Boteler, D. H., and R. J. Pirjola (2014), Comparison of methods for modelling geomagnetically induced currents, *Ann. Geophys.*, *32*, 1177–1187.

Boteler, D. H., R. J. Pirjola, and H. Nevanlinna (1998), The effects of geomagnetic disturbances on electrical systems at the earth's surface, *Adv. Space Res.*, *22*(1), 17–27, doi:https://doi.org/10.1016/S0273-1177(97)01096-X.

Bozoki, B. (chair) et al. (IEEE power system relaying committee) (1996), The effects of GIC on protective relaying, *IEEE Trans. Power Delivery*, *11*, 725–739.

Bui, U., T. J. Overbye, K. Shetye, H. Zhu, and J. Weber (2013), Geomagnetically induced current sensitivity to assumed substation grounding resistance, *2013 North American Power Symposium (NAPS)*, doi:https://doi.org/10.1109/NAPS.2013.6666893

Dixon, L. H. (2003), Eddy current losses in transformer windings and circuit wiring, *Texas Instruments*. Available from: http://www.ti.com/lit/ml/slup197/slup197.pdf

DOE (2015), United States electricity industry primer, *Tech. rep.*, Office of Electricity Delivery and Energy Reliability, U.S. Department of Energy. Available from: https://www.energy.gov/indianenergy/downloads/united-states-electricity-industry-primer

Dong, X., Y. Liu, and J. G. Kappenman (2001), Comparative analysis of exciting current harmonics and reactive power consumption from GIC saturated transformers, *Proceedings of IEEE 2001 Winter Meeting*, 318–322, doi:https://doi.org/10.1109/PESW.2001.917055.

FERC (2013), Order no. 779: Reliability standards for geomagnetic disturbances, Docket No. RM12-22-000, Federal Energy Regulatory Commission.

FERC (2016) Order no. 830: Reliability standard for transmission system planned performance for geomagnetic disturbance events, Docket No. RM15-11-000, Federal Energy Regulatory Commission.

Gannon, J. L., A. B. Birchfield, K. S. Shetye, and T. J. Overbye (2017), A comparison of peak electric fields and GICs in the Pacific Northwest using 1-D and 3-D conductivity, *Space Weather*, *15*, 1535–1547, doi:https://doi.org/10.1002/2017SW001677.

Glover, J. D., T. J. Overbye, and M. S. Sarma (2017), *Power System Analysis and Design*, Cengage Learning.

Horton, R., D. H. Boteler, T. J. Overbye, R. J. Pirjola, and R. Dugan (2012), A test case for calculation of geomagnetically induced currents, *IEEE Transactions on Power Delivery*, *27*(4), 2368–2373, doi:https://doi.org/10.1109/TPWRD.2012.2206407

IEEE (1991), IEEE guide for measurement of impedance and safety characteristics of large, extended or interconnected grounding systems, *IEEE Standard*, 81.2, doi:https://doi.org/10.1109/IEEESTD.1992.106970

IEEE (2012), IEEE guide for measuring earth resistivity, ground impedance and earth surface potentials of a ground systems, *IEEE Standard*, 81-2012, doi:https://doi.org/10.1109/IEEESTD.2012.6392181

Kappenman, J. G. (1996), Geomagnetic storms and their impact on power systems, *IEEE Power Engineering Review*, *16*, 5.

Lehtinen, M., and R. Pirjola (1985), Currents produced in earthed conductor networks by geomagnetically-induced electric fields, *Ann. Geophys.*, *3*, 479–484.

Love, J. J., G. M. Lucas, A. Kelbert, and P. A. Bedrosian (2018), Geo-electric hazard maps for the mid-Atlantic United States: 100 year extreme values and the 1989 magnetic storm, *Geophysical Research Letters*, *44*, doi:https://doi.org/10.1002/2017GL076042.

NERC (1990), March 13, 1989 geomagnetic disturbance, North American Electric Reliability Corporation. Available from: https://www.nerc.com/files/1989-quebec-disturbance.pdf

NERC (2012), 2012 Special reliability assessment interim report: Effects of geomagnetic disturbances on the bulk power system, Tech. rep., North American Electric Reliability Corporation.

NERC (2013), Application Guide: Computing geomagnetically-induced current in the bulk-power system, *Tech. Rep.*, North American Electric Reliability Corporation.

NERC (2014), TPL-007-1: Transmission system planned performance for geomagnetic disturbance events, *Standard*, North American Electric Reliability Corporation. Available from: http://www.nerc.com/pa/Stand/Reliability%20Standards/TPL-007-1.pdf

NERC (2017), TPL-007-2: Transmission system planned performance for geomagnetic disturbance events, *Standard*, North American Electric Reliability Corporation. Available from: http://www.nerc.com/pa/Stand/Reliability%20Standards/TPL-007-2.pdf

NERC and DOE (2010), High-impact low-frequency event risk to the North American bulk power system, *A Jointly-Commissioned Summary Report of the North American Electric Reliability Corporation and the U.S. Department of Energy's November 2009 Workshop*, June 2010.

Overbye, T. J., T. R. Hutchins, K. Shetye, J. Weber, and S. Dahman (2012), Integration of geomagnetic disturbance modeling into the power flow: A methodology for large scale system studies, *North American Power Symposium (NAPS), 2012*, 1–7, doi:https://doi.org/10.1109/NAPS.2012.6336365.

Overbye, T. J., K. S. Shetye, T. R. Hutchins, Q. Qiu, and J. D. Weber (2013), Power grid sensitivity analysis of geomagnetically induced currents, *IEEE Transactions on Power Systems*, *28*(4), 4821–4828, doi:https://doi.org/10.1109/TPWRS.2013.2274624

Prabhakara, F. S., L. N. Hannett, R. J. Ringlee, and J. Z. Ponder (1992), Geomagnetic effects modeling for the PJM interconnection system part ii – Geomagnetically induced current study results, *IEEE Transactions on Power Systems*, *7*(2), 565–571, doi:https://doi.org/10.1109/59.141760.

Ramos, R. (2018), On the development of GIC system models – A transmission planner's perspective, NERC Geomagnetic Disturbance Task Force, Available from: http://www.nerc.com/comm/PC/Geomagnetic%20Disturbance%20Task%20Force%20GMDTF%202013/Agenda_presentations_jan_30.pdf

Rezaei-Zare, A., L. Marti, A. Narang, and A. Yan (2016), Analysis of three-phase transformer response due to GIC using an advanced duality-based model, *IEEE Transactions on Power Delivery*, *31*(5), 2342–2350, doi:https://doi.org/10.1109/TPWRD.2015.2505499.

Schultz, A. (2010), EMScope: A continental scale magnetotelluric observatory and data discovery resource, *Data Science Journal*, *8*, IGY6–IGY20.

Shetye, K. S., J. L. Gannon, A. B. Birchfield, R. H. Lee, and T. J. Overbye (2018), Impact of 1D vs 3D Earth conductivity based electric fields on geomagnetically induced currents, *Presented at Innovative Smart Grid Technologies (ISGT) Europe*, October 2018, Sarajevo, Bosnia, and Herzegovina.

Shetye, K. S., and T. J. Overbye (2015), Modeling and analysis of GMD effects on power systems: An overview of the impact on large-scale power systems, *IEEE Electrification Magazine*, *3*(4), 13–21, doi:https://doi.org/10.1109/MELE.2015.2480356.

Torta, J. M., S. Marsal, and M. Quintana (2014), Assessing the hazard from geomagnetically induced currents to the entire high-voltage power network in Spain, *Planets and Space*, *66*:87, doi:https://doi.org/10.1186/1880-5981-66-87.

Walling, R. A., and A. H. Khan (1991), Characterisitcs of transformer exciting current during geomagnetic disturbances, *IEEE Transactions on Power Delivery*, *6*(4), 1707–1714, doi:https://doi.org/10.1109/61.97710

Xu, T. (2016), Electric grid test cases, Available from: https://electricgrids.engr.tamu.edu

Zheng, K., D. Boteler, R. J. Pirjola, L. Liu, R. Becker, L. Marti, S. Boutilier, and S. Guillon (2014), Effects of system characteristics on geomagnetically induced currents, *IEEE Transactions on Power Delivery*, *29*(2), 890–898, doi:https://doi.org/10.1109/TPWRD.2013.2281191.

11

Geomagnetically Induced Currents from Extreme Space Weather Events

L. M. Winter

ABSTRACT

In this chapter, we discuss the drivers and measured effects of extreme/severe geomagnetic storms. We focused on the eight modern storms and two historical storms with the highest disturbance storm time (Dst) index ($\lesssim$ –300 nT). The geomagnetically induced currents (GICs) from these modern storms at their most extreme instantaneous effects caused power blackouts lasting hours. Predicting GICs is difficult since many physical systems interact to produce each of the storms discussed, including solar drivers (flares and coronal mass ejections, CMEs), interplanetary magnetic fields, the magnetosphere, ionosphere, and ground conductivity. Every storm is unique and a number of factors from sudden storm commencement components, Dst index, auroral electrojet direction and strength, and time variation of the magnetic field along with the geospatial extent of the storm, local ground conductivity, and technological specifications of the system all contribute to whether GIC effects will occur. All of these effects should be taken into account for modeling GIC effects of extreme storms.

Key Points

- A historical review of the largest known geomagnetic storms is given along with their solar drivers and known GIC impacts.
- Extreme geomagnetic storms are rare and all unique in their drivers and impacts.
- GIC modeling and benchmarks need to account for this uniqueness with a range of input storm parameters from different types of geomagnetic storms.

11.1. INTRODUCTION

High-energy plasma storms in the space environment surrounding Earth can affect ground level technology. Specifically, movements of charged particles in space induce currents on the ground, which are known as geomagnetically induced currents (GICs). Due to the rarity of extreme geomagnetic storms the effects of GICs are mainly deduced based on the properties of the extreme storms, including the most severe (the 1 September 1859 solar storm; Carrington, 1859), recent lower level impactful storms, and modeling.

The 1859 storm, referred to as the Carrington event after the British astronomer Richard Carrington who recorded observations of the intense solar flare, occurred at a time when human technology was less susceptible to GICs. However, it provides a benchmark for the solar and geomagnetic conditions of an extreme event. Hand-drawn diagrams of the solar conditions and saturated magnetometer measurements from the geomagnetic latitudes where GICs are expected to be strongest leave gaps in our understanding of the impacts. Therefore, we must also rely on measurements from geomagnetic storms that were more recent. In Section 11.2, we discuss how extreme events are generated and the properties of the most severe historical and recent geomagnetic storms (Dst (min) below about −300 nT). In Section 11.3, we discuss how extreme storms are defined for GIC studies. In Section 11.4, we discuss how often extreme events are expected to occur.

Los Alamos National Laboratory, Los Alamos, NM, USA

Geomagnetically Induced Currents from the Sun to the Power Grid, Geophysical Monograph 244, First Edition.
Edited by Jennifer L. Gannon, Andrei Swidinsky, and Zhonghua Xu.

Studies from academic, industry, and government agencies identify infrastructure affected by GICs during past geomagnetic activity. Of utmost concern for extreme GIC effects are power grid systems. In at least one case, solar storm linked GIC events caused damage to electricity infrastructure that resulted in geographically localized black outs (e.g., Quebec Blackout Storm of 1989). Additional systems that rely on long wires conducting current are also likely to be affected during an extreme GIC event. We discuss these systems and impacts from past extreme GMDs in Section 11.5. Our conclusions are presented in Section 11.6.

11.2. EXTREME GEOMAGNETIC STORMS

This section reviews the conditions leading to geomagnetic disturbances (GMDs) – including solar drivers, interplanetary fields, and the Earth's magnetospheric and ionospheric response. Severe geomagnetic storms from the past ~30 years and the two largest historical storms are highlighted. The past storms used for extreme impact studies are discussed.

Geomagnetic disturbances (GMDs) refer to changes in the magnetic field with respect to time (i.e., dB/dt). GMDs drive GICs, which consequently result from the interplay between solar wind, interplanetary magnetic field (IMF), and the Earth's geomagnetic field. In particular, solar events including coronal mass ejections (CMEs), intense plasma mass loss events typically occurring close in time to a solar flare, and co-rotating interaction regions, produced from coronal holes, are major drivers of GMDs. Both CMEs and solar flares are related to complex sunspot groups or active regions on the solar surface. Solar flares result from magnetic reconnection events from these active regions and their intensity is often measured through their X-ray emission. Intense flares, with X-ray class of M- or X- based on a peak 1–8 Angstrom X-ray flux above 10^{-5} or 10^{-4} W/m^2, are often followed by strong CMEs. Geo-effective CMEs, which cause GMDs, result from fast CMEs with speeds typically above 800 km/s that drive shocks in the solar wind (e.g., Richardson and Cane, 2012). These CMEs also carry embedded magnetic fields that change the direction and intensity of the Earth's magnetic field. CMEs occur as often as two to three per day during solar maximum (St. Cyr et al., 2000) and range in speed up to ~3000 km/s (Yashiro et al., 2004).

A phenomenon known as "preconditioning" can occur when multiple CMEs happen in quick succession. The earlier CMEs change the ambient solar wind and IMF conditions so that more complex IMF structures arise from the combined CMEs. The resultant activity is longer, more intense geomagnetic storms. For example, such preconditioning has led to the high solar wind speed of 2246 ± 110 km/s and strong ejecta magnetic field of 109 ± 1 nT observed by the STEREO A spacecraft in July 2012 (Liu et al., 2014). This event missed the Earth, but is comparable to the inferred CME speed from the extreme 1859 Carrington event, which had an estimated speed of ~2400 km/s (Gopalswamy et al., 2005a; Riley, 2012). Extreme GICs are more likely to occur in conjunction with fast CMEs that are more probable through a preconditioning event, but the magnetospheric response to a CME driven shock is dependent on more factors than speed alone. See for example, Green and Baker (2015) for a summary of CME drivers related to severe space weather.

Geomagnetic activity is measured through geomagnetic disturbance indices, such as the hourly averaged disturbance storm time (Dst) index. Dst is a global measurement of the ring current in the inner magnetosphere (Dessler and Parker, 1959). The ring current is part of the inner magnetosphere (altitude ~15–30 000 km) consisting of ions and electrons flowing westward around the Earth in the equatorial plane. During a geomagnetic storm, the intensity of the ring current increases resulting in a diminished geomagnetic field and a negative Dst index.

The geomagnetic disturbance indices are shown to vary in response to CME interaction with the Earth's magnetosphere (Newell et al., 2007). GMDs are often associated with southward Bz, the component of the IMF northward and parallel to the Earth's magnetic field (Thomson and Wild, 2010). Estimates for Dst during the 1859 Carrington event range from −1760 to −850 nT (Tsurutani et al., 2003; Siscoe et al., 2006). The most severe storm with measured minimum Dst of −589 nT was the 13 March 1989 solar storm, which resulted in GICs that caused large-scale power grid outages in Quebec.

Extreme geomagnetic storms, with minimum Dst < −300 nT, are rare. Further, the current solar cycle 24 has been weaker than previous cycles with respect to sunspot number and thereby the solar drivers of GMDs were weaker than in the past decade. Uniform Dst measurements only began in 1957, despite the establishment of worldwide geomagnetic observatories in the early 1830s (Barraclough, 1991). Consistent CME observations enabled by the coronagraph instrument on NASA's Solar and Heliospheric Observatory only existed since 1996. Cliver et al. (2009) compiled a list of the 25 largest geomagnetic storms from 1905 to 2004 based on Dst and/or an extension of the Dst index for time periods before 1957 (from Love (2007)). These storms range from peak minimum Dst of −328 to −670 nT, including only five storms occurring after the 1980s. Prior to this time, both the Carrington event and the great storm of 1921 exceed the minimum Dst of these storms. No intense storms in this Dst range have occurred after 2004.

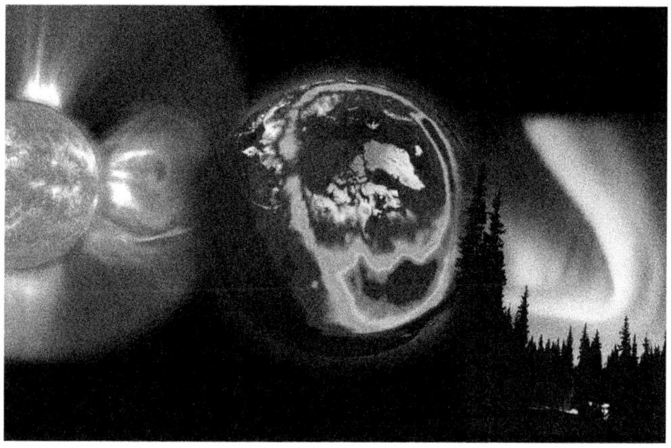

Figure 11.1 Composite image from the SOHO image gallery of the 14 July 2000 CME, which led to a strong GMD event at Earth. The first panel shows the extreme UV 304 Å image of the Sun superimposed on LASCO's coronagraph image of the CME. Next, the Polar VIS image shows the location and intensity of charged particles producing aurora and the final panel shows the view of aurora from the ground. Credit: SOHO (ESA & NASA). *(See electronic version for color representation of this figure.)*

The ionospheric response to the GMD is likely a more direct indicator of GICs (illustrated in Figure 11.1 as the chain between flares and CMEs at the Sun to the auroral currents seen on the ground). In particular, auroral electrojets are currents in the D and E regions of the auroral ionosphere created from the interaction between the solar wind and magnetosphere. During GMDs, they expand to lower latitudes and their strength increases due to an increase in energetic particle flux associated with CMEs. The Carrington event as an extreme event resulted in aurora seen as far toward the equator as Colombia (Cardinas, 2016). Both the strength and time variability in this current, which flows at about 100 km above the surface, can induce changes in geo-electric fields on the ground resulting in GICs. For this reason, the auroral electrojet indices (i.e., AE, AU, AL, AO) are useful in determining the strength of ionospheric currents that can lead to GICs.

In Table 11.1, we include a list of the largest geomagnetic storms from the past 30 years along with the two most extreme storms recorded (the Carrington event and the great storm of 1921). We also include the March 1991 storm, which corresponds with one of the largest GICs measured in the United States (Kappenman, 2003). Our list includes details of associated solar drivers of GMDs including sunspot number (SSN), X-ray flare class, and CME speed. For the magnetospheric response – we list the minimum Dst, maximum AU, and minimum AL – the upper auroral electrojet indices from the upper and lower envelope of the auroral electrojet are included as an estimate of the strongest current intensity in the eastward and westward auroral electrojets. For the recent storms, SSN, Dst, and AU are obtained from the NASA Near-Earth Heliosphere Data (OMNI) and X-ray flare measurements are from NOAA. The monthly sunspot number for the 1859 and 1921 events is from the Royal Observatory of Belgium's SILSO sunspot network (Available at: http://sidc.be/silso). For storms since 1996, CME speed estimates are the apparent speed from the Solar and Heliospheric Observatory's Large Angle and Spectrometric Coronagraph (e.g., Yashiro et al., 2004). References/notes include sources for historical storm data.

All geomagnetic storms are unique and "geoeffectiveness" cannot be measured solely by the strength of the solar storm or geomagnetic indices (Lanzerotti, 1992). As shown in Table 11.1, even among the extreme storms listed there is a great deal of variability in their solar drivers and magnetospheric and ionospheric response. For instance, the 1859 storm is believed to be associated with both the most intense X-ray flare in known history and a very fast CME. However, while the March 1989 coincided with a high intensity X15 X-ray flare, the associated CME was relatively slow (i.e., had a lower Sun–Earth transit time), but was exceptionally bright and thought to have a high dynamic pressure when it reached the magnetosphere (Feynman and Hundhausen, 1994). The 9 November 1991 event produced a severe geomagnetic storm that resulted from the eruption of a quiescent filament and not an active sunspot region (Cliver et al. 2009). The magnetic cloud released was highly inclined to the ecliptic plane (Gopalswamy et al., 2005b), showing that the geometry of events with respect to Earth is also a major factor in GMDs. Also of note is the importance of storm sudden commencements (SSCs), magnetic shocks associated with abrupt increases or decreases in the northward component of the geomagnetic field. SSCs likely caused the largest GIC measurements in the United States, associated with the March 1991 event (Kappenman, 2003).

The range of sunspot numbers for the extreme storms in Table 11.1 shows that not all GMDs are produced during solar maximum or peaks in the 11-year solar activity cycle. From this list, we also find that the auroral electrojet and magnetospheric responses, indicated by AU, AL, and Dst, are not directly correlated. The strongest auroral electrojet currents were seen with lower Dst index (i.e., October 2003, July 2000, and March 1991 GMDs). Intensity variations in the geomagnetic field (Kappenman, 2006), measured through local changes in the horizontal component of the magnetic field at ground magnetometer stations (dBH/dt), are another important factor in producing GICs (e.g., Viljanen et al., 2001). This is also dependent upon geology/local ground conductivity (e.g., Beamish et al., 2002) and

Table 11.1 Historical and Recent Large Geomagnetic Storms.

Date	SSN	X-ray flare class	CME speed (km/s)	Min Dst (nT)	Max AU (nT)	Min AL (nT)	Notes/references
1–2 September 1859	(147)	X45-100	2400	−850 to −1750	?	?	Largest known storm (SSN – SILSO; Dst – Turutani et al., 2003; Siscoe et al., 2006)
13–15 May 1921	(71)	?	?	−825 to −900	?	?	(SSN – SILSO; Dst – Kappenman, 2006)
13 March 1989	235	X15	770	−589	1884	−1892	Feynman and Hundhausen (1994); Ranked 3 in (Cliver et al., 2009) C09
20 November 2003	111	M5	1660 (11/18), 669 (11/20)	−422	1698	−1790	Ranked 5 in C09
31 March 2001	300	X1	942 (03/29)	−387	1524	−1420	Ranked 20 C09
8–10 November 2004	92	X2	2000 (11/09), 3387 (11/10)	−374	1912	−1586	Ranked 21 C09
9 November 1991	195	–	725	−354	1390	−1204	Ranked 15 in C09; eruption of quiescent filament (Cliver et al., 2009)
30 October 2003	250	X17	2459 (10/28), 2029 (10/29)	−353	2241	−2203	Ranked 12 in C09
16 July 2000	283	X5.7	1674 (07/14)	−301	2023	−1089	
24–25 March 1991	203	X9	?	−298	3195	−1643	Largest US GIC measurement from Kappenman (2003)

geolocation (e.g., Ngwira et al., 2015; Woodroffe et al., 2016) of the potentially affected technological systems.

To illustrate differences in the magnetic field measurements from modern severe storms, we downloaded magnetometer observations of all of the storms in Table 11.1 from the Ottawa station from INTERMAGNET. Only data since 1991 are available. In Figure 11.2, we plot the horizontal component of the magnetic field. The strongest storms in terms of Dst index, which is a global indicator, are often not the strongest (e.g., November 2003 GMD) at a specific geolocation. Additionally, storm profiles differ greatly.

11.3. EXTREME STORM DEFINITIONS FOR GIC STUDIES

For extreme GIC studies, the 1859 Carrington event, the 1989 Quebec Blackout storm, or 10 times the 1989 March storm (Bolduc, 2002) are often used as benchmarks. In particular, we next discuss selected examples of extreme storm models. Pulkkinen et al. (2012) define 1-in-100 year reference events through the maximum 100-year amplitude of the 10-s resolution horizontal geoelectric field from magnetometer data spanning January 1993 to December 2006. However, the data used

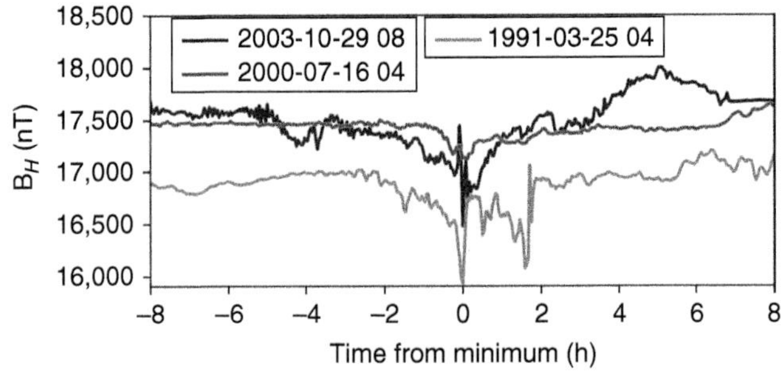

Figure 11.2 Horizontal component of the magnetic field measured in Ottawa, Canada for the modern storms listed in Table 11.1. The storms are ordered by minimum Dst index and the legend shows the time of minimum B_H. As shown, the profiles of each storm differ significantly, those believed to be most "geoeffective" exhibit a sudden storm commencement. (*See insert for color representation of the figure.*)

to construct the model are limited to a time frame of ~13 years, over which the most extreme storms did not occur. They also use uniform electric field models, not accounting for spatial variations. Responding to the FERC Order No. 779, NERC developed standards for GMD modeling (NERC, 2014). The NERC-defined benchmark GMD event is based upon the March 1989 event, using magnetometer observations from Finland and Japan with ground conductivity models from Quebec. It also does not account for differences in local conductivity and geomagnetic latitude. Further, while NERC defines this standard as a 1-in-100 year event, the 28 March 1946 storm had a minimum Dst of −569 nT and prior to that the 25 September 1909 storm had a minimum Dst of −670 nT (Cliver et al. 2009). This suggests instead a 1-in-40 year event.

Beggan et al. (2013) produced a model of an extreme geomagnetic field using the October 2003 event from magnetometer observations in the United Kingdom and North Sea. They determined spatial variations across the entire UK region using the Spherical Elementary Current Systems method (e.g., McLay and Beggan, 2010). These were scaled by a factor to account for both a 100 year and 200 year extreme event, which they define as having dBH/dt and BH of 1000–4000 nT/min and 2000–5000 nT (100 year) and 1000–6000 nT/min and 3000–6500 nT (200 year) from Thomson et al. (2011). A "thin-sheet" approximation is used to determine the electric field. They also compare simulated results from auroral electrojet models with the scaled up October 2003 storms. The Beggan et al. (2013) techniques have significant advantages in that they account for local conductivity and spatial variations. They also use magnetic field responses from a real storm.

Recently, Winter et al. (2017) presented a new technique for modeling an extreme Carrington-like electric field using a scaling relationship for geolatitude derived from local magnetometer measurements of moderate to severe storms. Like Beggan et al. (2013), this technique for defining an extreme event has the advantage of using local conductivity models and magnetometer observations to define the geomagnetic and geoelectric field directly for a specific location. The benefit of this model is that it incorporates higher frequency (1-min and 1-s scale) variations in the magnetic field at specific geolocations with the direct 1-h magnetometer observations of the extreme Carrington event from the Colaba Magnetic Observatory.

To summarize, a number of techniques and extreme event scenarios exist in the literature with a focus on characterizing a 1-in-100 year event. For accurate models of GIC effects, the geomagnetic latitude and local conductivity of the specific system of interest should be accounted for. Given the differences in dBH/dt profiles for all known severe storms, a range of extreme events should also be tested for systems vulnerable to GICs. Techniques like those from Beggan et al. (2013) and Winter et al. (2017) can be applied to additional historical storms to provide a range of extreme event electric field models for GICs.

Estimates for levels associated with severe GMDs range from geoelectric fields of 4–20 V/km. Pulkkinen et al. (2008) calculated the upper limit of GIC magnitude is on the order of 1000 A, consistent with Campbell, 1980 results from analyses of GIC effects in the Alaskan pipeline. Applying the Pulkkinen et al. (2008) to Carrington event effects at high latitude, a peak geoelectric field of 4 V/km is possible with induced currents of 2000 A. The NERC guidelines in TPL-007-1 for the 1-in-100 year reference event (the 1989 Quebec storm) is a maximum of 8 V/km in the United States. Beggan et al. (2013) and Winter et al. (2017) find similar Carrington-level estimates of 9–10 V/km for the United Kingdom. Yet, Kappenman (2003) calculate 20 V/km for a 1-in-100 year event. These estimates depend on geomagnetic latitude, definition of a severe storm (e.g., Carrington event, March 1989 event, or other), and ground conductivity models. However, accurate modeling for the expected GIC affects from a severe geomagnetic storm needs to account for the specific properties of the system of interest and should include a range of sample local geomagnetic field models with different time profiles, sudden storm components, and intensities.

11.4. OCCURRENCE RATE FOR EXTREME GMDs

In order to be prepared and to mitigate the effects of GICs, we need to know how likely they are to occur. This is difficult since the extreme space weather drivers are rare and the ground effects have a large dependence on the local ground conductivity, the geomagnetic latitudes affected by the particular storm, and the configurations of systems potentially affected are unique. GIC studies therefore focused on the occurrence rate of observable GMD-related indices/measurements associated with GICs.

Extreme value statistics (EVS) is a technique used in several studies to make inferences about the distribution followed by extreme cases (e.g., Coles, 2004). Tsubouchi and Omura (2007) took into account maximum sunspot number from solar cycle and Dst index in their EVS analysis of storms from 1957 to 2001. They predict that the March 1989 storm occurs as a 1-in-60-year event. Love et al. (2015) also found a 1 per-60 year frequency from a study of peak Dst for storms < −63 nT. However, as Cliver et al. (2009) pointed out, severe storms with similar Dst index occurred in 1909 and 1949 (i.e., every ~40 years over the last century).

Thomson et al. (2011) used EVS with one-minute resolution magnetometer observations from 28 European observatories to predict B_H (horizontal component of the magnetic field), declination (D), and their time series (dB_H/dt, dD/dt) for 100 year and 200 year GMDs. Pulkkinen et al. (2008) calculated occurrence statistics and conditional probabilities for GICs through analysis of peak daily geoelectric field measurements. They predict that large geoelectric fields of 10 V/km occur 10–100 times in 100 years.

For a Carrington-like event, Riley (2012) used a 55 year record of Dst index to find a 12% probability within the next decade. Kataoka (2013) used an 89 year dataset of dB_H, finding a 6% chance for the current solar cycle 24. Other studies estimate this event as a 1-in-150 year event (Lloyd's, 2013).

The statistics of how often extreme GMDs occur is heavily reliant on the dataset used. The limit of magnetometer records is 130 years, and even then it is difficult to compare modern, higher-resolution observations to older data from periods when more severe geomagnetic storms occurred more commonly. The question of how solar cycle level affects these estimates is an important point raised by Kataoka (2013). However, the NASA STEREO-A spacecraft observed a July 2012 CME with properties (i.e., speed ~2500 km/s) consistent with estimates of the Carrington storm directed on the opposite side of the Sun as the Earth (Baker et al., 2013). This shows that despite small probabilities of creating a severe GMD during a weak solar cycle, we may have narrowly escaped such a storm. Also, the danger of defining a storm as, for example, 1-in-60 or 1-in-100 years belies the fact that these studies really refer to probability of occurrence in a given year or decade and not that a severe storm can only occur once in this time frame.

11.5. GIC IMPACTS TO INFRASTRUCTURE

In this section, we focus on known GIC impacts from the 10 severe storms listed in Table 11.1. Knowing the impacts of a Carrington-like or higher super storm is difficult to assess since no storm of this level occurred with our modern technology infrastructure. The economic impacts could be severe, depending on the areas affected, with estimates of a $10 billion cost for a March 1989-like event impacting the Northeastern US (NRC, 2008; Baker, 2009).

11.5.1. The Extreme Carrington event of 1859 (Dst < −850 nT)

At high geomagnetic latitudes, the Carrington event exceeded the magnetometer scale at observatories across northern Europe. It occurred during a period of heightened geomagnetic activity with extended aurorae observed in contemporary storms on 24 October 1847 (B_H ~ −1100 nT in Greenwich), 25 October 1870, 9 April 1871, and 4 February 1872 (Cliver and Dietrich, 2013). Telegraph networks, consisting of long conducting lines, were the important communication technology of the time disrupted by GICs from the Carrington event. These systems were unable to directly measure the induced currents, as the equipment only had a needle display to show activity on the wire and a clock to announce the beginning of a new message.

Disruptions to the telegraph network were noted worldwide. For instance, Brussels was unable to communicate widely for an hour and a half on 29 August and the Ostend–Dover line lost communication for 3 h, until battery power was increased to regain capability (Muller, 2014). In Melbourne, Australia, a telegraph operator was able to transmit without any batteries (McGowan, 1860). Sparks and fires were observed in the Northeast US (Shea and Smart, 2006). Based on a quantitative assessment from the Shea and Smart (2006) review, Rivera et al. (2016) estimate a geoelectric field from 2 to 10 V/km.

11.5.2. The Great Railway Storm of 1921 (Dst < −825 nT)

The May 1921 storm likely resulted from a CME "preconditioning" event, where an earlier CME swept the interplanetary environment to lead to a more intense shock from a second CME arriving at 22:20 UT on 14 May (Lundstedt et al., 2015). By this time, technological systems evolved to include telephone and cable systems, in addition to telegraph networks. As a result, the effects from GICs were stronger than those during the Carrington event.

Fires and outages in service were experienced in Europe and the United States. For instance, a fire began at a telegraph station in Karlstad, Sweden (Em, 1921). Telegraph and telephone lines in the United Kingdom were disrupted, US wire transmission stopped for a day and a half, and Western Union transplants cables were affected (Odenwald, 2007). The storm got its name, however, from fires in Northeastern US railroad stations. The Central New England Railroad station was destroyed when a telegraph system caught on fire and spread to the building. A large outage occurred to the signal and switching system of the New York Central railroad system and a control tower caught fire from a short circuit. Rivera et al. (2016) estimated the geoelectric field in the Northeastern US to be 15 V/km based on measured voltages by telephone engineers and consistent with the Kappenman (2004) estimate of 20 V/km.

11.5.3. The Quebec Blackout Storm of 1989 (Dst −589 nT)

The March 1989 GMD was the most severe storm, in terms of Dst index, of the past 30 years. GIC effects to power grids primarily include half-cycle saturation of transformers (e.g., Subudhi et al., 1994). In severe cases, this leads to excessive localized heating and increases in harmonics. Changes in the harmonics from a transformer can lead to overloading of capacitor banks, misoperation of relays, and higher recovery voltages for circuit breakers. Additional issues from power transmission lines occur when there are unusual swings in power flow or limitations in the power generation production that lead to voltage collapses, that is, blackouts.

During the March 1989 GMD, electrical surges led to power grid impacts. Most famously, the Hydro-Quebec transmission grid experienced large voltage fluctuations tripping five lines with a generation loss of 9450 MW. This led to an over 9-h blackout across Quebec. Overheating occurred to the Meadow Brook 500/138 kV power transformer on the Allegheny power system – caused by an estimated 80 A GIC (Gattens et al., 1989). A transformer at the Salem nuclear power system as part of the Pennsylvania, New Jersey, Maryland (PJM) power grid also showed evidence of damage from overheating (Balma, 1992). Capacitor banks were tripped from high harmonic content in the current at Three Mile Island's 500 kV substation (Subudhi et al., 1994). Negative sequence currents in the Hope Creek, NJ and Bruce, Ontario nuclear generators caused multiple alarm triggers, but no transformer overheating (Subudhi et al., 1994). As a result of the March 1989 blackouts, the US Federal Energy Regulatory Commission commissioned NERC to address impacts of GMDs through creation of reliability standards.

11.5.4. The Halloween Storm of 29–31 October 2003 (Dst −353)

The GIC effects of the Halloween 2003 storm led to blackouts in Sweden. The storm resulted in lost power to 50 000 people for an hour and delays in trains. This amounted to approximately half a million dollars in losses (Pulkkinen et al., 2005). Failures in the transmission system coincided with a storm sudden commencement. Transformer burn out led to a 20–50 min blackout in the South African city of Malmo, affecting 50 000 customers (Wik et al., 2009). Transformers in Scotland also exhibited overheating.

Figure 11.3 shows the time-derivative of the horizontal component of the magnetic field at the Ottawa magnetometer station for the same storms in Figure 11.2. Sharp changes or spikes in the time-derivative of magnetic field are thought to be linked to GICs. While Pulkkinen et al. (2005) note that GIC effects were not known in North America from this storm, the time derivative does show a sharper increase than the majority of other storms. The exception is the March 1991 storm, which did have large GICs measured in North America (discussed in the following sub-section).

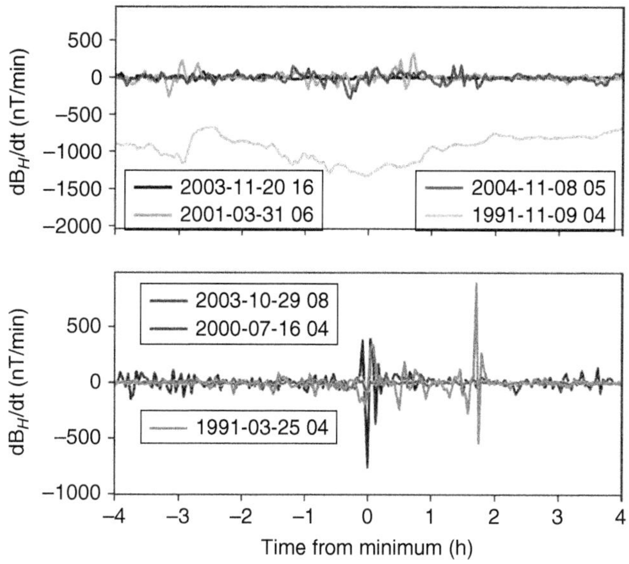

Figure 11.3 The time-derivative of the horizontal component of the magnetic field at the Ottawa, Canada station. The strongest GICs measured in the Northeastern US were from the March 1991 storm, which shows the largest changes in dB_H/dt. While the October 2003 storm has no known GIC effects, this storm also exhibited large differentials. (*See insert for color representation of the figure.*)

11.5.5. Additional Severe Storms

The GIC effects from other modern storms were far below the effects of the March 1989 and October 2003 GMDs. Wik et al. (2009) detail the known effects in Sweden. The March 1991 storm led to nine lines (220 kV) and a transformer being tripped. Kappenman (2003) also note that this storm corresponds with the highest GIC measurements in the United States of up to 130 A. The November 1991 storm tripped one high voltage line in Sweden and led to large pipe-to-soil voltages in a pipeline (Wik et al., 2009). In Japan, this storm also led to a >40 A GIC in transformers (Erinmez et al., 2002). The March 2001 storm led to tripping in a capacitor bank in New York at the time of the SSC (Kappenman, 2003). The November 2004 stop led to measured GIC's > 100 A in a transformer in southern Sweden. The most active GIC days in the European high-voltage power grid were presented in Viljanen et al. (2014). These included the 2003 Halloween storm as the highest (average over all nodes of 10 A). The July 2000 and November 2004 storms are also high, with average GICs of 7.5 and 5.8 A, respectively.

11.6. CONCLUSIONS

In this review, we discussed the solar drivers and measured effects of extreme/severe geomagnetic storms. We focused on the eight modern storms and two historical storms with the lowest Dst index. The GICs from these modern storms at their most extreme instantaneous effects caused power blackouts lasting hours. Predicting GICs is difficult since many physical systems interact to produce each of the storms discussed – including solar drivers (flares and CMEs), interplanetary magnetic fields, the magnetosphere, ionosphere, and ground conductivity. The GIC effects also depend upon the local system. For instance, the highest GIC measured did not result in blackouts despite being associated with the largest upper auroral electrojet index (>3000 nT) measured. The most intense historical storms (the 1859 Carrington event and the 1921 Railroad storm) also had limited effects (i.e., fires, telegraph disruption for hours to approximately a day), since the technological systems were less vulnerable to induced currents than today's systems.

A danger in current GIC modeling is use of terminology on the frequency of extreme storms. The 1989 storm is alternatively held up as a benchmark of a 1-in-100 year event or 1-in-60 year event. However, more extreme storms in terms of Dst index occurred about ever 40 years in the past century (Dst < −500) and the 1859 and 1921 storms with Dst ~ −900 nT were only ~60 years apart. Further, every storm is unique and a number of factors from sudden storm commencement components, Dst index, auroral electrojet direction and strength, and time variation of the magnetic field along with the geospatial extent of the storm, local ground conductivity, and technological specifications of the system all contribute to whether GIC effects will occur. Greater care should be taken in GIC effects modeling, such as including a range of realistic GMD storm properties used in benchmarks.

ACKNOWLEDGMENTS

This work used data from the SOHO/LASCO CME catalog. This CME catalog is generated and maintained at the CDAW Data Center by NASA and the Catholic University of America in cooperation with the Naval Research Laboratory. SOHO is a project of international cooperation between ESA and NASA.

REFERENCES

Baker, D.N. 2009, "What does space weather cost modern societies?," *Space Weather*, *7*, S02003, doi:https://doi.org/10.1029/2009SW000465.

Baker, D.N., et al. 2013, "A major solar eruptive event in July 2012: Defining extreme space weather scenarios," *Space Weather*, *11*, 585–591.

Balma, P.M. 1992, "Geomagnetic effects on a bank of single phase generator step-up transformers," Proc. Geomagnetically Induced Currents Conference, Millbrae, California, November 8–10, 1989, EPRI Report TR-100450, 20-1.

Barraclough, D. 1991, "150 years of magnetic observatories: Recent researches on world data," *The Observatory*, *111*, 148–154.

Beamish, D., et al. 2002, "Geomagnetically induced currents in the UK: Geomagnetic variations and surface electric fields," *Journal of Atmospheric and Solar-Terrestrial Physics*, *64*, 1779–1792.

Beggan, C.D., et al. 2013, "Prediction of extreme geomagnetically induced currents in the UK high-voltage network," *Space Weather*, *11*, 407–419, doi:https://doi.org/10.1002/swe.20065.

Bolduc, L. 2002, "GIC observations and studies in the Hydro-Quebec power system," *Journal of Atmospheric and Solar-Terrestrial Physics*, *64*, 1793–1802.

Campbell, W.C. 1980, "Observation of electric currents in the Alaska oil pipeline resulting from auroral electrojet current sources," *Geophysical Journal of the Royal Astronomical Society*, *61*, 437–449.

Cardinas, F.M. 2016, "The grand aurorae borealis seen in Colombia in 1859," *Advances in Space Research*, *57*, 257–267.

Carrington, R.C. 1859, "On a curious appearance seen in the Sun," *Monthly Notices of the Royal Astronomical Society*, *20*, 15.

Cliver, E.W., Balasubramaniam, K.S., Nitta, N.V., and Li, X. 2009, "Great geomagnetic storm of 9 November 1991: Association with a disappearing solar filament," *Journal of Geophysical Research*, *114*, A00A20.

Cliver, E.W. and Dietrich, W.F. 2013, "The 1859 space weather event revisited: Limits of extreme activity," *Journal of Space Weather and Space Climate*, *3*, A31.

Coles, S. 2004, "*An Introduction to Statistical Modeling of Extreme Values*," Springer, London.

Dessler, A.J. and Parker, E.N. 1959, "Hydromagnetic theory of geomagnetic storms," *Journal of Geophysical Research*, *64*, 2239–2252.

Em, A.A. 1921, "Jorstrommar sasom brandstiftare," *Brandskydd*, 77–79.

Erinmez, I.A., et al. 2002, "Application of modeling techniques to assess geomagnetically induced current risks on the NGC transmission system," paper presented at CIGRE, Council Int. Des Grands Res. Elect., Paris.

Feynman, J. and Hundhausen, A.J. 1994, "Coronal mass ejections and major solar flares: The great active center of March 1989," *Journal of Geophysical Research* *99*, 8451.

Gattens, P.R., et al. 1989, "Investigation of transformer overheating due to solar magnetic disturbances," Effects of Solar-Geomagnetic Disturbances on Power Systems, Special panel session report, pp. 29–32, IEEE PES Summer Meeting, Long Beach, CA, July 12, 1989

Gopalswamy, N.S., et al. 2005a, "Coronal mass ejections and other extreme characteristics of the 2003 October-November solar eruptions," *Journal of Geophysical Research*, *110*, A09S15.

Gopalswamy, N.S., et al. 2005b, "Solar source of the largest geomagnetic storm of cycle 23," *Geophysical Research Letters*, *32*, L12S09.

Green, L. and Baker, D. 2015, "Coronal mass ejections: A driver of severe space weather," *Weather*, *70*, 1.

Kappenman, J.G., 2003, "Storm sudden commencement events and the associated geomagnetically induced current risks to ground-based systems at low-latitude and midlatitude locations," *Space Weather*, *1*, 1016.

Kappenman, J.G. 2004, "The evolving vulnerability of electric power grids," *Space Weather*, *2*, doi:https://doi.org/10.1029/2003SW000028.

Kappenman, J.G. 2006, "Great geomagnetic storms and extreme impulsive geomagnetic field disturbance events – An analysis of observational evidence including the great storm of May 1921," *Advances in Space Research*, *38*, 188–199.

Kataoka, R. 2013, "Probability of occurrence of extreme magnetic storms," *Space Weather*, *11*, 214–218.

Lanzerotti, L.J. 1992, "Comment on 'Great Geomagnetic Storms' by Tsurutani et al.," *Geophysical Research Letters*, *19*, 1991–1992.

Liu, Y.D., et al. 2014, "Observations of an extreme storm in interplanetary space caused by successive coronal mass ejections," *Nature Communications*, *5*, 3481.

Lloyd's, 2013, "Solar storm risk to the North American electric grid," https://www.lloyds.com/news-and-insight/risk-insight/library/natural-environment/solar-storm

Love, J.J. 2007, "A continuous long-term record of magnetic storm intensity and occurrences," *Eos, Transactions American Geophysical Union*, *88*(23), Jt. Assem. Suppl., Abstract SH54B-03.

Love, J.J., et al., 2015, "On the lognormality of historical magnetic storm intensity statistics: Implications for extreme-event probablities," *Geophysical Research Letters*, *42*, 16, 6544–6553.

Lundstedt, H., Persson, T., and Andersson, V. 2015, "The extreme solar storm of May 1921: Observations and a complex topological model," *Annales Geophysicae*, *33*, 109–116.

McGowan, S.W. 1860, "*Half-yearly report of the General Superintendent of Electric Telegraph relative to the advancement and present condition of the Department, up to 31st December, 1859*," Melbourne, Australia: Parliament of Victoria.

McLay, S. and Beggan, C. 2010, "Interpolation of externally-caused magnetic fields over large sparse arrays using Spherical Elementary Current Systems," *Annales Geophysicae*, *28*, 1795–1805.

Muller, C. 2014, "The Carrington solar flares of 1859: Consequences on life," *Origins of Life and Evolution of Biospheres*, *44*, 3, 185–195.

NERC, 2014, "Transmission System Planned Performance During Geomagnetic Disturbances," NERC TPL-007-1.

Newell, P.T., et al. 2007, "A nearly universally solar wind-magnetosphere coupling function inferred from 10 magnetospheric state variables, *Journal of Geophysical Research*, *112*, A01206.

Ngwira, C.M. et al. 2015, "Characteristics of extreme geoelectric fields and their possible causes: Localized peak enhancements," *Geophysical Research Letters*, *42*, 6916–6921.

NRC, 2008, "*Severe Space Weather Events - Understanding Societal and Economic Impacts: A Workshop Report*," The National Academies Press, Washington, DC.

Odenwald, S. 2007, "Newspaper reporting of space weather: End of a golden age," *Space Weather*, *5*, S11005.

Pulkkinen, A., Pirjola, R., and Viljanen, A. 2008, Statistics of extreme geomagnetically induced current events, *Space Weather*, *6*, S07001, doi:https://doi.org/10.1029/2008SW000388.

Pulkkinen, A., et al. 2005, "Geomagnetic storm of 29-31 October 2003: Geomagnetically induced currents and their relation to problems in the Swedish high-voltage power transmission system," *Space Weather*, *3*, S08C03.

Pulkkinen, A., et al., 2012, "Generation of 100-year geomagnetically induced current scenarios," *Space Weather*, *10*, 4, 1542–7390.

Richardson, I.G. and Cane, H.V. 2012, "Near-earth solar wind flows and related geomagnetic activity during more than four solar cycles (1963-2011)," *Journal of Space Weather and Space Climate*, *2*, A02.

Riley, P. 2012, "On the probability of occurrence of extreme space weather events," *Space Weather*, *10*, S02012.

Rivera, M.K., et al. 2016, "EMP/GMD Phase 0 Report, A Review of EMP Hazard Environments and Impact," LA-UR-16-28380.

Shea, M. and Smart, D. 2006, "Compendium of the eight articles on the Carrington event attributed to or written by Elias Loomis in the American Journal of Science, 1859-1861," *Advances in Space Research*, *38*, 2, 313–385.

Siscoe, G., Crooker, N.U., and Clauer, C.R. 2006, "Dst of the Carrington storm of 1859," *Advances in Space Research*, *38*(2), 173–179, doi:https://doi.org/10.1016/j.asr.2005.02.102.

St. Cyr, O.C., et al. 2000, "Properties of coronal mass ejections: SOHO LASCO observations from January 1996 to June 1998," *Journal of Geophysical Research*, *105*, 18169.

Subudhi, M., et al. 1994, "Effects of solar geomagnetically induced currents on electrical systems in nuclear power stations," U.S. Nuclear Regulatory Commission: Supt. of Docs., NUREG/CR-5990, BNL-NUREG-52359.

Thomson, A.W.P. and Wild, J. 2010, "When the lights go out," *Astronomy and Geophysics*, *51*, 5, 23–24.

Thomson, A.W.P., et al. 2011, "Geomagnetic extreme statistics for Europe," *Space Weather*, *9*, S10001.

Tsubouchi, K. and Omura, Y. 2007, "Long-term occurrence probabilities of intense geomagnetic storm events," *Space Weather*, *5*, S12003.

Tsurutani, B. T., Gonzalez, W. D., Lakhina, G. S., and Alex, S. 2003, "The extreme magnetic storm of 1–2 September 1859," *Journal of Geophysical Research*, *108*, 1268, doi:https://doi.org/10.1029/2002JA009504, A7.

Viljanen, A., et al. 2014, "Geomagnetically induced currents in Europe," *Journal of Space Weather and Space Climate*, *4*, A09.

Viljanen, A.H., et al. 2001, "Time derivative of the horizontal geomagnetic field as an activity indicator," *Annales Geophysicae*, *19*, 1107–1118.

Wik, M., et al. 2009, "Space weather events in July 1982 and October 2003 and the effects of geomagnetically induced currents on Swedish technical systems," *Annales Geophysicae*, *27*, 1775–1787.

Winter, L.M., et al. 2017, "Spectral scaling technique to determine extreme Carrington-level geomagnetically induced currents effects," *Space Weather*, *15*, 5, 713–725.

Woodroffe, J.R., et al. 2016, "The latitudinal variation of geoelectromagnetic disturbances during large (Dst <−100 nT) geomagnetic storms," *Space Weather*, *14*, 668–681.

Yashiro, S., et al. 2004, "A catalog of white light coronal mass ejections observed by the SOHO spacecraft," *Journal of Geophysical Research 109*, A07105.

12

The Challenge Posed by Space Weather to Electric Power Reliability: Evidence from the New York Electric Power Grid

Kevin F. Forbes[1] and O. C. St. Cyr[2]

ABSTRACT

During a geomagnetic storm on 21 January 2005, a reliability action was declared by the system operator of the New York power grid and the real-time reference price spiked to over US$800 per MWh. Modeling estimates based on hourly data over the period 1 January 2002 through 31 December 2004 are consistent with the hypothesis that geomagnetic activity contributed to reliability actions. The model's out-of-sample predictions for January 2005 indicate that the reliability action declared on 21 January 2005 would not have occurred in the absence of the geomagnetic storm.

Key Points

- During a geomagnetic storm on 21 January 2005, an economically significant reliability action was declared by the operator of the New York power grid.
- Modeling estimates indicate that space weather has adversely affected power grid operations in the New York power grid.
- The model's predictions of reliability actions are more accurate when the predictions include the estimated effects of space weather.

12.1. INTRODUCTION

An intense geomagnetic storm on 13 March 1989 contributed to the collapse of the Hydro-Quebec power system in Canada. This storm clearly demonstrated that geomagnetic storms have the potential to give rise to the large societal costs associated with electricity blackouts. Despite the information learned from this event, Eastwood et al. (2017) have pointed out that there is a significant research gap in terms of quantifying the technical impacts of space weather. This chapter attempts to narrow this gap in the research. Evidence is presented that space weather can give rise to economic costs even in the absence of a blackout. Operations by the New York Independent System Operator (NYISO) in North America are examined over the period 1 January 2002 through 31 January 2005, during the declining phase of Solar Cycle 23.

We first discuss the geomagnetic data employed in this study. We then provide an overview of power grid operations. Graphical evidence is then presented that suggests geomagnetic activity was a contributing factor to the declaration of what is known as a "reserve pickup" by the system operator in New York on 21 January 2005. During this event, the real-time reference price in NYISO spiked to over US$800 per MWh. This price was about seven times the day-ahead price for the same hour. The research question in this chapter is whether geomagnetic activity was a contributing factor to this anomalous event. This issue is examined using a logit econometric model which arguably makes it possible to disentangle the effects of geomagnetic activity on the electric power system operations from purely terrestrial factors. The model is

[1] *Department of Economics, The Catholic University of America, Washington, DC, USA*
[2] *Goddard Space Flight Center, National Aeronautics and Space Administration, Greenbelt, MD, USA*

Geomagnetically Induced Currents from the Sun to the Power Grid, Geophysical Monograph 244, First Edition.
Edited by Jennifer L. Gannon, Andrei Swidinsky, and Zhonghua Xu.

estimated using data over the period 1 January 2002 through 31 December 2004. Based on the parameter estimates, the model is then used to make two series of hourly predictions over the period 1–31 January 2005. The first ignores the statistically significant parameters associated with geomagnetic activity while the second incorporates those estimated coefficients. The difference in the accuracy of the two sets of predictions may yield some useful insights about the root cause of the 21 January 2005 reserve pickup and the resulting US$800 per MWh price spike.

12.2. THE GEOMAGNETIC DATA EMPLOYED IN THIS STUDY

Following previous literature (Bolduc et al., 1998; Coles et al., 1992; Mäkinen, 1993; Viljanen, 1998; Viljanen et al., 2001), the GIC proxy employed in this study is based on the time derivative of the horizontal component (dH/dt) of the geomagnetic field. For each one-hour market period, the rate of the change in the horizontal component of the geomagnetic field (dH/dt) was calculated using the largest change (in absolute value) in the one-minute values of horizontal component of the geomagnetic field over the hour in question. The raw data were downloaded from the Intermagnet website (www.intermagnet.org). Specifically, the study uses data reported by the geomagnetic observatory operated by Natural Resources Canada in Ottawa Canada (OTT). The descriptive statistics are reported in Table 12.1. The peak value of dH/dt over the sample period was 758 nT/min, which occurred during hour 8 on 29 October 2003. The peak value of dH/dt over the out-of-sample period was 238.9 nT/min during hour 20 UT on 21 January 2005.

12.3. AN OVERVIEW OF THE POWER GRID AND THE VALUE OF ELECTRIC POWER RELIABILITY

Regardless of the fuel, electricity is produced at relatively low voltages. A transformer is used to "step up" the power onto the high voltage transmission system (Fig. 12.1). A high voltage system, that is 230 kV or higher, is used because this reduces the losses in moving the power long distances. When the electric power arrives near to where it will be consumed a transformer is used again to "step down" the power onto the distribution system. In short, transformers are an essential component of the electric power system. Unfortunately, transformers are vulnerable to the geomagnetically induced currents (GICs) associated with space weather. Empirical evidence on this vulnerability has been presented by Forbes and St. Cyr (2010, 2012). These currents can increase the reactive power consumed by the transformers which in turn can destabilize the power system. The currents can also lead to overheating issues in the transformers which have consequences for balancing electricity supply with demand.

Table 12.1 Descriptive Statistics for the GIC Proxy, 1 January 2002 to 31 December 2004.

	Based on data reported by the Ottawa Observatory (nT/min)
25th percentile	1.3
50th percentile	2.3
75th percentile	4.3
95th percentile	11.5
99th percentile	28.7
Largest value	758.0

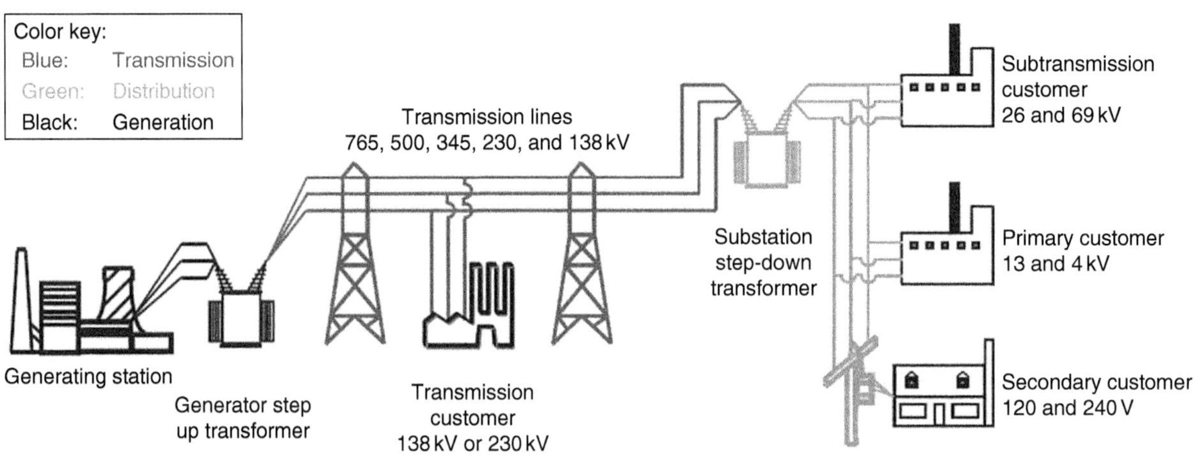

Figure 12.1 The basic structure of the electric power system. Source: U.S.-Canada Power System Outage Task Force. (*See electronic version for color representation of this figure.*)

The electricity that is ultimately delivered to consumers is extremely valuable to society. According to the United States Energy Information Administration, retail expenditures on electricity were approximately US$381 billion in 2016, the most recent year for which data are available, which represents approximately 2% of that year's Gross Domestic Product. These values, while consequential, significantly understate the economic contribution of this industry since they do not reflect the economic surplus that consumers receive from their purchases of electricity. The notion that individuals receive a surplus from consumption is strongly supported by microeconomic theory when individuals maximize their net levels of satisfaction subject to budget constraints (Henderson and Quandt, 1980, pp. 49–52; Willig, 1976). The rationale for this surplus is that the optimal level of consumption by an individual is based on the change in satisfaction from an additional unit relative to the price. At the optimum, the change in satisfaction from an additional unit will be less than average level of satisfaction because of the law of diminishing marginal utility and thus the total level of satisfaction received will be larger than the amount spent – that is there will be an economic surplus. Considering this surplus, the inability of consumers to purchase a good or service gives rise to an economic loss. The magnitude of this loss in the case of electricity is known as the value of lost load (VOLL). VOLL is generally considered to be very large because electricity is a key input in most production processes and is also a necessary component in many market transactions. The economic costs of an electricity blackout are not confined to the area directly affected but instead can spill over into other regions because of the interconnectedness of infrastructure (National Research Council, 2008). The economic costs can also spill over across the globe because of the interconnectedness of the global economy (Oughton et al., 2017). Electricity blackouts can also give rise to public safety issues. Consistent with these considerations, Mills and Jones (2016, p. 5) have reported that a leading modeler in the insurance industry has included a major blackout on a list of the "Top-10 Risks" even though the insured losses from an electricity blackout represent only a portion of total economic losses because of gaps in coverage.

The largest blackout over the past few decades was the 14 August 2003 blackout in the Northeast United States and Canada. The blackout has been attributed to terrestrial factors including inadequate system understanding by system operators, inadequate situational awareness, inadequate tree trimming in the transmission rights-of-way, and inadequate diagnostic support on the part of the reliability coordinators (U.S.-Canada Power System Outage Task Force, 2004). The blackout originated in northeast Ohio and cascaded into an area with an estimated 50 million people. Mills and Jones (2016, p. 17) have reported that the costs of the blackout was about US$7.7 billion when measured in 2014 dollars with only about 7 % this amount meeting the standards of an insured loss. The lowest published estimate of the economic costs to the United States of the August 2003 blackout in North America is US$4 billion (Electricity Consumers Resource Council, 2004) which is many times larger than what expenditures on electricity would have been in the absence of the blackout. While some might scoff at these estimates given that power was largely restored after about 32 h, Joskow (2012, p. 38) has reported that typical estimates of VOLL are in the range of US$5000–US$10,000 per MWh of electricity that is lost. A detailed literature review by Schröder and Kuckshinrichs (2015) yields similar estimates. We suspect that this range in the estimated costs can be explained by meteorological conditions, the geographic extent of the blackout, the blackout's duration, the effects on commerce, finance, and industrial production, the effects on the operations of critical infrastructure, and the magnitude of the challenge to health and public safety. Based on a VOLL of US$10,000 per MWh, the cost of the blackout on 14–15 August 2003 for just New York State was approximately US$3 billion even though our calculation of the foregone wholesale generation expenditures in New York yielded a value of about US$46 million.

The 2003 blackout was not New York's first major blackout. There were significant blackouts in 1959, 1961, 1965, and 1977. About half a million people were affected by the 1959 and 1961 blackouts. The 9 November 1965 blackout has been attributed to the "tripping", that is, a break in the circuit, of a high voltage transmission line in Canada at around hour 17 LT (Goodman, 2003, p. 8; Schewe, 2007, p. 147). It was subsequently discovered that the line tripped because the safety relay which was designed to protect the line from overloading by breaking the circuit, was set too low (Shanahan, 1965).

The 1965 blackout affected about 25 million people in parts of nine Northeastern states and two Canadian provinces (Gansberg, 1965). In New York City, about 800,000 rush-hour commuters were initially stranded in the subways. Thousands of people were stuck in elevators for hours. The theaters and movie houses were closed and none of the nine TV stations in NYC could broadcast (Schewe, 2007, pp. 137–138). All the traffic lights were inoperative. Patients at Bellevue Hospital were treated by candlelight. Despite these challenges, the 1965 blackout has been remembered as "largely characterized by cooperation and overriding good cheer" (Gottlieb and Glanz, 2003) with citizens helping the police by directing traffic (Pace, 1965). We cannot help but wonder if this event would be remembered differently if electric power had not been restored by the following morning.

The 25-h blackout that began on 13 July 1977 at about 21:30 LT has been attributed to lighting strikes that damaged transmission lines that brought electricity to NYC from the north (McFadden, 1977). The unmet demand for air conditioning was significant given that the heat index in Central Park had a peak value of about 34°C over the course of the blackout. Financial and retail activities were severely disrupted in all five of the city's boroughs (Jensen, 1977). All subways, the Long Island Rail Road, and Conrail were affected (McFadden, 1977). Tunnels from Brooklyn and Queens were closed because of the failure of the ventilation systems. The emergency backup generators failed at a number of hospitals including Bellevue, Metropolitan, New York University Medical Center, and Brooklyn Jewish (McFadden, 1977). In contrast to the 1965 blackout, the July 1977 blackout was marked by widespread looting and arson with more than 3330 individuals arrested (Raab, 1977). Ironically, this "night of terror," in the words of Mayor Beame, occurred just 3 days after Charles F. Luce, the chairman of Con Ed, indicated that no blackouts or brownouts were expected over the summer (McFadden, 1977).

While electricity load outages are costly to society, the losses cannot generally be avoided by storing the commodity. The hourly demand for electricity is highly variable. System operators manage this variability by forecasting demand and while forecast errors on average tend to be modest, the distribution of the forecast errors have "fat tails," that is more kurtosis than the Gaussian distribution. As a result, there will be occasions when the level of electricity consumption desired by consumers will exceed the forecasted level of consumption by a wide margin. Another challenge faced by system operators is that the generation of conventional electricity is subject to occasional outages because of temperature related equipment failures as well as operator error and other factors. Moreover, electricity flows on a transmission system follow the laws of physics and thus the actual flow at a given interface may be quite different from the desired flow. This may be an important consideration in the case of NYISO given that its actual electricity flows with Ontario over this period were poorly related with the scheduled flows. Specifically, Ontario was scheduled net exporter to New York over the period 1 May 2002 through 9 December 2003 but in terms of the actual flows, New York was actually a net exporter to Ontario, that is, on average the actual electricity flows were in the opposite direction from what was scheduled. This phenomenon is known as the Lake Erie Loop flows (NYISO, 2008). To control for these aberrant electricity flows, our econometric model includes explanatory variables that serve as proxies for expected transmission conditions. It also includes measures of load forecast errors.

12.4. THE NEW YORK ELECTRIC POWER SYSTEM

The New York Independent System Operator (NYISO) coordinates the generation and transmission of electricity in New York using an electricity market. NYISO was formed in 1998 as part of the restructuring of New York State's electric power industry. The NYISO is divided into 11 zones: The zones are Capital, Central Dunwood, Gennese, Hudson Valley, Long Island, Millwood, Mohawk Valley, New York City, North, and West (Fig. 12.2). In terms of electricity consumption, the most important of these zones are New

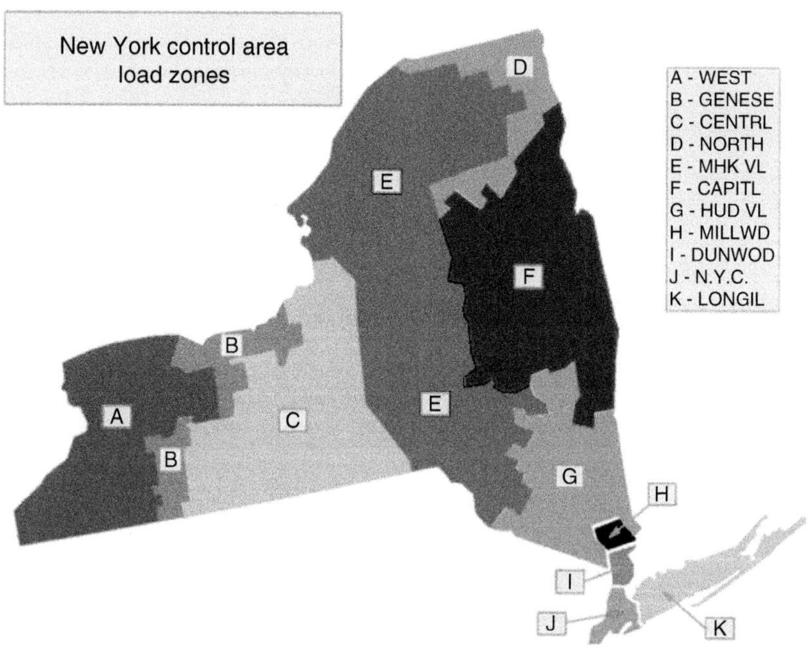

Figure 12.2 NYISO's geographic footprint. (*See electronic version for color representation of this figure.*)

York City and Long Island. One of the defining characteristics of the control area over the sample period is that the level of generating capacity in New York City and Long Island were incapable of meeting peak load (Table 12.2).

NYISO operates markets for capacity credits, ancillary services, and financial transmission rights. It operates two markets for energy, a day-ahead market and a real-time market. Real-time Locational Marginal Prices (LMPs) are calculated at 5-min intervals based on the actual operating conditions of the system. Load weighted average real-time prices are reported hourly. Real-time prices are determined for each of the 11 zones and for the four neighboring regions that have interconnections with the NYISO (New England, Hydro Quebec, Ontario Hydro and PJM). There is also a real-time reference price for the overall system. The day-ahead market is a forward market that market participants can use to avoid the price volatility in the real-time market. Nevertheless, events in the real-time market have consequences for prices in the day-ahead market given that there is a generally a price premium associated with reliance on the day-ahead market. For example, the average hourly day-ahead reference price over the sample period was about US$2.38 per MWh higher than the corresponding real-time reference price. In short, the two markets are linked such that risks in the real-time market ultimately get reflected in the day-ahead prices.

Table 12.2 Approximate Peak Load and Capacity within NYISO in 2002.

Zones	Peak load (MW)	Capacity (MW)
Western Zones (Zones A–E)	10,100	15,000
Hudson Valley (Zones F–I)	6500	8200
New York City (Zone J)	10,665	8850
Long Island (Zone K)	4800	4925

Source: NYISO.

Consistent with the peak load and generating capacity reported in Table 12.2, NYISO's high voltage transmission system is largely designed to move electricity from the northeast portion of the state to the southeast, that is New York City and Long Island, as well as from the west to the southeast (Fig. 12.3).

The reference price in NYISO is determined at the at the Marcy 345 kV substation in Marcy, New York which is about 300 km from New York City (Fig. 12.3). This location is the reference against which energy losses and congestion costs are calculated.

The LMP for each zone has three components: an energy component, a line loss component, and a congestion cost component. Specifically, for zone k,

$$\mathrm{LMP}_k = \text{Reference price} + \text{Losses}_k - \text{Congestion}_k \tag{12.1}$$

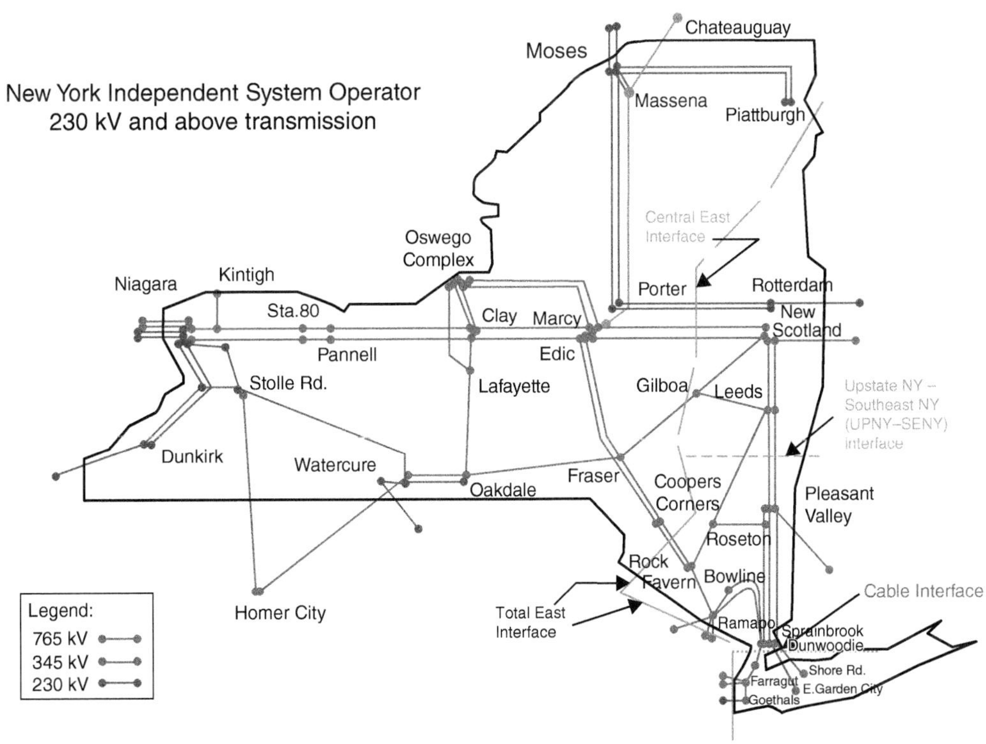

Figure 12.3 NYISO's high voltage transmission system. (*See insert for color representation of the figure.*)

Reported congestion costs for zone k can be positive, zero, or even negative. Ignoring line losses, $\text{Congestion}_k > 0$ indicates that there are transmission constraints such that LMP_k is less than the reference price. Likewise, $\text{Congestion}_k < 0$ indicates that there are transmission constraints that results in LMP_k being greater than the reference price. The LMPs for New York City (NYC) and Long Island are typically higher than the reference price and thus from (12.1) reported congestion costs for these two zones are typically negative. For example, over the sample period the day-ahead congestion costs for NYC ranged from approximately −US$165 per MWh to about US$79 per MWh with an average of −US$9.91 per MWh. These day-ahead congestion costs provide information about expected operating conditions on the transmission system and thus the econometric model in this chapter employs measures of these costs as explanatory variables.

12.5. SYSTEM ALERTS AND RESERVE PICK-UPS

Under normal circumstances, generators are dispatched on their economic merit, with the lowest cost generators being dispatched first. During a system alert, system reliability, not economic merit, is the priority. According to NYISO's system operations procedures, the conditions that prompt the system operator to declare an alert include "... thunder storms, hurricanes, tornadoes, solar magnetic flares or the threat of terrorism ..." During a system alert the system operator seeks to maintain the reliability of the system "... by directing the adjustment of the generator output levels in certain areas of the system to reduce power flows across transmission lines vulnerable to outages due to these adverse conditions" (NYISO, 2003, sect. 12, p. 1).

System alerts are relatively infrequent in the New York power grid. Over the period 1 January 2002 through 31 December 2004, there were 621 h (~2.4% of the total hours) in which a system alert was in effect. Consistent with the view that some of these alerts are space weather related, 33 of these hours occurred over the period 29–31 October 2003 (~46% of the 72 h), a period of well documented intense geomagnetic activity.

System alerts have economic consequences. On average, the differential between the real-time and day-ahead prices is larger during those periods in which an alert has been declared. For example, the average hourly reference real-time price during a system alert was US$8.12 per MWh higher than the day-ahead reference price. In contrast, the average real-time reference price was US$ 2.60 per MWh lower than the day-ahead price during other periods.

Another metric of adverse grid conditions on the New York grid is the incidence of a reserve pickup. As the name suggests, a reserve pickup represents a request by the system operator to generators to increase generation. On average, the differential between the real-time and day-ahead prices is larger during those periods in which a reserve pickup has been declared. For example, the average hourly reference real-time price during a reserve pickup was US$13.84 per MWh higher than the day-ahead reference price. In contrast, the average real-time reference price was US$2.90 lower than the day-ahead price during normal operating conditions.

The purpose of a reserve pickup is to recover from system events such as the loss of lines or loss of generators. They are also used to resolve sustained area control errors (ACE) where an area control error represents the difference between net actual and scheduled electricity flows with other electricity control areas. Specifically, a reserve pickup on the New York power grid may be declared if ACE is less than −100 MW (NYISO, 2003, sect. 0, p. 6). Finally, reserve pickups are sometimes declared in conjunction with a system alert (NYISO, sect. 0, p. 13).

Reserve pickups are also relatively infrequent in the New York power grid. Over the period 1 January 2002 through 31 December 2004, there were 777 h (~3.0% of the total hours) in which a reserve pickup was in effect. Three of these hours occurred over the period 29–31 October 2003 (~4.2% of the 72 h).

12.6. GICS AND RELIABILITY ACTIONS IN NYISO: IS THERE A RELATIONSHIP?

It is typically very difficult to make inferences about the possible effects of space weather on the real-time price of electricity by looking at a graph. The spike in both geomagnetic activity and the real-time reference price in NYISO in January 2005 may be an exception (Fig. 12.4). Before proceeding, it should be noted that the spike in the GIC proxy was the result of a significant eruptive event on the Sun (a large X-ray flare, a solar proton event, and a very fast coronal mass ejection) on the previous day, 20 January 2005. With respect to NYISO, a reserve pickup was declared at approximately the same time as the peak of the storm. We have been unable to obtain any evidence that the reserve pickup was declared in response to any space weather alert. The research question that this chapter attempts to address is whether the reserve pickup was induced by geomagnetic activity or whether there are some other drivers.

To address this issue, an econometric model to explain the incidence of reserve pickups and system alerts is estimated using data over the period 1 January 2002 through 31 December 2004. Based on the parameter estimates, the model will then be used to make two series of predictions over the period 1–31 January 2005. The first will ignore

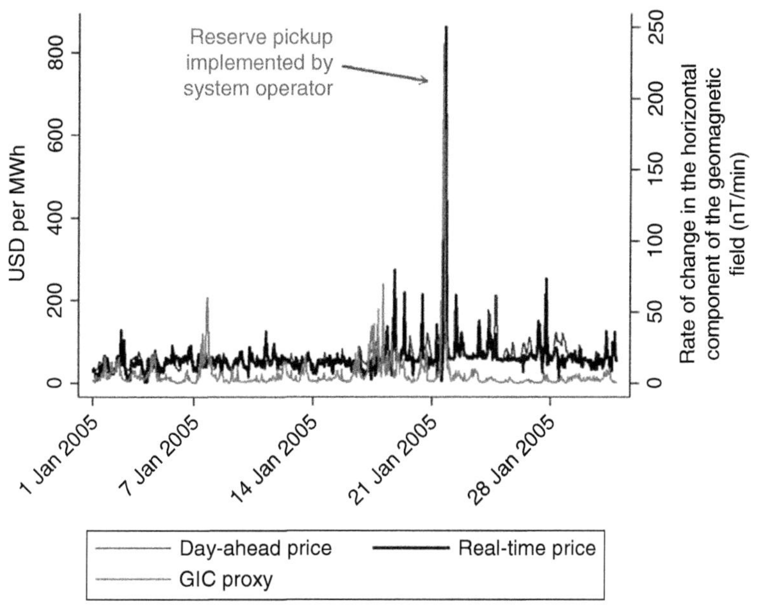

Figure 12.4 A GIC proxy, the day-ahead reference price in NYISO, and the real-time reference price in NYISO, 1 January 2005 to 31 January 2005. (*See electronic version for color representation of this figure.*)

the statistically significant parameters associated with geomagnetic activity while the second will incorporate those estimated coefficients. The difference in the accuracy of the two sets of predictions may yield some useful insights about the root cause of the approximately US$800 per MWh price spike on 21 January 2005.

Following from Forbes and St. Cyr (2012), a multivariate logit equation is employed to test whether GICs contribute to reserve pickups and system alerts (RPSA) in NYISO. This approach is a form of regression analysis that is used to model the incidence of binary outcomes. One advantage of this approach relative to a linear probability model is that the predicted probabilities are bounded between zero and one. Readers interested in learning more about this methodology may wish to consult Hosmer et al. (2013).

The explanatory variables in the analysis include temperature, the day-ahead hourly NYISO price relative to the prices of both natural gas and coal, the primary fuels used to generate electricity in NYISO, the day-ahead price in NYC relative to the day-ahead reference price in NYISO, forecasted load by region within NYISO, measures of the forecast errors by region within NYISO, and measures of day-ahead congestion costs relative to the day-ahead reference price for the following zones or interfaces: New York City (NYC), Long Island (LI), PJM, Ontario Hydro (OH), and Hydro Quebec (HQ).

The use of the day-ahead price variables follows from Forbes and Zampelli (2014) who have reported evidence that the day-ahead market for electricity in California is informationally efficient in the sense that the prices relative to fuel costs anticipate the operational outcomes. This approach has also been employed by Forbes and St. Cyr (2012). A variant of this approach using day-ahead prices and forecasted meteorological conditions is currently being explored using load data for NYC.

The list of explanatory variables also includes a GIC proxy, a one-hour lag of GIC proxy, and an interaction term between the GIC proxy and temperature. The one-hour lag of the GIC proxy is included as an explanatory variable to control for the possible cumulative effects of geomagnetic activity from 1 h to the next.

The price, congestion costs, system alert, reserve pickup, and electricity load data employed in this study were downloaded from NYISO web site (www.NYISO.com). The data representing the spot price of coal was obtained from Platts (www.platts.com). The data representing the spot price of natural gas was obtained from NGI Intelligence Press (www.naturalgasintel.com).

The temperature variable is based on the hourly temperature for LaGuardia Airport in Queens, New York. The data were obtained for the Weather Source (http://weathersource.com), a firm that previously specialized in the processing of raw data from the National Weather Service. We concede that our reliance on temperature data from a single weather station represents a shortcoming. In the ideal analysis, we would employ three geographically weighted average measures of temperature where the weights would include regional measures of scheduled generation, scheduled transmission, and forecasted load, respectively. This was not possible because of data limitations. Our analysis instead uses data from

LaGuardia airport because of its location in the New York City electricity zone. This zone is the most important load zone in NYISO. It also has substantial generating capacity (see Table 12.2). Finally, it has considerable amount of transmission capacity.

The form of the estimated equation is:

$$\begin{aligned}\mathrm{Ln}\left(\frac{p}{1-p}\right) = {} & \mathrm{constant} + \alpha_1 T + \alpha_1 \mathrm{GIC} + \alpha_2 \mathrm{GICLag} \\ & + \alpha_3 \mathrm{GICTEMP} + \alpha_4 \mathrm{NYCPR} \\ & + \alpha_5 \mathrm{PRGas} + \alpha_6 \mathrm{PRCoal} + \alpha_7\,\mathrm{FPR} \\ & + \alpha_8 \mathrm{FL}_{NYC} + \alpha_9\,\mathrm{FL}_{LI} + \alpha_{10}\,\mathrm{FL}_{Other} \\ & + \alpha_{11} \mathrm{NegFLE}_{NYC} + \alpha_{12} \mathrm{PosLFE}_{NYC} \\ & + \alpha_{13} \mathrm{NegLFE}_{LI} + \alpha_{14} \mathrm{PosLFE}_{LI} \\ & + \alpha_{15} \mathrm{NegLFE}_{Other} + \alpha_{16} \mathrm{PosLFE}_{Other} \\ & + \sum_{k=1}^{5} \delta_{YY}\,\mathrm{PosCC}_{YY} \\ & + \sum_{k=1}^{5} \delta_{YY} \mathrm{NegCC}_{YY}\end{aligned} \tag{12.2}$$

where:

p is the predicted probability of a RPSA.

T is ambient temperature in Kelvin.

GIC is the value of the GIC proxy.

GICLag1 is the value of the GIC proxy lagged by 1 h.

GICTEMP equals GIC*T.

NYCPR is the day-ahead hourly price in New York City relative to the NYISO's reference price.

PRGas is the day-ahead NYISO hourly reference price relative to the spot price of natural gas.

PRCoal is the day-ahead NYISO hourly reference price relative to the spot price of coal.

FPR is the energy equivalent spot price of coal relative to the spot price of natural gas.

FL_{NYC} is day-ahead hourly level of forecasted load for NYC.

FL_{LI} is day-ahead hourly level of forecasted load for Long Island.

FL_{Other} is day-ahead hourly level of forecasted load for the zones other than New York City or Long Island.

NegFLE_{NYC} is the absolute value of the error in the day-ahead hourly level of forecasted load for NYC when actual load for NYC is less than forecasted load for NYC. The variable has a value of zero otherwise.

PosFLE_{NYC} is the absolute value of the error in the day-ahead hourly level of forecasted load for NYC when actual load for NYC exceeds forecasted load for NYC. The variable has a value of zero otherwise.

NegFLE_{LI} is the absolute value of the error in day-ahead hourly level of forecasted load for Long Island when actual load for Long Island is less than forecasted load for Long Island. The variable has a value of zero otherwise.

PosFLE_{LI} is the absolute value of the error in the day-ahead hourly level of forecasted load for Long Island when actual load for Long Island exceeds forecasted load for Long Island. The variable has a value of zero otherwise.

NegFLE_{Other} is the absolute value of the error in the day-ahead hourly level of forecasted load for the zones other than New York City or Long Island when actual load for these zones is less than forecasted load. The variable has a value of zero otherwise.

PosFLE_{Other} is the absolute value of the error in the day-ahead hourly level of forecasted load for the zones other than New York City or Long Island when actual load for these zones exceeds forecasted load. The variable has a value of zero otherwise.

PosCC_{YY} is the hourly level of day-ahead congestion costs for zone/interface YY relative to the day-ahead NYISO hourly reference price when the day-ahead congestion costs for YY are positive, YY = NYC, LI, PJM, OH, and HQ. The variable has a value of zero otherwise.

NegCC_{YY} is the absolute value of the hourly level of day-ahead congestion costs for YY relative to the day-ahead NYISO hourly reference price when the day-ahead congestion costs for YY are negative, YY = NYC, LI, PJM, OH, and HQ. The variable has a value of zero otherwise.

The transformation on the left-hand side of (12.2), the logit, takes a number that is restricted to the [0,1] interval and converts it into a value that has no upper or lower bound. As a result, the marginal impact of any single independent variable on the probability of a RPSA is contingent on the values of the other explanatory variables. A mathematical demonstration of this point is provided by Greene (2008, p. 775). One could assume independence in the marginal effects by estimating the equation using ordinary least squares but this leads to the nonsensical result of negative predicted probabilities.

The estimation results using the logit specification are presented in Table 12.3. Reported standard errors are calculated using the Newey-West procedure and are therefore heteroskedastic and autocorrelation-consistent (Newey and West, 1987). Observe that the p values corresponding to the following "terrestrial" variables are less than 0.05: T, NYCPR, PRGas, FL_{LI}, FL_{Other}, PosFLE_{Other}, NegFLE_{Other}, NegCC_{NYC}, and PosCC_{OH}. The following geomagnetic variables are statistically significant as evidenced by their p values being less than 0.05: GIC, GICLag1, and GICTEMP. Observe that the estimated coefficients corresponding to T and GIC are both positive while the coefficient corresponding to GICTEMP is negative but small in absolute value. This

Table 12.3 Estimation Results.

Variable	Estimated coefficient	Standard error	*t*-Statistic	*P*-value
Constant	−6.9711	2.8806	−2.42	0.01600
T	0.0199	0.0074	2.70	0.00700
GIC	0.1731	0.0645	2.68	0.00700
GICLag1	0.0065	0.0016	4.03	<0.00001
GICTemp	−0.0006	0.0002	−2.61	0.00900
NYCPR	−4.7139	2.2851	−2.06	0.03900
PRGas	0.2984	0.0885	3.37	0.00100
PRCoal	−0.0031	0.0064	−0.49	0.62500
FPR	−1.1691	0.8700	−1.34	0.17900
FL_{NYC}	−0.0002	0.0001	−1.47	0.14000
FL_{LI}	−0.0009	0.0002	−4.47	<0.00001
FL_{Other}	0.0006	0.0001	6.87	<0.00001
$PosCC_{NYC}$	−66.1806	35.0033	−1.89	0.05900
$NegCC_{NYC}$	4.7986	2.2746	2.11	0.03500
$PosCC_{LI}$	2.3416	2.9500	0.79	0.42700
$NegCC_{LI}$	−0.0442	0.1021	−0.43	0.66500
$PosCC_{OH}$	3.5322	0.9070	3.89	<0.00001
$NegCC_{OH}$	1.5625	1.3545	1.15	0.24900
$PosCC_{PJM}$	−2.4745	1.2752	−1.94	0.05200
$NegCC_{PJM}$	−0.4025	0.6249	−0.64	0.52000
$PosCC_{HQ}$	−0.8096	0.8271	−0.98	0.32800
$NegCC_{HQ}$	−0.0377	1.4387	−0.03	0.97900
$NegFLE_{NYC}$	0.0001	0.0004	0.34	0.73400
$PosFLE_{NYC}$	0.0002	0.0003	0.61	0.53900
$NegFLE_{LI}$	−0.0010	0.0007	−1.42	0.15500
$PosFLE_{LI}$	0.0005	0.0005	1.13	0.25900
$PosFLE_{Other}$	0.0004	0.0002	2.36	0.01800
$NegFLE_{Other}$	−0.0012	0.0003	−4.72	<0.00001
Number of observations	26078			
McFadden R^2	0.0551			

suggests that the estimated probability of a RPSA increases with both temperature and geomagnetic activity but that the overall estimated effect moderates somewhat as both temperature and geomagnetic activity increase. It is not entirely clear why this occurs. One possibility is that the operator of NYISO is very familiar with contending with the adverse effects of extreme temperatures and that substantial levels of generation are available to be dispatched using standard operating procedures during these periods.

With respect to the model's explanatory power, it is not possible to calculate a conventional R^2 that is meaningful because the dependent variable is binary while the predicted values are probabilities. Several alternative scalar fit measures have been introduced. According to Greene (2002, pp. E15–E28), these measures "… share the flaw that none satisfactorily mimic the true measure of the proportion of variation explained given by R^2 in the linear regression context." McFadden's R^2 is one of the more commonly reported measures of scalar fit when the dependent variable is binary. Its value here is 0.0551. This value indicates that the log-likelihood function, the objective function whose maximization yields the estimated parameters, improves only by about 5.51%, compared to when the model is estimated with only a constant term as an explanatory variable.

The estimation of a receiver operating characteristic (ROC) curve provides a comprehensive method by which to evaluate the model's explanatory power. As explained by Pepe et al. (2009), a ROC curve is a plot that depicts the capability of a predictor to discriminate between two outcomes. It does so by plotting the detection rate versus the false-alarm rate as the discrimination threshold varies, that is by considering the detection and false-alarm rates at a multitude of discrimination thresholds, rather than only employing 0.5 as the cutoff. The curve is an increasing function in the unit square. The boundary points of the curve are (0,0) and (1,1). The horizontal axis represents the false-alarm rate (FAR), that is, the rate at which the model predicts a RPSA that does not occur. The vertical axis represents the detection rate (DR), that is, the rate at which the model predicts a RPSA that does occur. A perfect predictor of RPSA would have a ROC curve that rises along the left axis to the point (FAR = 0, DR = 1), while a completely useless predictor would have ROC curve that coincides with a 45° line emanating from (0,0) to (1,1). Moreover, the area under the curve (AUC) for a perfect predictor would equal one while the AUC for the completely useless predictor would be equal to 0.5. More specifically, the AUC equals the probability that a randomly selected pair of market periods, one being a RPSA and other being a normal operating outcome, will be correctly ranked by the test.

The ROC curve for the within sample predicted probabilities corresponding to Table 12.3 is presented in Figure 12.5. The curve lies well above the 45° line. The within sample AUC in this case equals 0.67646 indicating that the model's within sample predicted values, while very far from perfect, have significant informational content.

With respect to the model's out-of-sample predictive accuracy, the AUC is 0.65145 when the predictions are based on all of the estimated parameters. Consistent with causality between geomagnetic activity and the incidence of RPSA, the out-of-sample AUC declines to 0.61880 when the coefficients corresponding to the variables GIC, GICLag1, and GICTEMP are set equal to zero when making the out-of-sample predictions.

Table 12.4 presents the out-of-sample predicted probabilities for the January 2005 RPSA events. Observe that predicted probability of a RPSA during hour 21 UT on 21 January 2005 is 0.892 when the coefficients corresponding to GIC, GICLag1, and GICTEMP are equal to their estimated values; the probability is 0.078 when the coefficients for these variables are set equal to zero.

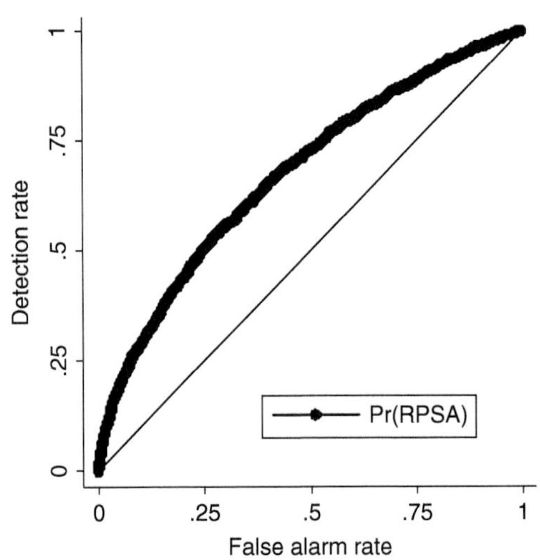

Figure 12.5 A receiver operating characteristic curve for the within-sample predicted probabilities.

12.7. GEOMAGNETIC ACTIVITY AND THE PREDICTED PROBABILITY OF A RPSA

In this section, we employ the estimated parameters and the within-sample data to explore the relationship between the GIC proxy and the incremental predicted probability of a RPSA. To simplify the analysis, we will ignore the contributions of GICLag1 and GICTEMP. The issue we want to address is whether the value of the GIC proxy is a sufficient indicator of the power system's vulnerability to geomagnetic activity.

We proceed by first using all the estimated coefficients to calculate the predicted probabilities of a PRSA for each hour of the sample period. We then calculate the predicted probabilities for each hour of the sample period under the assumption that the coefficient corresponding to the GIC proxy is equal to zero. The difference between the two predicted probabilities in the predicted incremental probability that can be attributed to geomagnetic activity if one ignores the contributions of GICLag1 and GICTEMP.

One might expect that the incremental predicted probability would monotonically increase as the value of the GIC increases. This is not the case (Fig. 12.6). Observe that some of the predicted incremental probabilities are greater than 0.6 when the GIC proxy is relatively modest in value while other predicted incremental probabilities are less than 0.2 even when the value of the GIC proxy exceeds 300 nT/min. This occurs because the predicted change in the probability of a RPSA from a change in the GIC proxy is conditional on the values of the other explanatory variables such as *T*, NYCPR, and PRGas. This variability in the incremental predicted probabilities illustrates the complexity of the reliability challenge posed by geomagnetic activity. In short, these results indicate that being able to prepare for a geomagnetic storm requires a forecast of the storm's intensity as well as a forecast of the power system's vulnerability.

12.8. CONCLUSION

There is no evidence of a space weather induced blackout in NYISO during Solar Cycle 23. Yet, the multivariate econometric analysis presented in this chapter presents strong evidence that NYISO's markets and operations were affected by space weather over this period. Moreover, the model's out-of-sample predictions indicate that the geomagnetic storm on 21 January 2005 presented a challenge to NYISO's reliability. The results reported here are consistent with Forbes and St. Cyr's (2008) analysis of space weather and real-time metrics of power grid operations in NYISO, the power system considered in this paper; in Eirgrid, the power grid that serves the Republic of Ireland; Scottish and Southern Electricity, the power

Table 12.4 Out-of-Sample Predicted Probabilities.

Date	UT	Predicted probability when the estimated effects of geomagnetic activity are incorporated into the predictions	Predicted probability when the estimated effects of geomagnetic activity are ignored	Value of the GIC proxy (nT/min)
1 January 2005	21:00	0.053	0.050	4.3
1 January 2005	22:00	0.067	0.055	19.0
2 January 2005	21:00	0.080	0.069	9.0
4 January 2005	19:00	0.074	0.060	18.6
6 January 2005	4:00	0.038	0.037	0.6
6 January 2005	13:00	0.096	0.093	2.2
7 January 2005	4:00	0.035	0.034	0.8
7 January 2005	12:00	0.116	0.104	8.1
7 January 2005	21:00	0.099	0.069	19.3
8 January 2005	5:00	0.048	0.035	16.0
8 January 2005	14:00	0.063	0.060	1.5
10 January 2005	3:00	0.030	0.029	1.3
10 January 2005	11:00	0.085	0.083	1.2
17 January 2005	14:00	0.081	0.047	26.2
18 January 2005	23:00	0.100	0.084	5.2
19 January 2005	0:00	0.086	0.078	3.6
19 January 2005	2:00	0.064	0.050	9.7
19 January 2005	13:00	0.108	0.072	16.4
19 January 2005	14:00	0.086	0.070	5.7
20 January 2005	2:00	0.047	0.046	0.8
20 January 2005	11:00	0.087	0.074	8.1
20 January 2005	15:00	0.082	0.062	13.6
20 January 2005	16:00	0.073	0.057	10.9
20 January 2005	21:00	0.083	0.052	20.9
21 January 2005	11:00	0.102	0.093	4.3
21 January 2005	21:00	0.892	0.078	158.3
21 January 2005	22:00	0.411	0.087	49.5
24 January 2005	11:00	0.093	0.088	1.7
24 January 2005	12:00	0.000	0.000	1.9
29 January 2005	17:00	0.046	0.042	4.5
30 January 2005	14:00	0.058	0.053	4.6
30 January 2005	15:00	0.053	0.050	2.1
30 January 2005	22:00	0.062	0.059	3.7

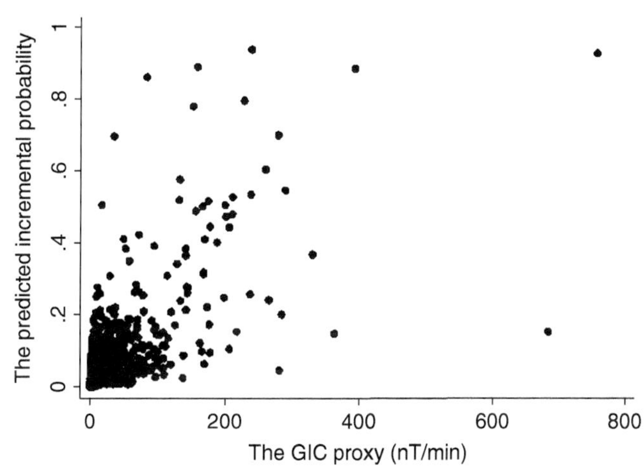

Figure 12.6 The GIC proxy and the geomagnetically induced predicted incremental probability of a RPSA over the sample period.

grid in northern Scotland until April 2005; Scottish Power, the power grid in southern Scotland until April 2005; the power grid that serves the Czech Republic; E.ON Netz, the transmission system operator in central Germany (now known as TenneT); the electric power system in England and Wales; the power grid in New Zealand; the power grid that serves the vast proportion of the population in Australia; ISO New England, the electric power system that serves New England; PJM, the electric power system directly south of NYISO, and the power grid in the Netherlands. In short, the findings presented here do not seem to be an outlier. Moreover, the results indicate that the challenge that geomagnetic activity poses to reliability does not increase monotonically with increases in the GIC proxy. Instead, the geomagnetic vulnerability is conditional on electric power system conditions at the time of geomagnetic storm. Improved understanding of this terrestrial component of the vulnerability would seem to be essential in mitigating the potentially large societal costs posed by geomagnetic activity.

ACKNOWLEDGMENTS

This research was made possible by grants from the U.S. National Science Foundation (Awards 0318582 and 0921964). We thank Jennifer Gannon and two anonymous reviewers for their helpful comments and suggestions. We wish to thank Michael A. Forbes for writing the programs that manipulated the geomagnetic data used in this study. We thank Gregory D. Forbes for his processing of the raw RPSA data. We have also benefited from discussions with Mark Lively, Antti Pulkkinen, and Ernest Zampelli. We thank NGI Intelligence Press for providing us with spot gas price data. We thank Platts for providing us with the coal pricing data. We thank NYISO for making its data available for analysis. The results presented in this chapter rely on the data collected at the Ottawa geomagnetic observatory. We thank Natural Resources Canada for supporting the operations of this observatory and INTERMAGNET for promoting high standards of magnetic observatory practice (www.intermagnet.org). Any errors are the full responsibility of the authors.

REFERENCES

Bolduc, V., Langlois, P., Boteler, D., and Pirjola, R. (1998). A study of geoelectromagnetic disturbances in Quebec. I. General results, *IEEE Transactions on Power Delivery*, *13*, 1251–1256, doi:https://doi.org/10.1109/61.714492.

Coles, R. L., Thompson, K., and Jansen van Beek, G. (1992). A comparison between the rate of change of the geomagnetic field and geomagnetically induced currents in a power transmission system, *Report TR-100450*, Electric Power Research Institute, Palo Alto, CA.

Eastwood, J. P., Biffis, E., Hapgood, M. A., Green, L., Bisi M. M., Bentley, R. D., ... Burnett, C. (2017). The economic impact of space weather: Where do we stand? *Risk Analysis*, *37*: 206–218. doi:https://doi.org/10.1111/risa.12765.

Electricity Consumers Resource Council (2004). The Economic Impacts of the August 2003 Blackout. Available from https://www.nrc.gov/docs/ML1113/ML111300584.pdf (Accessed 22 September 2017).

Forbes, K. F., and St. Cyr, O. C. (2008). Solar activity and economic fundamentals: Evidence from 12 geographically disparate power grids. *Space Weather*, *6*, S10003, doi:https://doi.org/10.1029/2007SW000350.

Forbes, K. F., and St. Cyr, O. C. (2010). An anatomy of space weather's electricity market impact: The case of the PJM power grid and the performance of its 500 kilovolt transformers, *Space Weather*, *8*, S09004, doi:https://doi.org/10.1029/2009SW000498.

Forbes, K. F., and St. Cyr, O. C. (2012). Did geomagnetic activity challenge electric power reliability during solar cycle 23? Evidence from the PJM regional transmission organization in North America, *Space Weather*, *10*, S05001, doi:https://doi.org/10.1029/2011SW000752.

Forbes, K. F., and Zampelli, E. M. (2014). Do day-ahead electricity prices reflect economic fundamentals?: Evidence from the California ISO, *The Energy Journal*, *35*(3), 129–144, doi: https://doi.org/10.5547/01956574.35.3.6.

Gansberg, M. (1965). Power blackout affects nine states in the northeast and three Canadian Provinces, *New York Times*, 10 November.

Goodman, J. (2003). *Blackout*, North Point Press, New York.

Gottlieb, M., and Glanz J. (2003). The blackout of 2003: The past; The blackouts of '65 and '77 became defining moments in the City's history, *New York Times*, 15 August.

Greene, W. H. (2002). *LIMDEP 8.0 Econometric Modeling Guide*, Vol. *1*, Econometric Software Inc., Plainview, NY.

Greene, W. H. (2008). *Econometric Analysis*, 6th ed., Prentice Hall, Upper Saddle River, NJ.

Henderson, J. M., and Quandt, R. E. (1980). *Microeconomic Theory: A Mathematical Approach*, 3rd ed., McGraw Hill, New York.

Hosmer, D. W., Lemeshow, S., and Sturdivant, R. X. (2013). *Applied Logistic Regression*, 3rd ed., Wiley., Hoboken, NJ.

Jensen, M. C. (1977). Stores, banking and wall street badly disrupted, *New York Times*, 15 July.

Joskow, P. L. (2012). Creating a smarter U.S. Electricity Grid, *Journal of Economic Perspectives*, *26*(1): 29–48.

Mäkinen, T. (1993). *Geomagnetically Induced Currents in the Finnish Power Transmission System, Geophys. Publ. 32*, 101 pp., Finnish Meteorological Institute, Helsinki.

McFadden, R. D. (1977). 100 POLICEMEN HURT: Thousands of false alarms add to problems – Service for commuters snarled, *New York Times*, 15 July.

Mills, E., and Jones, R. (2016). An insurance perspective on U.S. Electric Grid Disruption Costs, *The Geneva Papers on Risk and Insurance – Issues and Practice*, *41*(4), 555–586.

National Research Council (2008). Severe space weather events-understanding societal and economic impacts workshop report. Available at https://www.nap.edu/catalog/12507/

severe-space-weather-events-understanding-societal-and-economic-impacts-a (Accessed 22 September 2017).

New York Independent System Operator (2008). Lake Erie loop flow mitigation: A report from the New York Independent System Operator. Available at http://www.nyiso.com/public/webdocs/markets_operations/documents/ARCHIVE/white_papers/ARCHIVE/lerie_loopflow_report_11182008.pdf (Accessed 22 September 2017).

Newey, W. K., and West, K. D. (1987). A simple positive semi-definite heteroskedasticity and autocorrelation consistent covariance matrix, *Econometrica*, *55*, 703–708.

NYISO (2003). *NYISO System Operation Procedures*, NYISO.

Oughton, E. J., Skelton, A., Horne, R. B., Thomson, W. P., and Gaunt, C. T. (2017). Quantifying the daily economic impact of extreme space weather due to failure in electricity transmission infrastructure, *Space Weather*, *15*, doi:https://doi.org/10.1002/2016SW001491.

Pace, E. (1965). Citizens help police guide traffic flow, *New York Times*, 10 November.

Pepe, M. S., Longton, G., and James, H. (2009). Estimation and comparison of receiver operating characteristic curves, *The Stata Journal*, *9*(1), 1–16. Available at http://www.stata-journal.com/article.html?article=st0154 (Accessed 22 September 2017).

Raab, S. (1977). Ravage continues far into day; Gunfire and bottles beset police, *New York Times*, 15 July.

Schewe, P. F. (2007). *The Grid: A Journey through the Heart of Our Electrified World*, Joseph Henry Press, Washington, DC.

Schröder, T., & Kuckshinrichs, W. (2015). Value of lost load: An efficient economic indicator for power supply security? A literature review, *Frontiers in Energy Research*, *3*, 55, doi: https://doi.org/10.3389/fenrg.2015.00055.

Shanahan, E. (1965). F.P.C. criticizes power systems in Nov. 9 failure, *New York Times*, 7 December.

U.S.-Canada Power System Outage Task Force (2004). Final report on the August 14th blackout in the United States and Canada. Available at https://reports.energy.gov/BlackoutFinal-Web.pdf (Accessed 22 September 2017).

Viljanen, A. (1998). Relation of geomagnetically induced currents and local geomagnetic variations, *IEEE Transactions on Power Delivery*, *13*, 1285–1290, doi:https://doi.org/10.1109/61.714497.

Viljanen, A., H. Nevanlinna, K. Pajunpää, and A. Pulkkinen (2001). Time derivative of the horizontal geomagnetic field as an activity indicator, *Annales de Geophysique*, *19*, 1107–1118.

Willig, R. (1976). Consumer's surplus without apology, *The American Economic Review*, *66*(4), 589–597.

13

Mitigating Power System Response to GICs in Known Networks

Maryam Kazerooni[1] and Thomas J. Overbye[2]

ABSTRACT

This chapter focuses on mitigating the geomagnetic disturbances (GMDs) negative impacts through corrective control actions. Generation redispatch, transmission line switching, and transformer blocking device placement are recognized as three effective mitigation techniques. A generic reliability-based optimization problem is designed to obtain the power system optimal operation during GMDs. The coupling between the ac power flow solution and the geomagnetically induced current (GIC) flows is modeled and heuristic strategies are designed to maintain the system security measures in terms of both GICs and ac power flow. Moreover, the computational complexity of the algorithms are analyzed for large-system implementations.

Key Points

- A mathematical model is developed which relates the geoelectric field to geomagnetic induced currents (GICs) and calculates the AC power flows incorporating GICs.
- Generation redispatch, line switching and blocking devices are presented as three remedial action schemes to minimize the negative impact of GIC on AC power systems.
- The effectiveness of the proposed techniques on large scale power systems is evaluated through simulation.

13.1. INTRODUCTION

Several techniques have been investigated in the literature to mitigate the geomagnetic disturbances (GMDs). One technique is to install geomagnetically induced current (GIC) blocking devices with the capacitive circuits at the transformer neutral. These devices are expensive and optimization techniques are developed to find their optimal placement with minimum cost. The existing controls such as generator redispatch, load shedding, and line switching can mitigate GMDs with no additional employment cost. This chapter builds on the existing literature on the GMD mitigation with more focus on the Transformers safety and also scalability to large systems. The goal is to develop different types of corrective actions to satisfy the transformer thermal constraints in large-scale power systems while providing sufficient ac-related security measures.

The chapter is organized as follow: the background on GMD respose strategies is presented in Section 13.2. The GIC model is introduced in Section 13.3 and solving power flow including GICs is described in Section 13.4. Mitigating GMDs through generation redispatch is presented in Section 13.5. Section 13.6 consideres line switching as a remedial action to mitigate GMDs whereas Section 13.7 investigates blocking device placement as the corrective strategy. Section 13.8 presents direction for future work.

13.2. BACKGROUND ON GMD RESPONSE PROCEDURES

Remedial action schemes (RAS) also known as special protection schemes (SPS) are effective solutions to protect

[1] *Department of Electrical and Computer Engineering, University of Illinois at Urbana-Champaign, Urbana, IL, USA*
[2] *Department of Electrical and Computer Engineering, Texas A&M University, College Station, TX, USA*

Geomagnetically Induced Currents from the Sun to the Power Grid, Geophysical Monograph 244, First Edition.
Edited by Jennifer L. Gannon, Andrei Swidinsky, and Zhonghua Xu.

the network from GMDs. RAS is an automatic protection system designed to detect abnormal or predetermined system conditions and then to take corrective actions to restore the power system's safe operational mode (Anonymous, 2013b). RAS actions may include simple isolation of faulted components to maintain system reliability or changes in demand, generation (MW and Mvar) or system configuration to maintain system stability, acceptable voltage, and/or power flows. The economic and security benefit offered by RAS has urged many utilities to implement them in their systems (Varghese et al., 2009). RAS are widely deployed by Southern California Edison (Wen et al., 2014), Bonneville Power Administration (BPA) (Ramanathan et al., 2013) and British Columbia transmission corporations (BCTC) (Pai and Sun, 2008). Existing RAS designs may be classified based on their type of control action, for example, generation redispatch (Genc et al., 2010; Fouad et al., 1986; Ruiz-Vega and Pavella, 2003), generation tripping (Ota et al., 1996; Zhang and Tomsovic, 2004), load shedding (Taylor et al., 1981; Rovnyak et al., 2003), and line switching (Shao and Vittal, 2005; Gou and Zhang, 2013; Mazi et al., 1986).

There are several challenges associated with GMD mitigation: First, there is little real-world operational experience with respect to GMDs for the simple reason that they happen rarely. Moreover, they are caused by uncontrollable external forces. For example, vegetation contact with transmission lines may be avoided by proper control actions, yet no action can reduce the probability of geomagnetic storms. Finally, their risk assessment is difficult since historical events may not reflect the potential future impacts (Anonymous, 2010).

Enhancing power system resiliency to GMDs comprises four components (Anonymous, 2012b). The first building block is damage prevention prior to the event, that is, the ability to keep the system operational in the case of a disturbance. The second component is resourcefulness during the event, which entails the ability to manage a disaster as it unfolds. It includes identifying the options, prioritizing them, and communicating the decisions to the corresponding entities. Resourcefulness depends mostly on people, not technology. The third component is the rapid recovery that is the capability to move back to normal condition as quickly as possible after an event. The last building block is adaptability, which is learning new lessons from the event. The knowledge obtained from each event can help to revise the existing procedures and develop new ones. Adaptability is performed at all times and improves damage prevention, resourcefulness, and recovery. This chapter mainly focuses on damage prevention for improving resiliency against GMDs.

Various strategies may be used for damage prevention: First, risk assessment may be employed to analyze the probability of a solar storm and evaluate their impacts on the grid. This includes developing different scenarios and their associated probabilities (e.g., severe storm: once in 100 years, serious storms: once in 10 years). Second, the grid should be strengthened against GMDs. Enhancing relay technology by replacing electromechanical relays with solid state electronic relays is an effective hardening strategy against GMDs which reduces the relay misoperation and unintended disconnection of the reactive power providers such as static VAR compensators. The third approach is to improve the grid robustness. Installing GIC blocking devices, building more transmission lines, increasing the reactive power support and employing advanced technologies in communication and building controls can improve the system robustness. The fourth approach is situational awareness. Utilities are installing GIC monitoring devices at different locations in the network which contribute to the system awareness during solar storms. The outage notification system provided by the smart meters and the time-synchronized visibility capabilities of the phasor measurement units (PMUs) can increase the situational awareness. Corrective control through automated feeder switching contributes to enhanced resiliency as well. Other examples of damage prevention techniques include enhancing cooling strategies and health sensors for transformers (Anonymous, 2013a).

The term "new normal" is introduced in Anonymous (2012b) to describe the degraded planning and operating condition during catastrophic events. Once the event occurs, the mitigation procedure is employed to minimize the impacts and maximize the service to consumers. This might take days since resources such as reserve capacity, spare equipment, and personnel are limited. In the restoration phase, the system is restored to the new normal and may stay there for months or even years before it returns to pre-event reliability level. Studies on the 1989 Quebec blackout has provided some insight on the restoration procedures after GMDs, yet more in-depth analysis is required in this area.

13.2.1. Existing GMD Procedures Utilized by ISOs

Major utilities and ISOs have developed preliminary operating procedures to protect the network during severe storms. Different procedures are designed based on the severity of the storm. Storms with K-index smaller than K4 do not trigger any action. If K-index is larger than K4 and smaller than K7, a warning state is declared and the ISO notifies the transmission and generation owners. For storms with K-index of K7 and larger, alert state is declared and the following emergency procedures are executed with small variations among different ISOs (Anonymous, n.d., 2012a; System Operations Division, 2019):

• Out-of-service transmission facilities are restored if possible and taking long transmission lines out of service are avoided.

• The planned outages are reviewed, their impact on the system is evaluated and the problematic outages are canceled if possible.

• The MVA limit of the inter-area and internal transmission lines and transformers are reduced to a maximum of 90% of the normal rating where appropriate.

• The excitation of generators is adjusted to maintain the voltages within the acceptable range and protect the network from voltage swings.

• Generation redispatch is utilized to provide sufficient reactive power reserve, for example, synchronous condensers are dispatched, Maintain Thorium Molten Salt Reactor (TMSR) are maintained at minimum 50% and generators are possibly operated at their Eco Max state.

• SVC and capacitor banks are added if available to avoid voltage drop.

• Real-time contingency analysis is performed with a focus on voltage contingencies and the impact of tripping large capacitor banks and SVC is considered.

• Reclosing tripped capacitor banks and SVCs during severe storms are considered since they probably tripped due to relay malfunction rather than an actual damage.

• The GMD monitoring equipment is inspected to ensure that it is in service and personnel are possibly sent to those locations.

• The SCADA systems are monitored for unusual voltage and line flow variations.

13.2.2. Existing GMD Mitigation Techniques in Literature

Several techniques have been investigated in the literature to mitigate GMDs (Kappenman, 2010; Kappenman et al., 1991). One technique is through the installation of GIC blocking devices, consisting of the capacitive circuits, at the transformer neutral. The blocking device disconnect the network from the substation neutral for the dc flow without affecting the ac flow. If placed at proper locations, this technique can reduce the GIC-saturated reactive power loss (Zhu and Overbye, 2015; Arajärvi et al., 2011). Installing a capacitor in the neutral may compromise the ground fault detection system or cause insulation hazards and safety risks. Possible solutions to reduce such risks are using parallel resistors or spark gaps, vacuum switching or interrupting the protection circuit (Bolduc et al., 2005; Kovan and de León, 2015). The tradeoff among these solutions involves complexity, bulkiness, and deployment cost. Series capacitors (Arajärvi et al., 2011), polarizing cells (Kappenman, 1983), and neutral linear resistance (Kappenman, 1983; Hussein and Ali, 2016) are other types of blocking devices investigated in literature for GMD mitigation.

Most of the existing GIC mitigation programs focus only on the dc analysis of the system and reducing the GIC flows. However, the ac power flow solution is coupled with the GIC flows and it is desired that the mitigation framework integrates some aspects of the ac analysis along with the already existing dc ones. This is especially important when line switching or series capacitors are considered as the control action. Line switching changes the ac flows and if not performed correctly, may cause overloads and voltage violations. An effective GMD mitigation should properly model the effect of GICs on the ac power flow solution and develop a strategy that provides sufficient security measures in terms of both GIC flows and ac analysis.

In the conventional RAS designs, offline calculations are performed to obtain the best control action. These actions are stored and executed in real time when a contingency occurs. The optimal control action depends on the real-time state of the system, for example, the network topology and the amount of generations and loads. The offline calculation considers different scenarios for the system state and stores the best action for each scenario in the form of an arming table. For example, BPA RAS includes a 2D arming table that specifies the amount of generation drop at a particular generator based on the real-time flows across two major transmission lines (Ramanathan et al., 2013). A similar arming table is used for CBCT RAS (Pai and Sun, 2008). Creating and storing such arming tables require lots of data management and may still not capture all possible scenarios that might happen in the case of a real contingency. In the case of GMDs, real-time E-field has a significant impact on the state of the system and the RAS program should capture the time extended E-field in its design. To tackle this, online RAS design may be utilized, which calculates the best action in real time. Online RAS designs should be computationally efficient as it is desired to execute the control action as quickly as possible.

13.3. GIC MODELING

Within the footprint of an interconnected power system, the geoelectric field is often assumed to be uniform as represented by $\mathbf{E} := [E^N, E^E]^T$, with E^N and E^E denoting its northward and eastward components, respectively.

To this end, consider a transmission network with the set of buses $\mathcal{N}^B$ and the set of substations $\mathcal{N}^S$, which in together form the set of nodes $\mathcal{N} := \{1,\ldots,N\} = \mathcal{N}^B \cup \mathcal{N}^S$. As all the generators are grounded via the nearest substation, the buses directly connected to generators can be excluded from $\mathcal{N}^B$ in the GIC modeling. Furthermore, the node connectivity can be represented by the set of edges $\mathcal{E} := \{(n,m)\} \subseteq \mathcal{N} \times \mathcal{N}$, which includes the set of transmission lines $\mathcal{L} := \{(n,m)\} \subseteq \mathcal{N}^B \times \mathcal{N}^B$, as well as the pairs (n, m) if and only if substation n is directly connected to bus m by a transformer. For any $(n,m) \in \mathcal{E}$, let $g_{nm} = g_{mn}$ denote the equivalent conductance between n and m, which will be the transmission line conductance if $(n,m) \in \mathcal{L}$. Otherwise if $n \in \mathcal{N}^S$ and $m \in \mathcal{N}^B$, g_{nm} corresponds to the conductance due to transformer windings to the substation neutral and autotransformers' serires and common windings. Similarly, let g_{nn} represent the grounding conductance of the substation neutral for $n \in \mathcal{N}^S$. Notice all the conductance quantities have accounted for the three phases in parallel. With these notations, an $N \times N$ real symmetric matrix $\mathbf{G}$ can be constructed with nonzeros only at its diagonal and entries corresponding to $\mathcal{E}$, with the (n, m)-th entry given by

$$G_{nm} := \begin{cases} -g_{nm}, & \text{if } (n,m) \in \mathcal{E} \\ \sum_{v \in \mathcal{N}_n} g_{nv}, & \text{if } n = m \in \mathcal{N}^B \\ g_{nn} + \sum_{v \in \mathcal{N}_n} g_{nv}, & \text{if } n = m \in \mathcal{N}^S \\ 0, & \text{otherwise.} \end{cases} \quad (13.1)$$

The conductance matrix $\mathbf{G}$ serves as the counterpart of the bus admittance matrix for ac power network analysis, as detailed soon.

13.3.1. Input Voltages as Current Injections

The input of GIC modeling is the aforementioned induced voltage along the transmission lines. The effect of geoelectric field on the line $(n,m) \in \mathcal{L}$ is equivalent to having a dc voltage source in series with that line, with the voltage level given by

$$U_{nm} = E^N L_{nm}^N + E^E L_{nm}^E \quad (13.2)$$

where L_{nm}^N and L_{nm}^E denotes the northward and Eastward distance for line (n, m). Using the Norton equivalent, all the induced voltage sources can be converted to dc current injection to every bus n, as given by

$$I_n^{inj} = \sum_{(n,m)} I_{nm}^{inj} = \sum_{(n,m)} g_{nm} U_{nm} = \left(\sum_{(n,m) \in \mathcal{L}} L_{nm}^N g_{nm} \right) E^N + \left(\sum_{(n,m) \in \mathcal{L}} L_{nm}^E g_{nm} \right) E^E \quad (13.3)$$

Notice for all the substations that are not directly connected by transmission lines, the injected current is simply $I_n = 0$, $\forall n \in \mathcal{N}^S$. Concatenating (13.3) across all N nodes gives rise to

$$\mathbf{I} := [I_1, \ldots, I_N]^T = \mathbf{HE} \quad (13.4)$$

where the n-th row of matrix $\mathbf{H}$ is given by $\left[\sum_{(n,m) \in \mathcal{L}} L_{nm}^N g_{nm}, \sum_{(n,m) \in \mathcal{L}} L_{nm}^E g_{nm} \right]$.

To obtain the GIC flows, stack the dc voltage V_n at any bus or substation neutral in $\mathbf{V}$: = $[V_1, \ldots, V_N]^T$, which follows the dc power flow model as

$$\mathbf{V} = \mathbf{G}^{-1}\mathbf{I} = \mathbf{G}^{-1}\mathbf{HE}. \quad (13.5)$$

Note that V_n is the dc node voltage, different from the induced line voltage U_{nm}. The GIC flow between any two nodes is given by

$$I_{nm}^T = g_{nm}(V_n - V_m), \forall n \in \mathcal{N}^B, m \in \mathcal{N}^S. \quad (13.6)$$

13.3.2. Effect of Line Switching on GIC Flows

Next, the effect of opening a line on the GIC flows is studied. This will be used later on to develop line switching strategies for GMD mitigation. Opening line (n, m) modifies the $\mathbf{G}$ matrix by deducting the terms related to the disconnected pairs:

$$\mathbf{G} \leftarrow \mathbf{G} - g_{nm} e_{nm} e_{nm}^T \quad (13.7)$$

where vector e_{nm} of length N has all zero entries except for the n-th and m-th being $+1$ and -1, respectively. Similarly, the $\mathbf{H}$ matrix is modified by

$$\mathbf{H} \leftarrow \mathbf{H} - e_{nm} \left[L_{nm}^N, L_{nm}^E \right] \quad (13.8)$$

Opening a line does not affect the $\mathbf{\Phi}$ matrix. Opening multiple lines requires updating $\mathbf{G}$ and $\mathbf{H}$ successively for each opened line.

13.3.3. Effect of Blocking Device on GIC Flows

Placing a blocking device on transformer t which connects high-side bus m to low-side bus n at substation s, disconnects both side windings from the substation and hence, modifies $\mathbf{G}$ by deducting the terms related to the disconnected pairs:

$$\mathbf{G} \leftarrow \mathbf{G} - g_{ns} e_{ns} e_{ns}^T g_{ms} e_{ms} e_{ms}^T \quad (13.9)$$

To update $\mathbf{\Phi}$, its t-th row is set to zero and the other rows remain unchanged. Placing a blocking device does not affect $\mathbf{H}$.

13.3.4. GIC-saturated Reactive Power Loss

The GIC passing through the transformer increases its reactive power loss. A simplified model is presented in Overbye et al. (2012) which relates the additional GIC-saturated reactive power loss (GRPL) to the effective GICs as expressed in:

$$Q^{GIC} = K\,|V|\,I^{GIC} \tag{13.10}$$

where Q^{GIC} is the GRPL. K is the transformer loss coefficient which mostly depends on the core type, for example, single phase, three-legged three phase, and five-legged three phase. This parameter may be calculated by simplifying the transformer magnetizing curve into linear segments and determining the slope and knee point of each segment (Dong et al., 2001). $|V|$ is the ac voltage magnitude of the transformer's high-voltage side. Note that obtaining this value requires the ac power flow solution. The voltage profile of the base case or a flat voltage profile may be used instead to decouple the GIC-saturated loss from the ac analysis and simplify the calculations. Any reference to losses throughout the paper indicates the GIC-saturated reactive power loss.

I^{GIC} is the effective GIC and is calculated based on the transformer type and winding configuration. For a grounded wye-delta transformer, I^{GIC} is simply the current in the grounded coil. For transformers with multiple grounded windings (-autotransformers), the effective current is a function of the current in both coils as expressed in Overbye et al. (2013):

$$I^{GIC} = \frac{aI_H + I_L}{\alpha} \tag{13.11}$$

where I_H and I_L are respectively the per phase dc current passing through the high side winding (-series winding) and low side winding (-common winding) and α is the transformer turn ratio. Calculating I_H and I_L from (13.6), the effective current at transformer t is given by:

$$I_t^{GIC} = g_{sh}\left(V_h - V_s\right) + \frac{1}{\alpha} g_{sl}\left(V_l - V_s\right) \tag{13.12}$$

where h and l are the high and low side bus nodes for transformer t, and s is its substation node. Concatenating (13.12) across all transformers gives rise to:

$$\mathbf{I}^{GIC} = \mathbf{\Phi V} = \mathbf{\Phi G}^{-1}\mathbf{HE} = \mathbf{CE} \tag{13.13}$$

where $\mathbf{C} = \mathbf{\Phi G}^{-1}\mathbf{H}$ is the coefficient matrix and $\mathbf{\Phi}$ is a $T \times (N + S)$ matrix with the (t, m)-th entry given by

$$\Phi_{tm} := \begin{cases} g_{sm}, & \text{if } m = h \\ \alpha g_{sm}, & \text{if } m = l \\ -g_{sh} - \alpha g_{sl}, & \text{if } m = s \\ 0, & \text{otherwise.} \end{cases} \tag{13.14}$$

and T is the number of transformers.

13.4. POWER FLOW SOLUTION INCLUDING GICs

To solve power flow including the GICs, the GIC-saturated reactive power loss of each transformer is modeled as a constant current source. Adding these current sources changes the reactive power injections at the high voltage side of the transformers by:

$$Q_i \leftarrow Q_i - K\,|V|_i\, I_i^{GIC} \tag{13.15}$$

where Q_i is the reactive power injection at bus i. The power balance equations are nonlinear and the most common technique for solving them is Newton–Raphson solution. This technique uses the first order Taylor series to linearize the power balance equations as expressed in:

$$\begin{bmatrix} \Delta\theta \\ \Delta|\mathbf{V}| \end{bmatrix} = -\mathbf{J}^{-1}\begin{bmatrix} \Delta\mathbf{P} \\ \Delta\mathbf{Q} \end{bmatrix} \tag{13.16}$$

where $\Delta\mathbf{P}$ ($-\Delta\mathbf{Q}$) is the vector containing all the real (-reactive) power imbalances and $\mathbf{J}$ is Jacobian matrix defined as:

$$\mathbf{J} = \begin{bmatrix} \dfrac{\partial\Delta\mathbf{P}}{\partial\Delta\theta} & \dfrac{\partial\Delta\mathbf{P}}{\partial\Delta|\mathbf{V}|} \\ \dfrac{\partial\Delta\mathbf{Q}}{\partial\Delta\theta} & \dfrac{\partial\Delta\mathbf{Q}}{\partial\Delta|\mathbf{V}|} \end{bmatrix} \tag{13.17}$$

The algorithm starts with an initial guess, uses (13.16) to update the states in each iteration and continues the process until the power mismatches are smaller than a threshold. Adding the GIC-related constant current sources modifies the entries of the Jacobian that correspond to the partial derivative of the reactive power to voltage magnitudes of the same bus:

$$\frac{\partial\Delta Q_i}{\partial\Delta|V_i|} \leftarrow \frac{\partial\Delta Q_i}{\partial\Delta|V_i|} - KI_i^{GIC}. \tag{13.18}$$

The other entries of the Jacobian matrix stay unchanged.

13.5. GENERATION REDISPATCH THROUGH RELIABILITY-BASED OPF

In this section, generation redispatch is considered as a remedial action to mitigate GMDs. As mentioned earlier, the existing operating procedures employed by major utilities in the United States, consider some basic generation redispatch strategies to increase the reactive power reserve. This includes dispatching synchronous condensers, maintain Thorium Molten Salt Reactor (TMSR) at minimum 50% and operating generators at their Eco Max state. This section reformulates the optimal power flow in the context of security control during GMDs (Hug-Glanzmann and Andersson, 2009). This new formulation is termed as reliability-based optimal power flow (ROPF) and determines the optimal generation dispatch during GMDs.

First, we give a quick review on optimal power flow (OPF). OPF minimizes the operation cost subject to the power flow and other constraints:

$$\begin{aligned} &\min f(x,u) \\ &\text{s.t. } g(x,u)=0 \\ &\qquad h(x,u)\le 0 \end{aligned} \tag{13.19}$$

where u is the control variable and x is the state variable, which includes the voltage phasors (magnitudes and phase angles) on individual power buses. The control variables are the generator MW output set-points, settings of the flexible alternating current transmission system (FACTS) devices, phase shifting transformers, and static VAR compensators. For simplicity and without loss of generality, only the generator MW output may be considered as control variable and the objective function may be written as:

$$f(z)=\sum_{i\in\mathcal{U}_{PV}} C_i\left(\overline{P}_i\right) \tag{13.20}$$

where $C_i\left(\overline{P}_i\right)$ is the cost of operating generator i with the MW output of $\overline{P}_i$, and $\mathcal{U}_{PV}$ is the set of PV buses (generators).

The equality constraint $g(x, u)$ corresponds to the power flow equations and ensures that the active and reactive power at each bus match their given values. The inequality constraint $h(x, u)$ may include the line flow limits, the voltage magnitude limits, and the generators output limit as given by:

$$\begin{aligned} &V_i^{\min}\le V_i\le V_i^{\max} && i\in\mathcal{U}_{PQ} \\ &P_i^{\min}\le \overline{P}_i\le P_i^{\max} && i\in\mathcal{U}_{PV} \\ &Q_i^{\min}\le \overline{Q}_i\le Q_i^{\max} && i\in\mathcal{U}_{PV} \\ &P_{ij}\le P_{ij}^{\max} && (i,j)\in\mathcal{I} \end{aligned} \tag{13.21}$$

where $|V|_i$, $\overline{P}_i$ and $\overline{Q}_i$ are respectively the voltage magnitude, the active power and the reactive power at bus i; and $\mathcal{U}_{PQ}$ is the set of PQ buses. P_{ij} is the active power on the line between buses i and j, $P_{ij}^{\max}$ is the flow limit of this line, and $\mathcal{I}$ is the set of all (i, j) for which there is a line connecting bus i to bus j. Note that the generator output limit is a physical constraint and cannot be violated at any time. On the other hand, the voltage and line flow limits are operating constraints that relate to system reliability, and may be formulated as soft constraints.

Conventional OPF minimizes the operation cost subject to the power flow and other constraints. During the emergency state, retaining system security becomes the first priority rather than the operation cost. Hence, the objective function of ROPF optimizes the security instead of cost. Similar to the conventional OPF, the equality constraints of ROPF correspond to the power flow equations and the inequality constraints relate to the generator output limits, voltage constraints, and line flow limits. The GIC saturated loss at the individual transformers are included as additional constraints to ensure the transformers safety from overheating:

$$Q_i^{GIC}\le Q_i^{\text{Max}} \qquad \forall i\in\mathcal{T} \tag{13.22}$$

where Q_i^{Max} is the maximum permissible GIC saturated loss for transformer i which is determined based on its thermal ratings.

The voltage constraints and line flow limits may be formulated as soft constraints since they are operating constraints, and unlike the physical constraints, they can be violated during the emergency state. These constraints are modeled by the following,

$$\begin{aligned} &V_i\le V_i^{\max}+t_i && i\in\mathcal{U}_{PQ} \\ &-V_i\le -V_i^{\min}+r_i && i\in\mathcal{U}_{PQ} \\ &P_{i,j}\le P_{i,j}^{\max}+s_{ij} && (i,j)\in\mathcal{I} \\ &0\le t_i, r_i, s_{ij} \end{aligned} \tag{13.23}$$

where r_i and t_i are respectively the slack variables for the voltage upper and lower limits at bus i and s_{ij} is the slack variable for the line between buses i and j. The slack variables are to formulate the soft constraints and are penalized in the objective function. The objective function enforces the voltage constraints and the line flow limits as well as minimizing the generation cost:

$$\begin{aligned} f(x)=&\sum_{i\in\mathcal{U}_{PV}} C_i\left(P_i\right)+\sum_{i\in\mathcal{U}_{PQ}} \nu\overline{V}\left(2t_i+t_i^2\right)+\sum_{i\in\mathcal{U}_{PQ}} \nu\underline{V}\left(2r_i+r_i^2\right) \\ &+\sum_{(i,k)\in\mathcal{I}} \nu I\left(2s_{ik}+s_{ik}^2\right) \end{aligned} \tag{13.24}$$

where $\mathcal{V}\bar{V}$, $\mathcal{V}\underline{V}$, and $\mathcal{V}I$ are the weigthing parameters chosen with respect to the desired importance of each term.

The equality constraints including GICs are as follow:

ac Power Flow Equations

$$Q_i - Q_i^{GIC} = \sum_{k=1}^{N} |V_i||V_k|\left(G_{ik}\sin\theta_{ik} - B_{ik}\cos\theta_{ik}\right) \quad \forall i \in \mathcal{N} \tag{13.25a}$$

$$P_i = \sum_{k=1}^{N} |V_i||V_k|\left(G_{ik}\cos\theta_{ik} + B_{ik}\sin\theta_{ik}\right) \quad \forall i \in \mathcal{N} \tag{13.25b}$$

GIC Power Flow Equations

$$\sum_{k=1}^{N} g_{ik}\left(V_i - V_k + V_{ik}\right) + g_{ii}V_i = 0 \quad \forall i \in \left(\mathcal{N} \cup \mathcal{S}\right) \tag{13.25c}$$

$$Q_i^{GIC} = \begin{cases} K|V_i|I_i^{GIC} & i \in \mathcal{T}^{HKV} \\ 0 & \text{otherwise} \end{cases} \tag{13.25d}$$

where G_{ik} and B_{ik} are respectively the real and imaginary parts of the entry at the ith row and kth column of the admittance matrix and θ_{ik} is the phase angle between the voltages of bus i and k. N is the total number of buses. P_i and Q_i are respectively, the net real and reactive power injections at bus i.

To maximize the system reactive power reserve during GMDs, the generators cost function $C_i\left(\bar{P}_i\right)$ may be modified to capture the reactive power reserve of individual generators. For example, the generators with higher reactive power reserve may be assigned lower operating cost than their actual cost to be dispatched first and the ISO compensates for the cost mismatch. Details of reactive power support and its cost component in OPF is presented in Gross et al. (2002).

13.6. GMD MITIGATION THROUGH TRANSMISSION LINE SWITCHING

Line switching has been studied as an effective control strategy to improve power system reliability (Rolim and Machado, 1999; Shao and Vittal, 2005). Line outage distribution factors (LODFs) are utilized in Gou and Zhang (2013) to rank the candidate switching actions. Similar types of sensitivity factors have been employed in Mazi et al. (1986) and Shao and Vittal (2005) to determine the best switching actions. In general, line switching can modify the network-wide flows so that the overloads and voltage violations are relieved. Using the same concept, the GIC flows could be redirected through the line switching to reduce the negative impacts.

First, GMD-resilient OPF is modified to account for line switching as the remedial action. The generation dispatch is assumed to remain unchanged in this formulation. Hence, the generation output limits are excluded from the inequality constraints. Moreover, the cost term is removed from the objective function as given by:

$$\begin{aligned} f(x) = &\sum_{i\in\mathcal{U}_{PQ}} \mathcal{V}\bar{V}\left(2t_i + t_i^2\right) + \sum_{i\in\mathcal{U}_{PQ}} \mathcal{V}\underline{V}\left(2r_i + r_i^2\right) \\ &+ \sum_{(i,k)\in\mathcal{I}} \mathcal{V}I\left(2s_{ik} + s_{ik}^2\right) \end{aligned} \tag{13.26}$$

The equality constrains are given by:

$$Q_i - Q_i^{GIC} = \sum_{k=1}^{N} y_{ik}|V_i||V_k|\left(G_{ik}\sin\theta_{ik} - B_{ik}\cos\theta_{ik}\right) \quad \forall i \in \mathcal{N} \tag{13.27a}$$

$$Pi = \sum_{k=1}^{N} y_{ik}|V_i||V_k|\left(G_{ik}\cos\theta_{ik} + B_{ik}\sin\theta_{ik}\right) \quad \forall i \in \mathcal{N} \tag{13.27b}$$

GIC Model

$$\sum_{k=1}^{N} y_{ik}g_{ik}\left(V_i - V_k + V_{ik}\right) + g_{ii}V_i = 0 \quad \forall i \in \left(\mathcal{N} \cup \mathcal{S}\right) \tag{13.27c}$$

$$Q_i^{GIC} = \begin{cases} K|V_i|I_i^{GIC} & i \in \mathcal{T}^{HKV} \\ 0 & \text{otherwise} \end{cases} \tag{13.27d}$$

where y_{ik} is a binary variable which indicates the state of the line, for example, y_{ik} is 0 (-1) if line i is closed (-open). y_{ik} is 0 if there is no line between bus i and k.

Solving the proposed ROPF is computationally expensive and might not be practical for a larger system. Lu et al. (2017) employs a similar optimization problem on a 97-bus system which takes up to 34,000 s (Lu et al., 2017). In the case of a credible contingency, it is crucial to take an effective control action as quickly as possible. Hence, finding a solution that is less accurate, but faster to compute is preferable. Motivated by this, fast line switching algorithms based on linear sensitivities are developed to reduce the computation complexity for large-scale power systems application.

13.6.1. Iterative Line Switching based on Linear Sensitivity Analysis

In this section, an iterative algorithm is proposed based on linear sensitivities which minimizes the GIC-saturated reactive power loss. The coupling between the ac power flow solution and the GIC flows is modeled and heuristic strategies are designed to maintain the system security measures in terms of both GICs and ac power flow. Moreover, the computational complexity of the algorithm is analyzed for large-system implementations. The principles of the algorithm are presented in the following, techniques to reduce the computational complexity are designed in Section 13.6.1.1 and heuristic strategies are incorporated in Section 13.6.1.2 to provide sufficient ac-related security measures.

The transformer LODF (TLODF) can be expressed as:

$$TLODF = \left[s_{ij}\right] = \left[Q_i^{GIC,(j)} - Q_i^{GIC,(0)}\right], \qquad i \in T,\ j \in L \tag{13.28}$$

where s_{ij} is the variation of GIC-saturated loss at transformer i caused by opening line j, $Q_i^{GIC,(j)}$ is the GIC-saturated loss at transformer i when line j is opened and $Q_i^{GIC,(0)}$ is the initial loss. An analytical technique may be developed to derive TLODFs as a function of the network parameters. Alternatively, one can follow the sensitivity definition to calculate the TLODFs.

TLODFs are used to identify the best line switching strategy that minimizes the GIC-saturated loss. The total loss reduced from opening a line is obtained by taking the sum of TLODF matrix along the corresponding column. Let $Q^T = Sum(TLODF, 2)$ be the sum of the TLODF matrix along the columns. The best lines are obtained by sorting Q^T in descending order and selecting the lines with the lowest Q^T values.

Similar to LODFs, TLODFs consider single line outages. The TLODF for multiple-line outages is not equal to the sum of the single-line TLODFs of the corresponding lines. The notion of generalized LODF was introduced in Guler et al. (2007) which considers multiple-line LODFs. A similar concept may be considered for TLODF and analytical techniques can be developed to calculate multiple-line TLODFs. A rather simple approach for calculating TLODFs under multiple-line outages is to consecutively open one line and calculate the single-line TLODFs for the new system until all the desired lines are opened. The line switching algorithm can thus be improved using a similar approach. First, the TLODF is calculated and the best line to be opened is selected accordingly. Next, the TLODF matrix is calculated for the new system and the second line to be opened is selected. This process is repeated until the number of opened lines reaches a user-defined threshold, M. This technique is illustrated in Figure 13.1.

13.6.1.1. Reducing the Computational Complexity

The computational complexity of the proposed algorithm is $\mathcal{O}(LMT)$ where L is the number of lines, M is the maximum number of lines that can be opened and T is the computation time for solving for GIC flows. The number of lines for a power grid is typically slightly larger than the number of buses. Solving GIC flows requires calculating the inverse of $\mathbf{G}$ and then multiplying it by $\mathbf{\Phi}$. $\mathbf{G}$ is a sparse matrix roughly in the order of $O(N^2)$ and the computation time of taking its inverse is $\mathcal{O}(N^{2.2})$ (Alvarado, 1976). $\mathbf{\Phi}$ is a sparse matrix of roughly the same order and multiplying it by $\mathbf{G}^{-1}$ requires $\mathcal{O}(N^{2+\mathcal{O}(1)})$ (Yuster and Zwick, 2005).

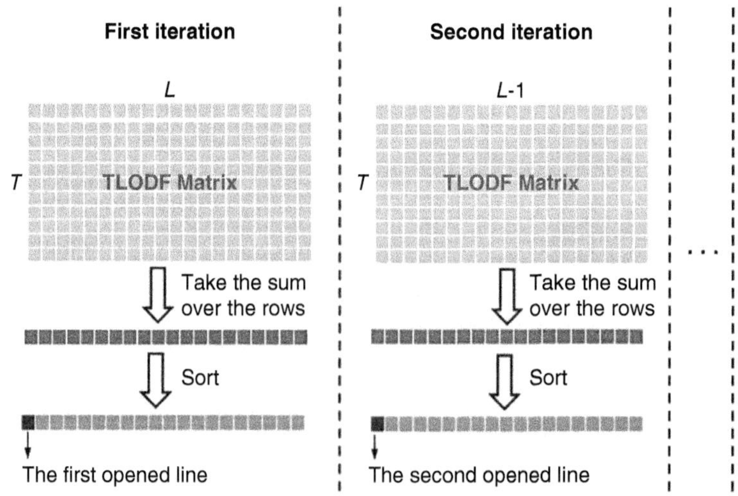

Figure 13.1 The line selection process in each step of the iterative line switching algorithm.

To reduce the running time, first, the TLODF matrix is calculated and Q^T is calculated for each line by taking the sum of the TLODF matrix over its columns. The first C lines in the sorted list are selected as critical lines and are investigated for further analysis. C is a user-defined parameter which controls the complexity. After selecting the critical lines, the iterative line selection algorithm is applied to only these lines; that is, the TLODFs are calculated for only the critical lines and the optimal lines are obtained based on the Q^T values. The critical lines may be updated after each U iterations by recalculating the full TLODF matrix. U is again a user-defined parameter which controls the computation time. Some insights on how to select suitable C and U values are presented later through simulation. Details of the fast line selection algorithm are illustrated in Figure 13.2.

13.6.1.2. Incorporating AC Analysis into the Algorithm

Opening a line changes the structure of the Jacobian matrix and the new system may not have a power flow solution. The line selection algorithm should guarantee that the switching strategy provides a power flow solution. One possible procedure to ensure this is as follows: The algorithm starts with the first line in the sorted list. The candidate line is opened and the Y-bus of the new system is calculated. If the new Y-bus is full rank, the candidate line is selected. If not, it is discarded and the second line in the sorted list is considered as the new candidate. The process is continued until a line that provides a full-rank Y-bus is found as detailed in Algorithm 4. Note that calculating the new Y-bus after opening a line is not computationally expensive as it can be derived from the initial Y-bus in negligible computational time.

Factors other than TLODFs may be used as a criterion for selecting the best lines. The power system experiences reactive power shortage during GMDs and reducing the GIC reactive loss is desired. However, opening a line changes the ac flows in the system and may compromise the system security. Hence, a meaningful line selection algorithm should consider some aspects of the ac analysis along with the already existing GIC related criterion (GIC-saturated loss). Motivated by this, the ac line flow may be considered as an additional criterion for selecting the best lines:

$$F_i = \frac{Q_i^T}{\bar{P}_i} \tag{13.29}$$

Algorithm 1 Power Flow Solution Check

```
procedure Find the Line with Power Flow Solution
    for 1 ≤ c < C do
        Open Line c in the sorted list.
        Calculate the Y-bus for the new System.
        if new Y-bus is full rank then
            Return line c.
        end if
        Close line c.
    end for
end procedure
```

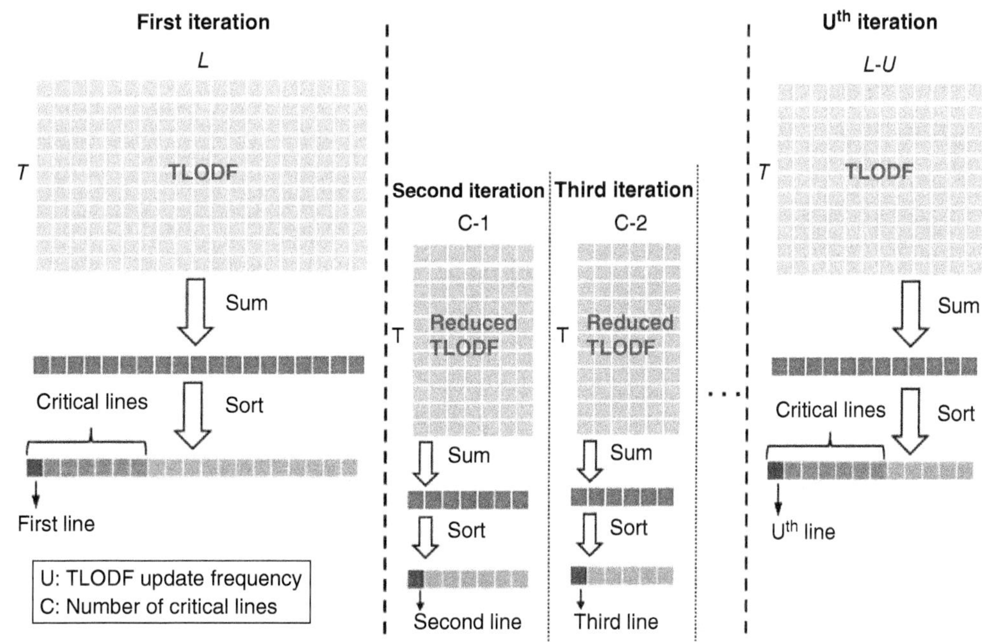

Figure 13.2 Reducing the computational complexity of the iterative line switching algorithm.

where $\overline{P}_i$ is the flow on line i. F is calculated for all the lines and the line with the highest absolute value of F is selected in each iteration. This criterion selects the line that reduces the GIC-saturated loss more and also has lower ac flow on it. Note that the ac flows change at each iteration and hence ac power flow needs to be solved to update them. This adds complexity to the algorithm and increases the running time. The ac flows may be updated after several iterations as opposed to every single iteration to reduce the running time associated with solving power flow. Moreover, the algorithm may be performed offline as part of the system planning in which case the running time would not be a constraint.

The proposed algorithm minimizes the total loss in the system but does not impose any limit on the loss of individual transformers. This can be demonstrated through numerical results on the 150-bus synthetic case in Birchfield et al. (2016). An electric field with 8 V/km magnitude and 26°N orientation is enforced to the system. This orientation results in the largest GICs for the system and hence is considered for the analysis. The GIC-saturated loss in the individual transformers before and after utilizing the line switching algorithm is illustrated in Figure 13.3. It is observed that the algorithm reduces the losses significantly in most of the transformers. The losses remain unchanged or even slightly increased for few of the transformers. In the following section, the algorithm is further refined to restrict the losses in each transformer while minimizing the total loss.

13.6.2. Multi-action Line Switching Algorithm

A maximum limit is defined for the GIC-saturated loss of each transformer based on its thermal rating and a transformer is considered overheated if its loss exceeds that limit. To find the optimal line switching strategy, a MILP is designed which minimizes the number of opened lines subject to satisfying the transformer loss constraints:

$$\begin{aligned} &\min \mathbf{a}^T \mathbf{y} \\ &\text{s.t. } \Psi \mathbf{y} < \mathbf{q}^{\max} - \mathbf{q}^0 \\ &\quad \mathbf{y} \in \{0,1\} \end{aligned} \tag{13.30}$$

where $\mathbf{q}^0$ is the vector containing the GIC-saturated loss of all the transformers prior to any action and $\mathbf{q}^{\max}$ is the vector of the transformers maximum permissible loss. $\mathbf{a} = \vec{1}_{L\times 1}$ is an all-one vector of size L where L is the number of lines. $\mathbf{y}$ is a binary vector of size L which indicates the state of the lines, for example, $\mathbf{y}_i$ is 0 (−1) if line i is closed (-open).

For a large system, performing mixed integer programming on all the lines is computationally expensive. To reduce the complexity, the critical lines are identified and the optimization is performed on the reduced system. Moreover, the loss constraints for only the overheated transformers are included in the formulation. Let $\mathbf{s}^0$ and C be the sets of overheated transformers and critical lines, respectively. The reduced-order MILP is given by:

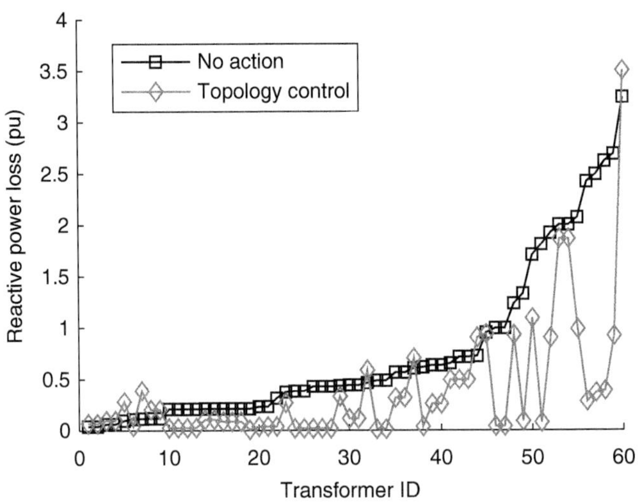

Figure 13.3 GIC-saturated loss in the individual transformers before and after utilizing the topology control. The 150-bus system in Birchfield et al. (2016) is used for the study. The enforced E-field has 8 V/km magnitude and 26°N orientation.

$$\min \mathbf{a}_s^T \mathbf{x}$$
$$\text{s.t. } \Psi\left(\mathbf{s}^0, \mathcal{C}\right) \times \mathbf{x} < \mathbf{q}^{\max}\left(\mathbf{s}^0\right) - \mathbf{q}^0\left(\mathbf{s}^0\right) \qquad (13.31)$$
$$X \in \{0,1\}$$

x is the binary state vector of size C where C is the number of critical lines and $\mathbf{a}_s = \vec{1}_{C\times 1}$ is an all-one vector of the same size. We use the indexing notation **a**(**b**) for vector **a** and **b** to indicate the subvector of **a** that contains the entries corresponding to **b**. For example, for **a** = [10, 11, 12] and **b** = [1, 3], we have **a**(**b**) = [10, 12]. Similarly, the notation **A**(**b**, **c**) for matrix **A** and vectors **b** and **c** indicates the submatrix of **A** that contains the rows corresponding to **b** and the columns corresponding to **c**.

The GIC-saturated losses may be significantly large during severe storms and no switching action may be able to reduce them to below the limit. To tackle this, we consider an alternating solution; a switching action is designed to satisfy half of the constraints and another action to satisfy the other half. Employing the first action for half of the time and the second solution for the remaining time can reduce the stress from all transformers. The transformers that are overheated under the first solution, can cool down while the second solution is executed and vice versa. This idea can be extended to multiple solutions instead of just two. Each solution is designed to relax constraints on certain transformers and an alternating schedule is developed to execute the actions periodically as illustrated in Figure 13.4.

The effectiveness of the multi-action strategy is illustrated through numerical results using the 20-bus system in Horton et al. (2012). An electric field with the magnitude of 8 V/km and the orientation of 124°N is enforced to the system. Again, the motivation for choosing this orientation is that it provides the highest GICs. It is assumed that the maximum permissible loss, $Q^{\max}$, is 100 MVAR for all transformers. Without any action, the losses at seven transformers exceed the limit whereas employing the multi-action algorithm generates two actions which can remove the stress from all the transformers. Figure 13.5 illustrates the impact of the actions

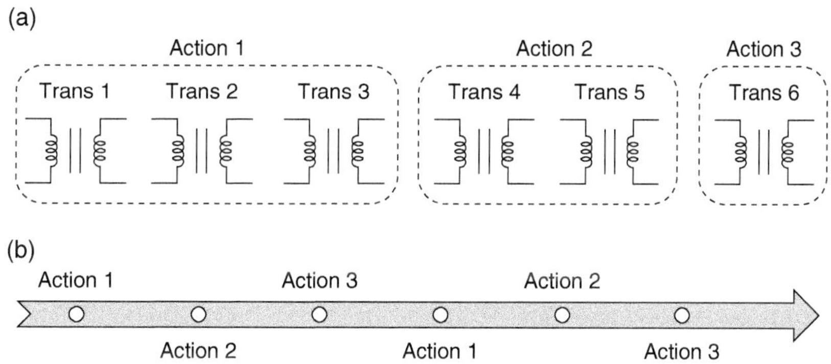

Figure 13.4 Multi-action line switching framework. (a) Assigning the actions to the overheated transformers. (b) Timeline of the alternating schedule.

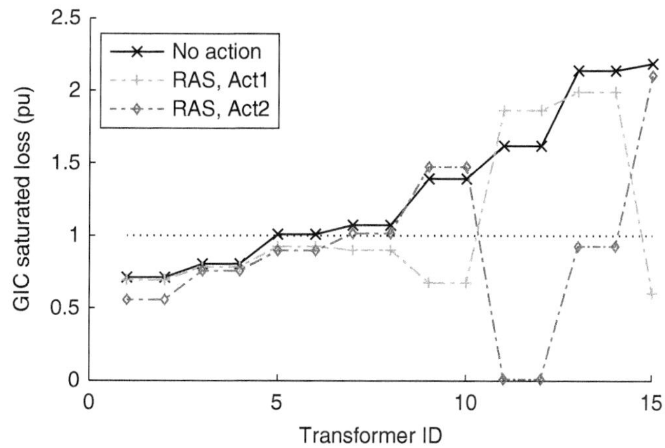

Figure 13.5 Impact of the proposed line switching on the GIC-saturated loss of the individual transformers for the 20-bus system. The magnitude of the enforced E-field is 8 V/km.

on the GIC-saturated loss of the individual transformers. For reference, the losses before taking any action are also presented in the figure. It is observed that each action target only part of the overheated transformers and the algorithm is effective only when a combination of both actions is utilized.

13.7. GMD MITIGATION THROUGH BLOCKING DEVICES

This section presents techniques for finding the optimal locations for GIC blocking devices. First, ROPF is modified to account for blocking device placement as the remedial action. Similar to the line switching formulation, the generation dispatch is assumed to remain unchanged and hence, the generation output limits are excluded from the inequality constraints. Instead, an additional constrain on the number of blocking devices is considered:

$$\sum_{i\in(\mathcal{N}\bigcup\mathcal{S})} z_i \le P \tag{13.32}$$

z_i is a binary variable which indicates whether substation i is selected for blocking device placement. P is the maximum number of blocking devices defined by the user based on the available budget. The objective function is similar to the line switching formulation with generation cost term excluded (the generation dispatch is constant and so is the cost):

$$f(x)=\sum_{i\in\mathcal{U}_{PQ}} \nu\overline{V}\left(2t_i+t_i^2\right)+\sum_{i\in\mathcal{U}_{PQ}} \nu\underline{\mathrm{V}}\left(2r_i+r_i^2\right) + \sum_{(i,k)\in\mathcal{I}} \nu I\left(2s_{ik}+s_{ik}^2\right). \tag{13.33}$$

The equality constrains are given by:

ac Power Flow Equations

$$Q_i-Q_i^{GIC}=\sum_{k=1}^{N}|V_i||V_k|\left(G_{ik}\sin\theta_{ik}-B_{ik}\cos\theta_{ik}\right)\quad\forall i\in\mathcal{N} \tag{13.34a}$$

$$P_i=\sum_{k=1}^{N}|V_i||V_k|\left(G_{ik}\cos\theta_{ik}+B_{ik}\sin\theta_{ik}\right)\quad\forall i\in\mathcal{N} \tag{13.34b}$$

GIC Power Flow Equations

$$\sum_{k=1}^{N}g_{ik}\left(V_i-V_k+V_{ik}\right)+z_i g_{ii}V_i=0\quad\forall i\in\left(\mathcal{N}\cup\mathcal{S}\right) \tag{13.34c}$$

$$Q_i^{GIC}=\begin{cases}K|V_i|I_i^{GIC} & i\in\mathcal{T}^{HighKV}\\ 0 & \text{otherwise}\end{cases}. \tag{13.34d}$$

Blocking device placement is part of the system planning and is performed before the actual storm. Hence, it needs to be independent of the system real time state. The proposed OPF formulation is state-dependent and requires the load and generation data along with the network topology as input. Using the historic data, a snapshot of the network during a typical day may be used to represent the system state. Alternatively, a blocking device formulation which is independent of the system state may be developed. Unlike line switching, blocking device placement does not affect the ac system directly and its impact is solely through the change of GIC flows and their resulting GIC saturated loss. Hence, incorporating the ac analysis into the blocking device placement problem is less critical. Motivated by this, a blocking device placement algorithm is developed which is independent of the system state and also ignores the ac analysis for simplicity.

13.7.1. Blocking Device Placement through Sensitivity Analysis

The transformer Blocking Distribution Factors (TBDF) can be expressed as:

$$TBDF=\left[s_{ij}\right]=\left[Q_i^{GIC,(\bar{S})}-Q_i^{GIC,(0)}\right],\qquad i\in T,\ j\in L \tag{13.35}$$

where s_{ij} is the variation of GIC-saturated loss at transformer i caused by placing a blocking device at substation s. $Q_i^{GIC,(\bar{s})}$ is the GIC-saturated loss at transformer i when blocking device is placed at substation s and $Q_i^{GIC,(0)}$ is the initial loss.

To find the optimal blocking device placement, a MILP is designed which minimzed the total GIC-saturated reactive power loss subject to satisfying the transformer loss constraints and number of blocking devices:

$$\begin{aligned}&\min\sum TBDF\mathbf{z}\\ &\text{s.t. } TBDF\mathbf{z}<\mathbf{q}^{\max}-\mathbf{q}^{0}\\ &\mathbf{b}^{T}\mathbf{z}\le P\\ &\mathbf{z}\in\{0,1\}\end{aligned} \tag{13.36}$$

where $\mathbf{b}=\vec{1}_{S\times 1}$ is an all-one vector of size S and $\mathbf{z}$ is a binary vector of the same size which indicates which substations are selected for blocking device placement.

13.8. FUTURE WORK

Corrective controls deal with the system during disturbances when the dynamics are significant and transient stability is crucial. Hence, the system dynamics are considered in the design of many corrective control actions (Fouad et al., 1986; Zhang and Tomsovic, 2004; Shao and Vittal, 2005). Genc et al. (2010) presents a generation redispatch strategy based on the maximum angle separation of the generators during the transient simulation. Transient energy analysis through a single machine equivalent (SIME) is used in Ruiz-Vega and Pavella (2003) for corrective control design. Transient stability analysis is computationally expensive, and getting a transient stability solution requires much more computation than solving the power flow in steady-state. Hence, performing RAS design in the context of steady-state analysis may simplify the calculations and improve the computational complexity. Moreover, transient stability analysis during GMDs has not been investigated well in the literature and most of the GMD studies are in the content of steady state analysis. The existing GMD mitigation programs do not consider system dynamics and future research may address this through understanding the system dynamics during GMDs and developing relevant transient stability tools.

Other matters with regard to GMD mitigation need to be considered in future research. First, more detailed transformer thermal models may be used to improve the mitigation technique. Presently, a maximum limit on the transformer GIC-saturated loss is considered to prevent its overheating. The duration of the GIC pulse affects the transformer temperature. A small yet consistent dc current at the transformer neutral may cause more overheating than a larger dc current that has a shorter duration. Future research tends to address this by considering simplified GIC profiles with a specific time-variant shape based on the historic measurements and performing detailed thermal modeling to evaluate the transformer safety during GMDs. Second, the current formulation assumes a constant E-field and magnitude whereas the E-field evolves over time during actual storms. The algorithm can be further refined to capture time-extended E-fields and possibly use real measurements from actual storms as input.

ACKNOWLEDGMENTS

This work was partially supported by a grant from the Spanish Ministry of Science and Technology.

REFERENCES

F. L. Alvarado, "Computational complexity in power systems," *IEEE Transactions on Power Apparatus and Systems*, vol. *95*, no. 4, pp. 1028–1037, Jul. 1976.

Anonymous, "High-impact, low-frequency event risk to the north American bulk power system," North American Electric Reliability Corporation (NERC), Tech. Rep., Jun. 2010.

Anonymous, "2012 special reliability assessment interim report: Effects of geomagnetic disturbances on the bulk power system," North American Electric Reliability Corporation (NERC), Tech. Rep., Feb. 2012a.

Anonymous, "Severe impact resilience: Considerations and recommendations," North American Electric Reliability Corporation (NERC), Tech. Rep., May 2012b.

Anonymous, "Economic benefits of increasing electric grid resilience to weather outages," *Executive Office of the President*, Aug. 2013a.

Anonymous, "Special protection systems (SPS)/remedial action schemes (RAS): assessment of definition, regional practices, and application of related standards," North American Electric Reliability Corporation (NERC), Tech. Rep., Apr. 2013b.

Anonymous, Transmission and Dispatching Operations Manual [Online], n.d. Available: http://www.nyiso.com/public/webdocs/markets_operations/documents/Manuals_and_Guides/Manuals/Operations/trans_disp.pdf

E. Arajärvi, R. Pirjola, and A. Viljanen, "Effects of neutral point reactors and series capacitors on geomagnetically induced currents in a high-voltage electric power transmission system," *Space Weather*, vol. *9*, no. 11, pp. 1–10, 2011.

A. B. Birchfield, K. M. Gegner, T. Xu, K. S. Shetye, and T. J. Overbye, "Statistical considerations in the creation of realistic synthetic power grids for geomagnetic disturbance studies," *IEEE Transactions on Power Systems*, 2016.

L. Bolduc, M. Granger, G. Pare, J. Saintonge, and L. Brophy, "Development of a dc current-blocking device for transformer neutrals," *IEEE Transactions on Power Delivery*, vol. *20*, no. 1, pp. 163–168, Jan. 2005.

X. Dong, Y. Liu, and J. G. Kappenman, "Comparative analysis of exciting current harmonics and reactive power consumption from GIC saturated transformers," in *IEEE PES Winter Meeting*, Jan. 2001.

A. A. Fouad, A. Ghafurian, K. Nodehi, and Y. Mansour, "Calculation of generation-shedding requirements of the B.C. hydro system using transient energy functions," *IEEE Power Engineering Review*, vol. *PER-6*, no. 5, pp. 31–32, May 1986.

I. Genc, R. Diao, V. Vittal, S. Kolluri, and S. Mandal, "Decision tree-based preventive and corrective control applications for dynamic security enhancement in power systems," *IEEE Transactions on Power Systems*, vol. *25*, no. 3, pp. 1611–1619, Aug. 2010.

B. Gou and H. Zhang, "Fast real-time corrective control strategy for overload relief in bulk power systems," *IET Generation, Transmission Distribution*, vol. *7*, no. 12, pp. 1508–1515, Dec. 2013.

G. Gross, S. Tao, E. Bompard, and G. Chicco, "Unbundled reactive support service: Key characteristics and dominant cost component," *IEEE Transactions on Power Systems*, vol. *17*, no. 2, pp. 283–289, May 2002.

T. Guler, G. Gross, and M. Liu, "Generalized line outage distribution factors," *IEEE Transactions on Power Systems*, vol. *22*, no. 2, pp. 879–881, May 2007.

R. Horton, D. Boteler, T. J. Overbye, R. Pirjola, and R. C. Dugan, "A test case for the calculation of geomagnetically induced currents," *IEEE Transactions on Power Delivery*, vol. *27*, no. 4, pp. 2368–2373, 2012.

G. Hug-Glanzmann and G. Andersson, "Decentralized optimal power flow control for overlapping areas in power systems," *IEEE Transactions on Power Systems*, vol. *24*, no. 1, pp. 327–336, Feb. 2009.

A. A. Hussein and M. H. Ali, "Suppression of geomagnetic induced current using controlled ground resistance of transformer," *Electric Power Systems Research*, vol. *140*, pp. 9 – 19, 2016.

J. G. Kappenman, "Mitigation of geomagnetically induced and dc stray currents," EPRI Rep. EL-3295, Tech. Rep., 1983.

J. Kappenman, "Low-frequency protection concepts for the electric power grid:geomagnetically induced current (GIC) and E3 HEMP mitigation," *Oak Ridge National Laboratory (ORNL)*, 2010.

J. G. Kappenman, S. R. Norr, G. A. Sweezy, D. L. Carlson, V. D. Albertson, J. E. Harder, and B. L. Damsky, "GIC mitigation: A neutral blocking/bypass device to prevent the flow of GIC in power systems," *IEEE Trans. on Power Delivery*, vol. *6*, no. 3, pp. 1271–1281, 1991.

B. Kovan and F. de León, "Mitigation of geomagnetically induced currents by neutral switching," *IEEE Transactions on Power Delivery*, vol. *30*, no. 4, pp. 1999–2006, Aug. 2015.

M. Lu, H. Nagarajan, E. Yamangil, R. Bent, and S. Backhausl, "Optimal transmission line switching under geomagnetic disturbances" [Online], 2017. Available: https://arxiv.org/pdf/1701.01469.pdf

A. A. Mazi, B. F. Wollenberg, and M. H. Hesse, "Corrective control of power system flows by line and bus-bar switching," *IEEE Power Engineering Review*, vol. *PER-6*, no. 8, pp. 53–53, Aug. 1986.

H. Ota, Y. Kitayama, H. Ito, N. Fukushima, K. Omata, K. Morita, and Y. Kokai, "Development of transient stability control system (tsc system) based on on-line stability calculation," *IEEE Transactions on Power Systems*, vol. *11*, no. 3, pp. 1463–1472, Aug. 1996.

T. J. Overbye, T. R. Hutchins, K. Shetye, J. Weber, and S. Dahman, "Integration of geomagnetic disturbance modeling into the power flow: A methodology for large-scale system studies," in *North American Power Symposium (NAPS)*, Sep. 2012.

T. J. Overbye, K. S. Shetye, T. R. Hutchins, Q. Qiu, and J. D. Weber, "Power grid sensitivity analysis of geomagnetically induced currents," *IEEE Transactions on Power Systems*, vol. *28*, no. 4, pp. 4821–4828, 2013.

S. C. Pai and J. Sun, "BCTCs experience towards a smarter grid - increasing limits and reliability with centralized intelligence remedial action schemes," in *Electric Power Conference, 2008. EPEC 2008. IEEE Canada*, Oct. 2008, pp. 1–7.

R. Ramanathan, B. Tuck, and J. O'Brien, "BPA's experience of implementing remedial action schemes in power flow for operation studies," in *2013 IEEE Power Energy Society General Meeting*, Jul. 2013.

J. G. Rolim and L. J. B. Machado, "A study of the use of corrective switching in transmission systems," *IEEE Transactions on Power Systems*, vol. *14*, no. 1, pp. 336–341, Feb. 1999.

S. M. Rovnyak, K. Mei, and G. Li, "Fast load shedding for angle stability control," in *Power Engineering Society General Meeting, 2003, IEEE*, Jul. 2003.

D. Ruiz-Vega and M. Pavella, "A comprehensive approach to transient stability control. ii. open loop emergency control," *IEEE Transactions on Power Systems*, vol. *18*, no. 4, pp. 1454–1460, Nov. 2003.

W. Shao and V. Vittal, "Corrective switching algorithm for relieving overloads and voltage violations," *IEEE Transactions on Power Systems*, vol. *20*, no. 4, pp. 1877–1885, Nov. 2005.

System Operations Division, PJM Manual 13 Emergency Operations [Online], 2019. Available: http://www.pjm.com/~/media/documents/manuals/m13.ashx

C. W. Taylor, F. R. Nassief, and R. L. Cresap, "Northwest power pool transient stability and load shedding controls for generation-load imbalances," *IEEE Transactions on Power Apparatus and Systems*, vol. *PAS-100*, no. 7, pp. 3486–3495, Jul. 1981.

M. Varghese, L. Jin, S. Ghosh, G. Lin, and B. Pek, "The caiso experience of implementing automated remedial action schemes in energy management systems," in *2009 IEEE Power Energy Society General Meeting*, Jul. 2009.

J. Wen, W. H. E. Liu, P. L. Arons, and S. K. Pandey, "Evolution pathway towards wide area monitoring and protection, a real-world implementation of centralized ras system," *IEEE Transactions on Smart Grid*, vol. *5*, no. 3, pp. 1506–1513, May 2014.

R. Yuster and U. Zwick, "Fast sparse matrix multiplication," *ACM Transactions Algorithms*, vol. *1*, no. 1, pp. 2–13, Jul. 2005.

Y. Zhang and K. Tomsovic, "Adaptive remedial action scheme based on transient energy analysis," in *Power Systems Conference and Exposition, 2004. IEEE PES*, Oct. 2004, pp. 925–931, vol.*2*.

H. Zhu and T. J. Overbye, "Blocking device placement for mitigating the effects of geomagnetically induced currents," *IEEE Transactions on Power Systems*, vol. *30*, no. 4, pp. 2081–2089, Jul. 2015.

INDEX

Geomagnetically Induced Currents from the Sun to the Power Grid, Geophysical Monograph 244, First Edition.
Edited by Jennifer L. Gannon, Andrei Swidinsky, and Zhonghua Xu.